CAMBRIDGE LIBRARY COLLECTION

Books of enduring scholarly value

Technology

The focus of this series is engineering, broadly construed. It covers technological innovation from a range of periods and cultures, but centres on the technological achievements of the industrial era in the West, particularly in the nineteenth century, as understood by their contemporaries. Infrastructure is one major focus, covering the building of railways and canals, bridges and tunnels, land drainage, the laying of submarine cables, and the construction of docks and lighthouses. Other key topics include developments in industrial and manufacturing fields such as mining technology, the production of iron and steel, the use of steam power, and chemical processes such as photography and textile dyes.

Treatise on Mills and Millwork

One of the great Victorian engineers, Sir William Fairbairn (1789–1874) had started his career as a millwright's apprentice, going on to become a civil engineer, a designer of industrial machinery and an expert on the failure of materials and structures. The present work distils a lifetime's experience of mechanical design into two highly illustrated parts. First published in 1861 and 1863, they are here reissued in a single volume. Part 1 gives a general overview of mechanisms such as gears, cranks and cams, and then moves on to the design of prime movers: waterwheels and turbines, steam engines and boilers, and windmills. Part 2 covers the design of mechanisms in more detail, and discusses power transmissions and their components: shafts, gears, bearings, couplings and so on. Lastly, Fairbairn gives overviews of the most important types of industrial mill – including cotton, wool, paper, iron and gunpowder – and their machinery.

Cambridge University Press has long been a pioneer in the reissuing of out-of-print titles from its own backlist, producing digital reprints of books that are still sought after by scholars and students but could not be reprinted economically using traditional technology. The Cambridge Library Collection extends this activity to a wider range of books which are still of importance to researchers and professionals, either for the source material they contain, or as landmarks in the history of their academic discipline.

Drawing from the world-renowned collections in the Cambridge University Library and other partner libraries, and guided by the advice of experts in each subject area, Cambridge University Press is using state-of-the-art scanning machines in its own Printing House to capture the content of each book selected for inclusion. The files are processed to give a consistently clear, crisp image, and the books finished to the high quality standard for which the Press is recognised around the world. The latest print-on-demand technology ensures that the books will remain available indefinitely, and that orders for single or multiple copies can quickly be supplied.

The Cambridge Library Collection brings back to life books of enduring scholarly value (including out-of-copyright works originally issued by other publishers) across a wide range of disciplines in the humanities and social sciences and in science and technology.

Treatise on Mills and Millwork

William Fairbairn

Cambridge University Press

CAMBRIDGE
UNIVERSITY PRESS

University Printing House, Cambridge, CB2 8BS, United Kingdom

Cambridge University Press is part of the University of Cambridge.
It furthers the University's mission by disseminating knowledge in the pursuit of education, learning and research at the highest international levels of excellence.

www.cambridge.org
Information on this title: www.cambridge.org/9781108070010

This edition first published 1861–3
This digitally printed version 2014

ISBN 978-1-108-07001-0 Paperback

MILLS AND MILLWORK

PART I.

LONDON
PRINTED BY SPOTTISWOODE AND CO.
NEW-STREET SQUARE

TREATISE

ON

MILLS AND MILLWORK

PART I.

ON THE PRINCIPLES OF MECHANISM

AND ON

PRIME MOVERS

COMPRISING THE ACCUMULATION AND ESTIMATION OF WATER POWER,
THE CONSTRUCTION OF WATER-WHEELS AND TURBINES, THE PROPERTIES OF STEAM,
THE VARIETIES OF STEAM-ENGINES AND BOILERS, AND WINDMILLS

BY

WILLIAM FAIRBAIRN, ESQ., C.E.

LL.D. F.R.S. F.G.S.

CORRESPONDING MEMBER OF THE NATIONAL INSTITUTE OF
FRANCE, AND OF THE ROYAL ACADEMY OF TURIN:
CHEVALIER OF THE LEGION OF HONOUR

LONDON

LONGMAN, GREEN, LONGMAN, AND ROBERTS

1861

PREFACE.

There is probably no department of practical science so generally useful, or so little studied of late years, as the machinery of transmission. The term "*millwork*," as applied to this class of machinery, is of modern origin, but "*millwright*" has long been a household word, and at no distant period conveyed the idea of a man marked by everything that was ingenious and skilful.

The millwright of former days was to a great extent the sole representative of mechanical art, and was looked upon as the authority in all the applications of wind and water, under whatever conditions they were to be used, as a motive power for the purposes of manufacture. He was the engineer of the district in which he lived, a kind of jack-of-all-trades, who could with equal facility work at the lathe, the anvil, or the carpenter's bench. In country districts, far removed from towns, he had to exercise all these professions, and he thus gained the character of an ingenious, roving, rollicking blade, able to turn his hand to anything, and, like other wandering tribes in days of old, went about the country from mill to mill, with the old song of "*kettles to mend*" reapplied to the more important fractures of machinery.

Thus the millwright of the last century was an itinerant engineer and mechanic of high reputation. He could handle the axe, the hammer, and the plane with equal skill and precision; he could turn, bore, or forge with the ease and despatch of one brought up to these trades, and he could set out and cut in the furrows of a millstone with an accuracy equal or superior to that of the miller himself. These various duties he was called upon to exercise, and seldom in vain, as in the practice of his profession he had mainly to depend upon his own resources. Generally, he was a fair arithmetician, knew something of geometry, levelling, and mensuration, and in some cases possessed a very competent knowledge of practical mathematics. He could calculate the velocities, strength, and power of machines: could draw in plan and section, and could construct buildings, conduits, or watercourses, in all the forms and under all the conditions required in his professional practice; he could build bridges, cut canals, and perform a variety of work now done by civil engineers. Such was the character and condition of the men who designed and carried out most of the mechanical work of this country, up to the middle and end of the last century. Living in a more primitive state of society than ourselves, there probably never existed a more useful and independent class of men than the country millwrights. The whole mechanical knowledge of the country was centred amongst them, and, wherever sobriety was maintained and self-improvement aimed at, they were generally looked upon as men of superior attainments and of considerable intellectual power. It, however, too frequently happened that early training, constant change of scene, and the temptation of jovial companions, led the young millwright into excesses which almost paralysed his good qualifications. His attainments as a mechanic, and his standing in the useful arts, were apt to make him vain; and with a rude independence he would repudiate the idea of working with an inferior craftsman, or even with another as skilful as himself, unless he was born and bred a millwright.* I remember an old millwright who, in palliation of an offence with which his employer charged him, urged that he ought not to forget, that he had condescended *to work even with carpenters to please him.*

The introduction of the steam engine, and the rapidity with

* This had reference to the young practitioner being the oldest son of a millwright, which circumstance in itself was, until of late years, considered a sufficient guarantee for skill and industry, whether he possessed them or not.

which it created new trades, proved a heavy blow to the distinctive position of the millwright, by bringing into the field a new class of competitors in the shape of turners, fitters, machine-makers, and mechanical engineers: and, notwithstanding the immense extension of the demand for millwork and the great stimulus which it afforded to the manufactures of the country, it nevertheless lowered the profession of the millwright, and levelled it in a great degree with that of the ordinary mechanic. He, however, retains his distinctive appellation, and I hope he will long continue the representative of a higher class of mechanical artisans, to whom the public are deeply indebted for many of our first and greatest improvements in practical science.

Serious, and perhaps not altogether unfounded, charges have been brought against millwrights as a class, but on examination I do not think that they are borne out to the extent some persons would wish us to believe. On the contrary, I am persuaded there is no class of mechanics so intelligent or who work harder than the millwright, or who exercise a sounder judgment in the performance of their varied duties in the perfect execution of their work. It is true that, in former times, they too frequently gave way to habits of dissipation, and neglected their work; but in this respect they were not alone, as the changes which have lowered their standing have proved of use in reforming their habits, and produced in the millwrights of the present day a highly moral and intellectual class of workmen. Taking them as a body, I believe there is not a more trustworthy or a more respectable class of men in existence. I make this statement from experience, and have great pleasure in doing so.

It used to be a custom, before the days of Mechanics' Institutes, for the millwrights to form one for themselves in every shop. Their meetings were generally held at a public house on Saturday evening; and many were the times when long discussions on practical science and the principles of construction were carried on between rival disputants with a fiery eagerness which not unfrequently ended in a quarrel, or effected a settlement by the less rational but more convincing argument of blows. It was a rough way of imparting knowledge but it was not worse than that practised in the schools and seminaries of the day, where the application of the rod was the general remedy for dull apprehensions and indocile minds. This was beginning at the wrong end, endeavouring to impart knowledge through the sensitive parts of the body, instead of appealing to the higher organs of the intellect. The principal difference between the Millwrights' Institute and such schools was, probably, that the former was the more ferocious of the two, as the rival disputants hit harder under the influence of potations than would now, fortunately, be tolerated. On more peaceful occasions, however, it was curious to trace the influence of these discussions on the young aspirants around, and the interest excited by the illustrations and chalk diagrams by which each side supported their arguments, covering the tables and floors of the room in which they were assembled. The great objection to these gatherings was, however, the angry feeling too frequently aroused, and the injurious influence of the place of meeting, which gave rise to prolonged debates under the encouragement of the landlord, who on most occasions was appealed to as referee in all matters in dispute.

The above is no overdrawn statement of the condition of the millwrights some fifty years ago. Their education and habits were those of the times in which they lived. There were then no schools for the working classes but those of the parish, nor any libraries or mechanics' institutes; and after the usual course of reading, writing, and accounts, the millwright was thrown upon his own resources in the attainment of the knowledge which might aid him in his profession. Hence his value and worth were most exhibited when away from home, and isolated from all assistance, where he was left to the construction and erection of work. In such a position his energies were frequently called into action, and on many occasions he displayed powers calculated to advance the interests of his employers, and to complete his task with accuracy and skill. Thus the genuine millwright became, to a fault, tenacious of his own views and position, and jealous of any interference or assistance from others. He would reconnoitre and survey the premises on which he was to work, rule and line in hand, and would stand for hours (much to the annoyance of his employers) before he could make up his mind as to what was best to be done. These preliminaries being settled, his decision was final, and he would fix his levels, stretch his lines, and in the course of a day or two commence work with an energy which generally led to the best and most satisfactory results.

Another feature of this class, which should not be lost sight of, was the kindly feeling and generous sympathy which generally belonged to them, and that exhibited especially towards those in declining years or in distress. It is in acts of charity and good will to those in want that the millwright of all times has shown his native goodness. He may frequently be reckless and dissipated, but he seldom fails in generosity, and I know of no other trade where a more hearty feeling of liberality and kindness exists.

Yet the millwrights, with all these good qualities, have been and are still subject to faults injurious to themselves and annoying to the public. Such are their frequent contests with their employers, either for an advance of wages or for some fancied privilege which they seek to maintain or establish. They are united in benefit societies for the relief of the old and indigent, and those who from sickness or other causes may be unable to work. Unfortunately these are connected with trade societies established for the purpose of maintaining what they consider their rights,—rights often of a very imaginary character, and ill calculated to advance their position or promote their individual interests. It is not my wish to enter here into the questions which these contests suggest. I am willing to forget bygone days and to look forward with sanguine hope to better times, when truer principles of freedom and social economy shall be acted upon, without destroying the independence and originality which have always been characteristic of an intelligent body of men for whom I entertain individually and collectively the highest veneration and respect.

I have deemed it necessary to give this brief account of the habits and character of a body of men whose skill and spirit of perseverance has done so much for the advancement of applied science, and whose labours have still a large influence on the industrial progress of the country. I am, perhaps, better qualified for this task than most others, from having been associated with them from early life, so that an experience of some fifty years must be my excuse for having imposed this narrative upon the reader.

For many years I have had it in contemplation to give an account of my own practical experience in millwright construction, but a multiplicity of engagements has combined with other causes to delay the work, and to modify considerably the original plan. This first volume, I hope, may contain reliable data and true principles for the successful guidance of the millwright in his professional duties.

The present portion of the work treats of the first principles of mechanism generally, and proceeds to the discussion of the various constructions of prime movers. I hope shortly to complete the work by a treatise on the new system of transmissive machinery, and on the arrangements necessary for imparting motion to the various descriptions of mills.

The accumulation, storage, and measurement of water has received attention; as well as the construction of prime movers depending upon this motive power, including the best forms of water-wheels, according to my own practice, and the more recently introduced varieties of turbines. In discussing the principles of the steam engine I have inserted a short treatise on the properties of steam, derived in part from researches carried on under my own superintendence, bearing on the density of saturated steam and the law of expansion of superheated steam. To this has been added a chapter on engines and boilers, their strength, powers, and principles of construction.

It is evident that, in the present improved state of mill machinery, steam and water are the chief agents on which we depend for motive power. In former times the wind was also looked to as a source of power, but it is now very little employed, except in Holland and the fenny districts of this country, where it is still used for pumping and other operations where constant uniformity of action is not required. Notwithstanding the changes effected by steam, as windmills are not yet obsolete, I have given a short chapter on their mode of construction.

In the prosecution of this work I have been ably assisted by my friend Mr. Thomas Tate, to whom I owe the chapter on the elementary principles of mechanism; as also to my assistant and secretary, Mr. William C. Unwin, to whose assiduous attention and love of science I am greatly indebted.

CONTENTS.

SECTION I.

INTRODUCTION.

CHAPTER I.

SECTION II.

THE PRINCIPLES OF MECHANISM.

CHAP. I.

SECTION III.

ON PRIME MOVERS.

CHAP. I.

CHAP. II.

On the Flow and Discharge of Water, and the Estimation of Water Power:—

CHAP. III.

On the Construction of Water-Wheels:—

CHAP. IV.

On the Undershot Wheel:—

CHAP. V.

On Turbines:—

CHAP. VI.

On the Properties of Steam:—

CHAP. VII.

VARIETIES OF STATIONARY STEAM ENGINES:—

CHAP. VIII.

ON BOILERS:—

CHAP. IX.

ON WINDMILLS:—

APPENDIX.

LIST OF PLATES.

ERRATA.

Page 171, line 13, *for* "44 cubic feet" *read* "90 cubic feet."
" 169, " 9, *for* "52 inches" *read* "46 inches."
" 240, " 12, *for* "Fig. 161" *read* "Plate VIII."

A TREATISE

ON

MILLS AND MILL-WORK.

SECTION I.

INTRODUCTION.

CHAPTER I.

EARLY HISTORY OF MILLS.

WE may search in vain for dates from which to calculate the earliest period at which the principles of accumulating power, and transmitting it for employment in mills, were first introduced, and it is equally impossible to trace consecutively the progressive developments that have taken place from the days of antiquity to the present improved state of the arts. Perhaps the earliest introduction of machinery was in the processes for the preparation of food, as we read of the Egyptians and Babylonians, and other nations in Europe and Asia, having mills for grinding corn at the earliest periods at which there are records of their history. Hesiod and Pliny both describe the most primitive method of the preparation of corn, a method still further illustrated among the pictorial remains of the Egyptians, viz. pounding in a mortar.

CORN MILLS.—When millstones were introduced is uncertain, although they boast of a high antiquity. Agatharcides (B.C. 113) mentions grinding stones employed in the reduction of gold ore in the mines on the Red Sea, and of the same kind, no doubt, were the early flour mills. Two round stones, with concave and convex fitting surfaces, roughened or notched like the pestle in Pliny's description, so as to distribute the grain introduced through a hole in the upper stone, and to throw off the flour at the edges. A pivot in either stone, fitting a recess in the other, would be necessary to guide the upper or running stone, which would be moved by simple manual labour. Millstones of this kind, or querns, as they are commonly called, are found not unfrequently amongst the foundations of Roman villas, and along the lines of Roman encampments. Fig. 1 is a representation of the nether stone of such a mill found at Gayton, near Northampton, and figured in the Archæologia, vol. xxx. The stone of which these mills were composed, was a sort of pudding stone, or rough lava, which from its varying hardness tended to retain a biting surface. Some of the querns retain traces of the notches or "work:" they vary from ten inches to twenty inches in diameter. Usually, as at the present time in some countries, these handmills were turned by women, but amongst the Romans male slaves were employed for this purpose, or it was reserved as a penal exercise for convicts.

Fig. 1.

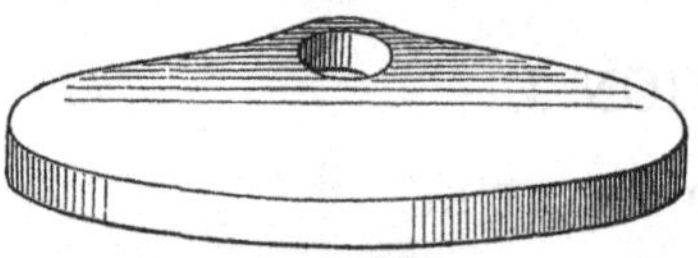

Roman Quern. Archæologia, xxx. 128.

The essential defects of this mode of grinding, the want of power, and the tediousness of the operation, led gradually to the employment of cattle mills, and these are sometimes mentioned by classical authors, although little was known of their construction until recently. In disentombing the baker's house at Pompeii, several of these large mills were found, in excellent preservation. Fig. 2 is a representation of one of them, and may be described as follows: The base A is a cylindrical stone, about five feet in diameter, and two feet high. Upon this, forming part of the same block, or else firmly fixed into it, is a conical projection, about two feet high, the sides slightly curving inwards; upon this rests another block C, externally resembling a dice-box, internally an hour-glass, being shaped into two hollow cones with their vertices towards each other, the lower one fitting the conical surface on which it rests, but not with any degree of accuracy. To diminish friction, however, a strong iron pivot was inserted in the top of the solid cone, and a corresponding socket let into the narrow part of the hour-glass shaped stone C. Four holes were cut through the stone parallel to this pivot, the narrow part was hooped on the outside with iron, into which wooden bars D D were inserted, by means of which the upper stone was turned on its pivot by the

Fig. 2.

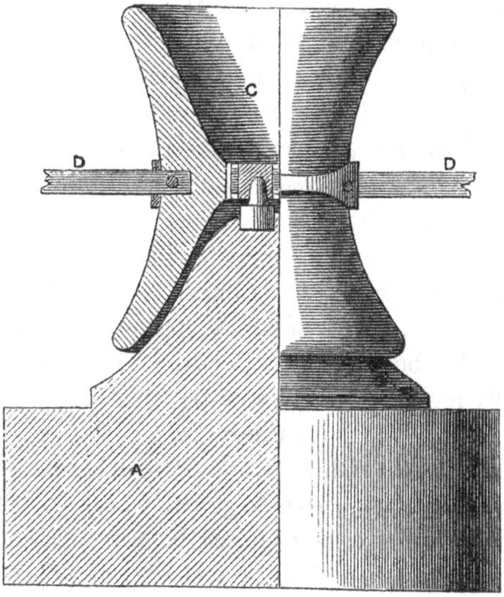

Pompeian Corn-Mill.

labour of men or asses. The upper hollow stone served as a hopper as well as a grinder, and was filled with corn, which fell by degrees through the four holes upon the solid cone, and was reduced to powder by friction between the two rough surfaces; of course it worked its way to the bottom by degrees, and fell out on the cylindrical base, round which a channel was cut to facilitate the collection. These machines are about six feet high in the whole, made of rough grey volcanic stone full of large crystals of leucite.*

Imperfect and tedious as the operation of grinding must still have been, cattle mills seem long to have held their ground

* Clarke, Pompeii, vol. ii. p. 136.

against further innovation, and even down almost to our own day in old works on machinery various contrivances for employing the labour of cattle in corn mills are described. However, before the Christian era a new power was beginning to be applied to corn mills, that of flowing or falling water. Probably the immense quantities of water required in Egypt and Assyria for the irrigation of the land first led to contrivances for turning to account the current of rivers as a motive power. Vitruvius describes water wheels employed both for raising water and for grinding corn *, the motion in the latter case being made available by a rude kind of gearing, in which we may trace the rudiments of our present transmissive machinery. Whittaker, in in his History of Manchester, describes a water mill ascribed to the Romans, of which traces were found in Manchester some years since. This mill served equally the purposes of the town and garrison, but was not alone sufficient, as the use of handmills remained very common in both, many having been found on the site of the station at Campfield. The Roman water mill at Manchester was placed upon the River Medlock, immediately below Campfield, and a little above an ancient ford. The sluice and conduit which actually regulated and conveyed the water to the mill was accidentally discovered about the middle of the last century. It was found at a place called Dyer's-croft, where a flood in the river swept away a dam with a large oak beam upon the edge of it, and disclosed a tunnel in the rock below. This, when excavated was found to be about three feet wide and three feet deep, gradually narrowing at the bottom, and upon the sides the marks of the tool were everywhere to be found. This ancient tunnel was bared to the extent of twenty-five yards, but it evidently had been continued in a direct line up to the commencement of a wide weir in the river above. From these discoveries it will appear that mills for grinding corn by power were of ancient date even in this country.

In our attempts to trace the progress of the mechanical arts, we are compelled to leave as a wide blank the period of intestine war which succeeded the decline of the Roman Empire. The conquest of Rome by Alaric and the spread of a race of barbarians over the whole of Europe, had the effect, for many

* Vitruvius, Architecture, book x. c. 10.

centuries, of obliterating almost every vestige of the arts which in the turmoil and tumult of war were either entirely lost or utterly neglected. Thus, for a long succession of years, during the middle ages and at the period of the Crusades, the industrial arts languished and retrograded, and it was not until the time of Michael Angelo and Galileo that mathematics, architecture, engineering and mechanics received the least encouragement or attention. The mathematician and natural philosopher had before then been looked upon with suspicion, and carefully watched as a person dangerous to society. During the rise of painting, sculpture, and architecture, the arts which rendered the republics of Italy so illustrious, mechanism began to attract notice, and to that age we may trace the introduction of water mills in many parts of Italy. Little or no progress, however, was made down to the close of the seventeenth century.

The Dutch, owing to the natural difficulties of their location, were urged, in their own defence, to take the lead in the field of mechanical appliance; and the vast embankments of that enterprising people, with their canals and docks, fully justify the remark that they were amongst the first to benefit mankind by the introduction of mills for grinding corn, which was chiefly imported, and of machines for draining the lands which their patient industry had reclaimed from the sea. As a prime mover the Dutch had no water power except what was obtained by impounding the tidal water and working it off during the reflux of the tide. At best this was an expensive and uncertain power which caused wind to come into more general use; and during the greater part of the seventeenth century we were chiefly indebted to the Dutch and Belgians for our improved knowledge of manufacture.

MILLS FOR THE MANUFACTURE OF TEXTILE FABRICS.—Woollen, cotton, and linen cloth was manufactured in this country from an early period, and the manufacture of silk was practised in Italy in the twelfth century. It was subsequently introduced into France and other parts of Europe, and we learn that James I. encouraged the manufacture, and made an attempt to grow the mulberry and produce silk in this country, which, however, as might have been expected, totally failed. During the reign of

Charles I., the Commonwealth, and the reign of Charles II. the manufacture of silk goods made great progress, and it is stated that in 1661 as many as 40,000 persons were employed in that branch of industry. In 1685, on the revocation of the Edict of Nantes, a large colony of skilful French weavers settled at Spitalfields, and from that time to the present have carried on the manufacture in that locality. The winding, throwing, and weaving was chiefly done by hand, and it was only from the construction of the large throwing mill at Derby, in 1719, that we date introduction of mill-machinery, technically so called, in the production of these fabrics.

Woollen mills have a much greater antiquity than either silk or cotton mills. Spinning and weaving processes were known in the time of Moses and are illustrated in ancient Egyptian monuments. Pliny attributes the discovery of the art of fulling cloth to Nicias of Megara (B.C. 1131). The origin of the woollen manufacture is evidently beyond the reach of tradition, though the process of felting was probably known before the art of spinning and weaving. Amongst the Romans the woollen manufacture attained considerable perfection, and several of their writers describe the different qualities of cloth as used for the tunic and common stuff garments.

From the time of the Romans until the Norman Conquest we have no record of the manufacture of woollens, and it is certain that amongst the Saxons, and, indeed, for several centuries after the Conquest, the costume of the peasantry was of leather, and there is reason to believe that the "buff-jerkin" retained its place as the ordinary dress of the labouring people of England until the time of the Commonwealth.

It is generally supposed that the woollen manufacture was introduced into this country in the reign of Edward III., but there is every reason to believe that it existed long before that time. Mr. McCulloch states that it was practised above a hundred years before that prince introduced improvements in the manufacture. What these improvements were is not known; probably they were neither more nor less than protective laws, which by giving an increased monopoly to guilds and corporations, seriously injured the freedom and restricted the extension of trade.

The whole of the woollen mills, from a very early period to the commencement of the present century, were driven by water, and this will account for the locations on the streams of the west of England and Yorkshire, where the woollen manufacture was carried on. The introduction of improved machinery for the manufacture of cotton gave to the woollen trade an entirely new character; and from that circumstance we may safely date the vastly increased production and the great extension that has taken place in that important branch of manufacture.

The next article of importance in an historical point of view is cotton; and to this production we may safely trace the advancement, prosperity, and power of the British Empire. The cotton manufacture had its origin in India, where the plant is indigenous, and where the climate renders a light absorbent fabric the most suitable clothing for the inhabitants. The manufacture of cotton in India may be dated from a period antecedent to the Romans; and the implements used in the different processes of the manufacture, from the cleaning of the wool to its conversion into muslin, are of a most simple kind and may be purchased for a few shillings.

The cotton manufacture of China is of the same character as that of India; and although of immense extent, the articles produced are chiefly employed for home consumption. The arts in that country, as far as we know from the accounts of the missionaries and the more recent expedition of Lord Elgin, are stationary; and the tools, implements, &c. are of the same primitive kind as those used in India. The chief description of cotton goods exported when the Chinese became famous for their manufacture were nankeens; but these have long since given way to the cheaper productions of Great Britain, and for years past we have supplied the Chinese with large quantities of yarn and cloth.

The first introduction of cotton into Europe and its manufacture were first attempted by the commercial states of Italy; and as early as 1560 cottons were exported from Venice to the different markets of Europe in the West. It was not, however, until the beginning of the seventeenth century that cotton was manufactured in this country; but we have records that the town of Manchester bought cotton wool in London which came from Cyprus and Smyrna, and worked the same into fustians, vermilions, and dimities. These goods were woven chiefly at Bolton, and finished by the Manchester dealers.

It is curious to trace the progressive increase of any description of manufacture, particularly that of cotton, which has attained to such colossal dimensions. In early times the weaver provided his own warp, which was of linen yarn, and cotton for his weft; buying these where he could best supply himself. In this way, every cottage formed an independent factory; the cotton was carded and spun by the female part of the family, and the cloth woven by the father and his sons.

Such was the state of the cotton manufacture before the introduction of power machinery, and the division of labour, and the separation of the different processes into distinct employments. At this time the workman had usually his residence in the country, where, with a little garden and perhaps grass for a cow, he carried on his trade and earned a comfortable subsistence. "How much more," says a philanthropic writer, "of the comforts of life and of the means of natural enjoyment belong to this state of manufacture than to the more advanced in which combined systems of machinery and a more perfect division of labour collect the workmen into factories and towns."

It will not be necessary to enumerate here the wellknown improvements of Arkwright, Hargreaves, and Crompton, or the changes which followed the introduction of machines for carding, roving, and spinning. Suffice it to observe, that these improvements inaugurated a new system of operations, and created a new demand for power and the means of transmitting it to the different machines required in the manufacture. It was about this time and at a rather later period that the improvements of the motive power and machinery of transmission were introduced.

To the steam engine in the first place, and subsequently to the improved machinery and mill-work, we may attribute the present gigantic extent of our manufactures. The factory system, which has supplanted the cottage manufacture, has enlarged the resources of the country far beyond those of any former period. This island stands pre-eminent in productive industry, and it is a source of pride and gratification to find that these blessings, springing out of the application of physico-mechanical science, have been attained by the skill and indomitable perseverance of our own countrymen.

To the immediate action, foresight, and intelligence of the Government of this country, the workers in coal, iron, and cotton are under no obligation; but they owe much to their own invention, skill, and industry in the prosecution and development of these pursuits, and the only merit that can be claimed by the Government is its non-interference and the protection it affords through the laws of the kingdom, which give security to property and to individual exertion in the varied departments of productive industry. Further, Dr. Ure, in his "Philosophy of Manufactures," argues that "the constant aim of scientific improvements in manufactures, is philanthropic, as they tend to relieve the workman either from niceties of adjustment, which exhaust his mind, or from painful repetition of efforts which distort or wear out his frame." Illustrations of this truth are presented every day in the remarkable extent to which labour is saved, with superior beauty and precision in the result, by self-acting machines, all of them within the domain of Automatic science.

The division of labour carried out by means of the factory system, is not exclusively applied in the manufacture of cotton, flax, silk, and woollen cloths; it pervades almost the whole of our manufacturing industry, and is beginning to show itself in mining and agriculture, and the time is probably not far distant, when we shall witness almost every operation of the human hand carried on by a system of divided activity, equally conducive to the interests of individual enterprise and to the public benefit.

The term *Factory*, according to Dr. Ure, designates, "the combined operation of many orders of work people, adult and young, tending with assiduous skill, a system of productive machines, continuously impelled by a central power. This definition includes cotton mills, flax mills, silk mills, woollen mills, and certain engineering works, but it excludes those in which the mechanisms do not form a connected series, or are not dependent upon one prime mover." The factory system is so much extended since these words were written as to change the relations of labour, and to affect almost every manufacturing process. It has created a much higher and more intelligent class of workmen than existed under the hand system, more respectable, better paid, better housed, and better clothed than heretofore.

Iron Manufactures.—We are at the present time in a state of transition in the manufacture of iron and steel, which is making rapid strides towards improvement. The inventive talent of the country has been directed to this object, and the production of homogeneous plates, having the elasticity and tenacity of steel, together with the improvements of Mr. Bessemer, Mr. Clay, and others, are likely to produce a complete revolution by a greatly increased economy in the production of iron. Mr. Bessemer is now proposing to roll plates in the form of a continuous web from liquid metal, run direct from the furnace to the rolls. We cannot vouch for the success of this enterprise, but we are most anxious to see its results realised, and there cannot exist a doubt from the number of able chemists and practical men at work, that the iron trade of this country is calculated to undergo a great change, and perhaps with as much benefit as was accomplished by Mr. Cort on the introduction of the puddling and rolling processes.

In the machinery department of iron manufacture there is nothing to boast of; it is still crude and rough in its character, perhaps necessarily so, on account of its liability to breakage in rolling, and other processes requiring great power. It is, however, possible, that the processes now in progress, may introduce new and more perfect machinery into the manufacture, and that the iron master may calculate with the same certainty of continued progress in his manufacture as now exists in other trades where machinery is employed.

Although much change has not been effected in the machinery of the iron manufacturer, considerable improvements have nevertheless been made in the smelting of the ores, and since the introduction of hot blast by Mr. Neilson the production of the furnaces has been more than doubled. Looking forward, therefore, to the improvements and changes now in progress, we may reasonably conclude that a new era is not only imminent, but has in great part been accomplished. The same progress, and even greater improvements, is observable in the conversion of iron into steel, and probably the time is not far distant when we shall be enabled to produce from the same furnace iron in either a cast or malleable state, or steel, as may best suit the requirements of the manufacturer. It is quite evident that our increasing knowledge of chemistry in iron manufactures lead to these results, and by a still closer adherence to chemical research, whereby impurities, such as phosphorus, sulphur, &c. are removed, the process just alluded to will be fully and satisfactorily realised.

SECTION II.

PRINCIPLES OF MECHANISM.

CHAPTER I.

GENERAL VIEWS.—LINK-WORK.—WRAPPING CONNECTORS.—WHEEL-WORK.—SLIDING CONTACT.

I. GENERAL VIEWS RELATIVE TO MACHINES.

Definitions and Preliminary Expositions.

1. MECHANISM may be defined as the combination of parts or pieces of a machine whereby motion is transmitted from the one to the other.

2. When a body, or any piece of mechanism, moves in a straight line it is said to have a *rectilinear* motion, and when it moves in a curved line it is said to have a *curvilinear* motion. When a point moves constantly in the same *path*, it is said to have a *continuous motion*, but if it move backwards and forwards it is said to have a *reciprocating motion*. We may have *reciprocating rectilinear motions* as well as *reciprocating curvilinear motions*.

If a body moves over equal spaces in equal intervals of time, it has a *uniform motion*; but if it moves over unequal spaces in equal intervals of time, it has a *variable motion*.

3. The velocity of a body is the *rate* at which it moves. In uniform motion the velocity is constant; but in variable motion the velocity continually changes. If the velocity of a body increase it is said to be accelerated, and if the velocity decrease it is said to be retarded.

The motion of a body is said to be *periodical* when it undergoes the same changes in the same intervals of time.

4. In order to express the velocity of a body, we must have a certain number of units of space passed over in a certain unit

of time. It is customary to take a foot as the unit of space, and a second as the unit of time.

In uniform motion, the space passed over is equal to the product of the velocity by the time. Thus, let s be the space in feet, t the time in seconds, and v the velocity per second; then

$$s = v\,t \ldots (1)$$

which expresses the general relation of space, time, and velocity, in uniform motions. Any two of these elements being given, the remaining one may be found; thus we have

$$v = \frac{s}{t} \ldots (2), \text{ and } t = \frac{s}{v} \ldots (3).$$

5. If the velocity in one certain direction be taken as positive, then that in the opposite or contrary direction will be negative.

6. If two wheels perform a revolution in the same time, their angular velocities are equal, whatever may be the dimensions of the wheels. The angular velocity of a revolving wheel or rod is the velocity of a point at a unit distance from the centre of motion. The wheel or rod will revolve uniformly when the angular velocity is uniform. If A be the angular velocity, r the radius of the wheel or length of the rod, v the velocity at this distance from the centre of motion; then

$$A = \frac{v}{r} = (1), \text{ and } v. = A\,r \ldots (2).$$

7. The motion of wheels is conveniently expressed by the number of rotations which they perform in a given time. Thus, let n be the number of revolutions performed per min., the other notation being the same as in Art. 6; then

$$v = \frac{1}{30}\,\pi\,n\,r \ldots (1), \text{ and } n = \frac{30\,v}{\pi\,r} \ldots (2).$$

Or substituting A for $\frac{v}{r}$. See formula (1), Art. 6,

$$n = \frac{30\,A}{\pi} \ldots (3), \text{ and } A = \frac{1}{30}\,\pi\,n \ldots (4).$$

Hence the number of turns performed in a given time varies as the angular velocity.

The number of turns which two wheels respectively make in the same time is called their *synchronal* rotations. Let Q and

q be the synchronal rotations of two wheels whose angular velocities are A and a, respectively; then $\frac{Q}{q} = \frac{A}{a}$; that is synchronal rotations are in the ratio of the angular velocities.

Example.—Let a wheel whose radius is 6 ft. perform 50 revolutions per min., required 1st, the velocity of its circumference, and 2nd, its angular velocity.

Here, by eq. (1), $n = 50$, and $r = 6$, then

$$v = \frac{1}{30} \times 3{\cdot}1416 \times 50 \times 6 = 31{\cdot}416 \text{ ft. per sec.}$$

And, by eq. (4), $A = \frac{1}{30} \times 3{\cdot}1416 \times 50 = 5{\cdot}236.$

8. If V and v be the velocities of two parts of a piece of mechanism, then $\frac{V}{v}$ is the *velocity ratio* of these parts. Let S and s be the corresponding spaces described in the same time, then when the motion is uniform

$$\frac{V}{v} = \frac{S}{s} = \text{a constant,}$$

that is, when the velocities are uniform, the velocity ratio is constant.

9. If the velocity ratio of the two parts remains constant, then however variable the velocities themselves may be, we still shall have $\frac{V}{v} = \frac{S}{s}$; where S and s are the entire spaces described in the same interval of time.

10. When a body moves with a variable motion, its velocity at any instant is determined by the rate at which it is moving at that particular instant, that is, by the space which it would move over in one second, supposing the motion which it then has to remain constant for that time.

Variable motions may be graphically represented, by taking the abscissa of a curve equal to the units of time, and the ordinates equal to the units of the corresponding velocities. Thus let A B be equal to the units of velocity at the commencement of the motion; A C the units in interval of time, C D the units in the corresponding velocity;

Fig. 3.

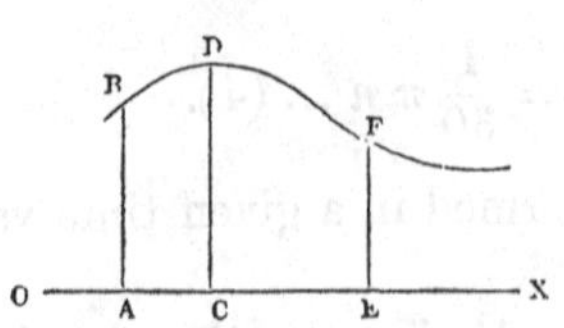

and so on; then the area of the curved space A B D F E will be equal to the space described in the interval of time represented by A E.

If the motion be uniform the curve B D F will become a straight line parallel to O X, and the space described in any given time will be represented by the area of a rectangle, whose length is equal to the units of time, and breadth equal to the units of velocity.

If the motion be uniformly accelerated or retarded, the curve B D F will become a straight line inclined to the axis O X, and the space described, in this case, will be represented by the area of a trapezoid, whose base is equal to the units of time, and parallel sides respectively to the velocity at the commencement and end of that time.

11. THE PARTS OF A MACHINE.—A machine consists of three important parts.

(1.) The parts which receive the work of the moving power—these may be called RECEIVERS of work.

(2.) The parts which perform the work to be done by the machine—these may be called WORKING PARTS, or more simply, OPERATORS.

(3.) The mechanism which transmits the work from the receivers to the working parts or operators—these pieces of mechanism may be called COMMUNICATORS OF WORK, or the TRANSMISSIVE MACHINERY.

The form of the mechanism must always be determined from the relation subsisting between the motions of the receivers and operators.

If there were no loss of work in transmission (from friction, &c.) the work applied to the receiver would always be equal to the work done by the operator. Thus, let P be the lbs. pressure applied to the receiver, and S the space in feet which it moves over in a certain time; P_1 the lbs. pressure produced at the working part, and S_1 the space in feet which it moves over in the same time; then, neglecting the loss of work by friction, we have—

Work applied to the receiver = work done upon the operator,

$$\text{or } P \times S = P_1 \times S_1 \ldots (1).$$

However, it must be borne in mind, that the actual or useful work done by a machine is always a certain fractional part of the work applied; this fraction, determined for any particular machine, is called the modulus of that machine. If m be put for this modulus, then we have from eq. (1)

$$m \times P \times S = P_1 \times S_1 \ldots (2).$$

In treating of the motion of these parts of a machine it is generally most convenient to find an expression for their proportional velocities. Thus, let V be the velocity of the receiver, and V_1 that of the operator; then $\frac{V}{V_1}$ is their velocity ratio. See Art. 8.

It must be observed, that this *velocity ratio* is not at all effected by the *actual velocities* of the parts, provided the velocity ratio of the mechanism be constant for all positions. In the more ordinary pieces of mechanism (such as common toothed wheels, wheels moved by straps, levers, &c.) the velocity ratio is constant, that is to say, it remains the same for all positions of the mechanism.

In eq. (1) S may be taken as the velocity of the power P, estimated in the direction in which it acts, and S_1 that of the resistance P_1; then this equality becomes—

$$P \times V = P_1 \times V_1 \ldots (3),$$

$$\text{or } \frac{P_1}{P} = \frac{V}{V_1} = \text{the velocity ratio, } \ldots (4).$$

Now $\frac{P_1}{P}$ is called the advantage gained by the machine, or the number of times that the resistance moved is greater than the power applied. Hence the advantage gained by a machine, irrespective of friction, &c. is equal to the velocity of the power divided by the velocity of the resistance, or the velocity ratio of the power and resistance.

This is called the *principle of virtual velocities.* Workmen express this dynamic law by saying, "What is gained in power is lost in speed."

12. The DIRECTIONAL RELATION of the motion of the receiver and the operator admits of every possible variation. It may be constant or it may be variable. By the intervention of me-

chanism rectilinear motion may be converted into curvilinear motion, and conversely; reciprocating rectilinear or circular motion may be converted into continuous circular motion, and conversely; and so on to the various possible combinations of which the cases admit. These directional changes are so important, in a practical point of view, that some eminent writers on mechanism have made them the basis of the classification of mechanism. But however eligible in a practical point of light such a classification may be, there is complexity in its application, which renders it less suitable for scientific purposes than that method of classification which is based upon the nature or mode of action of certain elementary pieces of mechanism which enter, more or less, into every mechanical combination.

Elementary Forms of Mechanism.

13. In analysing the parts of a machine we find motion transmitted by jointed rods or links, by straps and cords, by wheels rolling on other wheels, and by pieces of various forms sliding or slipping on other pieces. Hence we have the following elementary forms of mechanism:

(1.) Transmission of motion by jointed rods,—LINK-WORK.

(2.) By straps, cords, &c.,—WRAPPING CONNECTORS.

(3.) By wheels or curved surfaces, revolving on centres, rolling on each other,—WHEEL-WORK.

(4.) By pieces of various forms, sliding or slipping on each other,— SLIDING-PIECES.

14. The velocity ratio, as well as the directional relation, in an elementary piece of mechanism may be either constant or varying. The number of combinations of which these elementary pieces admit, is almost unlimited. The eccentric wheel is a combination of sliding pieces and link-work. The common crane is a combination of wheel-work, link-work, and wrapping connectors; and so on to other cases.

A train of mechanism must be supported by some frame work; the train of pieces being such, that when the receiver is moved the other pieces are constrained to move in the manner determined by the mode of their connexion. Revolving pieces, such as wheels and pullies, are so connected with the frame that every portion of them is constrained to move in a circle round the axis; and sliding pieces are constrained to move in straight lines by guides.

Mechanism is to a great extent a geometrical inquiry. The motion of one piece in a train may differ, both in kind and direction, from the motion of the next piece in the series: these changes are effected by the geometrical construction of the pieces, as well as by their mode of connexion. The investigation of the law of these changes constitutes one of the chief objects of the principles of mechanism.

II. ON LINK-WORK.

15. If a bent rod or lever A C B turn upon the centre C, the velocities of the extremities A and B will be to each other in the ratio of their distances from the centre of motion C, that is,

$$\frac{\text{velocity A}}{\text{velocity B}} = \frac{\text{circum. cir. A } Q_1}{\text{circum. cir. B Q}} = \frac{\text{A C}}{\text{B C}}$$

Fig. 4.

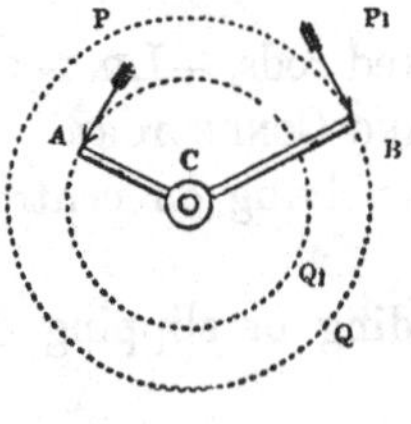

Fig. 5.

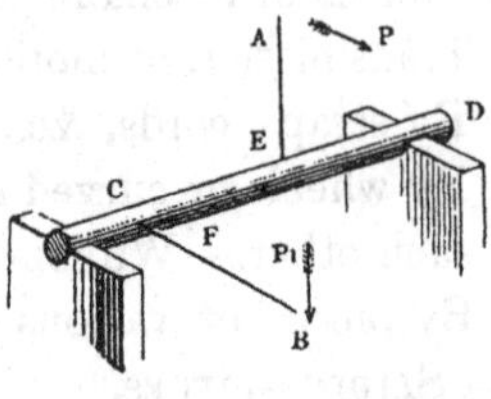

It is not necessary that the arms A C and B C should be in the same plane. Thus let C D be an axis round which the arms A E and B F revolve, then,

$$\frac{\text{velocity A}}{\text{velocity B}} = \frac{\text{perpend. dist. A from the axis}}{\text{perpend. dist. B from the axis}}$$

Fig. 6.

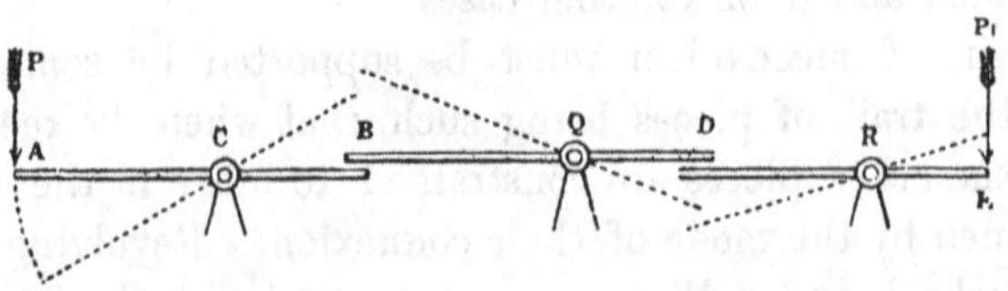

16. Let A B, B D, D E, be a series of levers turning on the fixed

centres C, Q, and R; then when the arcs, through which the extremities A and E are moved are small the velocity ratio will be expressed by the following equality:—

$$\frac{\text{velo. A}}{\text{velo. E}} = \frac{\text{A C. B Q. D R}}{\text{B C. D Q. E R}},$$

that is to say, *the velocity ratio of* P *and* P_1 *is found by taking the product of the lengths of the arms lying towards* P, *and dividing by the product of those lying towards* P_1.

17. To find *the velocity ratio of the rods* A B *and* C D, *turning on the fixed centres,* A *and* D; *and connected by the link* B C.

Fig. 7.

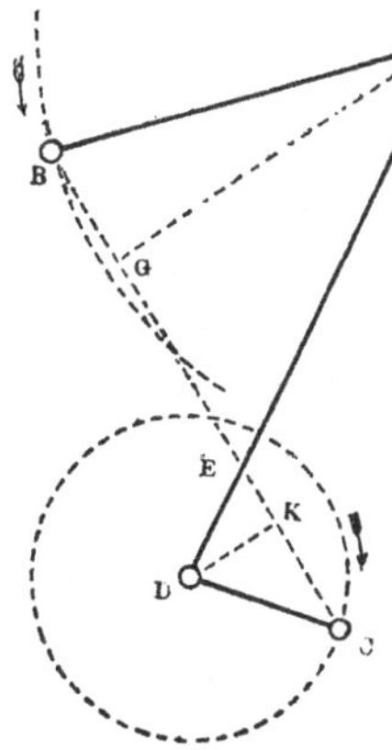

Through the centres A and D, draw the straight line D E A, cutting C B in E; and from A and D let fall the perpendiculars A G and D K upon C B, or it may be upon C B produced. Then

$$\frac{\text{ang. velo. D C}}{\text{ang. velo. A B}} = \frac{\text{A G}}{\text{D K}} \dots (1);$$

that is to say; *the angular velocities of the rods* D C *and* A B *are to each other in the inverse ratio of the perpendiculars let fall from their respective axes upon the direction of the link.*

Similarly we also have,

$$\frac{\text{ang. velo. D C}}{\text{ang. velo. A B}} = \frac{\text{A E}}{\text{D E}} \dots (2);$$

that is to say, *the angular velocities of the rods* D C *and* A B *are to each other in the inverse ratio of the segments into which the link divides the line joining their axes.*

These velocity ratios are obviously varying, depending upon the relative positions of the rods.

18. THE CRANK AND GREAT BEAM.—Let A B represent one half of the great beam of a steam engine, D C the crank, and B C the connecting rod. Putting β for the angle D C B, and β_1 for the angle A B C; then

$$\frac{\text{velo. crank}}{\text{velo. crank}} = \frac{\sin \beta_1}{\sin \beta} \ldots (1).$$

When the connecting rod B C is very long as compared with the length of the crank D C, then β_1 is nearly constant, being nearly equal to 90°, in this case, eq. (1) becomes

$$\frac{\text{velo. crank}}{\text{velo. beam}} = \frac{1}{\sin \beta} \ldots (2).$$

The crank must be in the same straight line with the connecting rod, at the highest and lowest points of the stroke of the beam, and then $\beta = 0$. In these positions the crank is said to be at its dead points.

The velocity ratio, expressed by eq. (2), will be a maximum when $\beta = 0$, that is, the velocity of the crank will be a maximum when it is in its dead points. When $\beta = 90°$, or when the crank is at right angles to the connecting rod, then the velocity of the crank is a minimum.

If R = A B, or one-half the length of the great beam; r = D C, the length of the crank; and A = the angular oscillation of the beam, or the whole angle described by the beam in one stroke; then

$$r = \text{R} \sin \frac{\text{A}}{2} \ldots (3)$$

which expresses the length of the crank in terms of the radius of the beam and angle of its stroke.

A *double oscillation* of the beam produces *one complete rotation* of the crank, or conversely, taking the crank as the driver, *each rotation* of the crank produces *a double oscillation* in the beam.

From eq. (1) it follows, that the velocity of the crank is equal to the velocity of the beam, when $\beta = \beta_1$ or angle D C B is equal to angle A B C; that is, when the position of the crank is parallel to that of the beam.

By this form of the crank the *reciprocating circular* motion of the extremity of the beam is changed into a *continuous circular motion*; and conversely a *continuous circular motion* is changed into a *reciprocating circular motion.*

19. *To determine the various relations of position and velocity of the* CRANK *and* PISTON *in a locomotive engine.*

Here the connecting rod, D E, is attached to the extremity of the piston rod, P D, and *the length of the stroke of the piston*

Fig. 8.

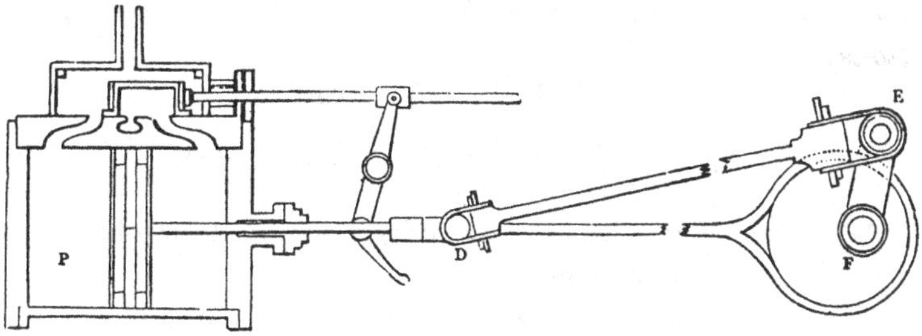

is equal to double the length of the crank, F E. Moreover, the centre, F, of the crank is in the same straight line with the axis of the cylinder or the direction of the piston rod.

Let l = D E, the length of the connecting rod;

l_1 = P D, the length of the piston rod;

r = F E, the length of the crank;

k = F D, the varying distance of the extremity of the piston rod from the axis of the crank;

h = the corresponding height of the stroke of the piston;

θ = the varying angle, F E D, which the crank forms with the direction of the connecting rod.

(1.) *The velocity ratio of the crank and piston* is expressed by the following equality:—

$$\frac{\text{velo. crank}}{\text{velo. piston}} = \frac{k}{l \sin \theta} \cdots (1), \text{ or}$$

$$= \frac{1}{\sin \beta} \cdots (2),$$

where β in eq. (2) is put for angle E F D; that is, the angle which the crank makes with the direction of the piston rod.

This latter form of the expression is the same as that given in eq. (2), Art. 18.

(2.) When the piston is at the bottom of point of its stroke, its distance from F = FE + ED + DP = $r + l + l_1$; also FD = FE + DE = $r + l$.

When the piston is at the middle point of its stroke, then FD = ED; that is to say, in this position of the piston DEF will be an isosceles triangle.

(3.) *The position of the crank at any point of the stroke of the piston, is determined by the two following general equations:*—

$$k = r + l - h \dots (3).$$

$$\cos\theta = \frac{r^2 + l^2 - (r + l - h)^2}{2\,r\,l} \dots (4).$$

When the piston is at the middle point of its stroke, then $h = r$, and eq. (4) becomes

$$\cos\theta = \frac{r}{2\,l} \dots (5).$$

When the crank is at right angles to the connecting rod, $\theta = 90°$, and then we find from eq. (4),

$$h = r + l - \sqrt{r^2 + l^2} \dots (6).$$

This expression is, obviously, less than r, or half the whole stroke of the piston. Hence it appears that the crank is at right angles with the connecting rod, before the piston has attained the middle point of its upward stroke.

20. Fig. 9 shows how a rotation of the axis A is transmitted to another C, by means of the two equal cranks AB and CD, connected by the connecting rod DB, whose length is equal to the distance AC, between the two axes. In all positions of the cranks, the figure ABCD will be a parallelogram, and the velocity of D will always be equal to the velocity of B, and the motion of the axis C will be exactly the same as that of the axis A.

Fig. 9.

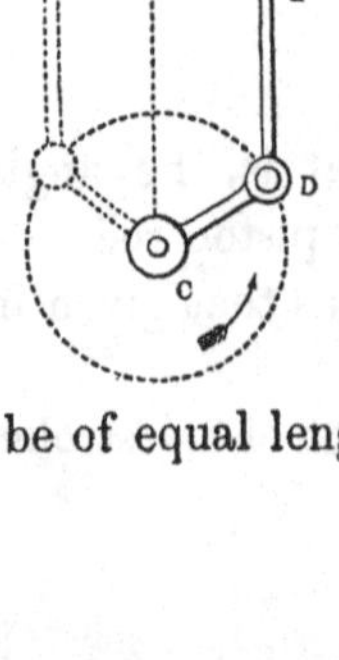

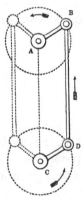

21. Two sets of cranks may be placed upon the axes, having the cranks on each axis at right angles to each other, as shown in fig. 10, where the cranks are formed by bending, or loops made in the axes. These axes must be parallel to each other, and the connecting rods must also be of equal lengths.

The advantage of this combination consists in maintaining a constant moving pressure, by which means an equable motion is sustained without the aid of the inertiæ of the machinery.

Fig. 10.

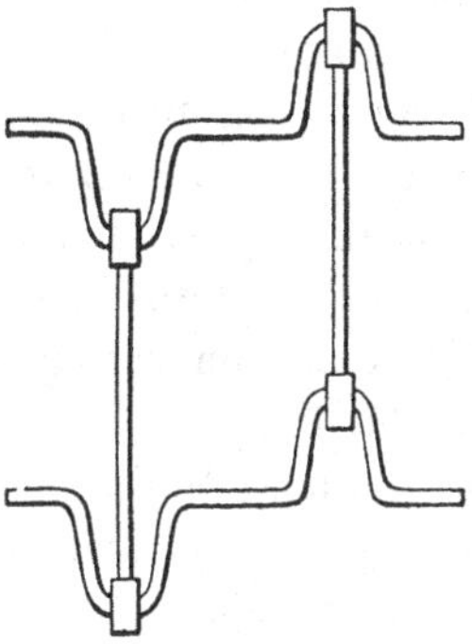

Fig. 11.

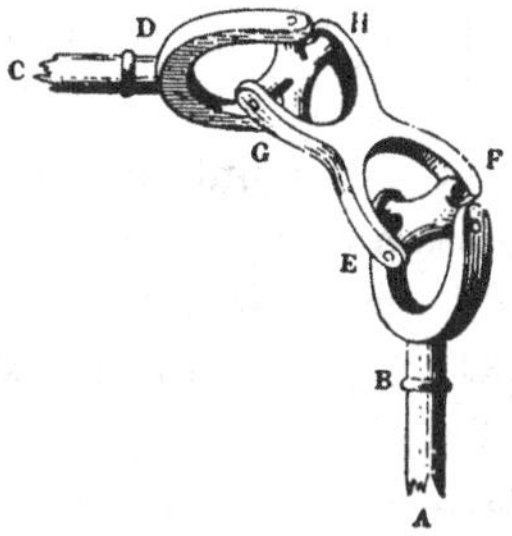

22. The double universal joint, represented in fig. 11, furnishes another example of link-work, for transmitting motion from one axis to another axis. This useful piece of mechanism should be constructed, so that the extreme axes, A B and C D, would meet in a point, if produced, and the angles which they respectively make with the central line of the intermediate piece, E F H G, shall be equal to each other.

Fig. 12.

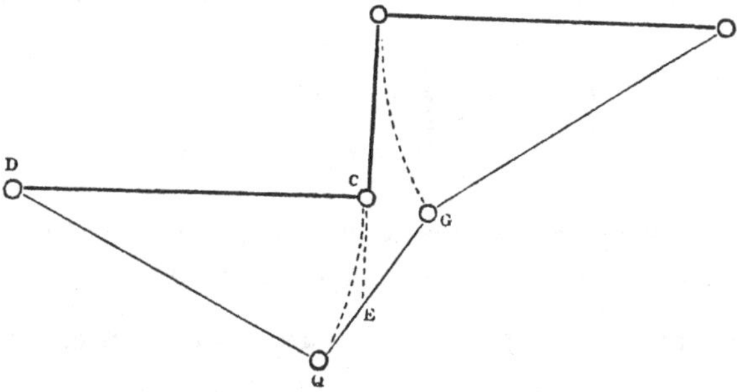

TO CONSTRUCT WATT'S PARALLEL MOTION.

23. This beautiful and useful piece of mechanism is formed by a combination of link-work.

Let A B and C D (see figs. 12 and 13) be two rods, turning on the fixed centres A and D, and connected together by the short link C B; then, when motion is given to the rods, there is a certain point, E, in the link C B, which will move, or very nearly move, in a straight line. In matter of fact the *path*, or *locus*, of this point is a curve of the fourth degree; but when the motion of the rods is limited, and their lengths are considerable, as compared with the length of their connecting link, this path becomes almost exactly a straight line.

Fig. 13.

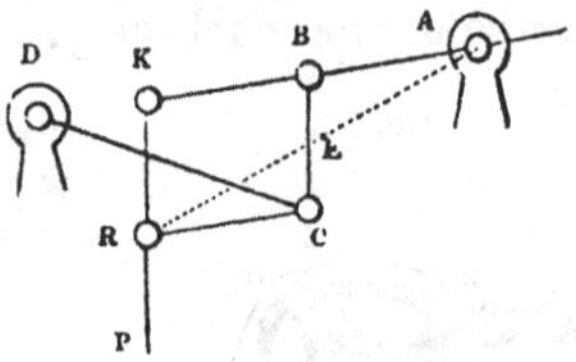

In fig. 13, C B K R is a parallel frame of links; to the joint R is attached the piston rod R P of the steam engine; and to the point E is attached the piston rod of the air-pump.

(1.) *To find the point* E (see fig. 12) *to which the air-pump rod must be attached, having given the radius rod* C D, *the link* C B *or* Q G, *and the rod* A B *or* A G *forming a part of the great beam.*

Let D Q, A G be an extreme position of the rods. Let the rods be moved to the position A B C D, where the link C B is perpendicular to A B and D C. Produce B C, meeting the link Q G in the point E; then E will be that point of the link which will most nearly move in a vertical straight line. The ratio of Q E to G E is generally expressed by the following equality,—

$$\frac{QE}{GE} = \frac{R}{r} \times \left(\frac{r \sin \frac{a}{2}}{R \sin \frac{A}{2}}\right) \dots (1);$$

where $R = AB$, $r = DC$, $a =$ angle C D Q, and $A =$ angle B A G.

Practically the link Q G or C B deviates very little from the vertical; and the angles a and A are small; hence, $r \sin \frac{a}{2} = R \sin \frac{A}{2}$ very nearly; in this case, therefore, eq. (1) simply becomes

$$\frac{QE}{GE} = \frac{R}{r} \dots (2);$$

and from this equality we readily find,

$$\text{G E} = \frac{\text{D Q} \times \text{G Q}}{\text{D Q} + \text{A G}} \ldots (3)$$

which gives the position of the point E, as required.

When D Q = A G, then $\text{G E} = \frac{\text{G Q}}{2}$, that is to say, in this case, the point E is at the middle of the link Q G or C D.

Example.—Let A B or A G = 5 ft., D C or D Q = 4 ft., and C B or G Q = 1·5 ft.; then by eq. (3) we have—

$$\text{G E} = \frac{4 \times 1{\cdot}5}{4 + 5} = \frac{2}{3}\text{ ft.}$$

(2.) *To find the length of the radius rod* D C (see fig. 13), *when the divisions,* A B *and* B K, *on the beam are given.*

In this case,

$$\text{The radius rod, D C} = \frac{\text{A B}^2}{\text{B K}} \cdots (4).$$

When A B = B K; then D C = A B; that is, in this case, the radius rod will be equal to the division A B on the beam.

Example.—Let A B = 6 ft., and B K = 4 ft.; then by eq. (4) we have—

$$\text{The radius rod, D C} = \frac{6^2}{4} = 9\text{ ft.}$$

To multiply Oscillations by means of Link-work.

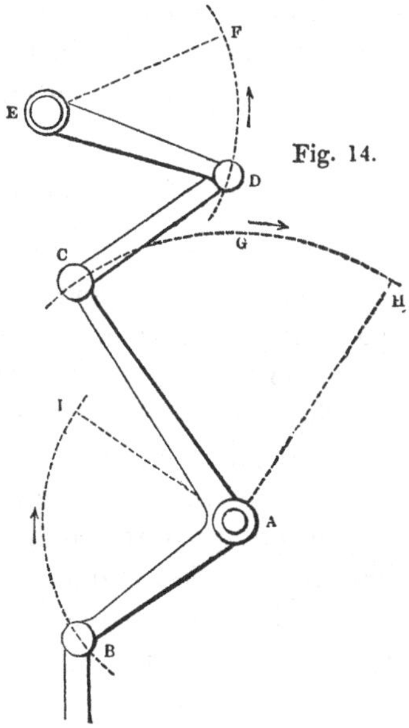

Fig. 14.

24. Fig. 14 represents a system of links B A C, C D, and D E, turning on the fixed centres A and E, and having the arms A B and A C united to the same centre A. The construction is such, that while the rod A B makes a *single* oscillation from B to I, the rod E D will make a *double* oscillation viz. from D to F, and back from F to D. The oscillations of A B are produced by the rotation of a crank (see Art. 17), or by any other means.

The conditions of the construction may be stated as follows:

Given the lengths of the arms A C and E D, the lengths or angles of their oscillations, and the length of the connecting link C D, to construct the mechanism, so that the rod E D shall perform two oscillations whilst A B makes one.

Let B A C be the position of the bent lever at the commencement of the upward oscillation. Draw A I and A H, making the angles B A I and C A H each equal to the angle of the oscillation. From A as a centre, with A B and A C as radii, describe the arcs B I and C H. Through A draw A G F bisecting the angle C A H cutting the arc C H in G. On A G F take A F equal to the sum of the rods A C and C D, and make F D equal to the given length of the oscillation of E D. From D and F as centres, with a radius equal to the length of the rod E D, describe circles, cutting each other in E; then E will be the centre of the rod E D, which will perform two oscillations, whilst the rod A B makes one.

When A B and A C are in the middle points of their oscillations, the rod E D will have the position E F, that is, it will have performed a complete upward oscillation. When A B and A C have performed the remaining halves of their oscillations, the rod E F will have returned to the original position, that is, it will have performed a complete downward oscillation.

In like manner the oscillations may be further multiplied, by connecting E D with another series of links.

To produce a Velocity which shall be rapidly retarded, by means of Link-work.

25. In fig. 15, R A C and E D represent two rods, turning on fixed centres A and E, and connected by a link C D; the rod E D is supposed to oscillate uniformly between the positions E D and E F. Now the construction is such as to produce a rapidly retarded motion of the rod R C in moving from the position R A C to the position S A B, and conversely.

The conditions of the construction may be stated as follows:

Given the rods E D and D C in position and magnitude, the angle of oscillation D E F, and the length of the rod A C to construct the mechanism.

Bisect the arc D F in G, and then bisect the arc F G in K; through the points K and E, draw the straight line K E C; from

Fig. 15.

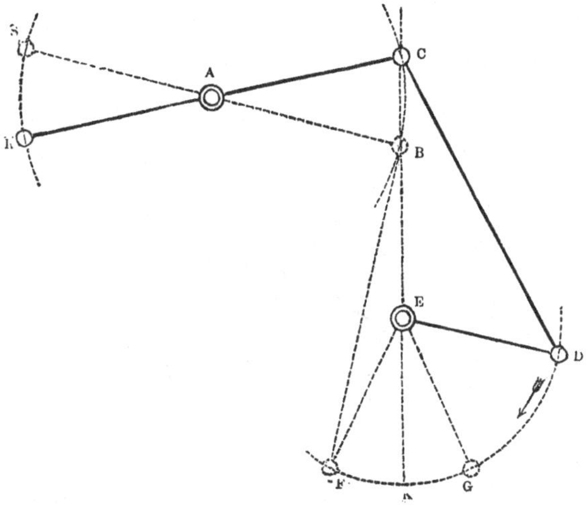

D and K as centres, with a radius equal to the length of the link D C, describe arc, cutting K E C in the points C and B; from B and C as centres, with a radius equal to the length of the rod A C, describe arcs cutting each other in the point A; then A will be the centre of the rod A C.

When the rod E D arrives at the position E G, the rod R A C will have the position S A B *very nearly*, and it will have moved with a rapidly retarded motion. During the remaining half of the oscillation G F, the rod S A B will remain, virtually, stationary.

This piece of mechanism was first employed by Watt for opening the valves of the steam engine.

To produce a Reciprocating Intermittent Motion by means of Link-work.

26. A B and C D (fig. 16) are two rods, turning on the fixed centres A and D, and connected by a link B C. The rod A B is made to oscillate between the positions A B and A I, by means of a crank and connecting rod. The construction of the mechanism is such, that the rod D C will oscillate between the positions D C and D F, but with an intermittent motion.

The conditions of the construction may be stated as follows:

Given the rods A B, B C, and C D in position and magnitude, to construct the mechanism.

Fig. 16.

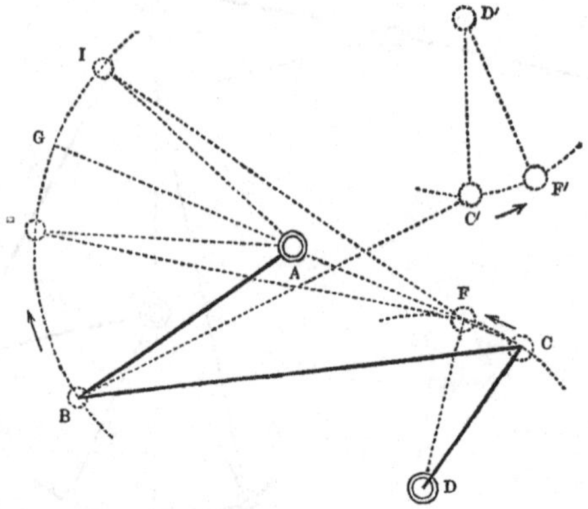

From A as a centre, with the radius A B, describe the arc B I; through C and A draw the straight line C A G, meeting the arc in G; make G E equal to one-third the arc G B, and on the arc take G I equal to G E; on the line G A C take G F equal to B C; then half the chord B I will give the length of the crank, and C F will be the arc through which the rod D C oscillates.

Bisecting the angle B A E, &c. the position of the rod D′ C′ is found, which being connected with B, by the link B C′, will oscillate exactly in a contrary manner to that of the rod D C, that is to say, when D C is stationary D′ C′ will be in motion, and conversely.

When the point B arrives at E, the rod D C will have completed, practically, its oscillation, and there it will remain stationary until the rod, turning on the centre A, returns from the position A I to A E.

The Ratchet-wheel and Detent.

27. In fig. 17, A represents the ratchet-wheel, and D the detent, falling into the angular teeth of the ratchet, thereby admitting the wheel to revolve in the direction of the arrow, but at the same time preventing it from revolving in the opposite direction.

Fig. 17.

In certain kinds of machinery, the action of the moving force undergoes periodic intermissions, in such cases the rachet and detent are used to prevent the recoil of the wheels, and sometimes to give an intermittent motion to the wheel as in the following example.

Intermittent Motion produced by Link-work connected with a Ratchet-wheel.

28. B E is a rod, turning on the fixed centre B, to which a reciprocating motion is given by the connecting rod C of a crank, or by any other means; E F is a click, jointed to the rod B E at its extremity, and gives motion to the ratchet-wheel A. At each upward stroke of the rod B E, the click E F, acting upon the saw-like teeth of the ratchet-wheel causes it to move round one or more teeth; and when the extremity F of the click is drawn back by the descent of the lever B E, it will slide over the beveled sides of the teeth without giving any motion to the wheel, so that at every upward stroke of the rod C the ratchet-wheel will be moved round and it will remain at rest during every downward stroke of the rod. Thus the reciprocating motion of the connecting rod, C, will produce an intermittent circular motion in the axis A.

Fig. 18.

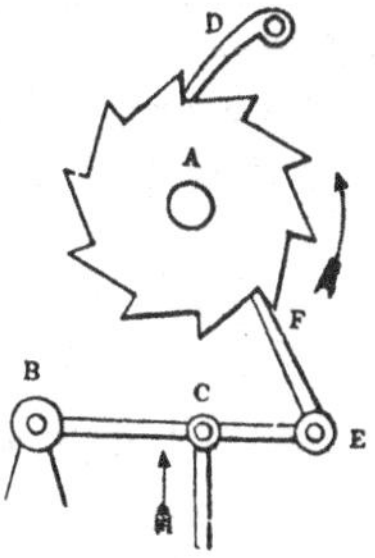

III. ON WRAPPING CONNECTORS.

29. When the moving force of the machinery is not very great, cords, belts, and other wrapping connectors, are most usually employed in transmitting motion from one revolving axis to another.

30. The *endless cord* or *belt* A B C D, represented in figs. 19 and 20, passes round the wheels, A B and C D, revolving on the parallel axes R K and Q F, and transmits motion from the axis Q F to the axis R K, with a constant velocity ratio. In all such cases the motion is entirely maintained by the frictional adhesion of the cord or belt to the surface of the wheel.

Fig 19.

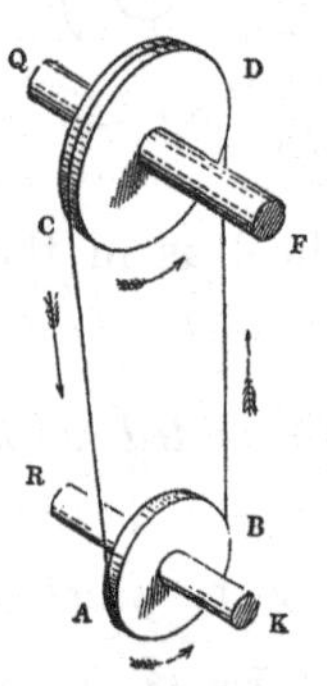

Fig. 20.

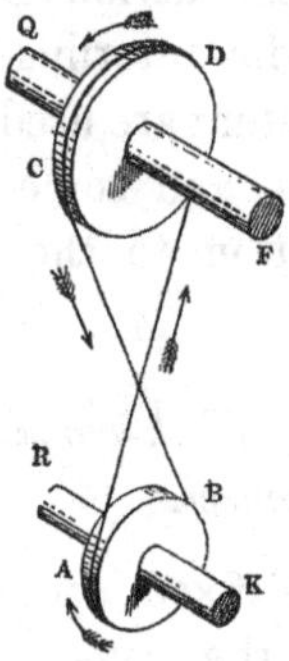

When the cord, passing round the wheels is *direct*, as in fig. 19, the motions of the wheels take place in the *same direction*, and when the cords cross each other, as in fig. 20, the motions of the wheels take place in *opposite directions*.

If the wheel C D makes one revolution, then

$$\text{No. revo. A B} = \frac{\text{circum. C D}}{\text{circum. A B}} = \frac{\text{radius C D}}{\text{radius A B}} \dots (1).$$

Or putting R and r for the radii of the wheels C D and A B respectively, and Q and q for their respective synchronal rotations, then

$$\frac{q}{Q} = \frac{r}{R} \dots (2).$$

Example.—If the radius of the wheel C D be 12 inches, and that of A B 9 inches, what will be the least number of entire revolutions which they must make in the same time?

Here, by eq. (2), we have

$$\frac{q}{Q} = \frac{R}{r} = \frac{12}{9} = \frac{4}{3}.$$

The fraction $\frac{12}{9}$ reduced to its least terms is $\frac{4}{3}$, therefore the least number of synchronal rotations are 4 and 3, that is to say, whilst the wheel C D makes 3 rotations, the wheel A B will make 4.

31. Fig. 21, represents a system of three revolving axes, in which motion is transmitted from one to the other, by means of a series of belts.

Fig. 21.

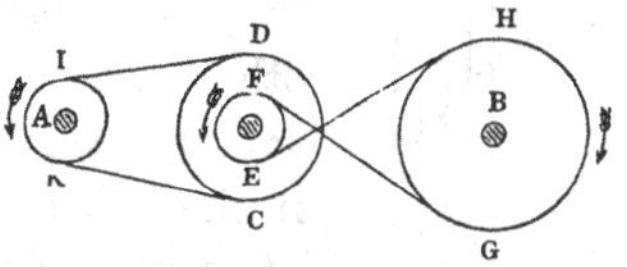

The belt being direct in the wheels A and D C, their axes will move in the same direction, but as the belt crosses in passing from D C to H G, their axes will move in opposite directions.

Here, whilst the axis B makes one rotation, the

$$\text{No. rotations A} = \frac{\text{rad. H G} \times \text{rad. D C}}{\text{rad. E F} \times \text{rad. I K}} \dots (1).$$

Or putting $R_1 = \text{rad. D C}$, $R_2 = \text{rad. H G}$, &c., $r_1 = \text{rad. I K}$, $r_2 = \text{rad. E F}$, &c., and putting q and Q for the synchronal rotations of the first and last axes respectively; then

$$\frac{q}{Q} = \frac{R_1 \times R_2 \times R_3 \times \&c.}{r_1 \times r_2 \times r_3 \times \&c.} \dots (2).$$

Example.—In the mechanism represented in fig. 21, let $R_1 = 8$, $R_2 = 15$, $r_1 = 5$, $r_2 = 4$; required the least number of entire rotations performed in the same time by the axes A and B.

Here, by eq. (2) we have,—

$$\frac{q}{Q} = \frac{8 \times 15}{5 \times 4} = \frac{6}{1}$$

that is, whilst the axis B makes one revolution, the axis A will make six.

32. In raising buckets from deep wells or from pits, a continuous cord coils round an axle or a drum wheel, as the case may be, the full bucket being attached to one end of the cord and the empty bucket to the other end, the rotation of the axle coils up the cord to which the full bucket is attached and at the same time uncoils the cord to which the empty one is attached so that whilst the former is ascending the latter is descending.

Speed Pulleys.

33. Fig. 22, represents an arrangement of speed pulleys; A B and C D are two parellel axes upon each of which is fixed a series of pulleys, or wheels, adapted for a belt of given length, so that it may be shifted from one pair of wheels to any other pair, say for example, from the pair $a\ a_1$ to the pair $c\ c_1$. In order to suit this arrangement, if the belt be crossed, *the sum of the diameters of any pair of pulleys must be a constant quantity*, that is to say, it must be equal to the sum of the diameters of any other pair. By this contrivance, a change in the velocity ratio of the two axes is produced by simply shifting the belt from one pair to another.

Fig. 22.

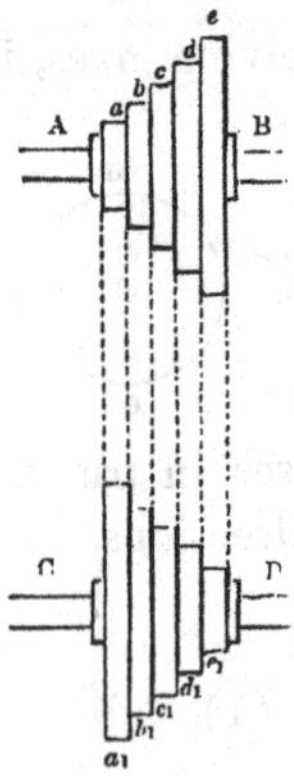

In practice it is customary to make the two groups of pulleys exactly alike, the smallest pulley of one being placed opposite to the largest of the other.

In a group of speed pulleys, let S = the constant sum of the diameters of the driver and follower, D = the diameter of the driver, d = the diameter of the follower, and Q, q the number of their synchronal rotations respectively; then $\frac{Q}{q} = \frac{d}{D}$, and

$$D = \frac{q \times S}{Q + q} \dots (1);$$

$$d = \frac{Q \times S}{Q + q}, \text{ or more simply,}$$

$$= S - D \dots (2).$$

Example.—Required the diameters of a pair of speed pulleys, when the sum of the diameters is 30 inches, and the driver makes 2 revolutions, whilst the follower makes 3.

Here S = 30, Q = 2, and $q = 3$; then by eq. (1) and (2) we have

$$D = \frac{3 \times 30}{5} = 18 \text{ in.; and } d = 30 - 18 = 12 \text{ inches.}$$

If the constant sum of the diameters of a group of 5 pairs of speed pulleys be 12 inches, and the diameters of the pulleys a_1, b_1, c_1, d_1, e_1, be 10, 8, 6, 4, and 2 inches respectively, then the diameters of the pulleys a, b, c, d, e, will be 2, 4, 6, 8, and 10 inches respectively; and as the strap is shifted from one pair of wheels to another, the relative velocities of the axes CD and AB will be as the numbers $\frac{1}{5}$, $\frac{1}{2}$, 1, 2, and 5.

34. It is customary to construct the pairs of speed pulleys so that the rotations of the follower may be increased or decreased in a certain geometric ratio. Thus, if r be this ratio, then for 5 pairs of speed pulleys we shall have the series of terms $\frac{1}{r^2}$, $\frac{1}{r}$, 1, r, r^2, for the different values of $\frac{Q}{q}$, the ratio of the synchronal rotations of each pair. Or, generally if n be the number of pairs, then $\frac{1}{r^{\frac{n-1}{2}}}$, $\frac{1}{r^{\frac{n-3}{2}}}$, ..., $r^{\frac{n-3}{2}}$, $r^{\frac{n-1}{2}}$, will be the different values of $\frac{Q}{q}$. In this case, let D_1, D_2, ..., D_n = the diameters of the 1st, 2nd, ..., and nth pulleys, respectively, on the driving axis; and these symbols, taken in a reverse order, will be the corresponding diameters of the pulleys on the driven axis; then

$D_1 = \frac{S}{1 + r^{\frac{n-1}{2}}}$, $D_2 = \frac{S}{1 + r^{\frac{n-3}{2}}}$, and so on: moreover we have

$D = S - D_1$, $D'_{n-1} = S - D_2$, and so on.

Example.—To find the diameters of a set of 5 pairs of speed pulleys, so that values of $\frac{Q}{q}$ (the ratio of the synchronal rotations of the different pairs) shall have the common ratio of $\frac{2}{3}$, the constant sum of the diameters of each pair being 26 inches.

Here $r = \frac{2}{3}$, $n = 5$, and $S = 26$, then from the foregoing formulæ we find—

$$D_1 = \frac{26}{1 + (\frac{2}{3})^2} = 18; \quad D_2 = \frac{26}{1 + \frac{2}{3}} = 15\tfrac{3}{5};$$

$$D_3 = \frac{26}{1 + (\frac{2}{3})^0} = 13; \text{ and so on.}$$

But the remaining diameters will be better found as follows:

$$D_5 = 26 - 18 = 8; \quad D_4 = 26 - 15\tfrac{3}{5} = 10\tfrac{3}{5}.$$

35. Two plain cones, having their axes parallel, as shown in

Fig. 23.

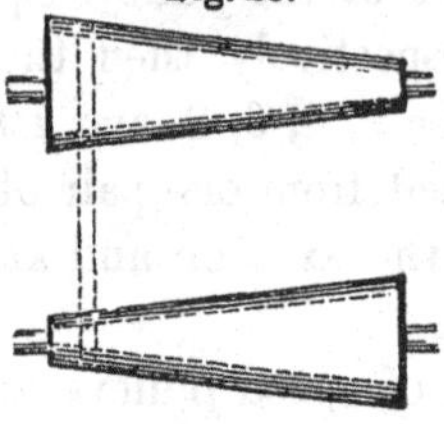

Fig. 23, will obviously answer the same purpose as the ordinary form of speed pulleys. The slant faces of the cones may be formed by any continuous curve; but with this condition,—that the sum of the diameters at every position of the band shall be a constant.

Guide Pulleys.

36. By the intervention of *guide pulleys* the direction of cords may be changed into any other direction. Thus, by means of the guide pulleys B and C, the motion of the cord in the direction CD is changed into the direction AB.

Fig. 24.

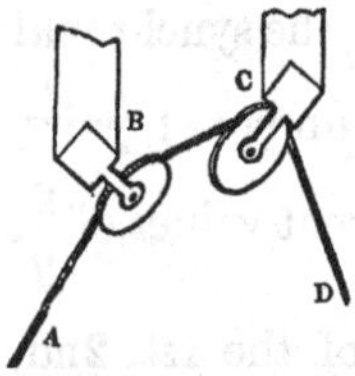

The cords D C and C B should be in the plane of the pulley C; and the cords C B and BA should be in the plane of the pulley B.

37. Two guide pulleys, E and H, may be employed to transmit motion from the wheel A to the wheel B, when the axes of these wheels have any given direction.

Fig. 25.

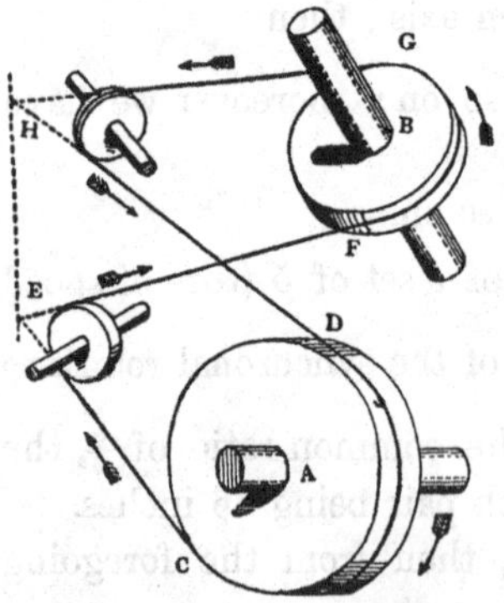

Let E H be the line where the planes, passing through the two wheels, intersect each other. In this line assume any two convenient points E and H; in the plane of the wheel A draw the tangents E C and H D; and in the plane of the wheel B draw the tangents E F and H G; then C E F G H D will be the path of the endless cord, which will be kept in this path by a guide pulley at E, in the plane of C E F, and another guide pulley at H, in the plane of D H G.

The relative velocities of the axes A and B depend entirely upon the ratio of the radii, A D and B G, of the two wheels. See Art. 30.

To prevent Wrapping Connectors from Slipping.

38. The slip of the band on the wheel, when it is not excessive, is in many cases rather an advantage than otherwise; but when motion is to be transmitted from one wheel to another according to some given exact ratio, *gearing-chains* of various forms are employed as the wrapping connectors.

39. In some cases the links of the gearing chain lay hold of pins or teeth formed upon the wheel, as shown in fig. 27. In other cases, the links of the gearing are joined together, some-

Fig. 26.

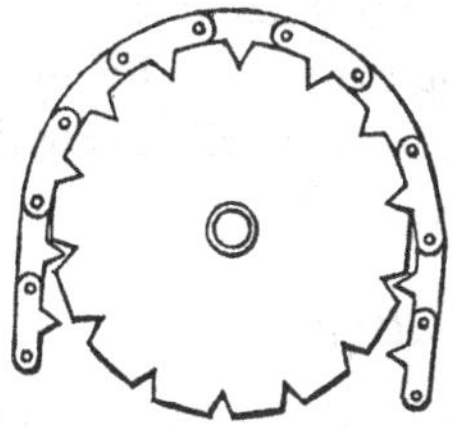

Fig. 27.

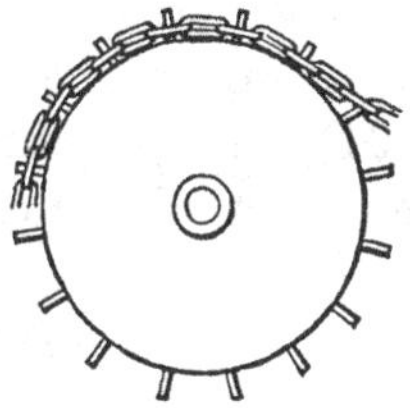

thing like a watch chain, and carry teeth which pass into certain notches made at corresponding distances on the edge of the wheel, as shown in fig. 26.

40. When a belt moves a conical wheel, it always happens that the belt gradually moves towards the broad end of the

Fig. 28.

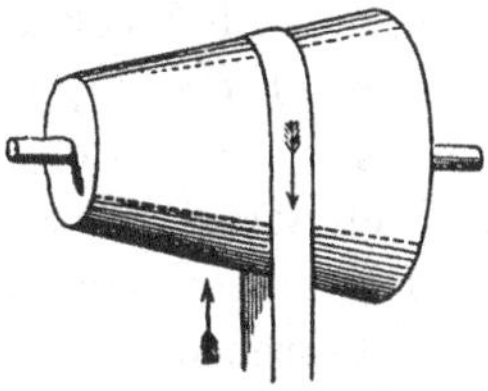

Fig. 29.

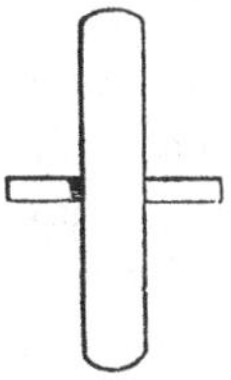

Fig. 30.

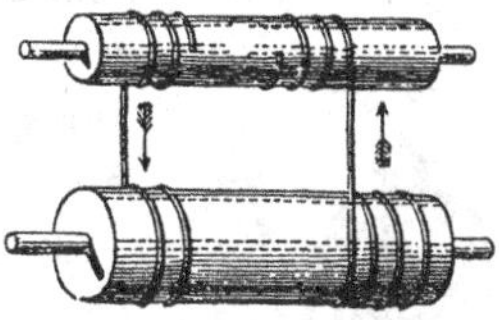

wheel: this is owing to the belt being more stretched on that side than it is on the other.

41. This property enables us to construct a wheel so that a belt shall not shift on its edge; this is simply effected by making the edge to swell a little in the middle, as shown in fig. 29.

42. When two rollers have to make only a limited number of revolutions in each direction, the slip of the cord may be prevented by having a cord coiled round each end of the rollers in opposite directions, so that while one cord is coiled on one extremity of the roller, the other cord is uncoiled from the other extremity, as shown in fig. 30.

Fig. 31.

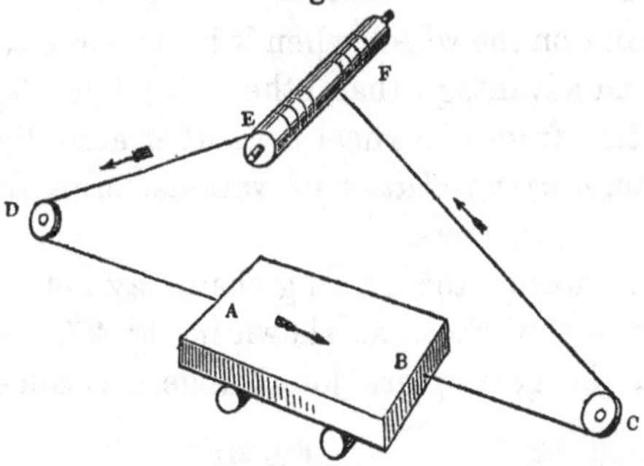

43. By a similar arrangement of cords on the cylinder E F (see fig. 31), a reciprocating motion of this cylinder will produce a back and forward motion of the carriage A B.

Systems of Pulleys.

44. A system of pulleys must at least contain one moveable pulley. When a wheel, forming a part of a system of wheels connected together by cords, has a progressive motion, it materially affects the velocity ratio of the receiver and the operator of the mechanism. There are a great many different systems of pulleys, but they all depend upon the different combinations of moveable and fixed pulleys, and the different modes of reduplication of a cord.

45. In this system of pulleys there is one moveable block and a single continuous cord with three duplications, so that whilst the moving force P acts by one cord, the moveable block with its load is suspended by six cords: if W ascend one foot, each of these cords will be shortened one foot, and therefore the cord P will be lengthened six feet; that is to say, the velocity of P will be six times that of W.

Fig. 32.

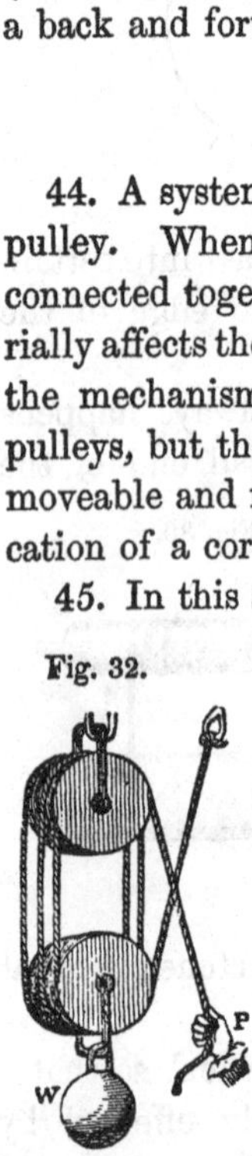

46. In the system of pulleys represented in fig. 33, there are two distinct cords and two moveable pulleys A and B, making two duplications of cord; then if A ascends one foot, B must ascend two

feet, and the cord at P must be lengthened four feet, that is, the velocity of P will be four times the velocity of W.

Generally if there are n moveable pulleys in such a system, then,

$$\text{velo. P} = 2^n \times \text{velo. W.}$$

Fig. 33.

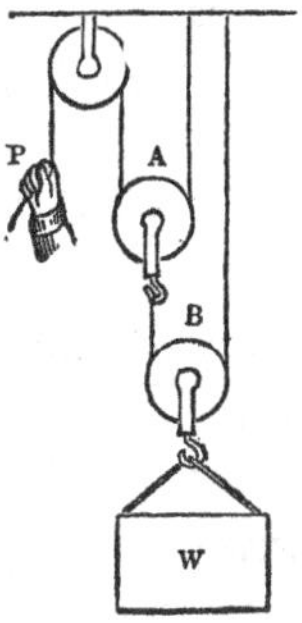

Fig. 34.

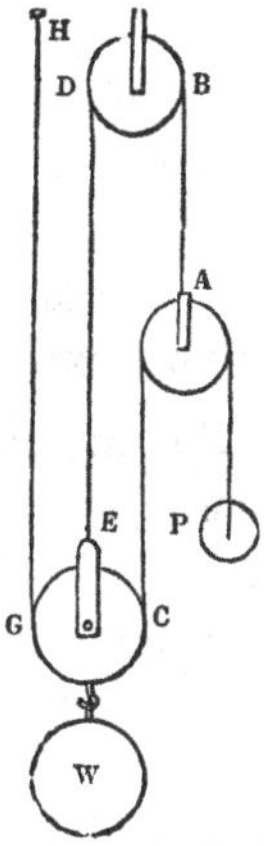

Fig. 35.

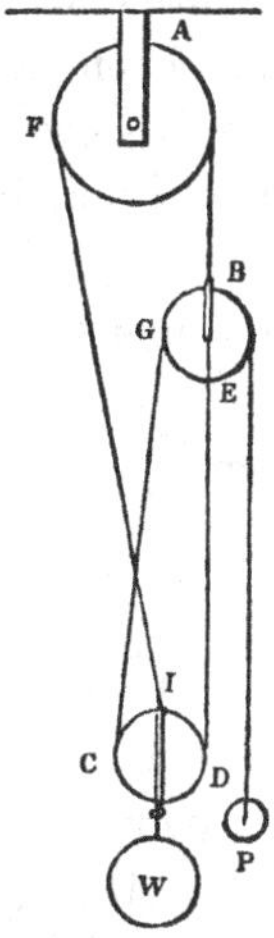

47. The system of pulleys represented in fig. 34, contains two moveable pulleys, one fixed pulley, and two single cords. In this case the velocity ratio of P to W is as four to one.

48. Fig. 35 represents a similar system of pulleys, in which the velocity ratio of P to W is as five to one.

In all these systems of pulleys the velocity ratios are constant.

49. In the compound wheel and axle, represented in fig. 36, the axle is made of different thicknesses as at A and B, and a continuous cord coils round these parts in different directions, and passes round the wheel of the moveable pulley D. In one revolution of the wheel C P the space moved over by the pulley D is equal to half the difference of the circumferences of the axles A and B. Putting R_1 for the radius of the wheel C P, R for

Fig. 36.

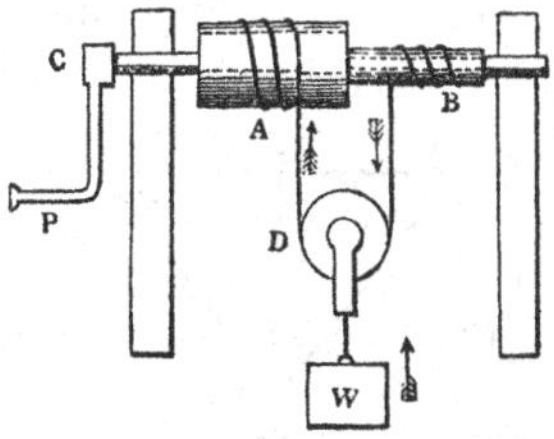

the radius of the axle A, and r for the radius of the axle B; then we have for the velocity ratio

$$\frac{\text{velo. P}}{\text{velo. W}} = \frac{2\,R_1}{R - r}$$

If $R_1 = 10$, $R = 4$, $r = 3\frac{3}{4}$; then $\frac{\text{velo. P}}{\text{velo. W}} = \frac{2 \times 10}{4 - 3\frac{3}{4}} = 80.$

This piece of mechanism belongs to a class which produces what has been called *differential motions*, their object being to produce a slow and definite motion in a body by the most simple and practicable means.

TO PRODUCE A VARYING VELOCITY RATIO BY MEANS OF WRAPPING CONNECTORS.

50. To find the ratio of the angular velocities of two eccentric wheels, moved by a cord wrapping over each.

Fig. 37.

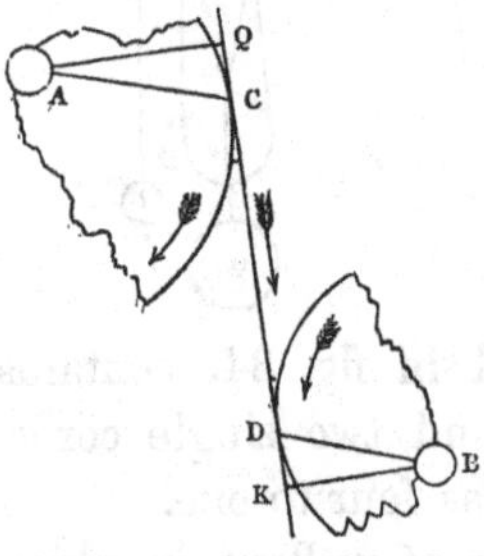

Let D C be a cord wrapping round the wheels, whose axes of motion are A and B; their line C D will be a tangent to the two curves forming the edges of the wheels. On D C produced let fall the perpendiculars A Q and B K; then the velocity of the cord, in this position of the wheels, will be equal to the velocity of the point Q, and at the same time it will also be equal to the velocity of the point K: hence we find,

$$\frac{\text{angular velocity A C}}{\text{angular velocity B D}} = \frac{\text{B K}}{\text{A Q}}, \ldots (1).$$

that is to say, the *angular velocities are inversely as the perpendiculars let fall upon the cord from the axes of motion.*

51. Let B be a moveable pulley suspended from the continuous cord P A B C, passing over a fixed pulley A, and attached to a point C in the same horizontal line with A. Let fall B D perpendicular to A C; then B C will always be equal to B A, and B will move in the vertical line B D. Hence we find,—

Fig. 38.

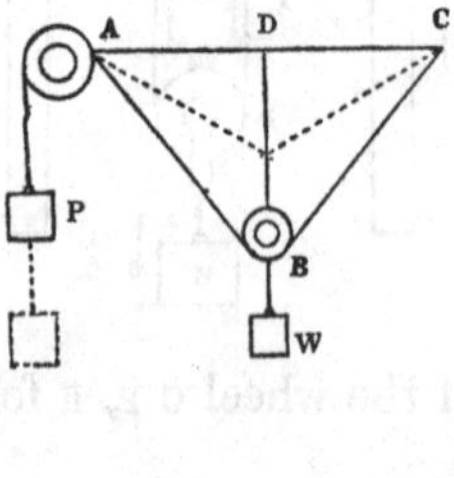

$$\frac{\text{velocity P}}{\text{velocity W}} = 2 \times \frac{\text{BD}}{\text{BA}} \dots (1).$$

This expression may be put in the following trigonometrical form :—

$$\frac{\text{velocity P}}{\text{velocity W}} = 2 \times \cos \text{PAB} \dots (2).$$

52. Fig. 39 represents a simple and ingenious contrivance for communicating a varying velocity to the axis B, by means of an endless band QKC, passing over an eccentric wheel A; a pulley BK, and a *stretching pulley* C. The curve of the eccentric wheel, A, must be such as to produce the varying velocity required. The weight W, attached to the stretching pulley C, keeps the band constantly stretched, so that whatever may be the velocity of the cord, upon leaving the eccentric wheel, it communicates the same velocity to the circumference of the pulley BK. From the axis A let fall AQ perpendicular to the cord QK; then by eq. (1) Art. 50, the velocity ratio may be expressed as follows:

Fig. 39.

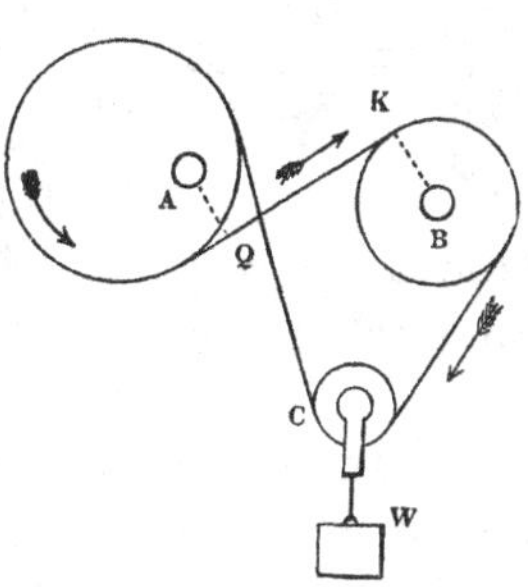

$$\frac{\text{ang. velo. axis A}}{\text{ang. velo. axis B}} = \frac{\text{BK}}{\text{AQ}}$$

Let the axis A revolve uniformly, and let the radius, BK, of the pulley be given; then

The ang. velo. axis B *will vary as the perpend.* AQ.

IV. ON WHEEL-WORK PRODUCING MOTION BY ROLLING CONTACT WHEN THE AXES OF MOTION ARE PARALLEL.

53. Two wheels E and F, in contact with each other, revolve on the parallel axes A B and C D; now if the wheels are in contact in any one position, they will also be in contact in every other position, and their circumferences will roll upon each other, so that if the driver F revolve on its axis C D it will communicate a rotatory motion to the follower E in a contrary direction, by the frictional adhesion of the parts successively brought in contact. The edges of these wheels must have the same velocity, and therefore their angular velocities will be inversely as their radii.

Fig. 40.

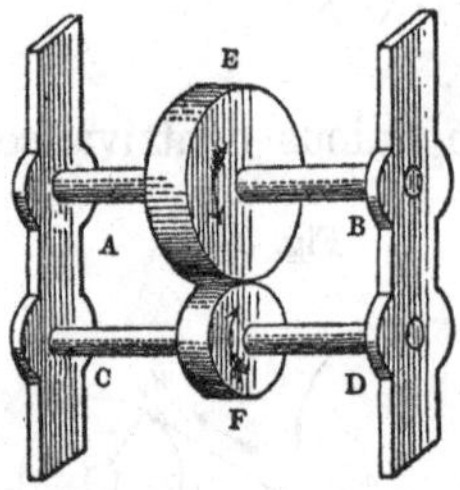

54. In order to render the transfer of motion perfectly exact, the edges of the wheels are formed into teeth, placed at equal distances from each other, so that when one wheel is turned, its teeth successively enter into the spaces formed on the edge of the other wheel. Thus, even with slight errors of construction, one wheel cannot escape from the other, which may happen in the case of simple rollers.

The numbers of teeth in the wheels, acting upon each in this manner, are in proportion to their radii. Thus, let the radius of the wheel A be 15 inches, that of B 6 inches, and let B contain 8 teeth; then

Fig. 41.

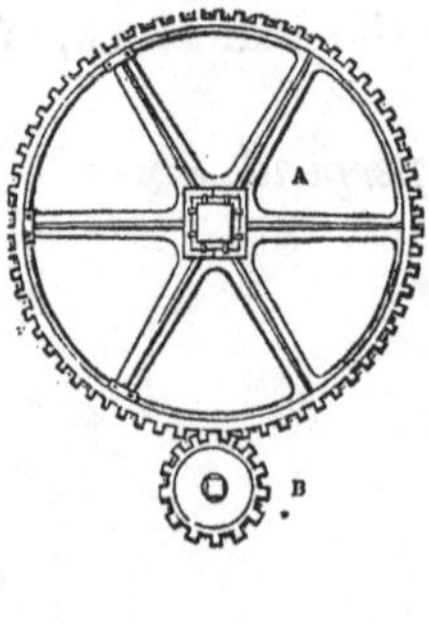

$$\text{No. teeth in A} = 8 \times \frac{15}{6} = 20$$

Or generally, if R and r be put for the radii of the wheels, and N and n the number of their teeth respectively; then

$$\frac{N}{n} = \frac{R}{r} \dots (1).$$

Hence angular velocities, as well as the synchronal rotations, of wheels may be expressed in terms of their numbers of teeth; thus we have

$$\frac{\text{ang. velo. A}}{\text{ang. velo. B}} = \frac{n}{\text{N}} \dots (2),$$

also, $$\frac{\text{synchronal rotation A}}{\text{synchronal rotation B}}, \text{ or } \frac{\text{Q}}{q} = \frac{n}{\text{N}} \dots (3).$$

Example.—Required the least number of teeth in the wheels A and B, so that B shall make 105 revolutions per min. and A only 40.

$$\text{Here by eq. (3);}\ \frac{n}{\text{N}} = \frac{40}{105} = \frac{8}{21};$$

that is, B will contain 8 teeth and A 21 teeth.

The form which must be given to the teeth of wheels, so as to maintain a perfect rolling contact, will be explained in another part of this work.

55. If the wheel A be the *driver* then B will be called the *follower*. Wheels acting this manner are sometimes called *spur-wheels*. Small toothed wheels are called *pinions;* thus B may be called a pinion in relation to A.

56. Toothed wheels are said to be *in gear* when their teeth are engaged together, and they are said to be *out of gear* when they are separated.

57. In the train of wheels represented in fig. 42, let. N_1, N_2, N_3, &c., be the number of teeth in the *driving* wheels, and n_1 n_2 n_3, &c., the number in the *driven* wheels; Q_1 = the no. of rotations of the first axis, Q_2 = the no. of the second axis, and so on, performed in the same time; then

Fig. 42.

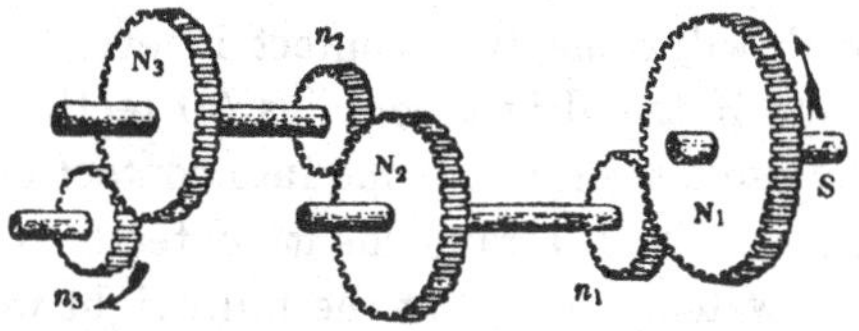

$$\frac{Q_{m+1}}{Q_1} = \frac{N_1 \cdot N_2 \cdot N_3 \cdot \dots N_m}{n_1 \cdot n_2 \cdot n_3 \cdot \dots n_m} \dots (1)$$

This equality may be expressed in language as follows:— *The ratio of the synchronal rotations of the last and first axes, is equal to the continued product of the number of teeth in the driving wheels divided by the continued product of the number of teeth in the driven wheels.*

Similarly we have,

$$\frac{Q_{m+1}}{Q_1} = \frac{Q_2}{Q_1} \times \frac{Q_3}{Q_2} \times \ldots \times \frac{Q_{m+1}}{Q_m} \ldots (2),$$

which may be expressed in language as follows:— *The ratio of the synchronal rotation of the first and last axes, is equal to the product of the separate synchronal ratios of the successive pairs of axes.*

The number of axes in this combination is always one more than the number of pairs of wheels.

It is evident, from eq. (1), that the drivers and followers may be placed in any order in a train of wheel-work without changing the velocity ratios of the first and last axes.

Example.—Let the number of pairs of drivers and followers be 3, that is, let $m = 3$, $N_1 = 16$, $N_2 = 15$, $N_3 = 14$, $n_1 = 7$, $n_2 = 6$, $n_3 = 5$; required the least number of synchronal rotations of the first and last axes in the train of wheels.

Here by eq. (1) we have—

$$\frac{Q_4}{Q_1} = \frac{16 \times 15 \times 14}{7 \times 6 \times 5} = \frac{16}{1};$$

that is, whilst the first axis makes one revolution, the last will make sixteen.

58. If the number of teeth in a driving wheel be some exact multiple of the number of teeth in the follower, then the same teeth will come into contact in every revolution of the driver. Thus if the driver contains 30 teeth and the follower 6, then the same teeth will come into contact at every revolution of the driver. This arrangement of teeth is preferred by the clock and watchmaker; but the millwright would add one tooth, called the HUNTING COG, to the large wheel, that is, he would have 31 teeth in the driver and 6 in the follower, because 31 and 6, being prime to each other, and at the same time nearly in the same ratio as 30 and 6, the same pair of teeth would not come again into contact until the large wheel had made 6 revolutions, and the small one 31.

59. Eq. (3), Art. 53, enables us readily to find the number of revolutions which the wheels must make in order that the same teeth may come again into contact with each other; for it is only

necessary to reduce the fraction $\frac{n}{N}$ to its least terms, and the denominator of this reduced fraction will give the number of revolutions of the driving wheel as required. Thus let $N = 144$, and $n = 54$, then $\frac{Q}{q} = \frac{54}{144} = \frac{3}{8}$; that is, the driver must make 3 complete revolutions, or the follower 8, before the same teeth can again come into contact.

60. In a combination of wheels, whose motions are expressed by the equality $\frac{Q_3}{Q_1} = \frac{N_1 \cdot N_2}{n_1 \cdot n_2}$, an indefinite number of values may be assigned to the numbers of teeth, which shall produce a given synchronal ratio of the first and last axes; but if n_1 and n_2 be given, and N_1 and N_2 be comprised within certain given limits; then a limited number of values may be found for N_1 and N_2.

Thus, for example, let $\frac{Q_3}{Q_1} = 60$, $n_1 = n_2 = 8$, and the values of N_1 and N_2 not to exceed 100 nor to be less than 40.

Here we have—

$$\frac{N_1 \cdot N_2}{8 \times 8} = 60;$$

$$\therefore N_1 \cdot N_2 = 60 \times 64;$$

hence, N_1 may be 60 and N_2 may be 64; but in order to determine all the combinations, we must put the product, 60×64, into prime factors, and then distribute these factors into different groups answering to the limiting values of N_1 and N_2.

Here, $60 \times 64 = 2^8 \times 3 \times 5$; hence we have—

1st combination, $(2^4 \times 3) \times (2^4 \times 5) = 48 \times 80$;

2nd combination, $(2^5 \times 3) \times (2^3 \times 5) = 96 \times 40$;

3rd combination, $2^6 \times (2^2 \times 3 \times 5) = 64 \times 60$.

61. When all the drivers contain the same number of teeth, and also the followers, then eq. (1), Art. 57, becomes

$$\frac{Q_{m+1}}{Q_1} = \left(\frac{N_1}{n_1}\right)^m \dots (1).$$

By means of this formula we may readily determine the least

number of axes requisite for producing a given synchronal ratio of rotation between the first and last axes, when the number of teeth in the drivers cannot exceed N_1 and the number in the followers cannot be less than n_1.

Find m, in eq. (1), equal to the highest whole number, which does not make the right member greater than the left; then the least number of axes will be $m + 2$. But if m, a whole number, can be found so as to make the right-hand member exactly equal to the left, then, in this case, the least number of axes will be $m + 1$.

Example.—Required the least number of axes in a train of wheels which shall cause the last axis to revolve 180 times as fast as the first axis, allowing that none of the drivers can contain more than 54 teeth, and none of the followers less than 9.

Here, we must find the greatest whole number for m, so that $\left(\frac{54}{9}\right)^m$ or $(6)^m$ shall not exceed 180. This value of m is obviously 2; and the least number of axes will be 4.

Idle Wheels.

62. The wheel C placed between two other wheels, A and B, does not affect the velocity ratio of these wheels; and hence the wheel C is called an idle wheel. This intermediate wheel, however, causes the wheels A and B to revolve in the *same direction*, whereas if A and B were in contact they would revolve in *opposite directions.*

Fig. 43.

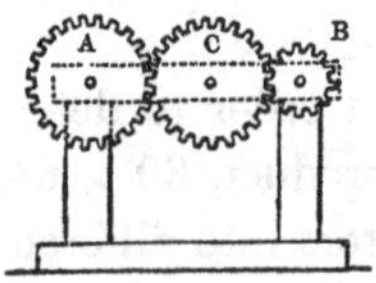

Annular Wheels.

Fig. 44.

63. Fig. 44 represents an annular wheel A, having its teeth cut on the internal edge of the annulus or rim. The toothed wheel B, revolving within the annular wheel A, causes it to revolve in the *same direction;* whereas two ordinary spur wheels revolve in *opposite directions.*

Concentric Wheels.

64. When two separate wheels revolve about the same centre of motion, they are called concentric wheels. The pinion D is fixed to the axis F E, whilst the concentric wheel C is fixed to a tube, or cannon, N, which revolves freely upon the axis F E. The driving wheels, A and B, fixed to the parallel axis H G, communicate the relative velocities to the axis FE and to the cannon N.

Fig. 45.

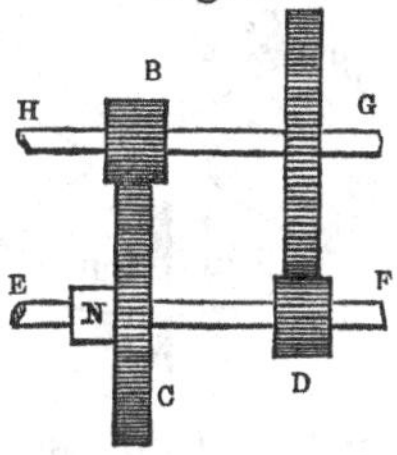

Wheel-work when the axes are not parallel to each other.

65. When the axes of two wheels are not parallel to each other, motion is generally communicated from the one to the other by *bevel wheels* or *bevel gear*. When the axes are perpendicular to each other, the *face wheel and lantern*, and the *crown wheel* are frequently employed.

Face Wheel and Lantern.

66. In fig. 46, F represents a *face wheel*, with its *lantern* L. Here motion is transmitted from the vertical axis A B to the horozontal axis A C. The teeth F on the face of the face wheel are called *cogs*, which are usually made of iron, whilst the round *staves* forming the teeth of the lantern, L, are made of hard wood. The axes A B and C D should, when produced, intersect in a point.

Fig. 46.

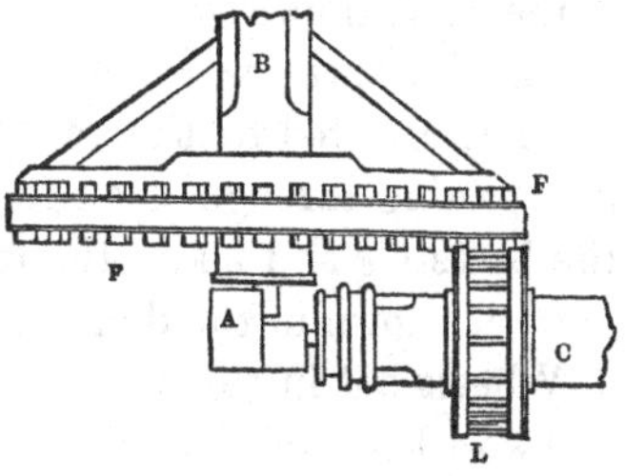

Crown Wheels.

Fig. 47.

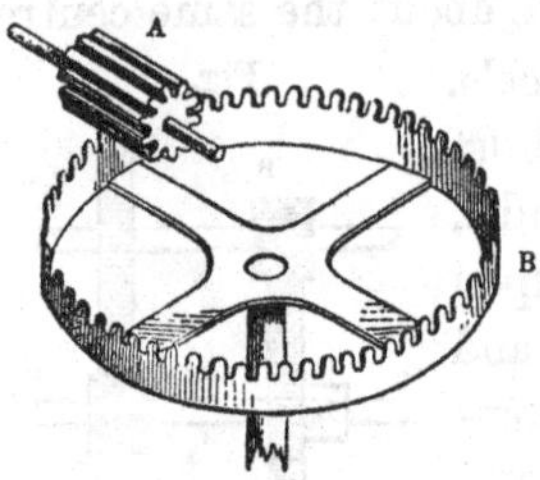

67. Fig. 47 represents a *crown wheel* B, with its pinion A, having their axes at right angles to each other. The teeth of the crown wheel are cut on the edge of a hoop, the plane of which is at right angles to its axis, and the pinion is thicker than wheels are commonly made.

CASE I. *To construct Bevel Wheels or Bevel Gear when the axes are in the same plane.*

68. Let A C and A B be two axes of rotation, in the same plane, and cutting each other in the point A. On these axes two right cones, A D F and A D E, may be formed, touching each other in the line A H D; and also two right frusta, D F G H and D H K E, of these cones may be formed.

Fig. 48.

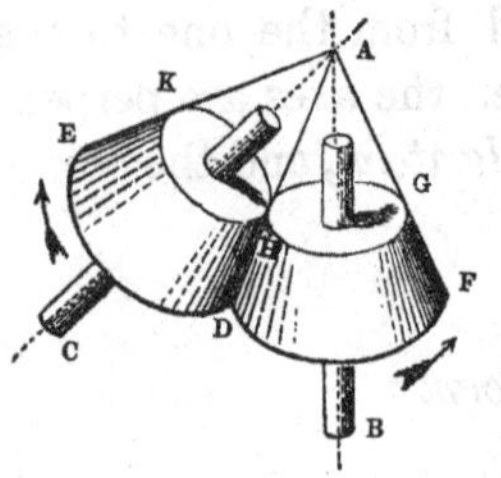

Now if the frustum D F G H revolve on its axis B A, it will communicate, by rolling contact, a rotatory motion to the frustum D H K E upon its axis C A.

These frusta of cones will obviously perform their rotations in the same time as the ordinary spur wheels previously described.

On the surfaces of these frusta a series of equidistant teeth are cut, directed to the apex A of the cones, so that a straight line passing through the apex to the outline of the teeth upon the bases D F and D E of the frusta shall touch the teeth in every part, as shown in the diagram.

Wheels cut in this manner are called *bevel gear*.

Two wheels of this construction will always transfer motion, with a constant velocity ratio, from one axis to the other, provided these axes meet each other in a point, which point being always made the apex of the frusta forming the bevel of the wheels.

69. *General problem.*—Given the radii of two bevel wheels, and the position of their axes, to construct the frusta forming the wheels, the two axes being in the same plane.

Let A B and A C be the position of the axes cutting each other in A. Draw I J parallel to A B at a distance equal to the radius of the wheel on the axis A B; and draw M L parallel to A C, at a distance equal to the radius of the wheel on the axis A C, cutting the line I J in the point D. From the point D, draw D B F perpendicular to A B, and D C E perpendicular to A C. Take B F equal to B D, and C E equal to C D. Join A E, A D, and A F. At a distance equal to the thickness of the wheel, draw H G parallel to D F, cutting A D in H; and through H, draw H K parallel to D E. Then D F G H and D H K E will be the frusta required.

Fig. 49.

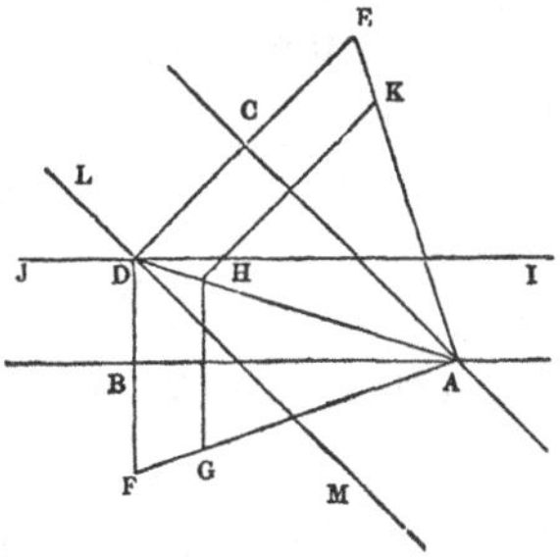

CASE II. *To construct Bevel Gear when the axes are not in the same plane.*

70. This is usually done by introducing an intermediate wheel with two frusta formed upon it, one frustum rolling in contact with the driving wheel, and the other frustum in contact with the driven wheel.

71. Let A B and C D be the direction of the given axes; take A D as a third axis, meeting the axes A B and C D at any convenient points, A and D; then A will be the vertex of two rolling frusta of cones G and H, and D will be the vertex of two other rolling frusta of cones I and K. Whilst the intermediate axis, with its two frusta of cones, revolves, the teeth of the frustum H will have a rolling contact with the teeth of the frustum G, and at the same time the teeth of frustum I will have a rolling contact with the teeth of the frustum K; and thus motion will be transmitted from the axis A B to the axis C D with a constant velocity ratio.

Fig. 50.

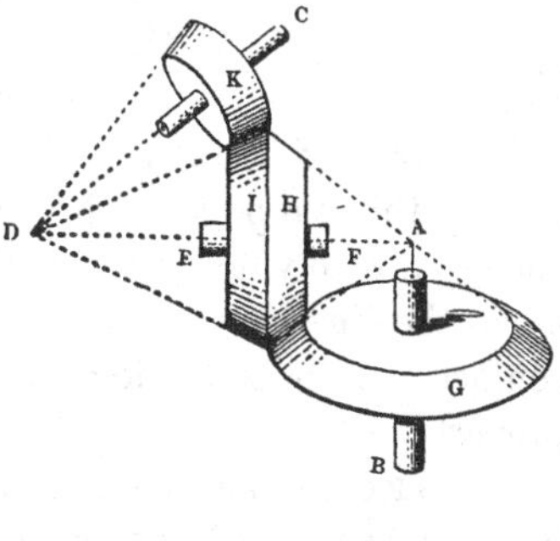

Let Q_1 and Q_3 be the number of rotations performed by the axes A B and C D respectively in the same time; N_1 = the number of teeth in the bevel wheel G; n_1 = the number in the edge H; N_2 = the number in the edge I; and n_2 = the number in the bevel wheel K; then,

$$\frac{Q_3}{Q_1} = \frac{N_1 . N_2}{n_1 . n_2} \dots (1),$$

which is similar to the expression given in eq. (1), Art. 57. When $n_1 = N_2$, then this equality becomes,

$$\frac{Q_3}{Q_1} = \frac{N_1}{n_2} \dots (2).$$

In this case the intermediate bevel wheel, I H, may be regarded as an idle wheel.

VARIABLE MOTIONS PRODUCED BY WHEEL-WORK HAVING ROLLING CONTACT.

72. Two curved wheels, E P and F P, having rolling contact, revolve on the axes A and B. In order that these wheels may roll on each other without slipping, or without producing any strain upon the axes A and B, these axes must always be in the line of contact A P B, and if the curve P E on the one wheel be equal to the curve P F on the other wheel, the sum of the lines A E and B F must always be equal to A B, the distance between the centres of motion. Various curves may be constructed, having this property. For example, two equal ellipses, E P and F P, revolving on their foci, A and B, and having A E and B F in the line of their major axes, will have a perfect rolling contact. Two equal logarithmic spirals have also the same property.

Fig. 51.

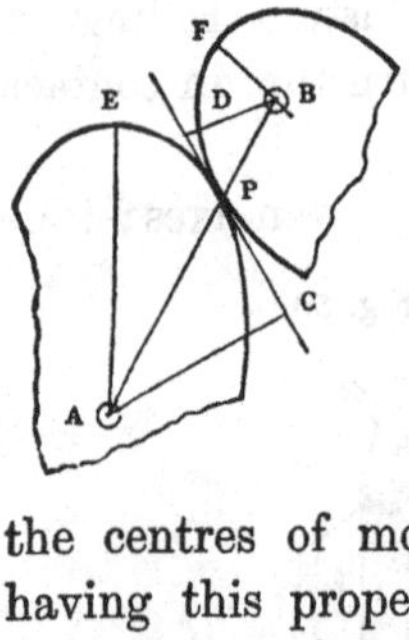

Let D P C be the common tangent to the point of contact P; from A and B let fall A C and B D perpendicular to D P C; then,

$$\frac{\text{angular velocity } AP}{\text{angular velocity } BP} = \frac{BD}{AC} \text{ or } \frac{BP}{AP} \dots (1).$$

This result may be expressed in language as follows: — *The angular velocities of the wheels are inversely as the perpendiculars let fall upon the common tangent from the centres of motion.*

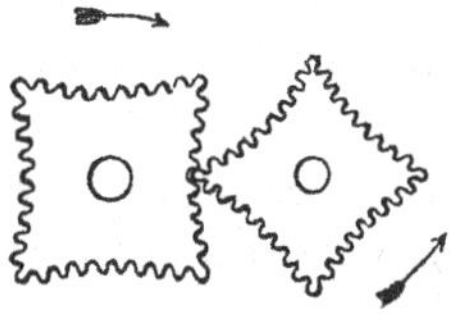

Fig. 52.

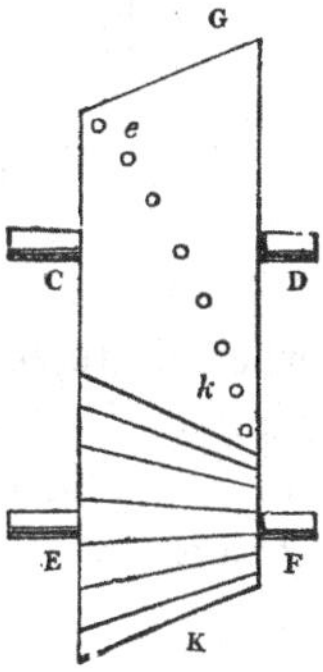

Fig. 53.

73. The form of wheels, represented in fig. 52, are used in silk-mills, and in the Cometarium. The curves may be indefinitely varied, but they must always be constructed to answer the conditions explained in Art. 72.

74. *Roëmer's Wheels.*—E F and C D are the axes of two conical wheels or bevel-wheels K and G, having their vertices turned in opposite directions; the teeth of K are formed like those of the ordinary bevel-wheel; but the teeth on G are formed by a series of pins *e k*, fixed on the surface of the frustum G. By varying the relative position of these pins, any given velocity ratio may be obtained.

75. Various combinations have been invented for producing a varying angular velocity; such as the eccentric crown wheel and broad pinion, the eccentric spur-wheel with a shifting intermediate wheel, and so on.

INTERMITTENT AND RECIPROCATING MOTIONS PRODUCED BY WHEEL-WORK, HAVING ROLLING CONTACT.

76. The following is an example of an intermittent motion produced by the continuous motion of a toothed wheel.

A driving wheel A, having sunk teeth on a portion of its edge, communicates an intermittent motion to the wheel B, which has a corresponding number of teeth on a portion of its edge. The portion D C of the wheel B, being a plain arc of a circle described from A as centre, allows the plain portion of the wheel A to revolve without any interruption. The wheels are brought into gear by a pin *p*, fixed to the wheel A, and a GUIDE-PLATE G *e*, fixed to the wheel B. Now when A revolves, in the direction of the arrow, the plain portion of its edge runs past D C without moving the wheel B, and at the same time keeps it from shifting; but when the pin *p* comes into contact with the guide-plate, the wheel B is moved round, and the teeth D E engage themselves with the teeth on B, and thus the wheel B is constrained to make a revolution; it then remains at rest until the pin *p* again comes round to meet the guide-plate.

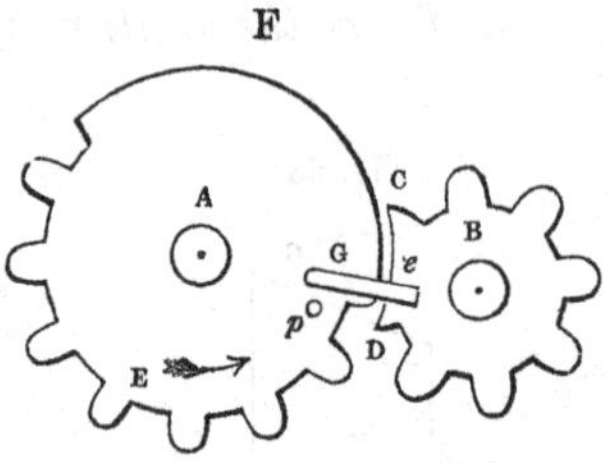

Fig. 55.

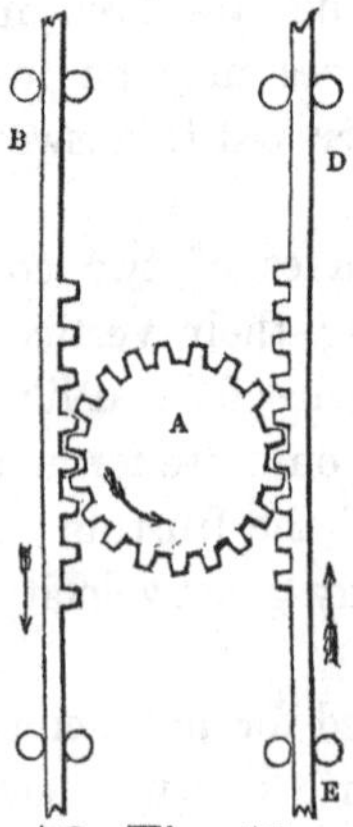

77. *The Rack and Pinion.*—By this combination a circular reciprocating motion is changed into a reciprocating rectilinear one. Teeth are cut upon the edge of the straight bars, B C and D E, so as to work with the teeth upon the pinion A. These toothed bars are called *racks*, and they are constrained to move in rectilinear paths by guides or rollers. The racks in this combination move in opposite directions.

78. Fig. 55 represents an application of the double rack, for converting a continuous circular motion of a wheel, A, into a reciprocating rectilinear motion, given to the frame B E.

Fig. 56.

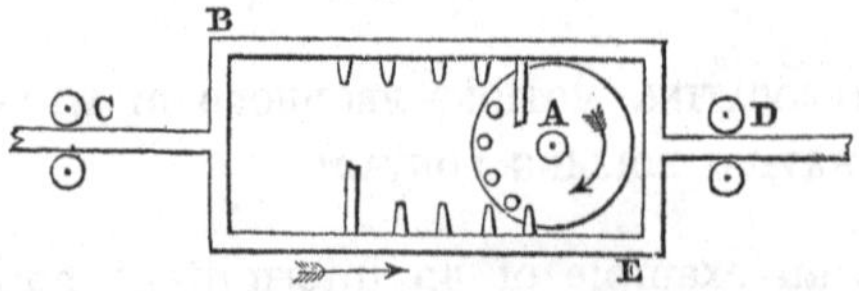

The teeth on A are formed by pins or staves placed about one quarter round the face of the wheel; these staves act alternately upon the racks formed on the upper and under sides of the frame. The tooth on each rack, which comes first into contact with the stave of the pinion, is made longer than the others, in order that the first stave should act obliquely upon it, thereby tending to lessen the shock. In this figure the lower stave is represented as leaving the last rack on the under side, and the upper stave as commencing its action on the elongated tooth of the upper rack.

V. ON SLIDING-PIECES, PRODUCING MOTION BY SLIDING CONTACT.

The Wedge or Moveable Inclined Plane.

79. Let A B C be a moveable inclined plane or wedge, sliding along the smooth surface D E, by a pressure P applied to the end B C, and producing a vertical motion in a heavy rod G P_1 resting on the plane A C, and constrained to move in a straight path by means of guide rollers. The velocity ratio of P and P_1 will be constant, being expressed by the following equality:—

Fig. 57.

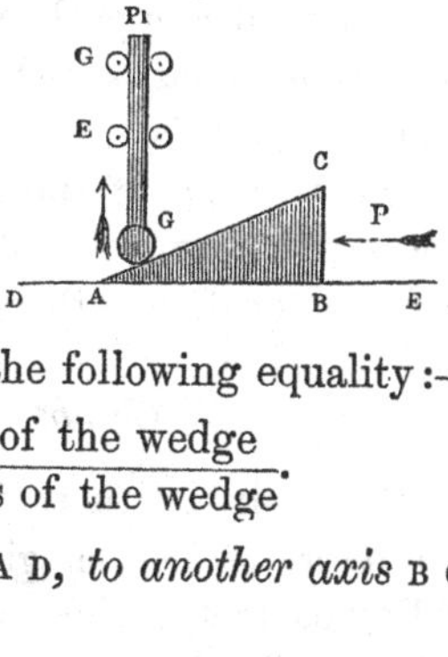

$$\frac{\text{velocity } P}{\text{velocity } P_1} = \frac{AB}{BC} \text{ or } \frac{\text{length of the wedge}}{\text{thickness of the wedge}}.$$

To transmit motion from an axis A D, *to another axis* B C, *parallel to it.*

80. The axis A D carries an arm A E, and a pin E F, which enters and slips freely in a slit made in the arm G B attached to the axis B C. When the axis A D revolves, it communicates a rotation, in the same direction, to the other axis B C, but with a varying velocity ratio, for the pin F continually changes its distance B F from the axis B C.

Fig. 58.

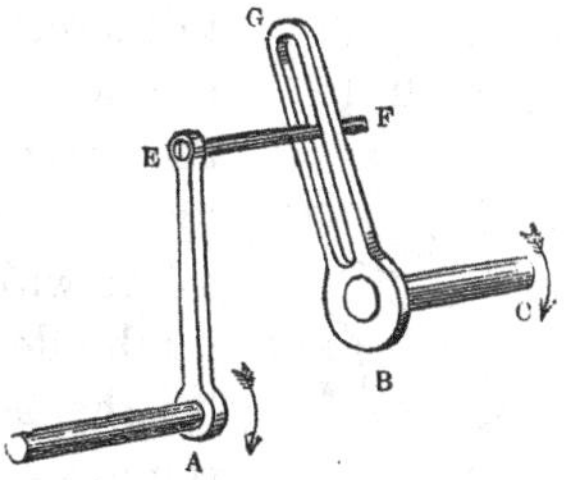

When the distance between the parallel axes is small, and the axis A D revolves uniformly, the angular velocity of the axis B C varies, very nearly, inversely as the distance, B F, of the pin from this axis.

The Eccentric Wheel.

81. This mechanism is usually employed to give motion to the slide-valve of the steam engine. In fig. 59, B represents the axis of the eccentric wheel; C the centre of the circle; E R F K a hoop which embraces the eccentric wheel, so that the one may revolve freely within the other; E F D a frame connecting this loop with the extremity D of the bent lever D L G, turning on the fixed centre L. Now when the eccentric wheel revolves in the direction of the arrow, shown in the figure, the frame with the pin D is pushed to the right, and when the lob side of the eccentric has passed the line of centres, B and D, the frame with the pin D is drawn to the left, and so on. Thus the continuous rotation of the axis B produces a reciprocating circular motion in the pin D. The stroke of the pin D will be equal to twice C B, or double the eccentricity of the wheel.

Fig. 59.

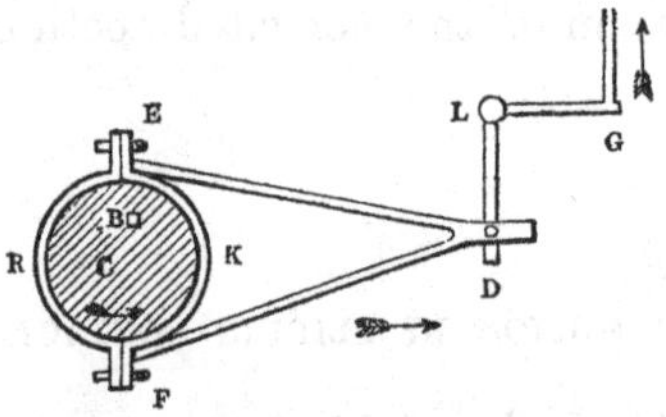

Cambs, Wipers, and Tappets.

82. Cambs are those irregular pieces of mechanism to which a rotatory motion is given for the purpose of producing, by sliding contact, reciprocating motions in rods and levers.

83. In fig. 60, B C D represents the camb, turning on its axis A, and giving a reciprocating rectilinear motion to the heavy rod E F, which is restrained to move in its rectilinear path by the guide rollers. The rotation of the axis A being in the direction of the arrow, the rod E F has an upward motion until the extreme point B of the camb comes in a line with the rod, then the portion B G of the camb allows the rod to fall, by its own weight or by the action of a spring, until the point G comes in a line with the rod, and so on; thus one revolution of the camb, here presented, will cause the rod

Fig. 60.

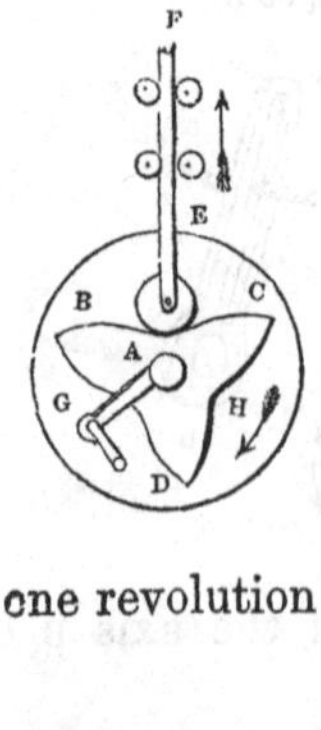

to make three upward and three downward strokes. By varying the curve of the camb, any law of motion may be given to the rod.

84. In fig. 61, the pin E of the rod is made to traverse a groove E G D, cut in the camb plate, so that the pressure of the camb upon the pin produces the downward stroke of the rod as well as its upward stroke. In this case the rod will only make one upward and one downward stroke in every revolution of the camb plate. The length of the stroke of the rod will be equal to the difference between A D and A G, where D is the point in the groove furthest from the centre A, and G is the point nearest to it.

Fig. 61.

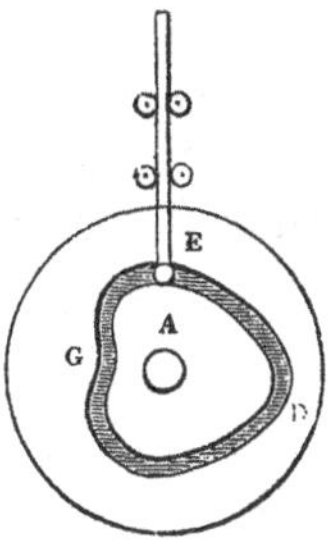

85. *To find the curve forming the groove of a camb, so that the velocity ratio of the rod and the axis of the camb may be constant.*

Let A be the centre of the camb, and C A B Q the direction of the rod. From A as a centre, with any convenient distance A C, describe the circle C E D B N. On B A take B a equal to the length of the stroke of the rod: divide it into any convenient number of equal parts, say five, in the points, b, c, d, e; and divide the semicircle B D E F G into the same number of equal parts by the radial lines A D, A E, A F. From A as a centre, with A b, A c, A d, A e, as radii, describe circles cutting A D, A E, &c., respectively in the points g, k, l, m; then through these points draw the curve $a\ g\ k\ l\ m$ C; and similarly in the semicircle B N C draw the other curve $a\ n\ p$ C.

Fig. 62.

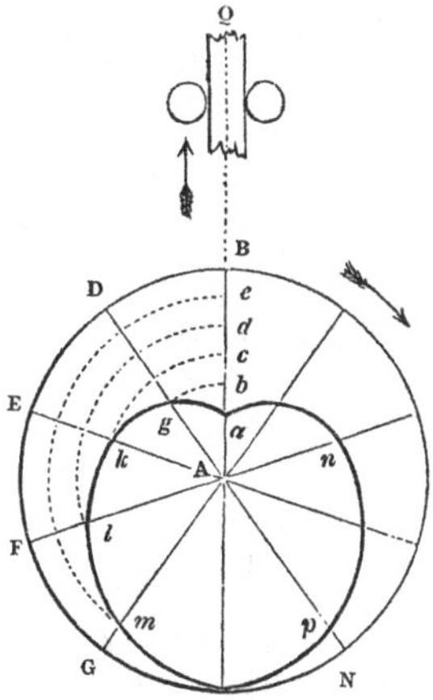

All lines drawn through the centre A of this curve are equal; thus a C $= l\,n = g\,p =$ &c. Hence if the rod had two pins, placed at a and C, the camb would revolve between them, and would cause the rod to make a downward as well as an upward stroke. This curve is the spiral of Archimedes.

By dividing the line B a into parts having a varying ratio to one another, any proposed law of velocity may be given to the rod.

86. In fig. 63, the continuous rotation of the camb A E C, revolving on the axis A, gives an oscillating motion to the rod or lever F a, turning on the centre F. In one revolution of the camb the rod makes a double oscillation in the arc a a_1.

Fig. 63.

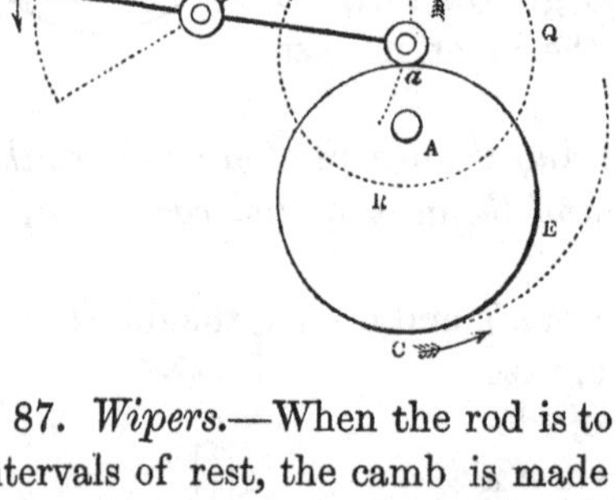

Fig. 64.

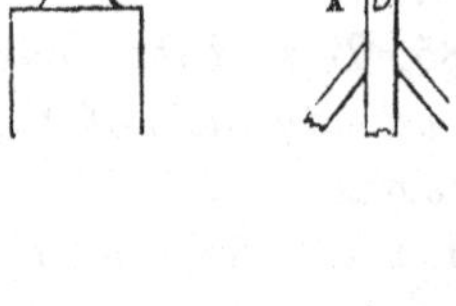

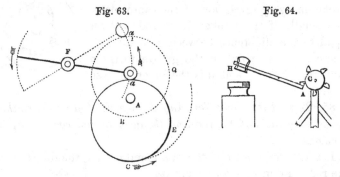

87. *Wipers.*—When the rod is to receive a series of lifts with intervals of rest, the camb is made into the form of projecting teeth which are commonly called *Wipers* or *Tappets.*

88. In fig. 64, the revolving cylinder C has five wipers upon its circumference, which give five downward strokes to the hammer, H, placed at the extremity of the lever A H, in each revolution of the cylinder.

Fig. 65.

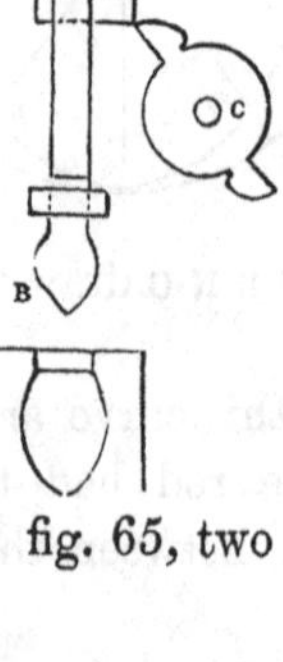

Fig. 66.

89. In fig. 65, two tappets, upon the revolving cylinder C,

give two downward strokes to the heavy bar or *stamper* A B, in each revolution of the cylinder. In this case the bar A B is constrained to move in a rectilinear path by means of guide rollers.

90. In fig. 66, a single wiper on the cylinder C gives an intermittent rotation to the ratchet wheel A with its detent D. At each revolution of C only one tooth in A is moved round, so that for the greater portion of the revolution of C the wheel B is at rest.

91. In fig. 67, the continuous rotation of three wipers *a*, *b*, *c*, communicates a reciprocating rectilinear motion to the frame A B C D. The wiper *a* is engaged with the *pallet e*, and at the instant of disengagement the wiper *b* becomes engaged with the pallet *g*, and then the frame starts its motion in a direction contrary to that of the arrows; and so on.

Fig. 67.

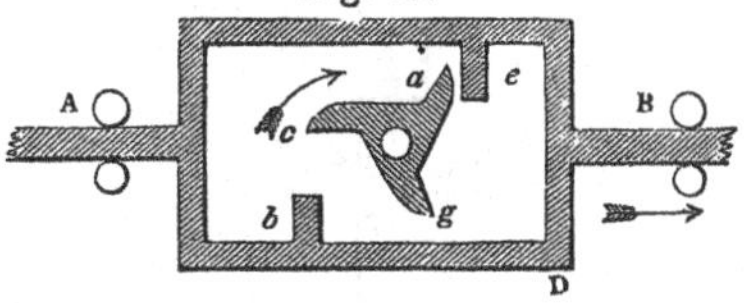

The Swash Plate.

92. By this mechanism, the continuous rotation of an axis produces a reciprocating rectilinear motion in a rod, in the direction of its length.

Here C E represents the revolving-axis, to the top of which is fixed the inclined circular flat plate A B, called the swash plate; A D F the rod to which a reciprocating motion is given in the direction of its length, having a frictional wheel A at its lower extremity resting on the swash plate. This rod is kept in contact with the plate by its own weight, or, if this be not sufficient, by means of a spring. Now as the swash plate turns round, the rod A F is alternately raised and depressed, so that at every revolution of the plate the rod performs an upward and a downward stroke. Supposing the rod, as represented in this figure, to be at the lowest point of its stroke; from C, the centre of motion of the plate, let fall C D perpendicular to A F; then A D will be equal to half the stroke of the rod. Moreover, let θ be any angle

Fig. 68.

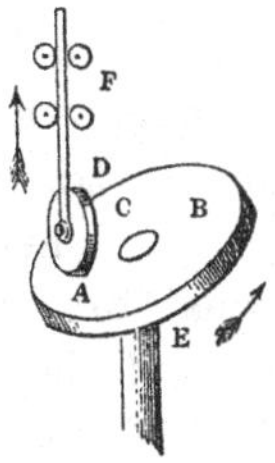

moved over by the axis, and let h be the corresponding space moved over by the extremity A of the rod; then

$$h = \text{AD} \times (1 - \cos\theta),$$

which gives the position of the rod at any point of the rotation of the plate.

93. There are an almost endless variety of combinations for producing reciprocating motions of this kind, by means of sliding contact.

94. In fig. 69, an eccentric revolving pin e, sliding or working

Fig. 69.

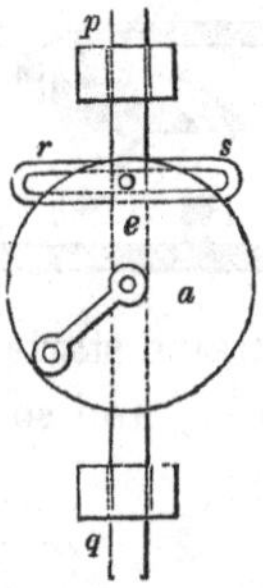

Fig. 70.

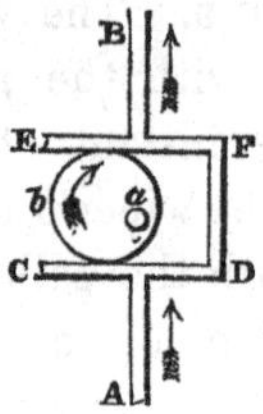

in the slit of the arm $r\,s$ gives a reciprocating motion to the rod $p\,q$ in the direction of its length.

95. In fig. 70, the same effect is produced by the rotation of an eccentric wheel, $a\,b$, on its axis a, within the frame C D F E.

SCREWS.

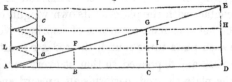

96. *Construction of a Helix or Screw.*—Let A a K be a cylinder, and A D E a piece of paper cut in the form of a right-angled triangle, having its height D E equal to the height A K of the cylinder. Now if this paper be wrapped round the cylinder, the slant edge A E of the paper will trace the helix or screw A a L b c K upon the cylinder. If A B = B C = C D be equal to the circumference of the cylinder, the edge of the paper will form four convolutions, and the perpendiculars

Fig. 71.

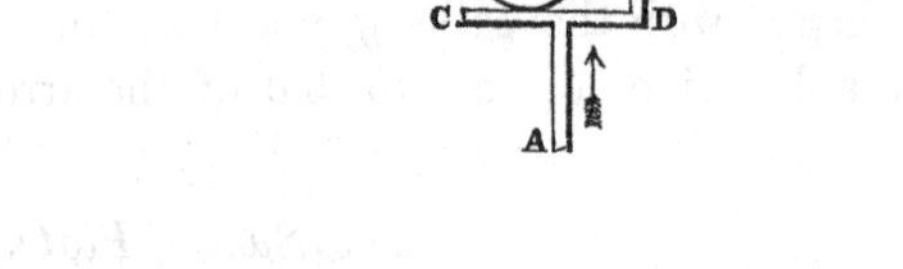

B F = I G = H E will be the distance between the threads of the screw.

97. *The pitch of a screw* is the distance B F between two successive convolutions. If t = B F, the distance between the threads of the screw, r = the radius of the cylinder, θ = angle B A F; then

$$t = \frac{2\,\pi\,r}{\tan\,\theta}.$$

98. We may also conceive the helix of the screw to be formed by the compound motion of a point. Suppose the cylinder to rotate uniformly upon its axis, whilst a point A upon its surface at the same time moves uniformly in the direction of its length; then, with this compound motion, the point A will trace the helix of a screw.

99. *Transmission of motion by the screw.*—Let $e\,a\,n\,c\,m\,g$ be a spiral groove cut upon a cylinder; A B the axis on which it turns; D E a rod parallel to the axis A B, and constrained to move in the direction of its length; e a tooth attached to this rod fitting the groove of the screw. Now when the wheel C is turned in the direction of the arrow, the tooth with the rod D E will be moved from left to right in the direction of its length, that is, parallel to the axis of the screw.

Fig. 72.

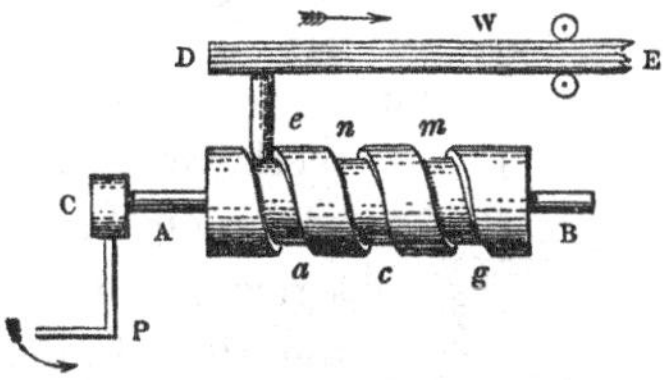

The velocity ratio of the wheel C and the rod D E will be constant, for we have

$$\frac{\text{velocity C}}{\text{velocity C D}} = \frac{\text{circum. described by C}}{\text{pitch of the screw}}$$

If R be the radius of the wheel C, r = the radius cylinder A B, V = velocity circum. C, v the velocity of the bar D E, and so on as in Art. 97; then the above equality becomes,—

$$\frac{\text{V}}{v} = \frac{2\,\pi\,\text{R}}{t} \;\ldots\; (1);$$

$$\therefore\; \frac{\text{V}}{v} = \frac{\text{R}}{r}\tan\,\theta \;\ldots\; (2);$$

and when R $= r$, then,—

$$\frac{V}{v} = \tan\theta \ldots (3);$$

that is, *the velocity ratio is equal to the tangent of the angle which the thread of the screw makes with the sides of the cylinder.*

100. It is obvious that the number of teeth in the bar D E will not at all alter its motion.

In fig. 73, the screw acts upon a series of teeth upon the *rack* D E. This arrangement, called *the rack and screw,* converts a circular motion into a rectilinear one.

Fig. 73.

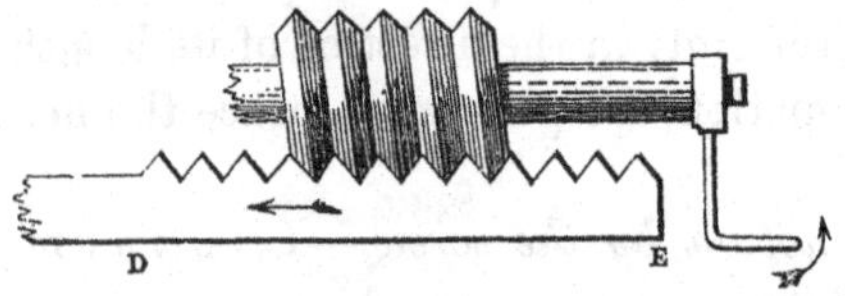

Solid Screw and Nut.

101. In general, the piece acted upon by the screw has its teeth, or rather its threads, formed in a cavity which embraces the whole circumferences of the screw, and the threads of the one exactly fitting the threads of the other. This modification is shown in fig. 63, where N is the hollow screw fitting the threads of the screw S. The solid piece S is called the male screw, and the hollow piece the female screw or *nut.*

Fig. 74.

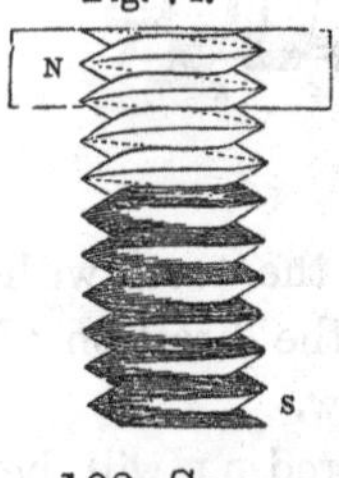

102. Screws are either left-handed, or right-handed, according to the direction of the threads.

103. It is important to observe that the following relations of motion subsist between the solid screw and the nut.

1. When the nut is fixed, the solid screw will have a motion in the direction of its length, upon being turned round.

2. If the nut revolves, without having any longitudinal motion, the solid screw will have a motion in the direction of its length, provided it is incapable of revolving.

3. If the solid screw revolves without having any motion in the direction of its length, the nut will have a longitudinal motion, provided it is incapable of revolving.

The first two cases are exemplified in the different forms which are given to the common press, and the last case is exemplified in the construction of the self-acting slide rest of the lathe, and in other kinds of mechanism.

The screw is usually employed for producing very slow uniform motions, and for exerting great pressure through a limited space.

The Common Press.

104. In fig. 75, S S is the solid screw, N the nut, N P the lever, B the lower press board which is constrained to move in an upward direction by means of the guide-frame.

Fig. 75.

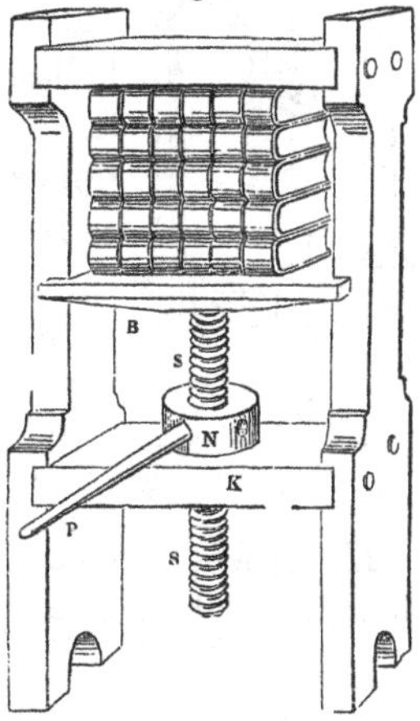

Case 2. In this case the nut N revolves, but does not move longitudinally, but the screw S S is incapable of revolving. Hence the press board B is moved upwards at every revolution of the nut, over a space equal to the pitch of the screw, or the distance between the threads, that is,

$$\frac{\text{velo. P}}{\text{velo. B}} = \frac{\text{circum. described by P}}{\text{distance between the threads}}$$

Example.— Let the distance between the threads = $\frac{1}{4}$ in., the length of the lever N P = $2\frac{1}{2}$ ft.; required the velocity ratio of the point P and the press-board B.

$$\frac{\text{velo. P}}{\text{velo. B}} = \frac{2 \times 2\frac{1}{2} \times 12 \times 3{\cdot}1416}{\frac{1}{4}} = 753{\cdot}984.$$

That is, the velocity of P is 753·984 times that of B.

Case 1.—In this case N is a perforated cylinder forming part of the solid screw S S, and therefore turns with it on a pivot which works in a socket placed on the under side of the press board B; the piece K fixed to the frame contains the hollow or female screw; so that the solid screw, S S, is capable of revolving and of moving longitudinally, whilst the nut K remains absolutely fixed.

Compound Screw.

105. This mechanism consists of two screws A and D, the smaller one D working within the larger one A. The screw A works in a fixed nut or female screw at K, and is capable of revolving and moving in the direction of its length; the small screw D is incapable of revolving, but is capable of moving in the direction of its length. In one revolution of the lever P, the screw A descends a space equal to the distance between its threads, but at the same time the screw D enters the hollow screw formed in A, a space equal to the distance between the threads on D, so that the extremity B will only descend a space equal to the difference between the thickness of the threads on A and the thickness of the threads on B; hence we have

Fig. 76.

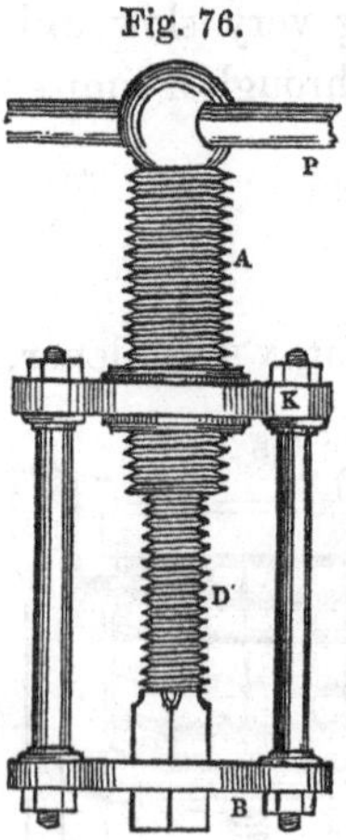

$$\frac{\text{velo. P}}{\text{velo. B}} = \frac{\text{circum. described by P}}{\text{dist. bet. threads on A} - \text{dist. bet. threads on D.}}$$

If the length of the lever P $= r$, the pitch of the screw A $= t$, and the pitch of D $= t_1$; then

$$\frac{\text{velo. P}}{\text{velo. B}} = \frac{2\,\pi\,r}{t - t} \;\ldots\; (1).$$

Example.—Let $r = 5$ ft., $t = \frac{1}{2}$ in., $t_1 = \frac{3}{8}$ in.; then

$$\frac{\text{velo. P}}{\text{velo. B}} = \frac{2 \times 5 \times 12 \times 3{\cdot}1416}{\frac{1}{2} - \frac{3}{8}} = 3015{\cdot}936$$

The same velocity ratio might be attained by making the pitch of a single screw A, equal to $t - t_1$, but the threads, in this case, might be too weak to stand the pressure; hence the advantage of the compound screw.

The Endless Screw.

106. When the threads or teeth of a revolving screw are made to act upon the teeth of a wheel, as in fig. 77, the mechanism is called the endless screw. Here, each rotation of the axis A B of the screw turns round one tooth of the wheel C, the pitch of the screw on the axis A B being equal to the pitch of the teeth on the wheel.

Fig. 77.

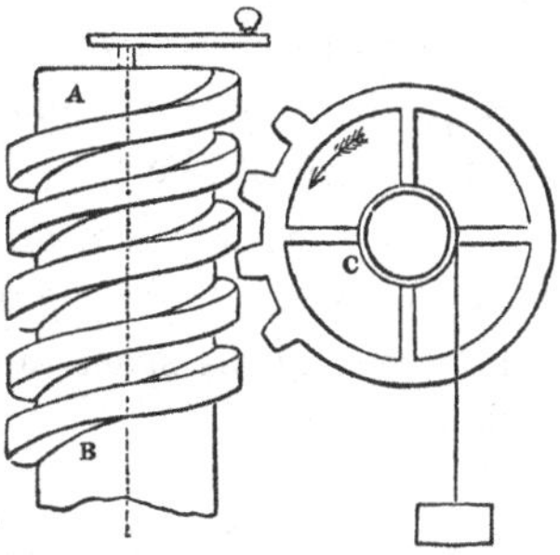

If Q and q be the synchronal rotations of the wheel and the screw respectively, and N the number of teeth in the wheel; then

$$\frac{q}{Q} = N \ldots (1).$$

If $N = 40$, then $\frac{q}{Q} = 40$; that is, for every revolution performed by the wheel the screw will make 40.

If R, r be the respective pitch-radii of the wheel and screw, θ being, as before, the angle which the thread of the screw makes with its axis; then

$$\frac{q}{Q} = \frac{R}{r} \tan \theta \ldots (2).$$

The Differential Screw.

107. A D is an axis on which are formed two screws, A B and B C, whose pitches are different. The screw A B passes through a fixed nut or female screw E, whilst B C passes through a nut N which is capable of moving longitudinally, but incapable of revolving from the intervention of the guides.

Fig. 78.

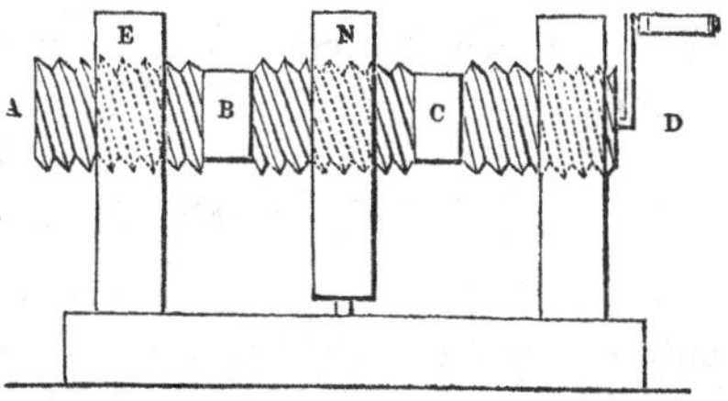

Let the screw make one turn so as to move the cylinder from right to left, then the screw A B will move through the fixed nut E a space equal to the distance between its threads; but, at the same time, the screw B C will move through the nut N a space equal to the thickness of the threads on B C; so that the nut N will only be moved through a space equal to the difference between the thickness of the threads on A B and B C, that is,

In one revolution of A, the space moved over by the nut N = pitch screw A B − pitch screw B C = $t - t_1$, where t is put for the pitch of the screw A B, and t_1 for that of B C.

If $t = t_1$, then nut N will remain at rest.

If the screw A B be right-handed, and B C left-handed; then $t + t_1$ will be the space moved over by the nut N in one revolution of A.

The Archimedian Screw Creeper.

108. This machine is used for conveying corn from one part of a corn-mill to another. It consists of a wooden cylindrical trough, A B C D, with which revolves a shaft, E F, having a deep

Fig. 79.

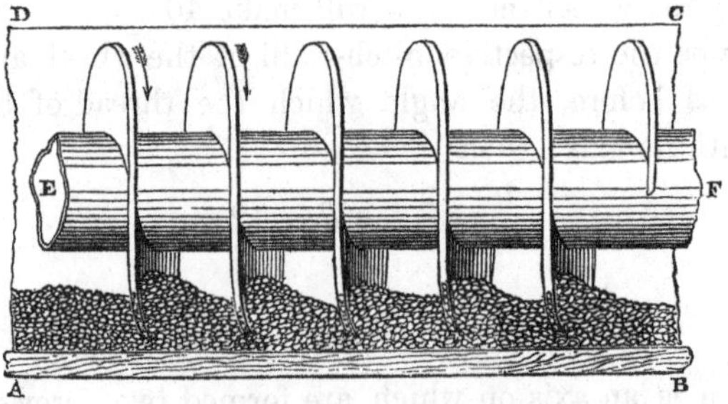

spiral thread formed upon its surface. The corn is dropped in at one extremity of the trough by a hopper, and by the revolution of the creeper the corn is pushed along towards the other extremity of the trough.

Mechanism for Cutting Screws.

109. C D is the cylinder, or axis on which the screw is to be cut, revolving with the mandril D of the lathe; A a toothed wheel revolving with the axis C D, and giving motion to the toothed wheel B, round its axis F E, on which is cut the parent

screw; this screw gives a longitudinal motion to the nut N, as in case 3, carrying the sliding table or saddle upon which is securely clamped the cutting tool P intended to cut the thread of the screw on the cylinder C D. In the place of the wheels A B, any combination of wheels may be used so as to produce any relative longitudinal velocity to the cutting tool P, and thereby to form a screw of any given pitch on C D with the same parent screw F E.

Fig. 80.

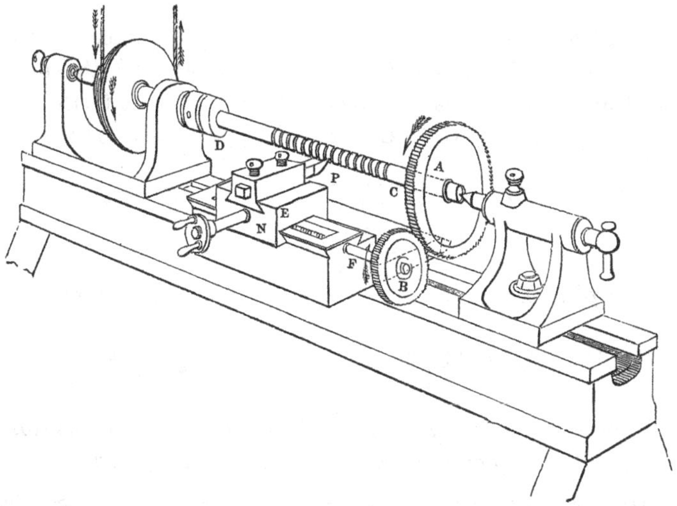

Let n = the no. teeth on the wheel A, n_1 = the no. teeth on B, t = the pitch of the screw on C D, t_1 = the pitch of the screw on F E; then

$$t = \frac{n}{n_1} \cdot t_1 \dots (1),$$

which expresses the pitch of the screw on C D.

From this equality we get,

$$\frac{t}{t_1} = \frac{n}{n_1} \dots (2),$$

that is to say, *the pitches of the screws are in the ratio of the number of teeth on their respective wheels.*

If n_1 and t_1 be constant, then

$$t \propto n,$$

that is to say, *the pitch of the screw on* C D *varies with the number of teeth on its wheel* A.

Let k and k_1 be the number of threads per inch on the cylinders C D and F E respectively, then

$$\frac{1}{k} = t \text{, and } \frac{1}{k_1} = t_1,$$

and eq. (2) becomes —

$$\frac{k}{k_1} = \frac{n_1}{n} \ldots (3).$$

Now, let there be an intermediate pinion and wheel, turning on the same axis, placed between A and B; and let the pinion (acted upon by A) contain e_1 teeth, and the wheel e teeth; then the velocity ratio of the axis F E will be increased by the ratio $\frac{e}{e_1}$, and hence eq. (3) becomes—

$$\frac{k}{k_1} = \frac{n_1 e}{n\ e} \ldots (4).$$

Example.—Let $n = 30$, $n_1 = 10$, $t_1 = \frac{1}{2}$ in.; required t.

Here by eq. (1), $t = \frac{n}{n_1} \cdot t_1 = \frac{30}{10} \times \frac{1}{2} = 1\frac{1}{2}$ in.

To produce a changing reciprocating rectilinear motion by a combination of the camb and screw.

110. E F is a conical shaped camb, turning on the eccentric axis A B, on which is cut the screw K B, working in the fixed nut or hollow screw N; D C a rod, resting on the camb, constrained to move in the direction of its length, and to which the varying reciprocating motion is to be given. Here, whilst the camb revolves, it has a continuous motion in the direction of the axis A B, so that the lower extremity, C, of the rod D C describes a spiral or screw curve upon the cone whose pitch is equal to the pitch of the screw K B. The effect of this is to make C D reciprocate in its path in such a manner that the stroke in one direction is shorter than that in the opposite direction.

Fig. 81.

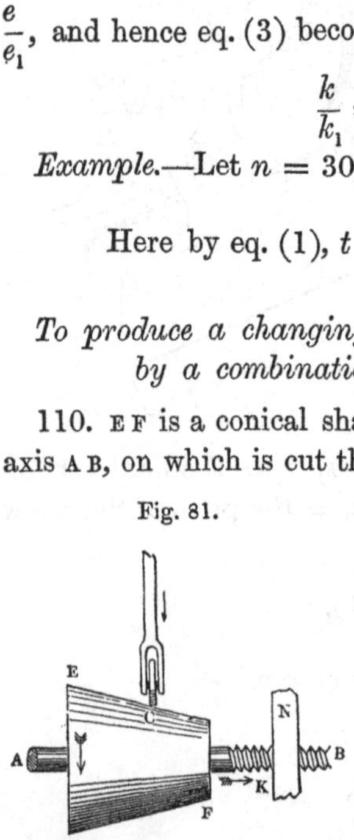

To produce a boring motion by a combination of the screw and toothed wheels.

111. Here it is required to produce a rapid rotation combined with a very slow motion in the direction of the axis.

The screw I B is cut upon a portion of the revolving axis A B; this screw passes through a nut K capable of revolving with the wheel G, but incapable of moving in the direction of its axis, as in case 2; the wheel G is driven by the pinion F revolving on the parallel axis D C; E is a long pinion, turning on this axis, and acting on the wheel L, which transmits a rotatory motion to the screw axis A B. Now the rotation of C D produces a rotatory motion in the axis A B, and at the same time causes it to advance, in the direction of its length, with a velocity determined by the following formula.

Fig. 82.

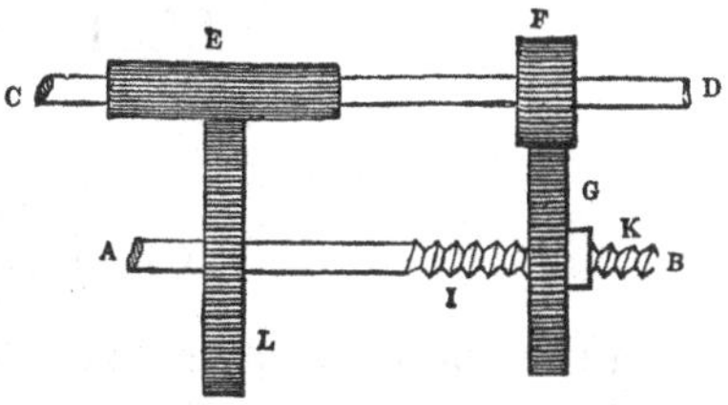

Let Q, Q_1, q_1 be the synchronal rotations of the axis C D, the nut K and wheel G, and the wheel and axis A B, respectively; N, N_1, n, n_1, the number of teeth in the wheels F, G, E, L, respectively; S the space moved over by A B in the direction of its length; and $t =$ the pitch of the screw I B.

Now Q_1 rotations of the nut K moves the screw A B through a space equal to $Q_1 \times t$; but q_1 rotations of L moves the screw through a space, in the *opposite* direction, equal to $q_1 \times t$; therefore in Q rotations of the axis C D, the screw A B will be moved through a space equal to the difference between $Q \times t$ and $q_1 \times t$, that is,

$$S = (q_1 - Q_1)\,t;$$

$$\text{but } \frac{Q_1}{Q} = \frac{N}{N_1}, \text{ and } \frac{q_1}{Q} = \frac{n}{n_1};$$

$$\therefore S = \left(\frac{n}{n_1} - \frac{N}{N_1}\right) Q\,t \;\ldots\; (1).$$

Now the difference $\dfrac{n}{n_1} - \dfrac{N}{N_1}$ may be very small as compared with Q, and consequently S may be made as small as we please as compared with Q, which is the condition required for the construction of a boring instrument. The boring tool is placed upon one extremity of the axis A B.

SECTION III.

ON PRIME-MOVERS.

CHAPTER I.

ON THE ACCUMULATION OF WATER AS A SOURCE OF MOTIVE POWER.

THE machinery of mills, as a whole, may be generally divided into three classes; — the *prime-movers,* from which the power is derived for keeping the machinery of the mill in motion; the *transmissive* machinery or *millwork* (shafting, gearing, &c.), by which the power obtained through the prime-mover is distributed over the different parts of the mill, so that it may be applied at the most convenient place and at the required velocity; and lastly, the *machines,* technically so called, by which the special operations of the mill in the preparation of its manufactures are carried out. It will be convenient to treat of these divisions in separate sections and in the order just named.

Prime-movers are those combinations of mechanism which *receive motion and force* directly from some natural source of power, and convert it into that condition in which it is applicable to the purposes of manufacture. Thus the water-wheel takes from the falling water a part of the *work* accumulated in it, and imparts it as a rotatory motion to the machinery of the mill; and, similarly in the steam-engine, the heat force of the fuel is converted through the medium of the pressure of the steam into motive power in a condition for producing work or mechanical effect. Also the force of currents in the atmosphere impinging upon the expanded sails of windmills, has been in former days extensively employed as a motive power. From these three sources, falling or moving water, the combustion of coal in the production of steam, and wind, we derive almost exclusively at the present time the motive power necessary for carrying on our immense mining and manufacturing systems.

It is only of late years that in this country the steam-engine has nearly superseded the use of air and water as a prime-mover. Until recently steam has been auxiliary to water, it is now the principal source of power, and waterfalls are of comparatively small value, except in certain districts. So long as water was depended upon, the mills of Great Britain and Ireland were necessarily circumscribed in their operations and diminutive in size; they have now become so colossal, that they require steam-engines of much greater power than the largest water-wheels, and there appears to exist no limit to the magnitude and importance to which they may yet attain.

Water-wheels, therefore, are those prime-movers which receive a certain portion of their energy from falling or flowing water, and their power or dynamic effect clearly depends upon the amount of water supplied and the height through which it falls, or its velocity at the point of application. Hence water-wheels are usually placed on the banks of rivers where a large body of water is at hand, and near some considerable natural or artificial fall in the bed of the stream.

In establishments where manufacturing processes are carried on, and a large body of men employed, it is essential to success that there should be no stoppages, and that there should be always at command a uniform power, equal to the requirements of the mill. Now, as the quantity of water in rivers varies considerably at different periods of the year and in different conditions of weather, it has been found necessary, in many instances, to impound the water by means of *reservoirs* placed at the sources or the higher portions of the river, so as to retain the waters of wet seasons, and to part with them again in periods of drought and deficiency. To a small extent this may be effected by weirs thrown across the river, so as to retain the water which comes down at night, for the use of the mill during the day. But in many instances large reservoirs of a hundred or more acres in extent, and containing when full several million cubic feet of water, have been constructed. In these the drainage from a large extent of country is collected during the rainy seasons, and remains stored for use, whenever the supply of water in the river becomes inadequate; in this way damage from floods is prevented on one hand, and on the other the supply for an indefinite period of time is equalised. Among the large works of this kind are the Shaws' waterworks at Greenock, and the Lough Island Reavy or Bann reservoirs in the county of Down, in the north-east of Ireland, together with later works of the same kind for the supply of water to the cities of Glasgow and Manchester, Melbourne, &c.

Reservoirs are best placed in billy districts, at the bottom of a valley into which the water drains from a considerable extent of country. In selecting a site for reservoirs regard must first be had to the value of the land. They should be placed in retired valleys, where the cost of the land does not bear a high ratio to the cost of construction, and should there exist a natural lake it may be converted into a reservoir with greatly increased economy. Regard must also be had to the nature of the site. The reservoir should be restrained as far as possible by the natural rise of the ground around it, in order that as few embankments as possible may be requisite for the retention of the water. Again, the geological structure of the country must be examined, as the quantity of water to be expected to flow from a flat country, well clothed with vegetation, will be very different from that which will pour in torrents down the steep declivities of uncovered mountains. In districts of limestone, abounding in vertical fissures and subterranean cavities, a very much smaller quantity of water will drain off the higher districts than from a non-absorbent formation of primitive rock; or where the beds are horizontal and impervious. The steeper the district and the more rapidly the water is discharged to the reservoir, the less will be lost by evaporation and absorption.

It is necessary in constructing reservoirs, to obtain some measure of the quantity of water which may be expected to accumulate annually, in order to provide sufficient storage. For this purpose it is most important to determine the area of land draining into the valley chosen for the formation of the reservoir, and the average annual rainfall of the district, with, if possible, the probable loss or waste arising from the re-evaporation and absorption by vegetation, &c.

To ascertain the drainage area it is sufficient to determine the *summit level* or *watershed*, i. e. the ridge surrounding the valley which marks the line at which the streams flow in opposite directions into contiguous valleys. This may be determined by a special survey, with a careful examination of an accurate chart like the Ordnance map, on which the contour of the country, brooks, &c. are plainly marked. The whole of the basin included within the watershed is termed the catchment basin. In the case of the Bann reservoirs it amounts to 3,300 statute acres in extent; in that of the Greenock reservoirs to 5,000 acres; at the Manchester waterworks to 19,000 acres.

An immense number of experiments have been made of late years on the rainfall in different parts of Europe, and with considerable success in determining the laws of rain distribution. For England the annual average rainfall amounts to about 36 in. in depth over the entire surface, distributed throughout the year as in the following table:—

TABLE OF MEAN RAINFALL AT LONDON AND MANCHESTER.

Month.	Greenwich.*			Manchester.†		
	Average for 34 Years in ins.	Greatest fall in one month.	Least fall in one month.	Average for 64 years in ins.	Greatest fall in one month.	Least fall in one month.
January . . .	1·68	4·83	0·30	2·4915	5·85	0·32
February . . .	1·58	3·69	0·04	2·4190	6·56	0·44
March	1·61	3·45	0·40	2·2588	6·03	0·18
April	1·73	4·79	0·06	2·0225	4·75	0·16
May	1·96	4·16	0·50	2·3746	8·00	0·09
June	1·83	4·26	0·59	2·8483	7·05	0·20
July	2·37	6·65	0·10	3·7231	11·48	0·29
August . . .	2·40	4·65	0·07	3·5715	8·74	0·73
September . .	2·40	4·79	0·40	3·1353	9·00	0·24
October . . .	2·67	5·37	0·53	3·8404	9·00	0·60
November. . .	2·53	4·33	0·85	3·5682	7·37	0·62
December. . .	2·02	4·72	0·08	3·3088	9·50	0·07
Mean annual depth	24·781			35·5620		

* J. H. Belville.

† Principally from Dr. Dalton; see Manchester Memoirs.

This would give a mean of 30 inches, but it must be borne in mind that in the lake districts and all along the west coast, there is an annual fall of rain greatly exceeding that amount, and in some places in the higher districts in Cumberland the returns have been as high as 180 to 200 inches; from this it will be seen that 36 inches is a fair average for the whole surface of Great Britain.

It is, however, important in the construction of reservoirs to have observations of the rainfall in the district in which they are to be placed. Local causes greatly influence the quantity of rain; thus the average fall in Essex is about 20 in., whilst at Keswick, in Cumberland, it is as much as 67·5 in, and at Seathwaite, in the same county, it averages the enormous quantity of 141·5 in.

The method of determining the rainfall is very simple. A cylindrical vessel, of the form shown in section in fig. 81, is placed on the ground or sunk into it in such a manner that its mouth is about 12 in. above the surface. Sometimes a permanent rod and float is added, by means of which the depth of rain received on the funnel and preserved in the vessel is read off at sight, and for ordinary purposes this is probably the best plan. But where the greatest accuracy is requisite it is necessary either to tie down the rod or to remove it altogether after making an observation, as otherwise the rain in driving obliquely impinges upon the rod instead of passing over the funnel, and a slight excess is in this way registered above the true rainfall upon the area of the vessel. It is most accurate, however, to draw off the rain and measure it in a graduated glass tube. By placing two or three of these rain-gauges at different elevations around the site of a proposed reservoir, and examining them at convenient intervals of a week or month, it is easy to estimate the exact quantity of rain which falls upon the catchment basin in the course of one year; which with certain deductions is the quantity to be

Fig. 83.

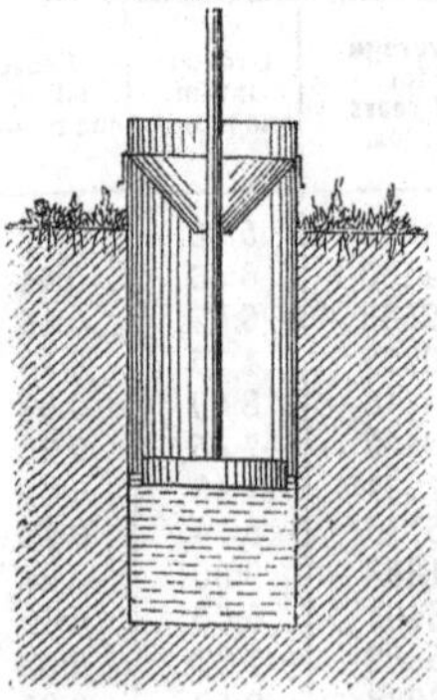

provided for in the reservoir. The precaution of placing the gauges within 5 in. or a foot of the ground is important, as, in accordance with an ill understood law, the quantity of rain rapidly decreases even at slight elevations from the ground, and it is also important to place the gauge where no artificial currents of air are created, as by the sloping side of the roof of a house. This subject was fully investigated several years ago by Mr. J. F. Bateman and a committee of the Manchester Philosophical Society. Observations had been made on and near the lines of the Ashton and Peak Forest Canals, about the accuracy of which, from their disagreement, doubts had arisen. The gauges in these observations were placed on the ridging of the roofs of the houses of the various lock-keepers, under the impression that, from the exposure of the position, all the rain which fell must there be caught. New gauges were placed in the same localities, but at the surface of the ground, and the results of these experiments were as follows:—

Locality.	Gauge on roofs.	Gauge on ground.	Excess per cent. on ground.
	in.	*in.*	
Near Middleton	18·14	28·8	58·76
Near Rochdale	20·50	30·3	47·8
Whiteholm Reservoir	22·64	35·1	55·0
Blackstone Edge	23·45	34·2	45·84
Blackhouse	24·89	35·9	44·23
Sowerby Bridge	16·77	23·8	41·92

This enormous difference, amounting to 50 per cent. on the average, fully proves the unfitness of the roofs of houses for registering the rainfall. The upward currents of wind created by the sloping roof appear to have carried the raindrops over the edge of the gauge.

Dr. Heberden found the annual fall of rain at the top of Westminster Abbey to be 12·099 in. On the top of a house close by of much inferior altitude 18·139 in.; on the ground 22·608 in.

Mr. Phillips, at York, found the total fall for three years at an altitude of 213 feet to be 38·972 in.; at 44 feet, 52·169 in.; and on ground 65·430 in.

Notwithstanding the explanations of these facts which have been offered, Sir J. F. W. Herschel has within the last year asserted that the cause is yet to seek. The raindrops certainly appear to increase in size in the moist lower strata of the atmosphere.

Mr. Phillips's explanation has been accepted by some Meteorologists, that this augmentation is caused by the deposition of moisture on the surface of the drop, in consequence of its temperature being lower than that of the moist strata of air through which it passes. But this does not appear to be consistent with the fact, that in the condensation of vapour a large amount of latent heat would be liberated. Mr. Baxendale, who pointed this out, estimates from Professor Phillips's observations that in the condensation of the amount of water which corresponds to the augmentation of the rain drop in a fall of 213 feet, sufficient heat would be liberated to raise the temperature of the drop to 434° F.

The quantity of rain which falls in twenty-four hours, is about 1 in. at the maximum in average districts in England, although in the remarkably exceptional district in Borrowdale, already alluded to, 6·7 in. have been known to fall in the same period. The western coasts generally receive a larger proportion of water than other districts. Mountainous districts in this country, to an elevation of 2000 feet, receive a larger proportion of rain than lowlands. According to the late Dr. Miller there fell in twenty-one months in the lake district:—

In the valley, 160 feet above the sea	. . .	170·55 inches.
Styehead, 1290	„	185·74 „
Scatoller, 1334	„	180·28 „
Sparkling Tarn, 1900	„	207·91 „
Great Gable, 2925	„	136·98 „
Scawfell, 3166	„	128·15 „

Mr. Bateman's observations agree with these results, in proving the increase of rainfall corresponding to increased elevation*,

* The increase of rainfall in passing from the valley to the mountain must be carefully distinguished from the decrease as we ascend upward into the atmosphere, as shown in Mr. Phillips's observations.

as shown by the following figures, representing the rainfall near Glossop in one year:—

Westerly foot of hills, 500 feet above the sea . .	45·0 inches.
„ edge of table-land, 1500 „	67·8 „
Easterly edge of Kinderscout, 1600 „	77·45 „
„ foot of hills	40·85 „

After having determined from these considerations the quantity of water annually falling on the drainage district of a proposed reservoir, it is necessary in the next place to ascertain the probable loss from evaporation and other causes during the transmission to the reservoir. The numerous experiments on evaporation made upon small surfaces of water and of earth may be dismissed as having afforded too inconsistent results to be of any practical value.* Dr. Dalton's experiments are accurate and valuable as far as they go, but they are deficient in points of application to practical investigations. The area from which evaporation takes place is identical neither with the area of the catchment basin nor with the reservoir surface; but is a variable quantity depending on the season, the climate and the locality. It appears to me that the evaporation from a surface of water in low flat land charged with moisture, or a level vegetated surface, is very different from the evaporation in mountainous districts where there are precipitous descents to the brooks. In the former case the waters are retained and remain for weeks more or less exposed to the solar rays and the drying influences of wind. In the latter the rain pours in torrents down the barren hill-sides, and is launched into the valley where the principal evaporation takes place upon a very limited area of surface.

So also in tropical countries; the evaporation from a surface of water is greater than the rainfall upon the same surface, but then the rain falls in torrents, and is rapidly carried away to its

* Dr. Dalton gives the annual evaporation from a surface of water as 25·158 inches; Dr. Dobson, 36·78 inches; Dr. Thomson, 32 inches. The above views in regard to these experiments I expressed in a report on the Bann reservoirs in 1836. Mr. Conybeare gives the evaporation from a surface of water at Greenwich Observatory 5 feet, at Bombay 8 feet, and at Calcutta 15 feet, per annum.

natural or artificial reservoirs, and then the evaporation takes place from a very small area of surface.

Since the establishment of reservoirs and the carrying out of large drainage operations, opportunities of estimating the relation of the rainfall to the discharge by rivers have been generally available, and several important experiments have been made in this way. The method of arriving at results is to ascertain the rainfall over a catchment basin the area of which is known. The whole of the water discharged by brooks, &c., is then conveyed over a rectangular weir or waste board, and the mean velocity of the current and its breadth and depth determined by observations made once or twice every day. The comparison of the amount of water discharged with the total fall will afford the data for ascertaining the amount of evaporation.

Observations of this kind were made by Mr. Bateman with great care in the years 1845, 1846, 1847, with reference to the construction of reservoirs for the supply of Manchester with water, from the Derbyshire hills beyond Staleybridge and Mottram. Gauges were placed at the bottom of the Swineshaw valley (through which flows a tributary of the Tame), and near the summit of Windyate edge, and for some time a gauge was placed midway between these places. Similar gauges were placed in Longendale valley, and the stream in each was measured two or three times a day. From these observations the following table is compiled:—

Locality.	Year.	Mean rain.	Mean discharge.	Waste or loss by evaporation.
		in.	*in.*	*in.*
Swineshaw Brook . .	1845	59·8*	40·70	19·10
	1846	42·6	33·24	9·36
	1847	49·3	37·10	12·20
Longendale Valley . .	1847	55·2	49·46	5·74

The first was a wet year, the second one of the dryest on record, the third an average year.

* The rainfall possibly somewhat too high. Manchester Memoirs, vol. ix. p. 17.

By uniting the observations at the Swineshaw and the Longendale valleys, we get the following general table of the monthly fall and flow for three years:—

Month.	Rain.	Discharge.	Difference.
	in.	*in.*	*in.*
January	2·36	2·85	−0·49
February	4·30	4·10	+0·20
March	1·70	1·30	+0·40
April	5·22	4·12	+1·10
May	6·48	4·75	+1·73
June	3·40	1·65	+1·75
July	1·52	0·99	+0·53
August	4·32	1·24	+3·08
September	7·38	5·12	+2·26
October	4·66	5·67	−1·01
November	5·48	6·25	−0·77
December	6·74	8·55	−1·81
	53·56	46·59	+6·97

In the following table I have collected the most reliable results on the relations of discharge, rainfall, and evaporation:—

District.	Year.	Area of country drained in acres.	Rain-fall in inches.	Discharge in inches.	Differences or loss by evaporation	Remarks.
Bute	1826–7	—	45·4	23·9	21·5	Dry year, Mr. Thom.
Greenock	1828	—	60·0	41·0	19·0	Mr. Thom.
Gorbals	1852	2,750	60·0	48.0	12·0	
Swineshaw Brook	1845–7	1,250	50·58	37·01	13·5	Mr. Bateman.
Rivington Pike	1847	10,000	56·5	44·0	12·5	Mr. Hawksley.
Lough Mask, Ireland	1851–2	70,000	49·34	28·59	20·75	Flat country, Mr. Betagh.

The above table shows a loss of from 12 to 20 in., or an average waste of 16 in. of rainfall arising out of re-evaporation and other causes of absorption.

"The storage requisite for equalising the supply of water between dry and wet years should be provided with a due reference to the continuance of drought, and the quantity of water which will flow off the ground: in extreme wet seasons no water should be allowed to run to waste. Experience has shown that in the regions of comparatively moderate rain in this country, the storage to effect this object should vary from 20,000 or 30,000 to 50,000 or 60,000 cubic feet for each acre of collecting ground, the smaller quantity being about sufficient for an available rainfall of perhaps 18 in. and the larger for one of about 36 to 40 in."* 80,000 cubic feet per acre of collecting ground are provided at Lough Island Reavy; 60,000 at the Gorbals reservoirs, Glasgow; 49,000 at Rivington Pike, and 34,000 at Manchester; at the last, the whole fall not being impounded.

I proceed, neglecting further details on this subject, which belongs rather to the province of the civil than the mechanical Engineer, to give an example of the carrying out of these views, of the utility and importance of reservoirs in districts abounding with waterfalls, and where mills are numerous and depending in whole or in part on a steady and regular supply of water.

In 1836 I was called upon to report upon the best means of regulating the water supply upon the river Bann, which from its excessive variations of flow was a source of great inconvenience to the manufacturers on its banks. The river Bann rises among the lofty bare summits of the Mourne mountains, in the north-east of Ireland, where there is a heavy rainfall, and in consequence devastating floods frequently poured down its channel, carrying bridges, embankments, and other obstructions before them. On the other hand, during the summer months, the ordinary supply of water was totally inadequate to the demands of the mills; whilst the flourishing state of the linen trade called for an extended application of power, in a district where steam was not available as a motive power unless at great cost. Hence, in co-operation with Mr. Bateman, the ground was surveyed, and two reservoirs erected in the upper part of the river, by which these evils were removed, and a continuous and adequate supply of water rendered available.

* Report of the British Association, "On the Supply of Water to Towns." By J. F. Bateman, C.E. 1858.

Lough Island Reavy, the site selected for the principal reservoir, was a natural lake, bounded on the north and south by land of considerable elevation, which although having a comparatively small extent of drainage (3,300 acres ultimately) was supplied by good feeders, which, united to the surplus waters of the river Muddock, would fill the reservoir at least once or twice a year. The original surface of the Lough, fig. 84, was 92½ acres in extent; on this it was proposed to raise a depth of 35 feet more water, by the aid of embankments, and to draw off at a depth of 40 feet under that height. The area thus enlarged would be 253 statute acres, and the capacity of the reservoir is 287,278,200 cubic feet.

Corbet Lough was the second site, and although at first abandoned from its proximity to the town of Banbridge, was afterwards adopted. At a small expenditure for embankments, Corbet Lough was raised 18 feet above its summer level. so as to cover 74½ acres, and to have a capacity of 46,783,440 cubic feet of water.

A third site was selected further up amongst the mountains, but at this part the works were never executed.

It is understood that 12 cubic feet of water per second falling one foot, will, in its best application on a water-wheel, afford a force equivalent to 33,000 lbs. raised one foot high per minute, or one horse power. Now supposing the reservoirs to discharge 40 cubic feet of water per second, the fall from the lowest point of outlet at Lough Island Reavy, to the tail water of the lowest mill on the Bann, being 350 feet, we have a total force of 1166 horses available for mill purposes, or in other words, the millowners will derive an average advantage of 3·3 horse power for every foot of fall. This, it must be observed, is not a supposed quantity, but the result of certain data, taken by calculation from the waters of the Bann. It must be noticed further that this supply of 2400 cubic feet per minute is not the whole power. The calculations are for one half, the river supplying the remainder, except in extremely low water, when the demands from the reservoir may be increased to meet the emergency.

From the estimates made at the time, the expenditure to secure this result would be

	£	s.	d.
For Lough Island Reavy	12,600	0	0
„ Corbet Lough	3,512	0	0

At Lough Island Reavy it was necessary to construct four embankments, marked A, B, C and D, in fig. 84.

The principal S.W. side	137,400	cubic yards,
Small do.	17,400	„
„ N.W. end . . .	5,200	„
„ E. end . . .	99,781	„
Total .	259,781	„

The substratum of the valley being water-tight, the footing for the puddle was easily obtained by sinking a trench into the water-tight stratum, whence the puddle wall was carried up vertically with the bank to the required height. It was 12 feet in width, at 40 feet below the top, diminishing to 8 feet wide at the summit. A layer of peat was brought up on the inside of the puddle, and a similar layer on the face of the slope. Above the peat a layer of three feet of gravel was laid, and on that the stone pitching forming the inner side of the bank. The inner slopes of the embankments were 2½ horizontal to 1 vertical, and 3 horizontal to 1 vertical. The outer slopes 2 horizontal to 1 vertical, and 2½ horizontal to 1 vertical. The discharge pipes, two in number, each 18 in. in diameter, were placed at the bottom of a stone culvert, at the lowest part of the embankment, with suitable discharge valves, &c. The rainfall for the district amounted to from 72 to 74 in. annually, of which at least 48 in. found its way to the reservoirs.

Fig. 84 is a plan of the original disposition of Lough Reavy and its feeders. The original area of the lake is shaded, and its present area is indicated by the dotted line connecting the embankments A, B, C and D. The diversions of roads and new feeders rendered necessary are also indicated.

Fig. 85 represents a section of the embankment of the Belmont reservoir, which will sufficiently explain the arrangement of the culvert and discharge pipe, *a a*, with the stop and discharge valves *v v*, in the valve house T, which in works of this

Fig. 84.

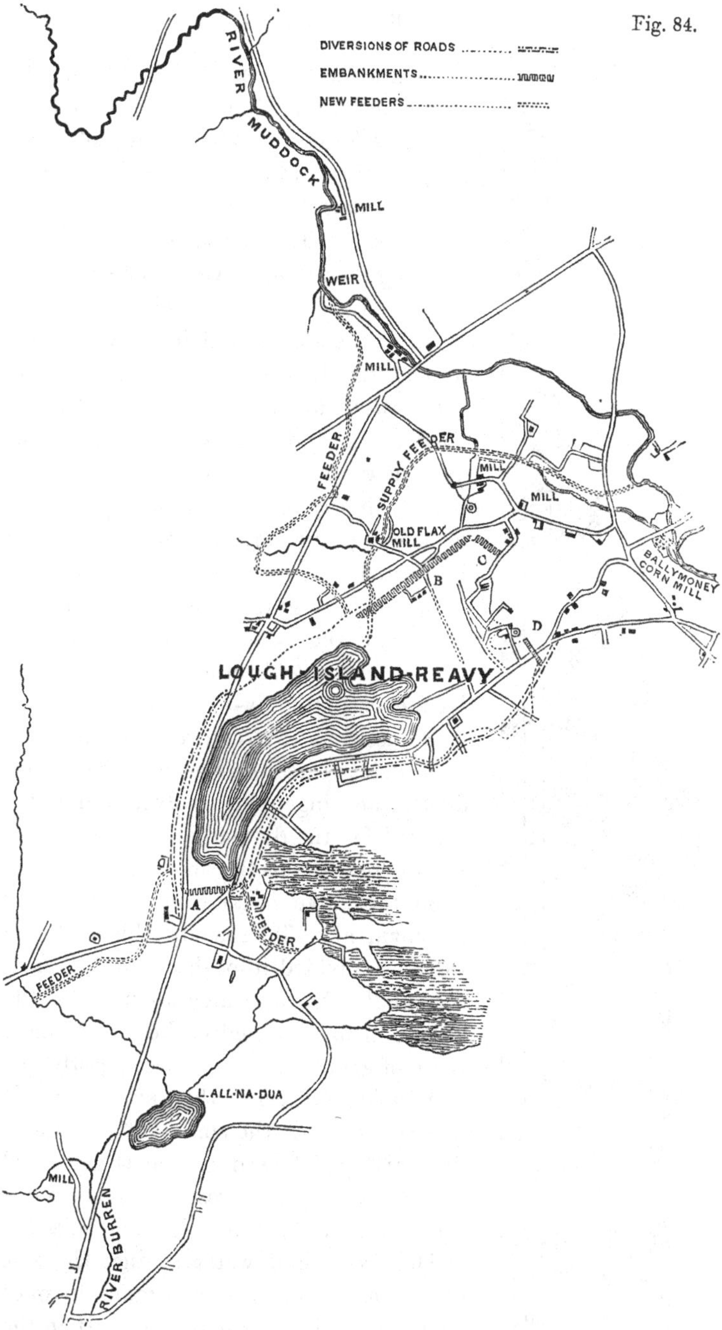

Map of Lough Island Reavy Reservoir.

Fig. 85.

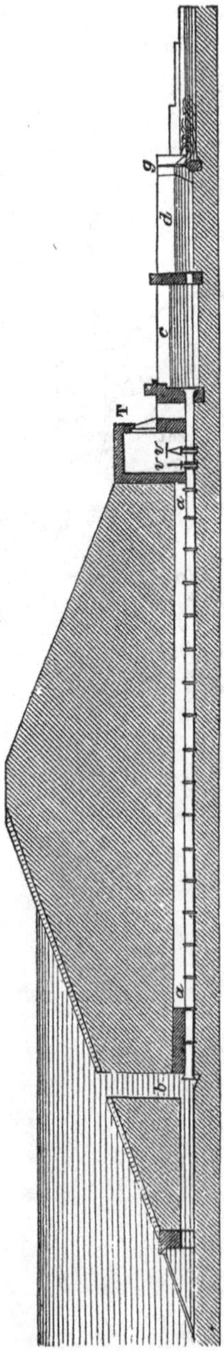

kind is always under lock and key. The water entering the pipe through the tunnel, *b b*, flows out into the well *c*, and gauge basin *d*, where, as it passes over the gauge or dam board *g*, its quantity may be ascertained. The construction of the regulating discharge valve is shown in figs. 86 and 87. Fig. 86 is an elevation of the valve at the side at which the water flows in, and fig. 87 a cross section. A is the valve case, closed below and fitted with a bonnet *b* at top; the valve *v*, works up and down in the valve box, against a brass facing *c*, and is confined by a guard *d* behind; the adjustment of the valve is effected by a valve spindle *f*, of wrought iron, cased in gun metal, so as to slide freely in the stuffing box *g*, and is worked by a fly-wheel *h*, and screw above. By means of this fly-wheel the valve may be adjusted to any required opening.

Of late years Mr. Bateman has introduced an ingenious valve, admirably adapted for the discharge of reservoirs of great depth, where the amount of pressure upon the valve is an impediment to its employment. To remedy this evil, the valve is divided into three parts; first, the small valve by which about $\frac{1}{10}$ of the area is opened; secondly, the intermediate valve of about $\frac{1}{4}$ the total area; lastly, the large valve unclosing the remainder. It will be seen that the small valve is drawn first, and is followed by the second, and ultimately by the largest, after the pressure is removed or partially neutralised.

The pressure of water against the side of an embankment is enormous in most instances, and varies upon any part in the ratio of its depth below the surface. Let h = the depth of water in a reservoir; A = the area in square feet of a vertical

Fig. 86.

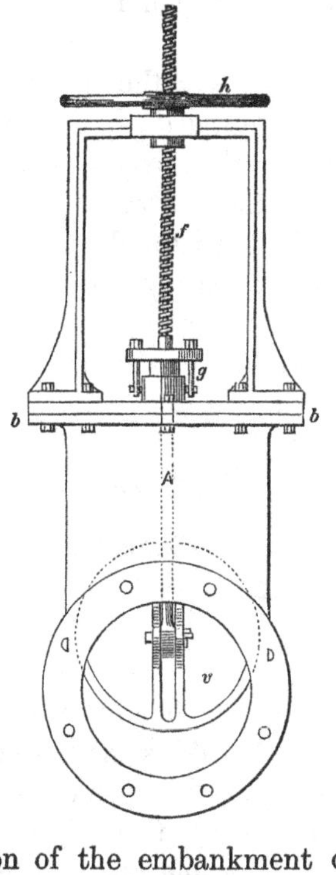

Fig. 87.

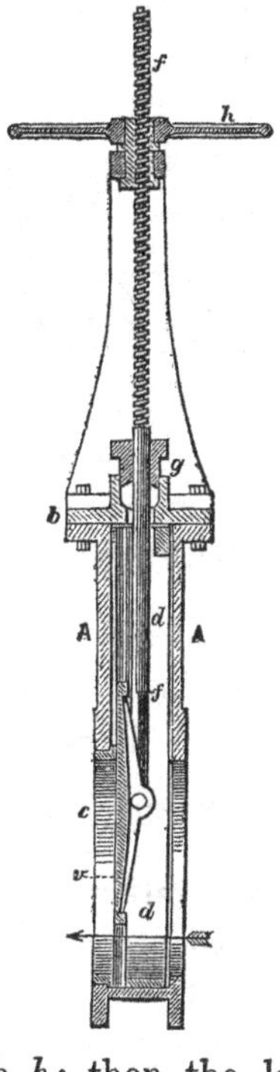

section of the embankment of the depth h; then the lateral pressure upon the embankment in a horizontal direction is, in lbs.

$$\text{P} = \tfrac{1}{2}\, h \times 62\tfrac{1}{2} \times \text{A} = 31{\cdot}25\ \text{A} \,.\, h,$$

a cubic foot of water weighing $62\frac{1}{2}$ lbs.

Or, generally, the whole pressure of water upon a submerged plane surface is equivalent to the area of the surface, multiplied by the weight per cubic unit of the fluid, and by the head of water measured from the centre of gravity of the submerged surface. That is, for water

$$\text{P} = 62\tfrac{1}{2} \,.\, \text{A}_1 \,.\, h_1.$$

Where h_1 = the depth of the centre of gravity below the level of the water in feet; A_1 the area of the surface in feet; P = the pressure on the surface in lbs.

And the whole pressure in any one direction is equal to the area of a section of the fluid vertical to that direction, multiplied by the weight of a cubic unit of the fluid and by the distance of the centre of gravity of the section from the level of the water.

Fig. 88.

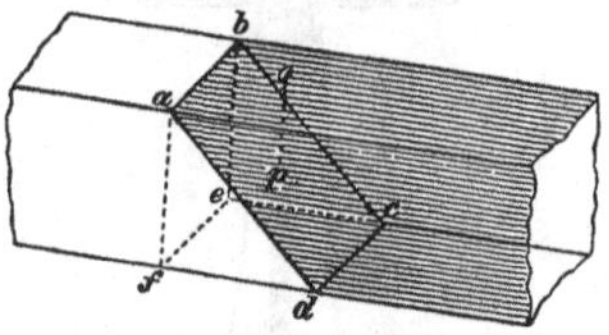

Putting l = the length $a\,b$; s = the slope $a\,d$; h = the height $a\,f$; k = the breadth $f\,d$, all in feet; p the centre of gravity of $a\,b\,c\,d$. Then the distance of the centre of gravity of $a\;b\;c\;d$ from the level of the water, $o\,p$, is equal to $\frac{1}{2}\,h$; and the distance of the centre of gravity of the plane $a\;b\;e\;f$ from the level of the water is also $\frac{1}{2}\,h$.

Therefore, the whole pressure upon

$$a\,b\,c\,d = \tfrac{1}{2}\,l\,s\,h \times 62{\cdot}5 = 31{\cdot}25\,l\,s\,h.$$

The horizontal pressure against the embankment

$$= \tfrac{1}{2}\,l\,.\,h^2 \times 62{\cdot}5 = 31{\cdot}25\,l\,.\,h^2.$$

The vertical pressure $= l \times k \times h \times 62{\cdot}5.$

To the statistics given above of the rainfall and evaporation in this country, it will be necessary to add some account of their amount in tropical climates, where the conditions are essentially different. In such climates for three quarters of the year the rain never falls, and the whole quantity for the annual consumption falls during the remaining quarter.

At the Bombay Water Works constructed by Mr. Conybeare, the annual rainfall is 124 inches, of which $\frac{6}{10}$ are assumed to be available for storeage. The area draining into the basin is 3948 acres, so that the supply is upwards of 6,600,000,000 gallons. The storeage capacity of the reservoir is 10,800,000,000 gallons, or 1,733,000,000 cubic feet.*

At the Melbourne Water Works constructed under the direction of Mr. Matthew Bullock Jackson, the area of the reservoir when full is 1303 acres, greatest depth 25 feet 6 inches, average depth 18 feet, and capacity 6,400,000,000 gallons. The area of the natural Catchwater basin is 4650 acres, together with

* Minutes of Proceedings of Institute of Civil Engineers, vol. xvii. p. 560.

600 acres drained by a water course. This area, however, may be increased if a larger supply is necessary. This watercourse at the same time opens a connection with the River Plenty, through which flows the water drained from an extent of 40,000 acres of country. This watercourse is opened during the winter to fill the reservoir from this source. The following table gives the detail of the rainfall and evaporation observed by Mr. Jackson during the construction of the works:—

TABLE, SHOWING THE AMOUNT OF SPONTANEOUS EVAPORATION AND RAINFALL FOR TWELVE MONTHS ENDING 31ST JANUARY 1858.

Months.	Rainfall at				Spontaneous Evaporation at Melbourne.
	Melbourne, 94¼ feet above the level of the sea.	Yan Yean.	Geelong, 96 feet above the level of the sea.	Ballarat, 1438 feet above the level of the sea.	
	inches.	*inches.*	*inches.*	*inches.*	*inches.*
February . .	3·98	1·33	2·39	0·23	8·14
March . .	3·80	3·61	1·99	3·75	5·10
April . . .	0·99	0·78	1·07	1·55	4·25
May . . .	2·00	2·05	1·72	1·85	1·97
June . . .	1·99	1·89	1·58	0·00	1·50
July . . .	1·16	2·39	1·15	0·00	1·86
August . .	1·69	2·42	1·14	2·42	2·57
September . .	3·83	3·70	3·19	2·68	3·76
October . .	5·28	4·70	2·63	4·63	4·23
November . .	2·12	1·80	3·15	2·27	5·84
December . .	0·83	1·76	0·33	0·73	11·01
January . .	0·88	1·07	0·00	0·00	11·23
Totals . .	28·55	27·50	20·34	20·11	61·46

It is to be presumed that the evaporation given above as nearly three times the rainfall is the evaporation from a surface of water such as that of the reservoir itself. The rain, however, is collected from a surface thirty-five times as great as that of the reservoir when at its maximum height.

Weirs or *Dams*, thrown across the beds of rivers, have always been employed in order to raise the head of water in the river bed, and to divert a portion of it for the purposes of the mill. We have now to consider how most economically to secure a sufficient fall, and to protect the dam from the destructive effects of floods.

There is hardly any department of engineering which requires more careful consideration than that of forming barriers to large quantities of moving water; and when the nature of rivers carrying off the drainage from a large area is considered, and the enormous power of suddenly accumulating floods, the nature of the resistance required from a dam may be easily conceived, and when all the care of the engineer has been exercised, it nevertheless sometimes occurs that the torrents tear up and destroy in a night the work which was intended to perform the quiet industrial duties of a mill for ages, leaving, in place of the well turned arch across the stream, only the horns of the abutments and an indistinguishable mass of rubbish mingled with the mountain debris of the flood.

Such is frequently the case with weir constructions, particularly those across the rapids of mountain torrents, and this not unfrequently causes the construction of a temporary dyke of boulder stones capable of withstanding the ordinary action of the river, and easily replaced when floods have caused its partial destruction. This description of weir is carried diagonally across the stream at *a* (fig. 89), and being considerably longer than its breadth forces part of the water into the conduit *b*, and passes the remainder over the top in a thin sheet, which does little or no damage to the banks below. In the above description of weir it seldom happens that much fall can be obtained, and they are

Fig. 89.

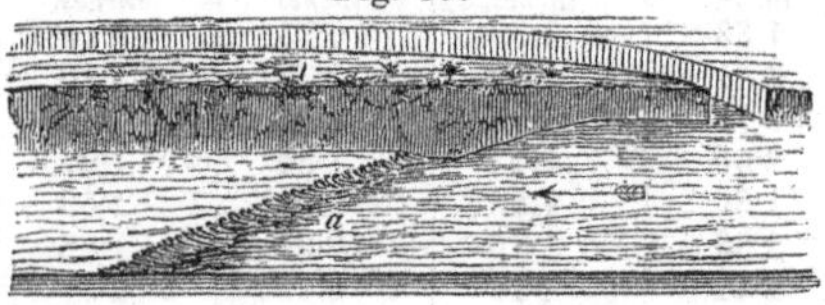

Fig. 90.

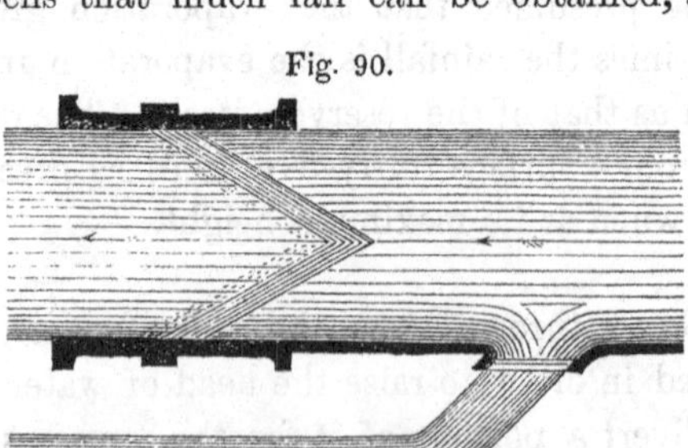

therefore adopted where there is a large supply of water employed upon an undershot wheel.

Another description of weir, which is generally employed on moderate sized rivers, is the V form constructed across the bed of the river, as shown in fig. 90, in plan. The object of adopting this form of weir is to increase its resisting powers, and by spreading the fall of water over a large surface, to diminish its destructive effects upon the apron below; the descending currents meeting in the angle of the V neutralise their effects on the foundations, and do less injury to the banks on either side. This weir is generally formed of piles (fig. 91), with an open

Fig. 91.

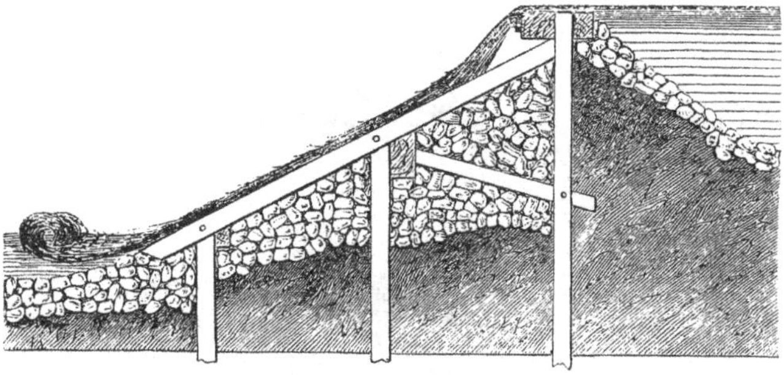

frame of timber, into which are inserted large boulder stones, forming a compact mass of boulder sheeting resting on gravel, and nearly impervious to water. Another weir, preferred to most others where timber is plentiful, is formed into a series of

Fig. 92.

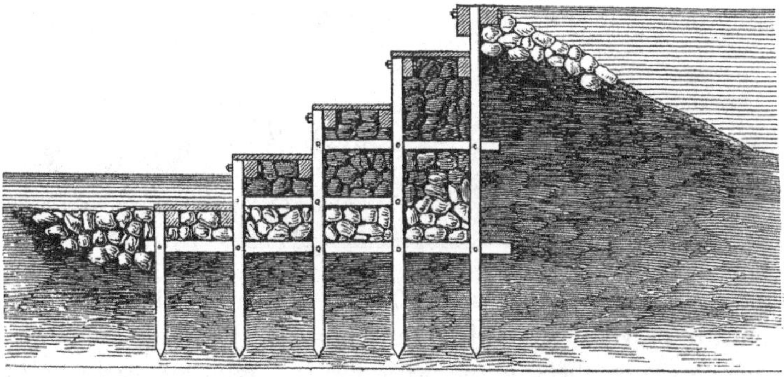

steps (fig. 92), over which the water falls in cascades, which destroys its injurious effect on the foundations; it is composed of piles placed at right angles with the direction of the stream, and placed in rows properly stayed and covered with planking firmly nailed to the horizontal and vertical timbers. When it is necessary to have the structure watertight a line of sheet piling is usually driven in, in the line of the weir across the whole breadth of the stream, and these again, supported by foot piles and stays at different distances, form a perfectly tight and very durable weir.

The most perfect weirs, however, are formed of stone, built of solid ashlar, and usually forming part of the segment of a circle across the breadth of the river (fig. 93). These are made,

Fig. 93.

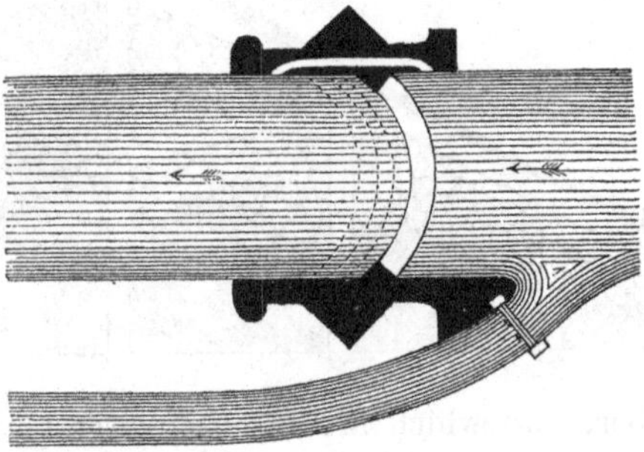

1st, with long inclined slopes on either side; 2nd, solid, with nearly perpendicular walls; or 3rd, with a curved apron to break the force of the fall.

Of the first kind we have a good example in the weir constructed by Smeaton at Carron (fig. 94), where *a*, *b*, represent

Fig. 94.

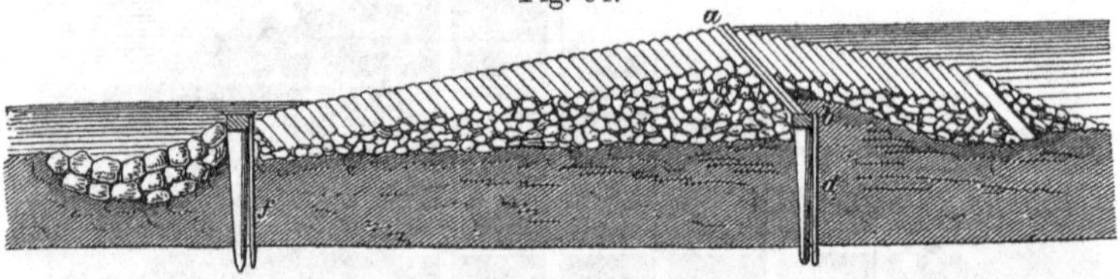

two courses of flag stones, breaking joint, and packed with live moss, to prevent the silt being driven through; these are footed upon grooved sheet piling with bearing piles and stringer *d*, the flags being supported on rubble; at the foot of the dam is another row of sheet piling *f*, similarly supported and protected by a fir plank at top from the action of the water. Over the rubble is placed a row of regular stones, laid endways so as to be perfectly secure from derangements by floods.

The second description of stone weir is a solid ashlar wall having its convex side to the current (fig. 93) and abutting upon heavy masses of masonry on each side of the stream. Fig. 95, exhibits this weir modified by having a curved apron, so as gradually to convert the vertical fall of the water into a horizontal flow in the direction of the stream.

It will suffice to observe further, that the head of water immediately over the crest is less than the head of water at some distance behind. It is usual to cut a channel, with a sluice gate in one of the wing walls of the weir, to draw off superfluous water, when requisite. The utmost caution is

Fig. 95.

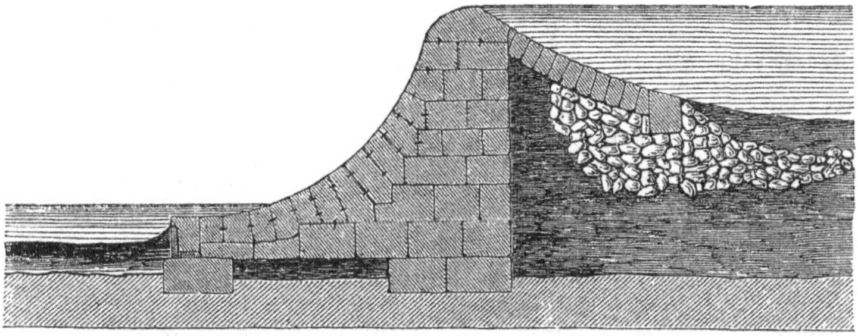

needed, both in observing the conditions of the river and the effects likely to result in times of flood from the increased head of water above the weir. Rapid rising of the waters and sudden changes in the state of the river are too often neglected with disastrous consequences to works of this kind, just on the eve of completion, or to the lands above the dam in consequence of flooding caused by the obstruction of the dam. In cases where this last danger is apprehended, a self-acting dam has sometimes been employed, consisting of a massive frame of planks carried across the river and attached by hinges to the crest of the dam. This plank is maintained in a vertical position in ordinary conditions of flow by balance weights attached or hung over wheels upon the wing walls, so as to retain the maximum desirable head of water. In floods the increased pressure of the overflowing water overcomes the balance weights and throws down the plank into a horizontal position, opening a free passage for the water.

Conduits.— Having thus considered the means of accumulating water power and regulating its supply by means of reservoirs and weirs, we have yet to consider the formation of conduits or lades, as they are called in some places, for the actual discharge of the water upon the water wheel or other machine by which its power is to be utilised. By the construction of a weir we may have dammed back the water half a mile or a mile, and formed the upper part of the stream into a reserve from which the supply of water can be drawn and two or three feet or more of fall gained; but unless the mill is built close up to the banks of the stream head courses, canals and tail races have to be cut in order to make the fall available, and these conduits are not unfrequently as difficult of construction and as expensive as the weir. In several large works with which I have been connected, the cost of conduits has extended to many thousands of pounds, as at the Catrine Works in Ayrshire, or the Deanston in Perthshire. In the former case a large tunnel, with retaining walls and embankments several hundred yards in extent, had to be constructed, and at the latter a wide and spacious canal, nearly a mile long, before the water reached the mills where it was turned to account in driving the different machines for spinning, weaving, &c.

The large expenditure in these and similar works, operates much against the economy of water power, and when the extremes of floods and droughts, including the interest of capital sunk, is considered, it will be seen that it frequently happens that steam power might have been purchased and maintained at as economical a rate. Let us take, for example, the Catrine Mills, at which there is a fall of forty-eight feet, and a power of 200 horses, nearly constant throughout the year. In this establishment there are two colossal water wheels, each fifty feet in diameter and twelve feet wide. Now taking the weir, the tunnel, the upper conduit, tail race, &c., arched to a distance of a third of a mile down the river, we may estimate the ultimate cost, approximately, as follows:—

Water privileges and land	£4,000
Cost of weir	1,000
Head race, tunnel, and canal	3,000
Archways, cisterns, sluices, &c.	1,000
Wheelhouse and foundations	1,500
Tail-race	1,500
Water-wheels and erection	4,500
Contingencies	1,500
Total	£18,000

The cost of power independent of mill-work equivalent to an annual rental for interest of capital, repairs, and wear and tear, at 7 per cent., amounting to 1260*l.*

This may be contrasted with steam power in a district where coal can be purchased at 7*s.* per ton, and we have,

Cost of engines of 100 nominal horse power	£4,000
Engine-house, foundations	1,500
Contingencies	500
Total	£6,000

This at 10 per cent. for interest of capital, repairs, and renewals, will be equivalent to

An annual rental of	£600
Add consumption of coal at 4lbs. per indicated horse power per hour, engineers' wages, &c.	900
Total	£1,400

Against the higher rental in the case of steam, must be set the cost of transit of the raw material and products of the mill, which must be transported to and from the market at a greatly increased cost, as in the case of the Catrine works, with the risk of stoppage also from want of water in long continued drought or frost. It is true that labour may be had cheaper in the country than in towns, but that is no counterpoise for want of skill amongst the operatives, or for the loss of those numerous conveniences which are to be obtained in the great foci of labour where the whole powers and energy of the country have been concentrated.

On the whole, there appears (in the present improved state of the steam engine and the price of coal) to be no advantage in this country in water power as applied to manufactures, and it is only at out districts, and where the mere wants of the inhabitants have to be supplied, that water mills can be used with profit. Before the introduction of the steam engine, water power was invaluable, but we now see that it cannot at all times be depended upon, and that in most cases where a large amount of power is required, the chief source from which it must be derived is steam.

CHAP. II.

ON THE FLOW AND DISCHARGE OF WATER, AND THE ESTIMATION OF WATER POWER.

In the present chapter it is proposed to enter only so far into those questions of Hydrodynamics which relate to the measurement of the discharge of water, and the estimation of water power, as it is necessary they should be understood by the practical millwright, in order that he may be at no loss in comparing the efficiency of various forms of water machinery, calculating their power, and proportioning them to their position and their work. For minute and accurate mathematical investigations, the reader is referred to special treatises on hydrodynamics, in which the subject is treated from another point of view.

From the nature of a non-elastic fluid such as water, in which the particles are free to move over one another without friction, the following relations hold between pressure, velocity and discharge.

1st. The *pressure* P upon a unit of area at the depth h beneath a fluid surface is equal to the weight of a column of the liquid h units high; that is, if W be the weight of a unit of volume,

$$\mathrm{P} = \mathrm{W}\, h \ldots (1).$$

And therefore the pressure on a horizontal surface of a units area $= \mathrm{W}\, h\, a$.

2nd. The *velocity* with which a fluid flows from a small orifice at the depth h beneath the surface, is the same as the velocity it would have acquired in falling freely the same distance under the action of gravity, if we neglect those causes of retardation to be considered presently. If we take $v =$ mean velocity of the effluent water; $h =$ mean depth of orifice beneath the surface, or in other words the *head* of fluid; $g = 32{\cdot}1908 =$ the velocity generated in a falling body in one second, then by the laws of accelerating motion,

$$v = \sqrt{2\,g\,h}^{*}, \ldots (2)$$

that is, the theoretical velocity of effluent water is equal to the square root of 64·38 times the mean head; understanding by mean head the head measured from the centre of the orifice. Thus we have in the following table the theoretical velocity at various heads.

TABLE I.—THEORETICAL VELOCITY OF EFFLUENT WATER.

Head.	Velocity per second.	Head.	Velocity per second.	Head.	Velocity per second.	Head.	Velocity per second.	Head.	Velocity per second.
ft.	ft.	ft.	ft.	ft.	ft.	ft.	ft.	ft.	ft.
1	8·02	11	26·6	21	36·8	31	44·6	41	51·3
2	11·34	12	27·8	22	37·6	32	45·4	42	52·0
3	13·90	13	28·9	23	38·5	33	46·1	43	52·6
4	16·04	14	30·0	24	39·3	34	46·8	44	53·2
5	17·93	15	31·1	25	40·1	35	47·4	45	53·8
6	19·64	16	32·1	26	40·9	36	48·1	46	54·4
7	21·21	17	33·1	27	41·7	37	48·8	47	55·0
8	22·68	18	34·0	28	42·4	38	49·4	48	55·6
9	24·09	19	34·9	29	43·2	39	50·1	49	56·1
10	25·38	20	35·9	30	43·9	40	50·7	50	56·7

3rd. The *quantity* of water which issues from an orifice at a depth h beneath the surface of a fluid is equal to the area of the orifice multiplied by the velocity of the effluent water, that is neglecting the diminution from the *venâ contractâ* to be mentioned shortly.

Let Q = units of volume discharged per second

a = area of orifice

v = velocity of effluent water

$$Q = a\,v = a\,\sqrt{2\,g\,h} \ldots (3).$$

And in t seconds $Q\,t = a\,v\,t$ will be discharged.

Where Q is called the *theoretical discharge*, and is found by multiplying the area of the orifice in feet by the velocity of the effluent water in feet per second found as above.

4th. If the orifice instead of opening freely into the air as supposed above, opens into another reservoir of fluid, we must substitute in the above equations the difference of level of fluid

* Or in feet $v = 8·03\,\sqrt{h}$.

in the two reservoirs for the head above the centre of the orifice. Let h' be the head above the centre of the orifice in the higher reservoir and h'' in the lower, then the effective head $h = h' - h''$ ... (4).

5th. If the water escape by a rectangular *notch* instead of an orifice, that is an aperture such that the upper level surface of the water does not come in contact with the sides of the vessel, falling freely in the air, the theoretical discharge is two thirds of the area of the effluent vein multiplied by the velocity of efflux, or more accurately if h = head of water, b = *breadth of notch*

$$Q = \tfrac{2}{3}\, b \,.\, h \,.\, \sqrt{2\, g\, h} \;\ldots\; (5).$$

We must next examine certain properties of fluid motion which cause the *actual* or *effective discharge* to differ materially from the theoretical discharge given in the above equations, although in a constant ratio, so that the one may always be calculated from the other.

1st. *Thick-lipped orifices* or mouth-pieces. For smooth orifices, the length of which is about twice or three times the smallest diameter, the *actual* does not widely differ from the *theoretical* discharge. The velocity of the effluent current is however never so great as that in equation (2), but is diminished for a constant ratio for each kind of orifice, and the discharge is less in the same proportion. For a simple cylin-

Fig. 96.

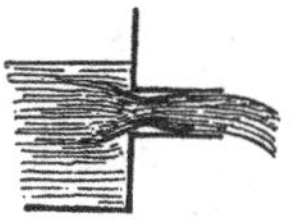

Fig. 97.

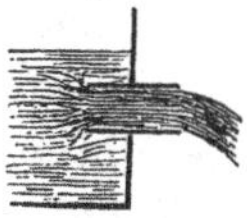

Fig. 98.

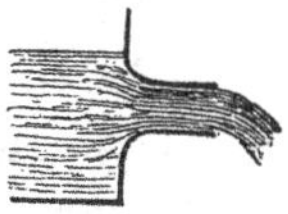

Fig. 99.

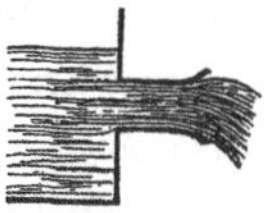

drical tube, fig. 94, of about $1\frac{1}{2}$ diameters in length, the velocity of the effluent water is equal to $0{\cdot}8\, v = 0{\cdot}8 \sqrt{2\, g\, h}$ and the actual discharge $= a \times 0{\cdot}8 \times v = 0{\cdot}8\,$Q. Where the interior angle of the tube is rounded, as in fig. 96, the velocity amounts to as much as $0{\cdot}96\, v$ to $0{\cdot}98\, v$, and hence the discharge to $0{\cdot}96\,$Q to $0{\cdot}98\,$Q, where Q as before is the theoretical discharge given by the above formulæ. This constant ratio is called the coefficient of velocity.

Hence we have this rule for determining the quantity of water discharged by a thick-lipped orifice; seek first in Table I. the velocity corresponding to the given head of water, measured from the centre of the orifice; multiply this velocity by the area of the orifice in square feet, and the product will be the theoretical discharge in cubic feet per second. For the actual discharge, this must be multiplied by 0·7, if the orifice be of the form of fig. 97; by 0·8, if the orifice be of the form of fig. 96, and by 0·97 if it be of the form of fig. 98. The importance of the form of orifice is manifest, and hence a trumpet mouth * should be employed in all water pipes, wherein a maximum discharge is desirable, the quantity being increased which ever way the trumpet mouth is turned, whether as in fig. 98 or 99, but most in the former case. For conical converging tubes, d'Aubuisson found the coefficient of efflux to vary from 0·829 to 0·946 as the lateral convergence increased from 0° 0′ to 13° 24′, and from 0·946 to 0·847 as the convergence increased from 13° 24′ to 48° 50′; the area of the orifice being measured at the small extremity. For tubes which at first converge and then diverge, so as to take the form of the fluid vein, the coefficient of discharge is 1·55, that is, of course, taking the minimum area of the tube.

2nd. *Thin-lipped orifices*, the fluid escaping freely into the air. With orifices of this nature, the fluid vein contracts very remarkably at a short distance beyond the orifice, and the discharge is diminished in the ratio of the least area of the vein to that of the orifice. This contraction amounts to five-eighths of the area of the orifice in most cases, and hence the actual discharge is scarcely more than five-eighths that estimated by equation (3). Putting $m =$ the coefficient of contraction, we have the actual discharge from an orifice $= m\,\mathrm{Q} = m\,a\,\sqrt{2\,g\,h}$.

The velocity of the effluent vein is also diminished in a slight degree, perhaps by three or five per cent., but it will be most convenient to combine the coefficients of contraction and velocity together, and to call m the coefficient of discharge or ratio of actual to theoretical discharge.

* As for instance in the reservoir drawing, fig. 84.

A very large number of experiments have been made upon the values of the co-efficient m for various forms of orifices, the most important of which we owe to Michelotti, Castel, Bidone, Bossut, Rennie and others. But by far the most important and complete are those conducted by MM. Poncelet and Lesbros under the auspices of the French government, and all interested in hydraulic investigations must feel indebted to them for the skill, perseverance and accuracy with which they have registered so large a body of results. These determinations go to show that the value of the coefficient of discharge ranges between 0·58 and 0·7 *, being greater for small orifices and small velocities and less for large orifices and high velocities.† For heads of three and four feet and upwards, the coefficient of discharge may be taken at 0·6.

Mr. Rennie's results give the following values of m.

	Head of 4 feet.	Head of 1 foot.
Circular orifices . . .	0·621	0·645
Triangular orifices . . .	0·593	0·596
Rectangular orifices . . .	0·593	0·616

For more accurate calculations I have abridged the following tables of M. Poncelet's results from the "Aide-Memoire" of M. Morin, reducing the measures to the English standard.

TABLE II. — COEFFICIENTS OF DISCHARGE OF VERTICAL RECTANGULAR ORIFICES, THIN-LIPPED, WITH COMPLETE CONTRACTION. THE HEADS OF WATER MEASURED AT A POINT OF THE RESERVOIR WHERE THE LIQUID WAS PERFECTLY STAGNANT.

Head or summit of orifice. ins.	Coefficients of discharge for orifices of a height of					
	7·9 ins.	3·9 ins.	1·9 ins.	1·18 ins.	0·78 ins.	0·39 ins.
0·79	0·572	0·596	0·615	0·634	0·659	0·694
1·9	0·585	0·605	0·625	0·640	0·658	0·679
3·9	0·592	0·611	0·630	0·637	0·654	0·666
7·9	0·598	0·615	0·630	0·633	0·648	0·655
11·8	0·600	0·616	0·629	0·632	0·644	0·650
15·7	0·602	0·617	0·628	0·631	0·542	0·647
39·4	0·605	0·615	0·626	0·628	0·633	0·632
59·1	0·602	0·611	0·620	0·620	0·619	0·615
78·7	0·601	0·607	0·613	0·612	0·612	0·611
118·1	0·601	0·603	0·606	0·608	0·610	0·609

* Rankine. † Weisbach.

TABLE III.—COEFFICIENTS OF DISCHARGE OF VERTICAL, THIN-LIPPED, RECTANGULAR ORIFICES, WITH COMPLETE CONTRACTION. THE HEADS OF WATER MEASURED IMMEDIATELY OVER THE ORIFICE.

Heads or summit of orifice in ins.	Coefficient of discharge for orifices of a height of					
	7·9 ins.	3·9 ins.	1·9 ins.	1·18 ins.	0·78 ins.	0·39 ins.
0·78	0·594	0·614	0·638	0·668	0·697	0·729
1·97	0·593	0·614	0·636	0·651	0·672	0·686
3·94	0·595	9·614	0·634	0·640	0·657	0·669
7·87	0·599	0·615	0·630	0·633	0·649	0·656
11·81	0·601	0·616	0·629	0·632	0·644	0·651
15·74	0·600	0·616	0·630	0·632	0·646	0·653
39·37	0·605	0·615	0·626	0·628	0·633	0·632
59·05	0·602	0·611	0·620	0·620	0·619	0·615
78·74	0·601	0·607	0·614	0·612	0·612	0·611
118·11	0·601	0·603	0·606	0·608	0·610	0·609

Thence we derive this rule for estimating the discharge of water from thin-lipped orifices; seek in Table I. the velocity corresponding to the head of water measured from the centre of the effluent vein; multiply this by the area of the orifice in square feet, and the product is the theoretical discharge. Five eighths of the theoretical discharge will give the actual discharge in cubic feet a second, if a rough approximation only is required. If the estimation is to be accurate, seek in Tables II. and III. the coefficient of discharge most nearly corresponding to the given head and area of orifice, and multiply the theoretical discharge by the coefficient so found. Thus the following table has been calculated.

TABLE IV.—THEORETICAL AND ACTUAL DISCHARGE FROM A THIN-LIPPED ORIFICE OF A SECTIONAL AREA OF ONE SQUARE FOOT.

h Head.	v Table I.	m Table II.	$a \times v$ Theoretical Discharge per second.	$m\,a\,v$ Actual discharge per second.	h Head.	v Table I.	m Table II.	$a \times v$ Theoretical discharge per second.	$m\,a\,v$ Actual discharge per second.
ft.	ft.	ft.	ft.	cub. ft.	ft.	ft.		cub. ft.	cub. ft.
1	8·02	0·60	8·02	4·812	11	26·6	0·60	26·6	15·96
2	11·34	0.60	11·34	6·804	12	27·8	0·60	27·8	16·68
3	13·90	0·60	13·90	8·340	13	28·9	0·60	28·9	17·34
4	16·04	0·60	16·04	9·624	14	30·0	0·60	30·0	18·00
5	17·93	0·60	17·93	10·75	15	31·1	0·60	31·1	18·66
6	19·64	0·60	19·64	11·78	16	32·1	0·60	32·1	19·26
7	21·21	0·60	21·21	12·73	17	33·1	0·60	33·1	19·86
8	22·68	0·60	22·68	13·61	18	34·0	0·60	34·0	20·40
9	24·10	0·60	24·10	14·46	19	34·9	0·60	34·9	20·94
10	25·40	0·60	25·40	15·24	20	35·0	0·60	35·9	21·54

The above table is given as a sample of the method of calculation according to the above rule. For other areas and different heads the calculation may very easily be performed.

3rd. *Discharge with incomplete contraction.*—It is very frequently the case in practice that one of the sides of a thin-lipped orifice is prolonged, so that the vein of fluid no longer contracts upon all sides, as in fig. 99A. In this case the coefficients in Tables II. and III. give too low a result. M. Morin gives the following rule for discharge with incomplete contraction:—

Fig. 99A.

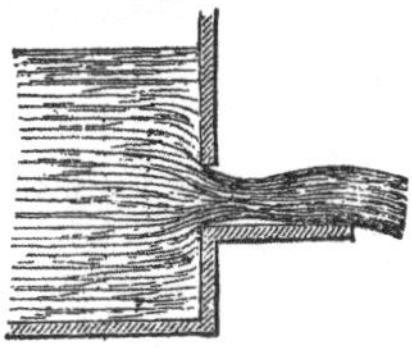

Multiply the coefficient of discharge for complete contraction found as above by

1·035	when the vein	contracts on	3 sides	only
1·072	"	"	2 sides	"
1·125	"	"	1 side	"

in order to obtain the true coefficient by which the theoretical discharge must be multiplied to give the actual discharge. Hence, for an approximate calculation, we may multiply the theoretical discharge ($= a \times v$) by 0·63, when the orifice is prolonged upon one side; by 0·66 when it is prolonged on two sides, and by 0·69 when it is prolonged on three sides. When all four sides are prolonged, the thick-lipped orifice, fig. 94, is formed of which the coefficient of efflux is 0·8.

4th. *Discharge from rectangular notches, waste-boards, and weirs.*—In this case the theoretical discharge =

$$\left.\begin{aligned} Q &= \tfrac{2}{3}\, b \,.\, h \,.\, \sqrt{2\,g\,h} \\ &= \tfrac{2}{3}\, a\, v \end{aligned}\right\} \ldots\ldots(5)$$

where Q = discharge in cubic feet per second.

b = breadth of notch or weir.

h = head of water, measured at some distance behind the crest of the weir.

v = velocity in feet per second = $\sqrt{2\,g\,h}$...(2)

a = area of effluent vein = $b \times h$.

The actual discharge is found by multiplying the theoretical discharge by the coefficient of efflux *m*, which varies under different circumstances. The millwright must select this coefficient from the following tables, so as to suit the particular case to which it is to be applied.

TABLE V.—COEFFICIENT OF DISCHARGE FOR WEIRS, FROM EXPERIMENTS ON NOTCHES 8 INCHES BROAD, BY PONCELET AND LESBROS.

Head of Water in inches.	0·89	0·78	1·18	1·57	2·36	3·15	3·93	5·90	7·86	8·65
Coefficient of discharge $= \frac{2}{3} m$.	0·424	0·417	0·412	0·407	0·401	0·397	0·395	0·393	0·390	0·385

This gives a mean value of 0·4 or $\frac{2}{5}$ths for $\frac{2}{3}\ m$, and hence we may approximately find the discharge from a waste board by multiplying the head in feet by the breadth of the notch in feet, and by the velocity due to the head found by equation (2) or Table I. Two-fifths of this product will be the discharge in cubic feet per second.

In 1852 the council of the Institution of Civil Engineers awarded to Mr. Blackwell a premium for a valuable series of experiments on the discharge of water from weirs made on a very large scale, and with various conditions of head and with different kinds of overfall bars. What constitutes the principal value of Mr. Blackwell's paper is the scale on which the experiments were made, and their close approximation to actual practice. It must be borne in mind, however, that in calculating the quantity of water discharged in the flow of rivers over weirs, reference must be had to the form of the top, in order to ascertain the state of the overfall as compared with those in the following table, from which the coefficient is taken for calculation.

TABLE VI.—COEFFICIENTS OF DISCHARGE FROM WEIRS, FROM EXPERIMENTS BY MR. BLACKWELL.

No. of Trials.	Description of Overfalls.	Head in ins.	$\frac{2}{3}$ m.
6	Thin plate 8 feet long	1 to 3 3 to 6	·440 ·402
11	Thin plate 10 feet long	1 to 3 4 to 6 6 to 9	·501 ·435 ·370
23	Plank 2 inches thick, with notch 3 feet broad .	1 to 3 3 to 6 6 to 10	·342 ·384 ·406
56	Plank 2 inches thick, with notch 6 feet broad .	1 to 3 3 to 6 6 to 9 9 to 14	·359 ·396 ·392 ·358
40	Plank 2 inches thick, with notch 10 feet broad .	1 to 3 3 to 6 6 to 9 9 to 12	·346 ·397 ·374 ·336
4	Plank 2 ins. thick, notch 10 ft. broad, with wings	1 to 2 4 to 5	·476 ·442
7	Overfall with a crest, 3 feet wide, sloping 1 in 12, 3 feet long like a weir . . .	1 to 3 3 to 6 6 to 9	·342 ·328 ·311
9	Overfall with a crest, 3 feet wide, sloping 1 in 18, 3 feet long, like a weir . .	1 to 3 3 to 6 6 to 9	·362 ·345 ·332
6	Overfall with a crest, 3 feet wide, sloping 1 in 18 and 10 feet long . . .	1 to 4 4 to 8	·328 ·350
14	Overfall with a level crest 3 feet wide by 6 long	1 to 3 3 to 6 6 to 9	·305 ·311 ·318
15	Overfall with level crest, 6 feet long by 3 broad	3 to 7 7 to 12	·330 ·310
12	Overfall with level crest, 3 feet wide by 10 long	1 to 5 5 to 8 8 to 10	·306 ·327 ·313
61	At Chew Magna, overfall bar 10 feet long, 2 inches thick	1 to 8 3 to 6 6 to 9	·437 ·499 ·505

The most important of the generalisations from this table are—

1st. That the discharge is decreased in proportion to the breadth and inclination of the crest, being least when the crest is level.

2nd. That converging wing walls, above the overfall, increase the discharge.

Where, as in the case of a river, the water approaches the weir with a certain velocity, it should be taken separately into account, the above coefficients being deduced from experiments on reservoirs so large that the water was approximately stagnant.

Let k = height of head due to velocity v of water as it approaches the weir; that is, let $k = \frac{v^2}{64\cdot38}$;

the effective discharge is then =

$$Q = \tfrac{2}{3} m \, . \, b \, . \, \sqrt{2g} \, [h+k)^{\frac{3}{2}} - k^{\frac{3}{2}}]^* \quad \ldots (6).$$

TABLE VII.—EXAMPLES OF ESTIMATION OF DISCHARGE FROM WEIRS.

Head of water.	v. Velocity due to head. Table I.	$\frac{2}{3}m$ Coefficient of discharge. Tables V. VI.	b. Breadth.	$\frac{2}{3}b.h.v.$ Theoretical discharge per second.	$\frac{2}{3}m.b.h.v.$ Actual discharge per second.	Remarks.
inches.				*cubic feet.*	*cubic feet.*	
1	2·32	0·412	1 ft.	1·283	·793	Thin lipped
6	5·67	0·393	1 ft.	1·890	1·114	waste
12	8·02	0·380	1 ft.	5·35	3·047	board.
6	5·67	0·350	10 ft.	18·90	9·922	Weir with crest.

According to the above formula I have computed the following table, showing at a glance the velocity in feet, the theoretical discharge in cubic feet per second, and the actual discharge of water over a thin-edged notch or weir for various heads from ½ an inch to 6 feet—the water approaching the weir with no perceptible current.

* Weisbach.

TABLE VIII.—DISCHARGE OF WATER OVER A THIN-EDGED NOTCH OR WEIR FOR EVERY FOOT IN BREADTH OF THE STREAM IN CUBIC FEET PER SECOND.

h. Head in feet.	*v.* Velocity per second.	$\frac{2}{3}m.$ Coefficient of discharge. Table V.	$\frac{2}{3}b.h.v.$ Theoretical discharge per second.	$\frac{2}{3}m.b.h.v.$ Actual discharge per second.	*h.* Head in feet.	*v.* Velocity per second.	$\frac{2}{3}m.$ Coefficient of discharge.	$\frac{2}{3}b.h.v.$ Theoretical discharge per second.	$\frac{2}{3}m.b.h.v.$ Actual discharge per second.
	ft.		*cub. ft.*	*cub. ft.*		*ft.*		*cub. ft.*	*cub. ft.*
·05	1·79	0·42	·0596	·0376	2·1	11·62	0·35	16·94	8·890
·1	2·53	0·41	·168	·1037	2·2	11·90	0·35	17·44	9·163
·2	3·58	0·40	·478	·286	2·3	12·17	0·35	18·66	9·796
·3	4·39	0·40	·878	·527	2·4	12·43	0·34	19·88	10·142
·4	5·07	0·40	1·352	·811	2·5	12·68	0·34	21·14	10·778
·5	5·67	0·39	1·890	1·105	2·6	12·94	0·34	22·42	11·439
·6	6·20	0·89	2·414	1·411	2·8	13·42	0·34	25·04	12·774
·7	6·71	0·39	3·130	1·832	3·0	13·89	0·34	27·78	13·751
·8	7·17	0·38	3·824	2·179	3·2	14·35	0·33	30·60	15·153
·9	7·31	0·38	4·386	2·500	3·4	14·79	0·33	33·52	16·593
1·0	8·02	0·38	5·34	3·047	3·6	15·22	0·33	36·53	18·081
1·1	8·42	0·38	6·18	3·519	3·8	15·64	0·33	39·62	19·612
1·2	8·82	0·37	7·04	3·914	4·0	16·04	0·33	42·78	21·173
1·3	9·15	0·37	7·98	4·399	4·2	16·24	0·33	46·03	22·786
1·4	9·47	0·37	8·84	4·906	4·4	16·82	0·33	49·33	24·422
1·5	9·79	0·37	9·78	5·430	4·6	17·20	0·33	52·74	26·109
1·6	10·11	0·36	10·78	5·821	4·8	17·57	0·33	56·22	27·829
1·7	10·67	0·36	12·09	6·529	5·0	17·93	0·33	59·76	29·564
1·8	10·75	0·36	12·90	6·966	5·2	18·29	0·33	63·40	31·385
1·9	10·99	0·36	13·92	7·516	5·4	18·64	0·33	67·11	33·216
2·0	11·34	0·36	15·12	8·164	6·0	19·64	0·33	78·56	38·887

Mr. Sang of Kirkcaldy has proposed a very ingenious arrangement for the approximate measurement of the flow of water over a rectangular notch in a waste board, particularly applicable in cases where the flow has to be frequently registered, as in the daily observations by which drainage is estimated. He employs a scale graduated variably, so as to give at once the number of cubic feet per, minute of water to every inch in breadth of the rectangular notch. Hence instead of employing a complex formula, nothing more is required than to observe at what number the water stands on the rule, and to multiply that by the number of inches of breadth of the notch, and we obtain at once the discharge in cubic feet per minute. He proposes that such a scale engraved on paper should be placed in a glass tube, hermetically sealed, and permanently fixed in a suitable position on the weir. The following table calculated by Mr. Sang shows the position of the divisions of the scale corresponding to cubic feet, and tenths of cubic feet discharge:—

Division.	Distance from Zero of scale in inches.	Division.	Distance from Zero of scale in inches.	Division.	Distance from Zero of scale in inches.	Division.	Distance from Zero of scale in inches.
·0	·000	2·6	3·327	5·1	5·214	7·6	6·803
·1	·379	2·7	3·412	5·2	5·282	7·7	6·862
·2	·602	2·8	3·496	5·3	5·350	7·8	6·921
·3	·789	2·9	3·579	5·4	5·417	7·9	6·980
·4	·955	3·0	3·661	5·5	5·483	8·0	7·039
·5	1·109	3·1	3·741	5·6	5·549	8·1	7·098
·6	1·252	3·2	3·822	5·7	5·615	8·2	7·156
·7	1·387	3·3	3·901	5·8	5·681	8·3	7·214
·8	1·516	3·4	3·979	5·9	5·746	8·4	7·272
·9	1·641	3·5	4·057	6·0	5·811	8·5	7·329
1·0	1·760	3·6	4·134	6·1	5·875	8·6	7·387
1·1	1·875	3·7	4·210	6·2	5·939	8·7	7·444
1·2	1·987	3·8	4·285	6·3	6·003	8·8	7·501
1·3	2·096	3·9	4·360	6·4	6·066	8·9	7·558
1·4	2·202	4·0	4·434	6·5	6·129	9·0	7·614
1·5	2·306	4·1	4·508	6·6	6·192	9·1	7·671
1·6	2·407	4·2	4·581	6·7	6·254	9·2	7·727
1·7	2·507	4·3	4·654	6·8	6·316	9·3	7·783
1·8	2·604	4·4	4·726	6·9	6·378	9·4	7·838
1·9	2·700	4·5	4·797	7·0	6·440	9·5	7·894
2·0	2·794	4·6	4·868	7·1	6·501	9·6	7·949
2·1	2·886	4·7	4·938	7·2	6·562	9·7	8·004
2·2	2·977	4·8	5·007	7·3	6·622	9·8	8·059
2·3	3·066	4·9	5·077	7·4	6·683	9·9	8·114
2·4	3·155	5·0	5·146	7·5	6·743	10·0	8·168
2·5	3·242						

5. *Friction of fluids in conduits and pipes.*—In long tubes an increased retardation arises, which must be ascribed to the friction of the fluid against the sides, and it has been ascertained that this element of retardation, whilst independent of the pressure of the fluid, increases in the ratio of the length of the tube, and decreases in the ratio of the width or diameter. It also increases nearly as the square of the velocity.

For pipes of uniform size and with no considerable amount of bending, it may be shown that the velocity of discharge

$$= v = \sqrt{\frac{2380\,h\,d}{l + 54\,d} - \frac{1}{12} \cdot \frac{l}{l + 54\,d}} \quad \ldots (7)$$

or if h be not very small, neglecting the last term,

$$\sqrt{\frac{2380\,h\,d}{l + 54\,d}} \quad \ldots (8)$$

and if the pipes be very long,

$$v = \sqrt{\frac{2380\,h\,d}{l + 54\,d}} - \frac{1}{12} \ldots (9)$$

where l = length of pipe in feet; d = diameter in feet; h = head in feet; and the constants have been derived from the experiments of Prony and d'Aubuisson.

Formula (8) very nearly coincides with that given by Poncelet, namely,

$$v = 47{\cdot}9 \sqrt{\frac{d\,h}{l + 54d}} = \sqrt{\frac{2300\,d\,h}{l + 54\,d}}$$

The most convenient way in practice of estimating the retardation of friction in the pipes is to measure the head of water which is requisite to overcome the friction, without increasing the velocity of the current. In calculations of quantity when the head h, necessary to generate the required velocity of exit, has been estimated by the rule for thick-lipped orifices already given, another head h_1 must be added as necessary to overcome the friction, if the orifice is prolonged into a tube.

The height to overcome friction may be calculated from the formula

$$h_1 = n\frac{l}{d} \cdot \frac{v^2}{2\,g} \ldots (10)$$

where n is the coefficient of friction derived from experiment. Hence, putting Q for the discharge,

$$v = \frac{4\,Q}{\pi\,d^2} \ldots (11)$$

when the pipe is cylindrical.

We may combine this formula with the preceding formula for the discharge from a thick-lipped orifice, putting m = coefficient of resistance for the portion of tube next the cistern, and n, coefficient of resistance for the remainder of the tube, we then have for the whole head of water

$$h = m\frac{v^2}{2\,g} + n.\frac{l}{d} \cdot \frac{v^2}{2\,g} + \frac{v^2}{2\,g} \ldots (12)$$

or

$$h = \left(1 + m + n\frac{l}{d}\right)\frac{v^2}{2\,g} \ldots (13)$$

and

$$v = \frac{\sqrt{2\,g\,h}}{\sqrt{1 + m + n\dfrac{l}{d}}} \ldots (14)$$

and putting a = area of orifice, the discharge =

$$Q = a\,v \ldots (15).$$

If there be bends in the tube an increased element of resistance is introduced; if we put p for the sum of the resistances due to this source,

$$h = \left(1 + m + n\frac{l}{d} + p\right)\frac{v^2}{2\,g} \ldots (16)$$

$$= \left(1 + m + n\frac{l}{d} + p\right)\left(\frac{4\,Q}{\pi}\right)^2\frac{1}{2\,g\,d^4} \ldots (17).$$

These formulæ we have from Weisbach's "Mechanics of Engineering," from which the following table has been reduced and adapted to English measures.

TABLE OF THE VALUE OF THE COEFFICIENT OF FRICTION n FOR DIFFERENT VELOCITIES.

Inches.	Velocity in Feet.											
	0	1	2	3	4	5	6	7	8	9	10	11
0	∞	·0316	·0265	·0244	·0229	·0220	·0215	·0209	·0205	·0202	·0198	·0195
4	·0443	·0293	·0257	·0239	·0226	·0218	·0213	·0208	·0204	·0201	·0197	·0195
8	·0355	·0277	·0250	·0233	0223	·0217	·0211	·0206	·0203	·0200	·0196	·0194
12	·0316	·0265	·0244	·0229	·0220	·0215	·0209	·0205	·0202	·0199	·0195	·0193

To use this table look in the horizontal line at top for the nearest velocity in feet, in the vertical column underneath and opposite the nearest number of inches will be found the value of n required. Thus for a velocity of 6 feet 8 inches per second, the coefficient n will be found to be 0·211, being under 6 and opposite 8.

From sixty-three experiments Weisbach deduces another general formula for the flow of water in tubes which is very convenient for calculation. It is based on the hypothesis that the resistance to friction increases simultaneously as the square, and as the square root of the cube of the velocity, and is of the form

$$h_1 = \left(0{\cdot}01482 + \frac{0{\cdot}017963}{\sqrt{v}}\right)\frac{l}{d}\cdot\frac{v^2}{2\,g}, \ldots (18)$$

which gives the head due to friction in feet. From this formula, Mr. James Thomson, M.A., C.E., and Mr. George Fuller,

C.E., have calculated the following very complete and convenient table.

In laying pipes the following directions are not unimportant; the mouth, both for ingress and egress, should be trumpet-shaped; bends should be as far as possible avoided, and especially sharp angular bends; at junctions the smaller pipe should be brought round in a curve to agree in direction with the main. And, lastly, where a pipe rises and falls much, air is apt to collect in the upper parts of the bends, and thus reduce the section at that part, and it is advisable to make provision by a cock or otherwise for drawing it off at intervals.

Flow of water in open channels.—This is a question of importance, and requires careful consideration on the part of the engineer, as it is a case of frequent occurrence in calculations of the flow of water. It is often, from various circumstances, impossible to throw even a temporary waste board or weir across a stream the quantity of discharge from which it is desirable to ascertain, and hence it becomes necessary to determine a formula which takes into account the friction of the river sides.

In estimating the velocity of a stream on a canal or river, by throwing in floating bodies of nearly the same specific gravity as the water, and estimating the time they require to pass a given measured distance, it must be borne in mind that the velocity is greatest in the centre of the stream and near the surface, and less at the bottom and near the sides. It is generally most convenient to ascertain the velocity at the centre, where the stream is fleetest, but it is essential in calculations to know the mean velocity, or the velocity of a stream of the same section, discharging the same quantity of water, but unaffected by friction at the sides. In practice it will be sufficient to assume that the mean velocity of a stream is equal to 0·83 per cent. or $\frac{4}{5}$ of the velocity at the surface.

Or we may use an empirical formula of Prony's, putting v for the mean velocity, and v for the surface velocity, measured by a floating body at the middle of the stream

$$\mathrm{v} = \frac{v\,(v + 7{\cdot}77)}{v + 10{\cdot}33} \;\ldots\; (19)$$

For small streams the most accurate method of measurement is the formation of a temporary weir by a vertical board thrown

FRICTION OF WATER IN PIPES.

Table calculated from a Formula of Julius Weisbach, to show, for Pipes 100 feet in Length, the relation between 1st. The Velocity of the Water, in feet per second.—2nd. The Internal Diameter of the Pipe.—3rd. The Head to overcome the Friction, in feet.—4th. The number of Cubic Feet of Water delivered per Minute; so that when any two of these four quantities are given, the remaining two can be found.*

By James Thomson, A.M., C.E., and George Fuller, C.E., Belfast.

Velocity of Water, in feet per second.	Head to produce velocity, in feet.	Internal Diameter of the Pipes in Inches. 3		3½		4		4½		5		6	
		Head to overcome friction, in feet.	Cubic feet per minute.	Head to overcome friction, in feet.	Cubic feet per minute.	Head to overcome friction, in feet.	Cubic feet per minute.	Head to overcome friction, in feet.	Cubic feet per minute.	Head to overcome friction, in feet.	Cubic feet per minute.	Head to overcome friction, in feet.	Cubic feet per minute.
2·0	·062	·659	5·89	·565	8·02	·494	10·4	·439	13·2	·395	16·3	·329	23·5
2·2	·075	·780	6·48	·669	8·82	·585	11·5	·520	14·6	·468	18·0	·390	25·9
2·4	·090	·911	7·07	·781	9·62	·683	12·5	·607	15·9	·547	19·6	·456	28·2
2·6	·105	1·05	7·65	·901	10·4	·788	13·6	·701	17·2	·631	21·3	·526	30·6
2·8	·122	1·20	8·24	1·03	11·2	·900	14·6	·800	18·5	·720	22·9	·600	32·9
3·0	·140	1·35	8·83	1·16	12·0	1·02	15·7	·905	19·8	·815	24·5	·679	35·3
3·2	·160	1·52	9·42	1·31	12·8	1·14	16·7	1·02	21·2	·915	26·2	·763	37·7
3·4	·180	1·70	10·0	1·46	13·6	1·27	17·8	1·13	22·5	1·02	27·8	·851	40·0
3·6	·202	1·89	10·6	1·62	14·4	1·41	18·8	1·26	23·8	1·13	29·4	·943	42·4
3·8	·225	2·08	11·2	1·78	15·2	1·56	19·9	1·39	25·2	1·25	31·0	1·04	44·7
4·0	·250	2·28	11·8	1·96	16·0	1·71	20·9	1·52	26·5	1·37	32·7	1·14	47·1
4·2	·275	2·49	12·3	2·14	16·8	1·87	22·0	1·66	27·8	1·50	34·3	1·25	49·5
4·4	·302	2·71	12·9	2·33	17·6	2·03	23·0	1·81	29·1	1·63	36·0	1·35	51·8
4·6	·330	2·94	13·5	2·52	18·4	2·21	24·0	1·96	30·4	1·76	37·6	1·47	54·1
4·8	·360	3·18	14·1	2·72	19·2	2·38	25·1	2·12	31·8	1·91	39·2	1·59	56·5
5·0	·390	3·43	14·7	2·94	20·0	2·57	26·2	2·28	33·1	2·05	40·9	1·71	58·9
5·2	·422	3·68	15·3	3·15	20·8	2·76	27·2	2·45	34·4	2·21	42·5	1·84	61·2
5·4	·455	3·94	15·9	3·38	21·6	2·96	28·2	2·63	35·8	2·37	44·2	1·97	63·6
5·6	·490	4·22	16·5	3·61	22·4	3·16	29·3	2·81	37·1	2·53	45·8	2·11	65·9
5·8	·525	4·50	17·1	3·85	23·2	3·37	30·3	3·00	38·4	2·70	47·4	2·25	68·3
6·0	·562	4·78	17·7	4·10	24·0	3·59	31·4	3·19	39·7	2·87	49·1	2·39	70·7
6·2	·600	5·08	18·2	4·36	24·8	3·81	32·4	3·39	41·0	3·05	50·7	2·54	73·0
6·4	·640	5·39	18·8	4·62	25·6	4·04	33·5	3·59	42·4	3·23	52·3	2·69	75·4
6·6	·680	5·70	19·4	4·89	26·4	4·28	34·5	3·80	43·7	3·42	54·0	2·85	77·7
6·8	·722	6·02	20·0	5·16	27·3	4·52	35·6	4·01	45·0	3·61	55·6	3·01	80·1
7·0	·765	6·35	20·6	5·45	28·0	4·77	36·6	4·24	46·4	3·81	57·2	3·18	82·4

Velocity of Water, in feet per second.	Internal Diameter of the Pipes in Inches. 7		8		9		10		11		12		13	
	Head to overcome friction, in feet.	Cubic feet per minute.	Head to overcome friction, in feet.	Cubic feet per minute.	Head to overcome friction, in feet.	Cubic feet per minute.	Head to overcome friction, in feet.	Cubic feet per minute.	Head to overcome friction, in feet.	Cubic feet per minute.	Head to overcome friction, in feet.	Cubic feet per minute.	Head to overcome friction, in feet.	Cubic feet per minute.
2·0	·282	32·0	·247	41·9	·220	53·0	·198	65·4	·180	79·2	·165	94·2	·152	110
2·2	·334	35·3	·293	46·1	·260	58·3	·234	72·0	·213	87·1	·195	103	·180	121
2·4	·390	38·5	·342	50·2	·304	63·6	·273	78·5	·248	95·0	·228	113	·210	133
2·6	·450	41·7	·394	54·4	·350	68·9	·315	85·1	·287	103	·263	122	·242	144
2·8	·514	44·9	·450	58·6	·400	74·2	·360	91·6	·327	111	·300	132	·277	156
3·0	·582	48·1	·509	62·3	·453	79·5	·407	98·2	·370	119	·339	141	·313	166
3·2	·654	51·3	·572	67·0	·508	84·8	·458	105	·416	127	·381	151	·352	177
3·4	·729	54·5	·638	71·2	·567	90·1	·510	111	·464	134	·425	160	·393	188
3·6	·808	57·7	·707	75·4	·629	95·4	·566	118	·514	142	·472	169	·435	199
3·8	·892	60·9	·780	79·6	·693	101	·624	124	·567	150	·520	179	·480	210
4·0	·979	64·1	·856	83·7	·761	106	·685	131	·623	158	·571	188	·527	221
4·2	1·07	67·3	·935	87·0	·832	111	·748	137	·680	166	·624	198	·576	232
4·4	1·16	70·5	1·02	92·1	·905	116	·814	144	·740	174	·679	207	·626	243
4·6	1·26	73·7	1·10	96·3	·981	122	·883	150	·803	182	·736	217	·679	254
4·8	1·36	76·9	1·19	100	1·06	127	·954	157	·867	190	·795	226	·734	265
5·0	1·47	80·2	1·28	105	1·14	132	1·03	163	·935	198	·857	235	·791	276
5·2	1·58	83·3	1·38	109	1·23	138	1·10	170	1·00	206	·920	245	·850	287
5·4	1·69	86·6	1·48	113	1·31	143	1·18	177	1·07	214	·986	254	·910	298
5·6	1·81	89·8	1·58	117	1·40	148	1·26	183	1·15	222	1·05	264	·973	309
5·8	1·93	93·0	1·68	121	1·50	154	1·35	190	1·22	229	1·12	273	1·04	321
6·0	2·05	96·2	1·79	125	1·59	159	1·43	196	1·30	237	1·19	283	1·10	332
6·2	2·18	99·4	1·90	130	1·69	164	1·52	203	1·38	245	1·27	292	1·17	343
6·4	2·31	102	2·02	134	1·79	169	1·61	209	1·47	253	1·35	301	1·24	354
6·6	2·44	106	2·14	138	1·90	175	1·71	216	1·55	261	1·42	311	1·31	365
6·8	2·58	109	2·26	142	2·01	180	1·81	222	1·64	269	1·50	320	1·39	376
7·0	2·72	112	2·38	146	2·12	185	1·90	229	1·73	277	1·59	330	1·46	387

Internal Diameter of the Pipes in Inches

Velocity of Water, in feet per second.	Head to produce this velocity, in feet.	14: Head to over-come friction, in feet.	14: Cubic feet per minute.	15: Head to over-come friction, in feet.	15: Cubic feet per minute.	16: Head to over-come friction, in feet.	16: Cubic feet per minute.	17: Head to over-come friction, in feet.	17: Cubic feet per minute.	18: Head to over-come friction, in feet.	18: Cubic feet per minute.	19: Head to over-come friction, in feet.	19: Cubic feet per minute.	20: Head to over-come friction, in feet.	20: Cubic feet per minute.
2·0	·062	·141	128	·132	147	·123	167	·116	189	·110	212	·104	236	·099	262
2·2	·075	·167	141	·156	162	·146	184	·138	208	·130	233	123	260	·117	288
2·4	·090	·195	154	·182	176	·171	201	·161	227	·152	254	·144	283	·137	314
2·6	·105	·225	167	·210	191	·197	218	·185	246	·175	275	·166	307	·158	340
2·8	·122	·257	179	·240	206	·225	234	·212	265	·200	297	·189	331	·180	366
3·0	·140	·291	192	·271	221	·255	251	·240	284	·226	318	·214	354	·204	393
3·2	·160	·327	205	·305	235	·286	268	·269	302	·254	339	·241	378	·229	419
3·4	·180	·365	218	·340	250	·319	284	·300	321	·283	360	·269	401	·255	445
3·6	·202	·404	231	·377	265	·354	301	·333	340	·314	382	·298	425	·283	471
3·8	·225	·446	243	·416	280	·390	318	·367	359	·347	403	·328	449	·312	497
4·0	·250	·489	256	·457	294	·428	335	·403	378	·380	424	·360	472	·342	523
4·2	·275	·534	269	·499	309	·468	352	·440	397	·416	445	·394	496	·374	550
4·4	·302	·582	282	·543	324	·509	368	·479	416	·452	466	·429	519	·407	576
4·6	·330	·631	295	·589	339	·552	385	·519	435	·490	488	·465	543	·441	602
4·8	·360	·682	308	·636	353	·596	402	·561	454	·530	509	·502	567	·477	628
5·0	·390	·734	321	·685	368	·642	419	·605	473	·571	530	·541	590	·514	654
5·2	·422	·789	333	·736	383	·690	435	·650	492	·614	551	·581	614	·552	680
5·4	·455	·845	346	·789	397	·740	452	·696	511	·657	572	·623	638	·592	707
5·6	·490	·903	359	·843	412	·791	469	·744	529	·703	594	·666	661	·632	733
5·8	·525	·964	372	·899	427	·843	486	·793	548	·749	615	·710	685	·674	759
6·0	·562	1·02	385	·957	442	·897	502	·844	567	·798	636	·755	709	·718	785
6·2	·600	1·09	397	1·01	456	·953	519	·897	586	·847	657	·802	732	·762	811
6·4	·640	1·15	410	1·08	471	1·01	536	·951	605	·898	678	·851	756	·808	838
6·6	·680	1·22	423	1·14	486	1·07	553	1·01	624	·950	700	·990	780	·855	864
6·8	·722	1·29	436	1·20	500	1·13	569	1·06	643	1·00	721	·951	803	·904	890
7·0	·765	1·36	449	1·27	515	1·19	586	1·12	662	1·06	742	1·00	827	·953	916

Velocity of Water, in feet per second.	22: Head to over-come friction, in feet.	22: Cubic feet per minute.	24: Head to over-come friction, in feet.	24: Cubic feet per minute.	26: Head to over-come friction, in feet.	26: Cubic feet per minute.	28: Head to over-come friction, in feet.	28: Cubic feet per minute.	30: Head to over-come friction, in feet.	30: Cubic feet per minute.
2·0	·090	316	·082	377	·076	442	·070	513	·066	589
2·2	·106	348	·097	414	·090	486	·083	564	·078	648
2·4	·124	380	·114	452	·105	531	·097	616	·091	707
2·6	·143	412	·131	490	·121	575	·112	667	·105	766
2·8	·164	443	·150	528	·138	619	·128	718	·120	824
3·0	·185	475	·170	565	·157	663	·145	770	·136	883
3·2	·208	507	·191	603	·176	708	·163	821	·152	942
3·4	·232	538	·213	641	·196	752	·182	872	·170	1001
3·6	·257	570	·236	678	·218	796	·202	923	·189	1060
3·8	·284	601	·260	716	·240	840	·223	974	·208	1119
4·0	·311	633	·285	754	·263	885	·244	1026	·228	1178
4·2	·340	665	·312	791	·288	929	·267	1077	·249	1237
4·4	·370	697	·339	829	·313	973	·290	1129	·271	1296
4·6	·401	728	·368	867	·339	1017	·315	1180	·294	1355
4·8	·434	760	·397	905	·367	1062	·341	1231	·318	1414
5·0	·467	792	·428	942	·395	1106	·367	1283	·343	1472
5·2	·502	823	·460	980	·425	1150	·394	1334	·368	1531
5·4	·538	855	·493	1018	·455	1194	·423	1385	·394	1590
5·6	·575	887	·527	1055	·486	1239	·452	1437	·422	1649
5·8	·613	918	·562	1093	·519	1283	·482	1488	·450	1708
6·0	·652	950	·598	1131	·552	1327	·513	1539	·478	1767
6·2	·693	982	·635	1168	·586	1371	·544	1590	·508	1826
6·4	·735	1013	·673	1206	·622	1416	·577	1642	·539	1885
6·6	·778	1045	·713	1244	·658	1460	·611	1693	·570	1943
6·8	·821	1077	·753	1282	·695	1504	·645	1744	·602	2003
7·0	·867	1109	·794	1319	·733	1548	·681	1796	·635	2061

* This Table also gives, by the first and second columns, the relation between the velocity and the head required to produce the velocity, calculated by the laws of falling bodies, independently of friction, contraction of the vein, or other retarding causes.

The formula of WEISBACH used for this Table is given in his "Ingenieur-und-Maschinen-Mechanik," vol. 1. p. 434, and when reduced to English measure is as follows:

$$H = \left\{ 0{\cdot}0144 + \frac{0{\cdot}01716}{\sqrt{V}} \right\} \frac{L}{D} \times \frac{V^2}{64{\cdot}4};$$

where H = head to overcome the friction, in English feet; L = length of pipe, in English feet; D = internal diameter of pipe, in English feet; and V = velocity of water, in English feet per second.

across the stream and carefully puddled at the edges. A rectangular notch of sufficient capacity to pass the water must be cut in the middle portion. The height of the water above the level of the notch should then be measured either at its crest, or better still at some distance behind, where the water is nearly still, and the constant for calculation will be found in Table V. or VI. as the case may be.

But if a waste board or weir cannot be employed, we may find the surface velocity, and from that obtain its mean velocity by the methods given above. If then we take the depth of the stream at various parts of its breadth and so calculate its sectional area, we may find the cubic feet of water discharged per second by multiplying the mean velocity (in feet per second) by the area so found (in square feet).

Thus, if a body floats along the surface of a stream 300 feet in a minute, its maximum velocity $= \frac{300}{60} = 5$ feet per second, and its mean velocity, according to Prony, $= \frac{5\ (5 + 7{\cdot}77)}{5 + 10{\cdot}33}$ $= 4{\cdot}16$. Now let the depth of the stream, 16 feet broad, measured at equal distances of two feet apart, be 0, 1, $2\frac{1}{2}$, 3, 3, $3\frac{3}{4}$, 3, $1\frac{3}{4}$, 0 feet respectively, then the area =

$$2 \times (1 + 2\tfrac{1}{2} + 3 + 3 + 3\tfrac{3}{4} + 3 + 1\tfrac{3}{4}) = 36 \text{ sq. ft.}$$

∴ Cubic feet of water discharged per second

$$= 36 \times 4{\cdot}16 = 149{\cdot}76 \text{ cubic feet.}$$

In rivers, the coefficient of friction n in formulæ (12) (13) and (14) may be taken at 0·0075; it varies according to Weisbach from 0·00811 to 0·00748 as the velocity increases from 0·1 to 1·0 metres, and from 0·00748 to 0·00743 as the velocity increases from 1 to 3 metres.

The following formulæ express the relations of velocity, fall, and discharge, when the flow of the stream is uniform:

$$h = \zeta \cdot \frac{l\,p}{F} \cdot \frac{c^2}{2\,g}, \ldots (20)$$

$$c = \sqrt{\frac{F}{\zeta\,l\,p} \cdot 2\,g\,h}, \ldots (21)$$

$$Q = F\,c \ldots (3).^*$$

* Weisbach, vol. i. p. 493.

Where h = whole fall, Q = discharge, F = transverse section of stream, c = mean velocity of stream, l = distance which the river flows for a fall h, $\frac{p}{F}$ the perimeter of the water profile, and ζ the coefficient of friction.

The best form of section must be that which presents the least resistance to a given quantity of water flowing through the channel. Now it has been shown that the resistance of friction varies directly as the wetted perimeter and inversely as the area of the section, and when the area is constant it will therefore vary directly as the wetted perimeter. Consequently the best form of section will be that with the least perimeter for a given area. Hence for open channels in which the upper water line is not part of the wetted perimeter, the half square is the best rectangular section, and the semihexagon the best trapezoidal section. For equal flows of water the semicircle will have less friction than the semihexagon, and the semihexagon than the semisquare. In designing conduits, for instance, the head race or tail race of water wheels, not only must the sectional form be attended to, but bends must be avoided as much as possible.

Estimation of Water Power.—Where a natural reservoir of mechanical power is employed through the medium of a prime-mover in overcoming resistances, in sawing, grinding, &c., we term the moving force the power, and the resistances overcome the work.

The dynamic unit by which we estimate force or resistance is the *foot-pound*, or the unit of force which is capable of lifting a weight of one pound one foot high. A second unit is employed when estimating large expenditure of force, namely, the *horse-power.* One horse-power, according to the estimate of Watt, was equivalent to 33,000 lbs. raised one foot high in a minute or 550 foot-pounds per second.

It is evident that a power exerted by a weight of water falling a given number of feet is capable of raising an equal weight the same number of feet. The power expended must equal the resistances overcome. In transmitting power through a prime-mover, however, a certain loss necessarily takes place, arising (1) from the loss or waste of the power by spilling, leakage, &c., and (2) from the absorption of a part of the power in overcoming the resistances of the prime-mover itself, friction, &c. Hence the work accomplished by a prime-mover is never equivalent to the power expended on it; the useful effect is always only a certain percentage of the power, and this percentage is called the efficiency or modulus of the machine.

Now for a water-wheel on which a stream of water acts by gravity alone:—

Let h = height of fall in feet.

W = weight of water delivered on the wheel per second.

n = the number of cubic feet per second.

P = the dynamic force of the falling water in foot-pounds.

P_1 = P reduced to horses power.

U = the useful effect of the machine in foot-pounds and $\text{U}_1 = \text{U}$ to reduced H. P.

u = the modulus of the machine.

Then for the total water power of the fall we have, in foot-pounds,

$$\text{P} = \text{W}\,h \ldots (1)$$

and water weighing 62·5 lbs. per cubic foot,

$$\text{P} = 62{\cdot}5\ n\ h$$

or, in horses power,

$$\text{P}_1 = \frac{\text{W}\,h}{550} = \frac{62{\cdot}5\,n\,h}{550} = \frac{n\,h}{8{\cdot}8}$$

Hence, for every foot of fall 8·88 cubic feet of water per second, or 1·47 tons per minute, theoretically afford an available force of one horse power.

But by definition,

$$\frac{u\,\text{P}}{100} = \text{U} \ldots (22).$$

$$\therefore \text{U} = \frac{u\,\text{W}\,h}{100} \text{ in foot-pounds.}$$

$$\therefore \text{U}_1 = \frac{u\ \text{W}\ h}{55000} \text{ in horses power.}$$

and,

$$\frac{100 - u}{100}\,\text{P} = \text{the sum of the resistances from friction, \&c.,}$$

and the loss from wasted water, in accumulating and transmitting the power.

CHAP. III.

ON THE CONSTRUCTION OF WATER WHEELS.

In the present age, the same importance is not attached to water power as before the introduction of steam, as has been already shown. Nevertheless, since water is still largely employed in some districts and for certain kinds of work, it is of importance that the machinery for rendering it useful should be constructed upon the best principles so as to secure a maximum effect. In numerous localities in Europe and America, water is the principal motive agent by which manufacturing processes are carried on; and the time has not yet arrived when it can be dispensed with even in our own country. We shall endeavour to point out the difference between the ordinary and improved forms of water wheels, and to lay down sound principles of construction, accompanied by examples for the guidance of the millwright.

CLASSIFICATION OF WATER MACHINES.

Water may be expended upon water machines, 1st. By gravitation, as in vertical wheels generally; 2nd. By pressure simply, as in the water pressure engine, where the water acts on a reciprocating piston; 3rd. By the impulse of effluent water striking float boards, as in the Poncelet wheel; 4th. By the reaction of effluent water issuing from an orifice, as in the Barker's mill and Whitelaw's turbine; or lastly, by momentum, as in the case of the water ram.

It is not, however, always possible in practice to classify water machines according to the mode in which the water expends its force, and hence it will be more convenient to divide them according to the point at which the water is applied, and the direction in which it passes through the wheel, as in the following summary:—

1st. *Vertical Water Wheels*, the plane of rotation being vertical and the water received and afterwards discharged at the same orifice on the external periphery. These may be subdivided into:—

a. Overshot wheels, where the water is applied over the crest, or near the upper extremity of the vertical diameter.

b. Breast wheels, where the water is applied below the crest at the side of the wheel.

c. Undershot wheels, where the water is applied near the bottom of the wheel, and acts, 1. By gravitation as in the improved undershot wheel; or 2. By impulse as in the ordinary undershot and Poncelet wheels.

2nd. *Horizontal Wheels,* the plane of rotation being horizontal and the water passing through the wheel from one side to the other. These may be subdivided into:—

a. Horizontal wheels strictly so called, in which the water passes vertically down through the wheel, acting as it passes on curved buckets.

b. Turbines, annular wheels in which the water enters the buckets at the internal periphery, and passing horizontally is discharged at the external periphery.

c. Vortex wheels, in which the water entering at the external periphery flows horizontally and is discharged at the internal periphery.

3rd. *Reciprocating Engines,* in which the water is applied upon a piston and regulated by valves on the same principle as the steam engine.

The improvements of the Vertical Wheel.—In the present chapter it will be convenient to enter on the consideration of the construction of vertical wheels. Since the time of Smeaton's experiments in 1759, the principle on which vertical water wheels have been constructed has undergone no important change, although considerable improvements have been effected in the details. The substitution of iron for wood has afforded opportunities for extensive changes in their forms, particularly in the shape and arrangement of the buckets, and has given a lighter and more permanent character to the machine than had previously been attained. A curvilinear form for the buckets has been adopted, the sheet iron of which they are composed affording great facility for being moulded into the required shape. It is not the object of the present treatise to enter into the dates of past improvements, but it will suffice to observe that the breast wheel has taken precedence of the overshot wheel, probably from the increased facilities which a wheel of this description affords for the reception of the water under a varying head. It is in most cases more convenient to apply the water of high falls on the breast at an elevation of about 30° from the vertical diameter, as the support of the pentrough is much less expensive and difficult than when it has to be carried over the top of the wheel. In cases of a variable head, when it is desirable to work down the supply of water, it cannot be accomplished without a sacrifice of power on an overshot wheel; but when applied at the breast, the water in all states of the river is received upon the wheel at the highest level of its head at the time, and no waste is incurred. On most rivers this is important, as it gives the manufacturer the privilege of drawing down the reservoir three or four feet before stopping time in the evening, in order to fill again during the night; or to keep the mill at work in dry seasons until the regular supply reaches it from the mills higher up the river. This becomes an essential arrangement where a number of mills are located upon the same stream, and hence the value of small regulating reservoirs behind the mill as a resource for a temporary supply.

Another advantage of the increased diameter of the breast wheel is the ease with which it overcomes the obstruction of back water. The breast wheel is not only less injured by floods, but the retarding force is overcome with greater ease, and the wheel works in a greater depth of back water.

Component parts of Water Wheels.—Vertical water wheels consist essentially of a main axis resting on masonry foundations, and together with arms and braces forming the means of support for the machine. Chambers for the reception of the water constructed of shrouding, sole-plate, and buckets. A pentrough with sluice for laying on the water, and a tail-race for conveying it away; and an internal or external geared spur wheel and pinion for transmitting the power. These parts we shall treat of successively, before describing the modifications of the vertical wheel.

The main axis is a large and heavy cast-iron shaft carried upon plummer blocks bolted to the masonry foundations of the wheel house. It sustains the weight of all the moving parts of the wheel, and in some cases the power is taken from it, when it is subjected to a force of torsion. It is usually cast with deep ribs or wings, calculated to resist the tensile and compressive strain to which they are alternately subjected as the wheel revolves. A section and elevation of the main-axis of a water wheel 20 feet in diameter and 22 feet wide are shown in figs. 100 and 101.* A A is one half the main-axis with its four deep ribs. The part *e* is the journal on which the wheel revolves, and *d* is left square, for the convenience of fixing a screw-jack should the wheel require raising. B, B, B are the recesses for the radial arms of 2½-inch round iron fixed by the keys *f*, *f*, *g*, *g* the corresponding recesses for the braces which pass diagonally across the wheel and alternate with the arms; *c*, *c* are the key beds on the main axis for fixing the main centre. It is difficult to estimate the strain on this shaft when the wheel is on the suspension principle, although the work it has to perform is trifling compared with what it would have to sustain in the event of the power being taken from the axle. In the latter case the wheel has to sustain not only the weight of the wheel and the water in the buckets, but also the force of torsion, as the power is transmitted from the periphery through the arms and axle to the main gearing of the mill.

Fig. 100.

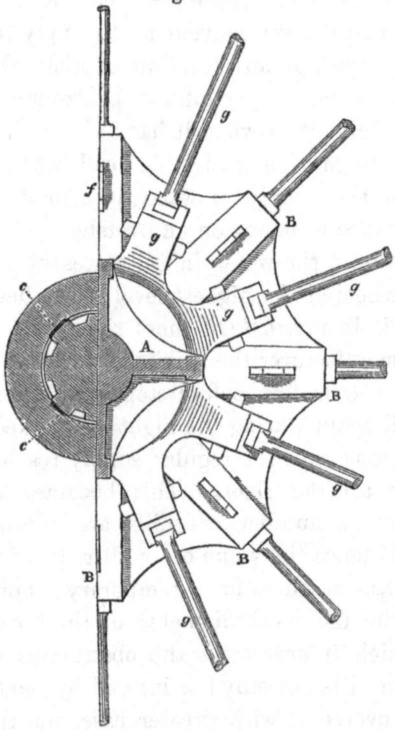

* The wheel is shown in Plate IV. Fig. 110 is also an enlarged detail drawing of this wheel.

Fig. 101. Elevation of half the main axis, and one main centre.

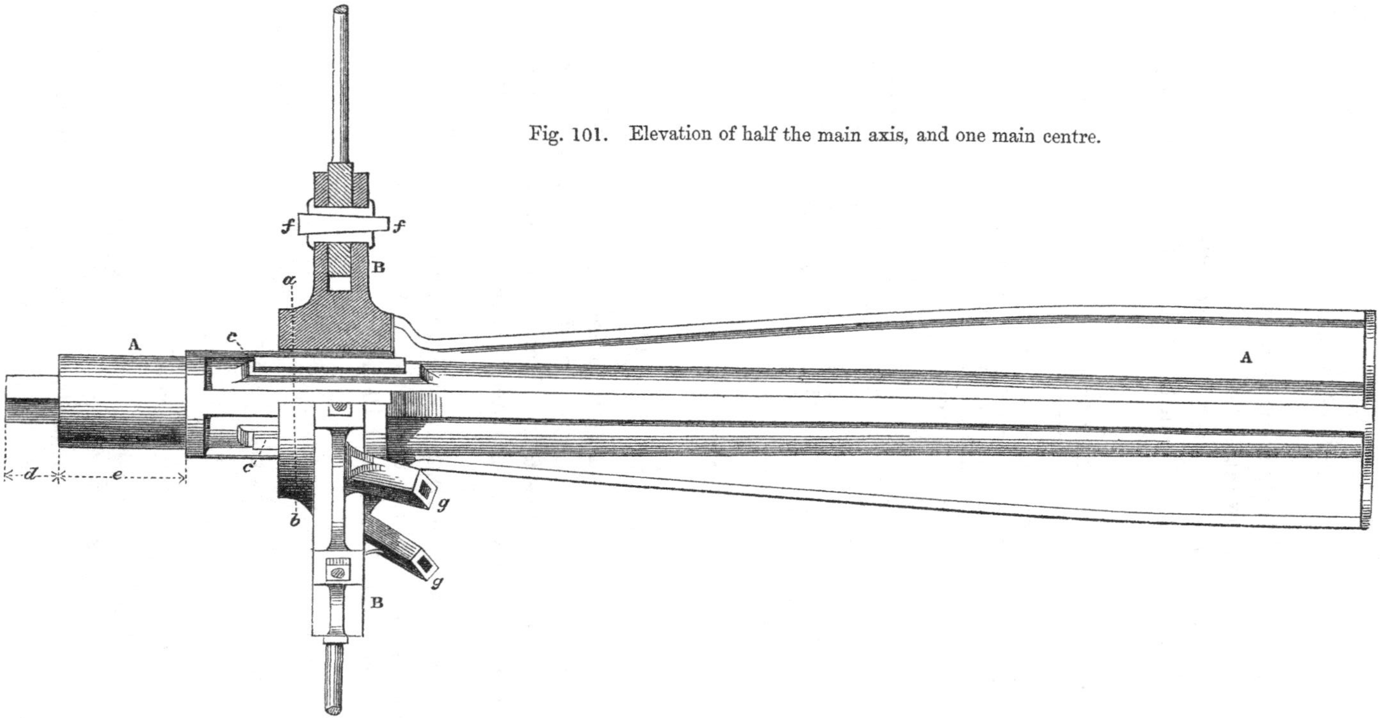

The following table exhibits the dimensions of the journals, which for high and low breast wheels, where the depth of the buckets is nearly the same, I have found effective, and is a summary of my own practice in this respect for the last forty years:—

TABLE OF DIAMETERS OF THE MAIN AXIS JOURNALS OF WATER WHEELS.

Diameter of Wheels in feet.	Diameter of Journal for a wheel				
	5 ft. broad.	10 ft broad.	15 ft. broad.	20 ft. broad.	
	inches.	*inches.*	*inches.*	*inches.*	
15	6	7	$8\frac{1}{2}$	10	The lengths of the bearings are usually equal to one and a half diameters of the journal.
18	$6\frac{1}{2}$	$7\frac{1}{2}$	9	11	
20	7	8	10	12	
25	$7\frac{1}{2}$	$8\frac{1}{2}$	11	$12\frac{1}{2}$	
30	8	9	$11\frac{1}{2}$	13	
40	$8\frac{3}{4}$	10	$12\frac{1}{2}$	$14\frac{1}{2}$	
50	$9\frac{1}{2}$	11	14	16	

Tredgold's rule for the diameter of water wheel journals is that,

$$d = \frac{1}{9} (l\ \text{w})^{\frac{1}{3}} \ldots (1)$$

where d = diameter of gudgeon in inches, l = its length in inches, and w = the maximum load placed on it in lbs.; or supposing the power to be taken off at the loaded side and the pinion to carry the weight of water, w = half the weight of the wheel.

Example.—A wheel 18 feet in diameter and 20 feet broad weighs 34 tons, required the diameter of the gudgeon of the main axis, taking its length at 10 inches.

$$\text{Here, } d = \frac{1}{9} (10 \times \frac{34}{2} \times 2240)^{\frac{1}{3}} = 8 \text{ in.}$$

Another rule which has been proposed is,

$$d = \frac{1}{25} \sqrt{\text{w}} \ldots (2).$$

Example.—Taking the same wheel as before,

$$d = \frac{1}{25} \sqrt{\frac{34}{2} \times 2240} = 7{\cdot}8 \text{ in.}$$

where the length is nearly equal to the diameter; but both these give a somewhat smaller journal than in the table above.

There exists a wide difference of principle amongst millwrights as to the mode of attaching the wheel to the axis. It may either be rigidly fixed by cast-iron arms which resist its weight, as a series of columns alternately exposed to a tensile and compressive strain, or it may be supported by tension rods on the principle now most generally practised in the construction of improved iron water wheels. In the former case the arms are of cast-iron fixed in recesses in a cast-iron main centre, to which they are accurately fitted on chipping strips, and then bolted as shown in fig. 102. Flat wrought-iron arms are sometimes riveted to the main centre in a somewhat similar manner.

Fig. 102.

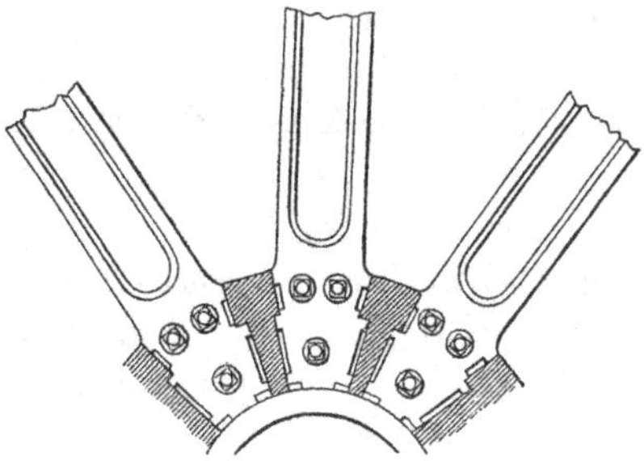

It was reserved for Mr. T. C. Hewes, of Manchester, to introduce an entirely new system in the construction of water-wheels, in which the wheels, attached to the axis by light wrought-iron rods, are supported simply by suspension. I am informed that a wheel on this principle in Ireland was actually constructed with chains, with which, however, from the pliancy of the links, there was some difficulty. But the principle on which this wheel was constructed was as sound in theory as economical in practice, and is due originally, it is said, to the suggestion of Mr. William Strutt, and was carried out fifty years ago by Mr. Hewes, whilst at the same time Mr. Henry Strutt applied the principle to cart-wheels, some of which, thus put together, were for a long time in use. Mr. Hewes employed round bars of malleable iron in place of the chains, and this arrangement has kept its ground to the present time, as the most effective and perfect that has yet been introduced.

Fig. 103.

In the earlier construction of suspension wheels the arms and braces were attached to the centre by screws and nuts, as shown in fig. 103. The arms *c, c* passed through the rim *b b*, and the braces *e, e* are set diagonally in the angle of the rim. This arrangement, although convenient for tightening up the arms and braces, was liable to many objections; the nuts were subject to become loose from the vibration in working, so as to endanger the wheel, and to create a difficulty in keeping it truly circular in form. To obviate this, in 1824, I substituted gibs and cotters, on the same principle as those which secure the piston rod of a steam engine, as shown in figs. 100 and 101: the ends of the arms are forged square, and are fixed in sockets in the cast-iron centre, and are there retained by the gibs and cotters *f, f* in perfect security from the danger of becoming loose.

The shrouds a, a consist of cast-iron plates cast in segments with curved flanches to receive the bucket plates which are attached to them by bolts or rivets (*d, d*, fig. 104), and round the inner periphery a projecting flanch (*b*, figs. 104 and 105) is formed for the reception of the sole plates (*c*). Fig. 104 is a side elevation, and fig. 105 a section of a large shrouding of this description, 15 inches deep. *a a* the cast-iron segmental plate of the shroud; *b* the flanch to which the sole plate *c c* is riveted; *d d* the curved flanches and bucket plates; B the bucket. The segments of the shrouds are bolted together by overlap joints, *j, j*, shown also in section in fig. 106. The overlap is placed on the bucket side of the shroud to preserve a smooth face on the outside of the wheel. The arms are attached to the shrouds either by riveting, or, according to my own practice, by dovetailing into recesses cast upon the inner face of the shroud. Fig. 107 * represents this arrangement in section, and fig. 108 in plan. The ends of

Fig. 104.

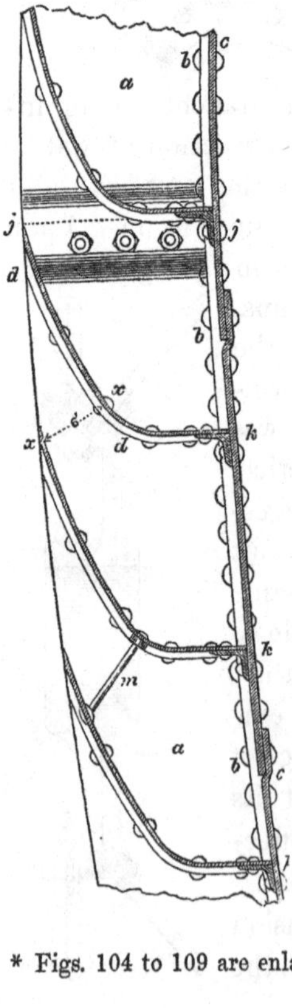

* Figs. 104 to 109 are enlarged details of the Catrine Wheels, Plates I. and II.

the arm *c c* are forged into a **T** form, and are fitted into a similar shaped recess on the shroud. To retain the arms in position, it

Fig. 105.

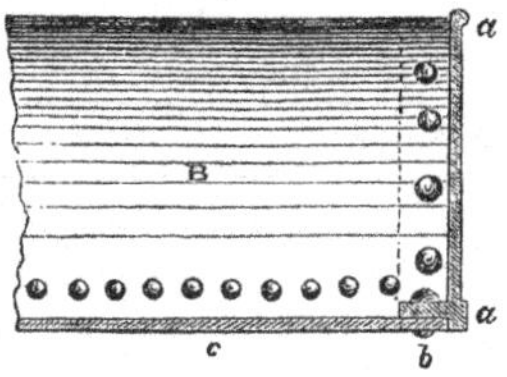

Fig. 106.

is only requisite to give to the recess and **T**-head a dovetail, as shown at *d*. The boss on the shroud must be tapered gradually

Fig. 107.

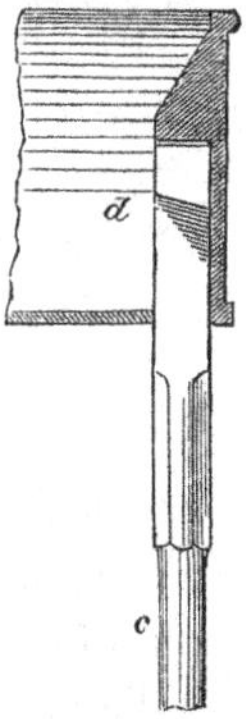

Fig. 108.

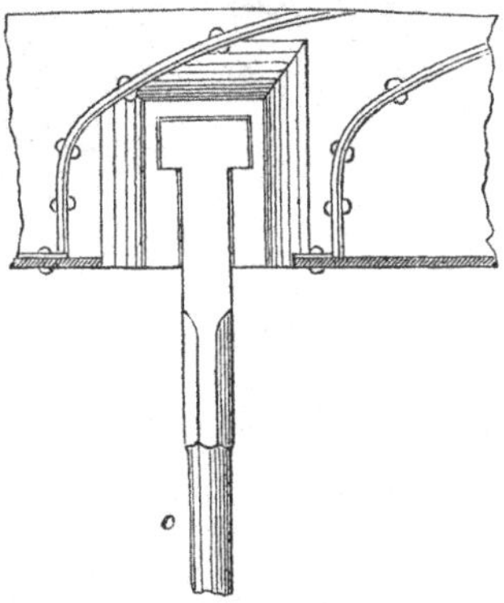

down, to avoid injury in casting from unequal contraction in cooling. The arms are usually 2 to $2\frac{1}{2}$ inches in diameter for almost all wheels, and the braces $1\frac{3}{4}$ to 2 inches.

To strengthen the wheel laterally, diagonal arms, called braces, are used (*g g*, *g g*, figs. 100 and 101), and where the wheel is not of great width, these braces pass from the main centre on one side to the shroud on the opposite side of the wheel, alternating with the radial arms and fixed in the same manner (fig. 109). Where the wheel is broad I prefer to attach the braces to a middle ring of cast-iron, riveted to the interior of the sole plates in their centre between the shroudings. This ring strengthens the wheel in an important degree, by supporting the bucket and sole plate at their weakest part, where they are liable to

Fig. 109.

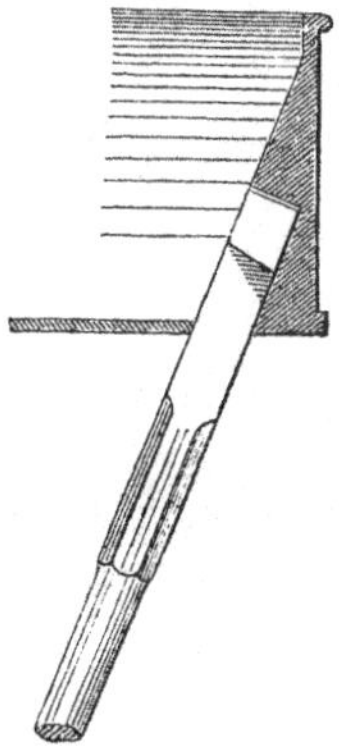

yield to the weight of the water. The middle ring is cast in

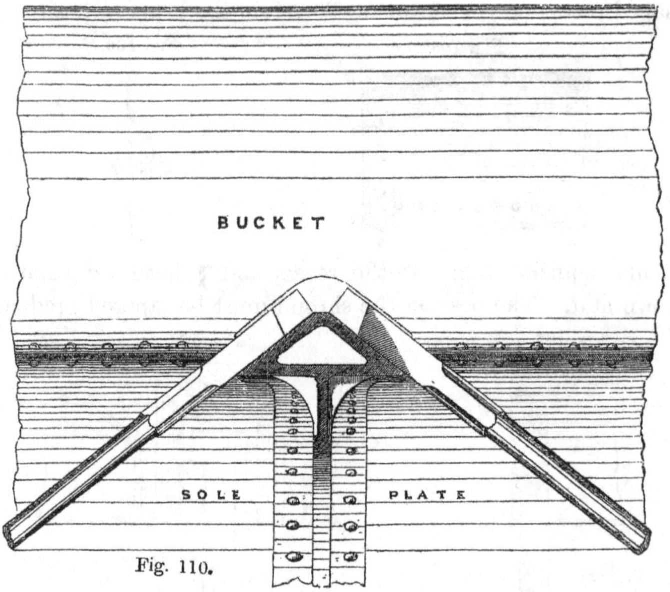

Fig. 110.

segments like the shrouding, and the braces are attached in the way already described. Fig. 110 shows the middle ring of a wheel 20 feet diam. and 22 feet broad.

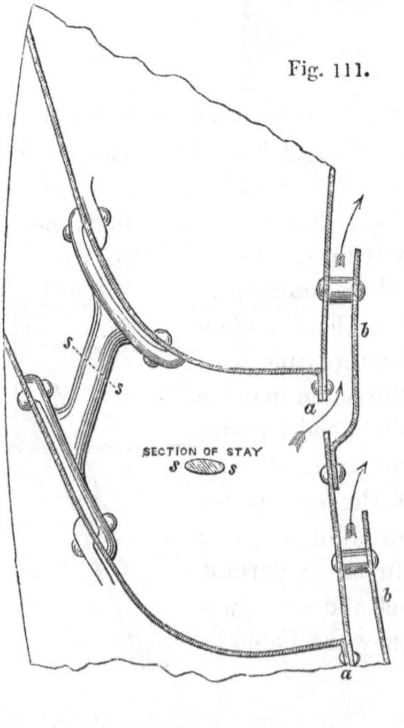

Fig. 111.

The sole plates are of wrought-iron, $\frac{1}{8}$th inch thick (No. 10 Wire Gauge) riveted together with lap-joints. The buckets are riveted throughout their whole length to the sole plate by a bend at the bottom, or in some cases by a small angle iron (*k k*, fig. 104). For the further support of the bucket plates, at every two feet of their length they are riveted to bucket stays forming a complete ring of auxiliary columns round the wheel at every two feet of its breadth. These bucket stays may be of wrought-iron, turned, with two collars, and riveted through each bucket plate, as at *m*, fig. 104, or else of cast-iron, as at *s s*, fig. 111.

The overshot water wheel.—By the overshot water wheel was originally intended that form of wheel in which the stream of water was led *over the summit* of the wheel, and thrown upon it

Fig. 112.

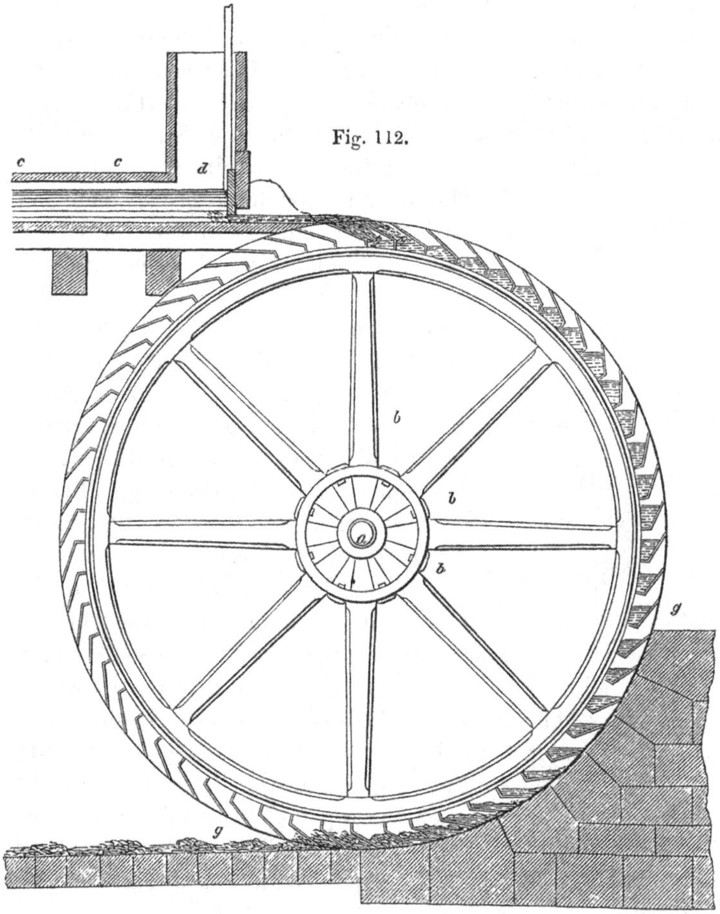

just beyond the extremity of the vertical diameter. The water is retained upon the wheel in troughs or buckets, and by its weight continuously depresses the loaded side of the wheel so as to create a motion of revolution. By a convenient modification of the mode of applying the water, however, the stream was laid on to the wheel upon the same side as it approached, by reversing the direction of the spout or sluice, and for this form the name of pitch-back over-shot wheel was employed. In present use the term overshot is no longer used strictly, but is arbitrarily applied to all wheels in which the water is laid on near the summit, although high-breast is perhaps a more correct and descriptive designation.

The form of the overshot wheel, as constructed about seventy to eighty years ago, is shown in fig. 112. The wheel revolves on a cast-iron shaft *a*, with broad flanches to which the wooden arms *b*, *b* are bolted, as shown in section fig. 113*, with wedges between them to retain them in place.

Fig. 113.

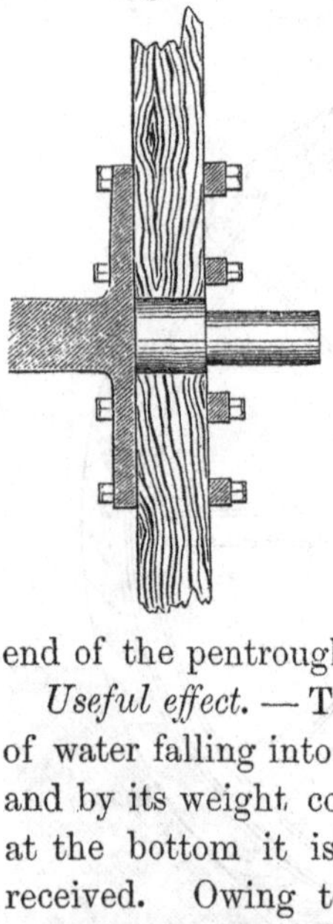

The water is brought from the dam and carried to the summit of the wheel in a wooden trough *c c*, which is nearly horizontal as in fig. 112, or has an inclined apron or spout over the wheel, that the water may flow with a velocity somewhat greater than that of the wheel, so as not to be struck by the back of the revolving float-boards, and thrown off the wheel. This apron is usually made to incline at an angle of about 15° with the course, and is 18 or 24 inches long. A sluice or shuttle *d* is generally placed at the end of the pentrough, to regulate the discharge on the wheel.

Useful effect. — Thus provision is made for a constant supply of water falling into the buckets at the summit of the wheel, and by its weight constantly depressing the loaded side, whilst at the bottom it is discharged with the same facility as it was received. Owing to the form of the buckets, however, the water begins to be discharged at a point considerably above the bottom of the wheel, and thus escapes before it has performed all the work due to the fall. The amount of this waste may be reduced —

1. By adopting a curvilinear form of bucket.

* In earlier wheels, in which the main axis is of wood instead of iron, the principal arms are usually placed in parallel pairs extending across the main axis to the shrouding on either side.

2. By only partially filling the buckets.
3. By a close-fitting breast to retain the water on the wheel.

But when decreased as far as possible, this waste is still an important item in the performance of the wheel, and hence the useful effect secured is never equal to the work of the water due to the space through which it falls. The fraction expressing the percentage of useful effect derived from a given quantity of power expended by the water is called the efficiency of the machine, and is found by the formula —

$$m = \frac{100\,\text{U}}{u} = \frac{100\,w\,h}{\text{W H}}$$

$$\text{U} = \frac{m\,\text{W H}}{100}$$

where m is the efficiency of the machine per cent.; u the work of the water employed per minute, or the weight W of the water in pounds multiplied by the fall H in feet, measured from the surface of the water in the pentrough to that in the tail-race; U the useful effect of the machine, or the pressure p in pounds moved by the working point of the machine multiplied by h, the space in feet through which this point is moved per minute, or the number of pounds raised one foot high by the machine per minute. In ordinary overshot water wheels, the useful effect amounts to about 60 per cent. of the power; or a supply of 12 cubic feet of water per second will give one horse-power for every foot of fall. In the improved iron high-breast wheels, as I have been in the habit of constructing them, the efficiency amounts to 75 per cent., in which case 10·8 cubic feet of water per second will give one horse-power per foot of fall. This is about a maximum effect for water machines, and hence the improved high-breast wheel may be considered as nearly perfect as a water machine.

The waste of water from spilling may to a certain extent be reduced by decreasing the opening of the buckets, but with the disadvantage of at the same time increasing the difficulty of the exit of the water at the bottom of the wheel, and of its entrance at the summit. The waste may be further lessened in an important degree by increasing the breadth of the wheel and the capacity of the buckets, and in general it is not advisable that the buckets should ever be more than two-thirds filled with the average supply of water. The buckets then reach a much lower position before they begin to discharge than when they have been nearly filled. The third means of preventing the spilling of the water is by a curved breast fitting closely to the wheel, as shown in fig. 112, *g g*, and serving, when accurately fitted, to retain the water on the buckets. With low falls this breast is of considerable importance, and secures a considerable increase of efficiency. But with large wheels for high falls, with small openings in the buckets, it is of no value, and does not compensate for its cost, when the buckets are made of the best form, so as to retain the water as long as possible upon the wheel; and in these cases the breast is invariably dispensed with.

The pitch-back wheel.—The most important modification of the old over-shot wheel is known as the pitch-back wheel, in which the course of the current of water is reversed in the pentrough, and laid on the wheel from the same side at which it approaches. In old wheels it was essential, as the wheel generally worked more or less frequently in back water, that the tail-race should always lie in the same direction as the revolution of the wheel. Hence when the position of the waste water culvert was fixed by other circumstances, it often happened that the millwright was driven to the use of the pitch-back wheel to meet the conditions of the case, and the advantages of this form of wheel were thus forced on his notice. It was perceived that by increasing the diameter of the wheel, the water might be laid on at a distance from the summit, and it was shown theoretically that a larger useful effect would be secured by laying it on at about 25° to 30° from the summit, than if it took the water over the top. And in this way, when the introduction of iron gave sufficient facility for the construction of wheels of large diameter, the high breast wheel was adopted, and has maintained its ground to the present time as one of the most perfect and economical machines.

Direction of tail-race.—It is no longer necessary that the flow of the tail water should be in the direction of the wheel's revolution. On the contrary, I frequently take it in the opposite direction or at the side, according as the circumstances of the case determine the position of the wheel, and the point of discharge. The old plan of setting the wheel parallel with the stream is no longer requisite, provided proper care is taken to give a sufficient outlet to the water. To effect that object it is essential to sink the bottom of the tail-race two or two and a-half feet beneath the bottom of the wheel, and that depth should be continued to the river, so as to form the tail-race into a canal, with the water flowing gently, and with a comparatively slow motion from the wheel. In this arrangement the bottom of the wheel, when standing in an ordinary condition of the river, is 8 or 9 inches above the water in the tail-race, so that its motion cannot be impeded, and there is left ample space for the rise occasioned by the continuous discharge from the buckets during the working of the wheel. To show how immaterial is the direction of the tail-race, I may add that I have in some cases formed the tail-race into an underground tunnel, in the shape of an inverted syphon. Fig. 114 shows this arrangement as adopted for a mill in 1832, to secure an increase of fall. A shows the wheel and wheel-house, in which originally the wheel was 24 feet in diameter, the fall 22 feet, and the tail-

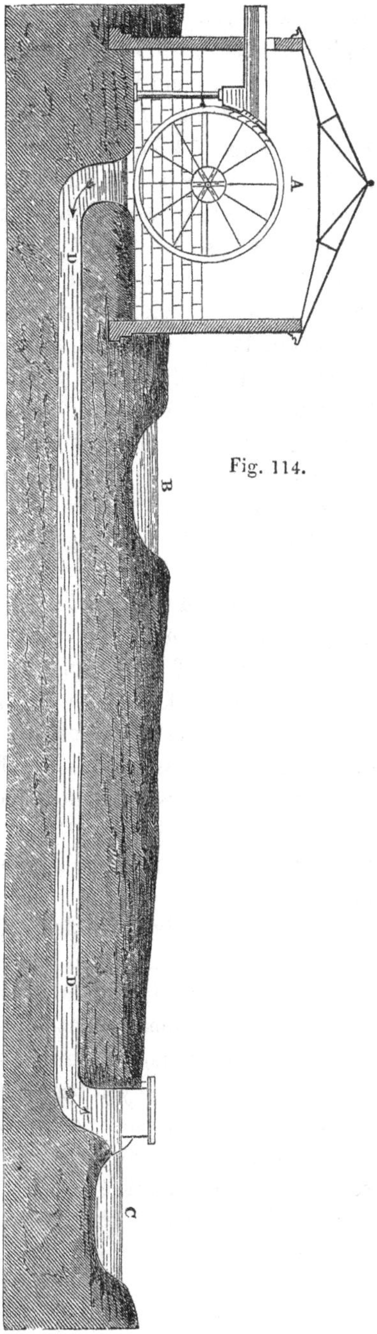

Fig. 114.

water conveyed direct into the river Eagley, at B. When replacing this wheel by a new one, it was found that by taking advantage of a bend in the river, and conveying the tail water to C, an increase of about 6 feet of fall could be obtained. Hence a wheel 32 feet in diameter was adopted, with a fall of 28 feet; and for the tail-race a tunnel D D was constructed, nearly a quarter of a mile long, and passing under the bed of the river at B, so as to meet the stream on the other side of the field at C. The substratum being composed of hard rock and shale, afforded every facility for the drifting of the tunnel, and when complete, the flow of water through it was so exceedingly sensitive, that only a few gallons falling from the wheel into the trumpet mouth at A, immediately caused a perceptible discharge into the river at C, at a distance already stated, of nearly a quarter of a mile. The perfect success of this arrangement caused its adoption in other cases, where the conditions were favourable for carrying it out.

The Catrine high-breast wheels.—Plates I. and II. illustrate the construction of the improved iron high-breast wheel as applied at the Catrine Works in Ayrshire, between the years 1825 and 1827, on a fall of forty-eight feet. Taking into consideration the height of the fall, these wheels, both as regards their power and the solidity of their construction, are even at the present day among the best and most effective structures of the kind in existence. They have now been at work upwards of thirty years, during which time they have required little or no repairs, and they remain nearly as perfect as when they were erected.

It was originally intended to erect four of these wheels at the Catrine Works, but only two have been constructed. Preparations were made, however, for receiving two others in the event of an enlargement of the reservoirs in the hill districts, and more power being required for the mills. This extension has not as yet been wanted, as these two wheels are working to 240 horses' power, and are sufficiently powerful, except in very dry seasons, when they are assisted by auxiliary steam-power, to turn the whole of the mills.

Plate I. is a plan of the wheel-house, showing the position of the wheels, and the arrangement of the main gearing. The

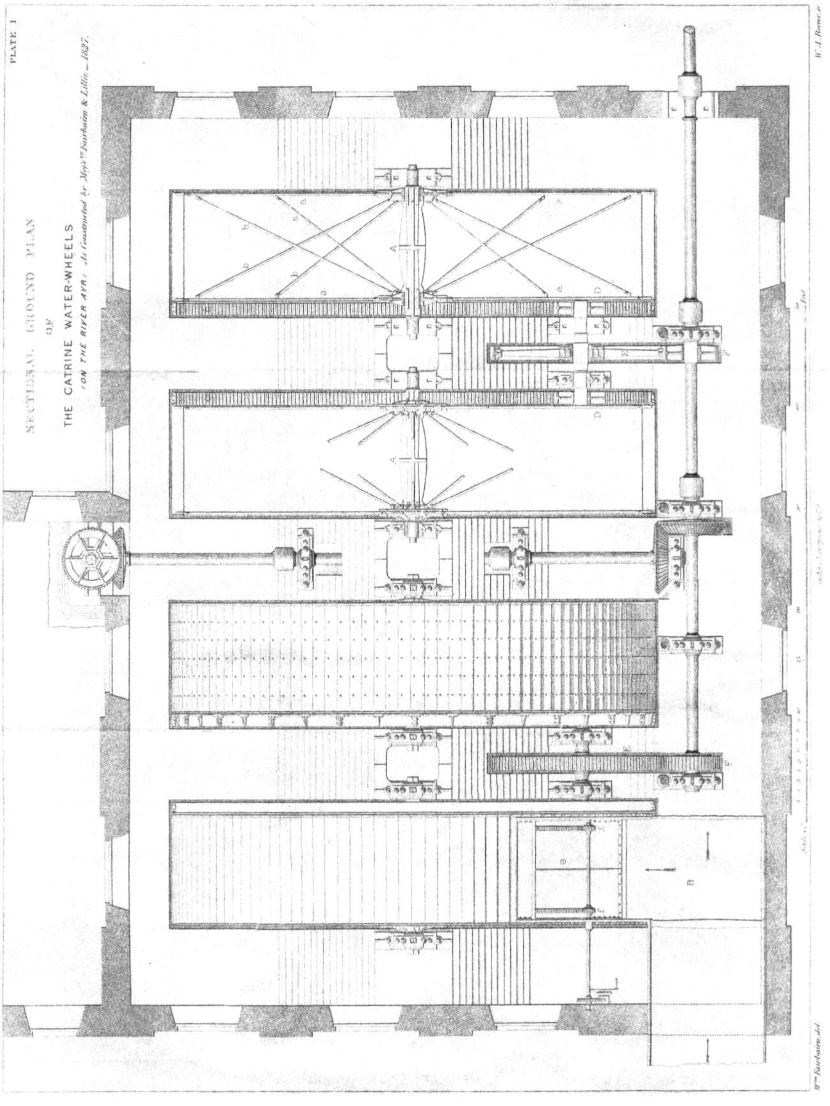

The material originally positioned here is too large for reproduction in this reissue. A PDF can be downloaded from the web address given on page iv of this book, by clicking on 'Resources Available'.

first pair of wheels is shown in section, to exhibit the main axle, arms, braces, spur segments, and pinions. The other pair are shown in plan, one exhibiting the buckets, and five rows of bucket-stays, while the pentrough, sluice, and regulating gear are shown on the other. It will be seen that the motion of each pair of wheels is transmitted through a common pinion shaft, and thence by another pinion and spur-wheel, by which the velocity is increased to the first motion shaft of one mill, whilst between the two pairs of wheels there is the first motion shaft of another mill geared into the preceding shaft by a pair of large bevil wheels.

Plate II. is an elevation of the wheel house, with the masonry for supporting the wheels, tail-race, tunnel, &c. The right half of the wheel is shown in section, and the left half in elevation, and there is a section of the pentrough, sluice, and plates, to guide the water into the buckets.

The following are the references to the different parts of the wheel:—

A, main axis.

a a a, arms. *c*, segments.

b b b, braces. *d*, joints of segments.

B, pentrough.

e, sluice with racks. *f*, pinion connected with governor.

C, tunnel running through the wheel-house, and acting as the tail-race.

D, pinion gearing into internal segmental spur-wheel on shrouds.

E, wheel on the same axis as D, and communicating the power to the pinion F on the first motion shaft.

G, galleries to obtain access to the pentrough and other parts of the wheel.

The water is brought from the reservoirs in a tunnel 10 feet in diameter through the hill part, and thence in a conduit 12 feet wide, and 5 feet deep, arched over. The reservoirs cover 120 Scotch acres, of an average depth of 8 or 10 feet, giving storeage room for a large supply of water; and the sill of the reservoir sluice, from which the aqueduct bottom is carried level to the pentrough, is 16 inches above the lowest overflow of the sluice on the wheel; hence in dry seasons the water may be drawn off to within 16 inches of the bottom of the lade. At the same time the pentrough is made of a depth of six feet, in order that in seasons of plentiful supply the water may be drawn off at the highest level, and the entire fall, as far as possible, rendered effective.

The total supply of water requisite to work the mills when the wheels were started was about 60 tons or 2150 cubic feet per minute, the wheels revolving at a circumferential velocity of four feet a second, or 182 buckets passing each sluice per minute. This gives $\frac{2150}{182 \times 2} = 5{\cdot}9$ ft. or 6 cubic feet of water nearly for each bucket of the wheels. The whole capacity of each bucket is $17\frac{1}{4}$ cubic feet, hence when thus working the buckets were just one-third filled.

When working to their full power of 240 horses, however, the fall being 48 feet, this pair of wheels would require,

$$\frac{100 \times 33000 \times 240}{75 \times 2240 \times 48} = 98{\cdot}2 \text{ tons of water per minute,}$$

if we suppose the useful effect to be 75 per cent. of the water power expended. Now if we take the circumferential velocity at 5 feet per second, at which the wheel should then run, this would give 7·7 cubic feet of water per bucket, or $\frac{7{\cdot}7}{17{\cdot}25} = \frac{10}{22}$ or one-half nearly, as the ratio of the quantity of water in the buckets to their capacity.

Between these limits these water wheels act effectively and economically.

The wheels are 50 feet in diameter, 10 feet 6 inches wide inside the bucket, and 15 inches deep on the shroud; the buckets are 120 in number, and have an opening of 6 inches; the internal spur segments are 48 feet 6 inches diameter, 3 inches and a quarter in pitch, 15 inches broad, and have 560 teeth. The pinions are the same width and pitch, and are 5 feet 6 inches in diameter. The intermediate wheel between the pair of segment pinions is 18 feet $3\frac{3}{4}$ inches in diameter, 16 inches broad, and $3\frac{1}{4}$ inches pitch; and the large bevil-

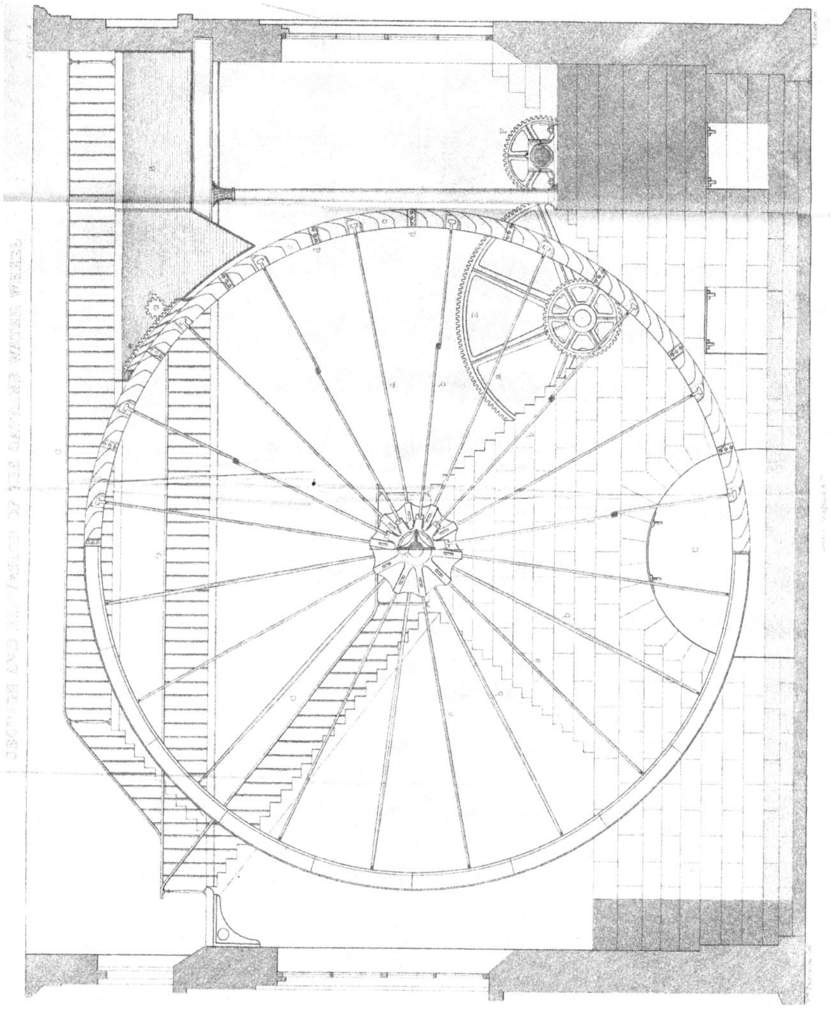

The material originally positioned here is too large for reproduction in this reissue. A PDF can be downloaded from the web address given on page iv of this book, by clicking on 'Resources Available'.

wheels are 7 feet in diameter, $3\frac{1}{2}$ inches pitch, and 18 inches broad on the cog, so as to be of sufficient strength to convey, if necessary, the united power of the four water wheels.

When viewed from the entrance, the two wheels already completed have a very imposing effect, from their elevation on stone piers. And as the whole of the cisterns, sluices, winding apparatus, galleries, &c., are considerably elevated, they are conveniently approached in every part. Under the wheels there is a capacious tunnel, terminating at a considerable distance down the river and conveying away the tail water from the wheels.

TABLE OF SPEEDS.

Water wheel 50 ft. 0 in. = 1·5 revolutions = 4 ft. per second.

	Diameter.		Revs.		Diameter.		Revs.
Segments,	48 ft. 6 in.	and	1·5	into wheel	5 ft. 6 in.	=	13·3 of shaft.
Wheels A,	18 ft. $3\frac{3}{4}$ in.	and	13·3	into pinion	5 ft. 6 in.	=	44 of main shaft to mill.
Wheels B,	7 ft. 0 in.	and	44	into wheel	7 ft. 0 in.	=	44 of shaft to new mill.
Wheels S,	5 ft. 9 in.	and	44	into wheel	4 ft. 0 in.	=	63 of upright in new mill.

The journals of the main axes of the water wheels and of the pinion shafts are 14 inches in diameter. The first motion shafts are $13\frac{1}{2}$ inches in diameter, and of an average length between the couplings of 19 feet.

The maximum fall may be estimated at 48 feet 9 inches. The distance from the bottom of the wheel to the floor of the tail-race is 3 feet 6 inches, the average depth of tail-water 2 feet, and the distance from the floor of the tail-race to the level of the water in the reservoirs is 50 feet 9 inches.

I have been more particular in describing these wheels, as they are the first erected upon the principle of concentration and combined action. In former cases it had been the custom to erect the wheels near where the work was required, so that it was not unusual to have three or four wheels at a short distance from each other, working independently. This was the case at the Catrine works before the large wheels were erected. It was found desirable, however, in extending the works to have the whole power concentrated in one wheel-house, with a uniform fall, so as to simplify the transmission of the power to the different parts of the mills. This was effected in the manner already described with great success, and the result has been a continuous and efficient supply of power from 1827 to the present time.

Immediately following the erection of the Catrine wheels, those of Deanston, belonging to the same proprietary, were commenced. The Deanston Works were designed upon a much larger scale than even those at Catrine, as it was intended to erect eight powerful water wheels instead of four, as in the

Fig. 115.

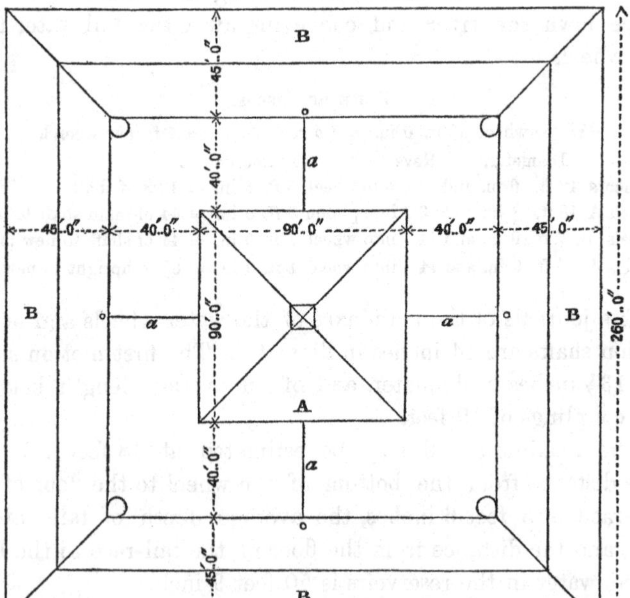

works in Ayrshire. The Deanston Works were erected with two water wheels in the bottom room of the factory about the year 1780, and came into the hands of their present proprietors about 1798 or 1800. After the completion of the alterations in Ayrshire, a similar concentration of the power was desired for Perthshire, and I was requested to prepare both for a renewal of the old wheels and the erection of new ones on a larger and more comprehensive scale. In obedience to these instructions, an entirely new site was selected for the water power, close to the old mills on the River Teith, and provision made for an increased fall, and an improved application of the water power.

The new wheels as then designed were eight in number, and were placed together in a rectangular building adjoining the old mill, but arranged to afford power to an entirely new establishment surrounding the wheel-house, according to the annexed plan (fig. 115), in which the centre building A is the wheel-house, and the buildings B B B B, surrounding it on all four sides, and three stories in height, contained the machinery driven by the wheels. From this design it will be seen how the power, amounting to 800 horses, was given out on each side by the shafting *a a a a*, radiating from the centre of the wheel-house at right angles to the mills on every side. Another shaft was extended in an underground tunnel to the old mill, where it still gives motion to the machinery in that portion of the works.

It is much to be regretted that this design was never carried out in its integrity; but the late Mr. James Smith, so well known as the inventor of the subsoil plough and many other ingenious contrivances, altered the plans after having raised one side of the new mill to a height of one story, when unfortunately it was abandoned for a much less convenient and less perfect structure.

As respects the water wheels, the two first were erected by myself — then in partnership with my much respected friend, Mr. James Lillie,—and the last two by Mr. Smith, who, with the Cotton Mills, has since carried on considerable engineering works. The remainder have never been erected. There were, however, several novelties in the arrangement of these wheels which it may be desirable to describe at greater length. The River Teith is the principal feeder, and falls into the Forth about a mile above Stirling. The supply in ordinary seasons is about 260 cubic feet per second, and for many months in the year more than double that quantity. The original fall was about 18 feet; but, by the erection of a weir higher up the stream, and the construction of a canal three-quarters of a mile long, it was increased to 33 feet, so as to afford, except in very dry seasons, nearly 800 available horses' power. Of late years this has been increased by a copious supply from Loch Vennaquar, the surface of which has been raised at the cost of the Corporation of Glasgow, as a compensation for the water taken from Loch Katrine, which falls into the Teith for consumption in the city. From Loch Vennaquar, therefore, there is a continuous supply at all seasons.

The augmentation of the fall from 18 to 33 feet nearly doubled the power for the mills, and also the supply of water which was conveyed direct from the weir to the new wheels in the rectangular building. The water flowed into a wrought-iron pentrough A, fig. 116, supported on iron columns, and delivered the water into the wheels on each side. The wheels were 36 feet in diameter, and of the same construction as

Fig. 116.

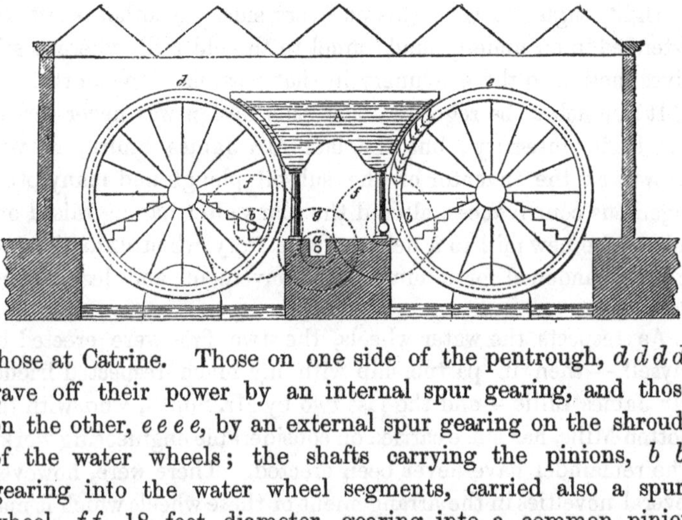

those at Catrine. Those on one side of the pentrough, *d d d d*, gave off their power by an internal spur gearing, and those on the other, *e e e e*, by an external spur gearing on the shrouds of the water wheels; the shafts carrying the pinions, *b b*, gearing into the water wheel segments, carried also a spur-wheel, *ff*, 18 feet diameter, gearing into a common pinion *g*. This last pinion was on the central shaft *a a*, passing along the centre of the wheel-house, and giving off motion to the shafts *á á* by the bevel wheels *k*, at the centre of the wheel-house (fig. 117).

Water wheel, 4·1526 ft. circumferential velocity = 2·203 revolutions.

	Ft.	in.		Revs.		Ft.	in.			Revs.	
Segments,	33	8½	and	2·203	into	5	6	pinion	=	13·59	cross shaft.
Wheels *b b*,	18	2½	and	13·59	into	5	6	pinion	=	45·3	of main shaft to mill.
Wheels	6	0	and	45·3	into	4	2½	wheel	=	65	of upright in mill.

It will be observed that these high-breast wheels have the

peculiar advantage of permitting the use of a sliding or folding sluice, for the admission of the water, which can be adjusted to a very variable fall. So that, at whatever height the water may stand, the velocity at which it enters the wheel will be the same, because it falls over the top of the adjusted sluice. But with this advantage they are apt to become liable to the defect of admitting the water with too much difficulty, a defect which was remedied by the principle of ventilation, which I first introduced in the year 1828, under the following circumstances:—

Ventilation of Water Wheels.—Shortly after the construction of the water wheels for the Catrine and Deanston Works, a breast wheel was erected for Mr. Andrew Brown of Linwood near Paisley. In this it was observed, that when the wheel was loaded in flood waters, each of the buckets acted as a water blast, and forced the water and spray to a height of 6 or 8 feet above the orifice at which it entered. This was complained of as a great defect, and in order to remedy it openings were cut in the sole-plates, and small interior buckets attached, inside the sole, as shown at *b* fig. 111. The air in the bucket made its escape through the openings *a*, *a*, and passed upwards as shown by the arrow, permitting the free reception of the water from the pentrough. The buckets were thus effectually cleared of air as they were filling, and during obstruction from back-water in the tail-race the same facilities were offered for its re-admission, and the free discharge of the water from the rising buckets. The effect produced by this alteration would scarcely be credited, as, in consequence of the freedom with which the wheel received and parted with its water, an increase of power of nearly one-third was obtained, and the wheel, which remains as then altered, continues, in all states of the river, to perform its duties satisfactorily.

This difficulty in the admission of the water had often been noticed by the early millwrights, and where it interfered with the working of the wheel, their remedy was to bore holes for the escape of the air in the sole-plate or the start of each bucket. Thus, in his "Mechanical Philosophy," Dr. Robison gives a similar instance to that of Mr. Brown; a wheel 14 feet in diameter and 12 feet wide was working in 3 feet of back-water and labouring prodigiously; three holes, each one inch diameter, were made in each bucket, when the wheel ceased to labour, and its power was increased one-fourth. The objection to holes in the sole-plate or buckets is a certain spilling of the water over the interior of the wheel, which cannot be avoided. But it must be remembered that air being 800 times rarer than water will escape through a hole at least thirty times faster with

Fig. 117.

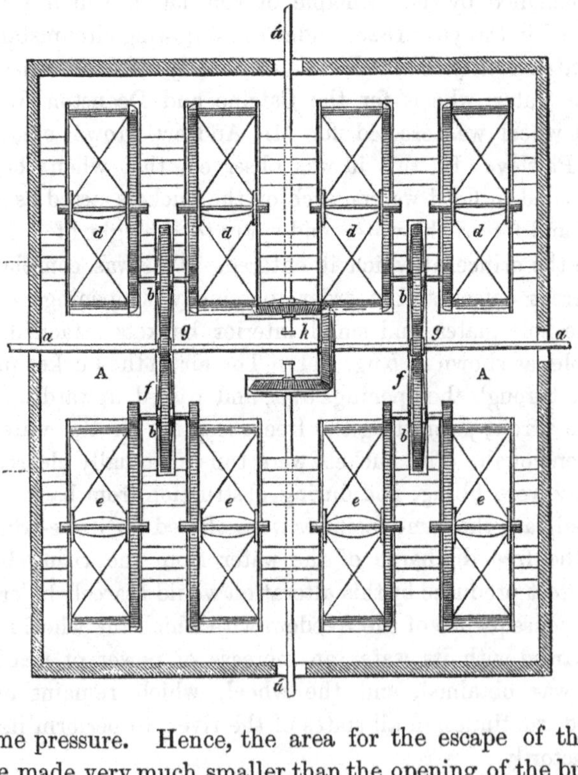

the same pressure. Hence, the area for the escape of the air may be made very much smaller than the opening of the bucket.

The amount of power gained, and the beneficial effects produced upon Mr. Brown's wheel, induced the adoption of the ventilating principle as a permanent modification of construction. The first wheel thus designed was erected at Wilmslow in Cheshire, and was started in 1828. It was identically the same with that shown in Plate III., and it was closely followed by a further improvement, as shown in Plate IV.

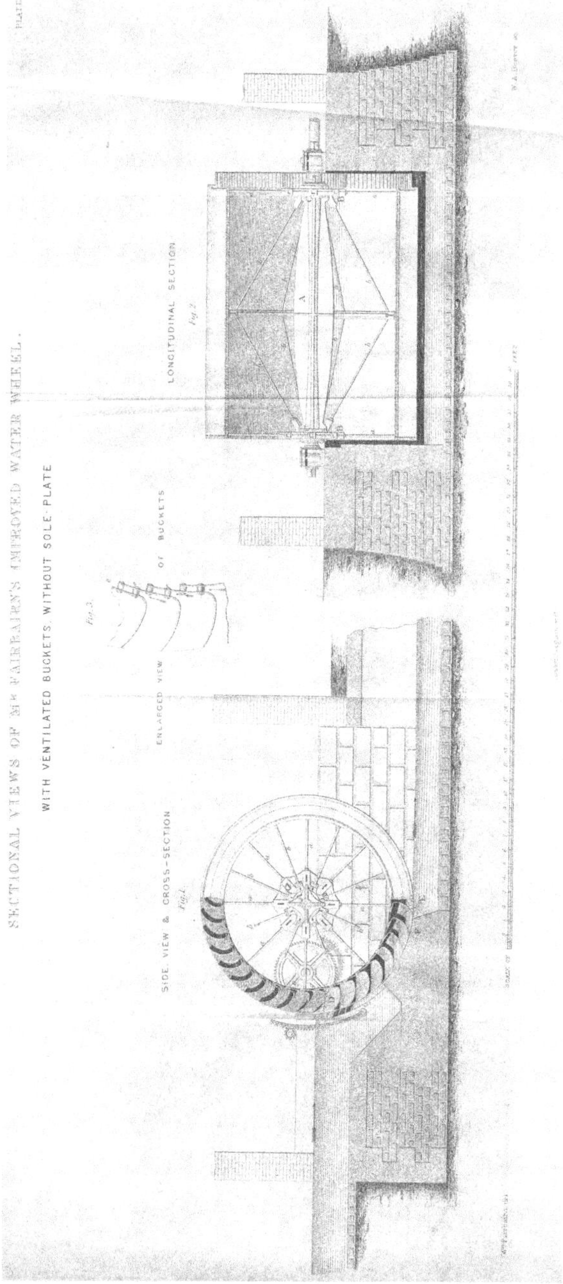

The material originally positioned here is too large for reproduction in this reissue. A PDF can be downloaded from the web address given on page iv of this book, by clicking on 'Resources Available'.

Close-bucketed wheels labour under great disadvantages when receiving the water through the same orifice at which the air escapes. When, as is frequently the case, the water is discharged upon the wheel in a sheet of greater depth than the opening between two buckets, the air is thus suddenly condensed in the bucket, and re-acting by its elastic force throws back the water upon the orifice of the cistern, and thus allows the buckets to pass imperfectly filled. A similar obstruction occurred whenever the wheel worked in backwater, the water being lifted in the rising buckets, the mouths of which being under water the entrance of air was effectually prevented; and the deeper the backwater the more completely they filled with water and the greater became the difficulty in discharging. Many millwrights to remedy this were in the habit of boring holes in the sole near the start of the bucket, and of narrowing the spout or sluice so as to leave room on each side of the buckets for the escape of the air, means which to some extent remedied the evil of the spilling and sputtering of the water, but in most cases occasioned considerable waste of power, from the water being driven through the openings and falling over the interior of the wheel.

Other remedies have been attempted, such as circular tubes and boxes attached to the sole-plates; but these plans have been generally unsuccessful, owing to the complexity of their structure and the inadequate manner in which they attained the object contemplated. In fact, in wheels of this description it has been found more satisfactory to submit to acknowledged defects, than to incur the trouble and expense of partial and imperfect remedies. In the ventilated wheels about to be described, the perfect escape of the air is effected by very simple means, and great success has attended their application in situations where interruptions frequently arise from excess of backwater or a deficiency of supply.

Low-breast ventilated Wheel.—Plate III. represents a front and side view of a water wheel with ventilated buckets. Portions of the shrouding and segments are removed in order to show a section of the buckets, and the position in which they receive the water.

A is the axle or ribbed shaft, supporting the two main axes, C C, from which the wheel is suspended; B B are the projecting sockets into which the ends of the malleable arms *a a* and the diagonal braces *b b* are keyed. The arms are 2 inches, and the braces $1\frac{3}{4}$ inches in diameter. D represents the buckets, with the shuttle which regulates the admission of the water, and which is made to slide downwards. F the termination of the stone breast, and E the tail-race. This wheel, it will be seen, is arranged as a low-breast.

The principle of the construction of the buckets is more clearly shown on an enlarged scale in fig. 3, Plate III., the sole-plate being abandoned and the bucket plates bent round and prolonged upwards so as to overlap one another, leaving an opening, indicated by the arrows, for the escape of the inclosed air. The bucket plates are connected together by tubular ferules, or stays, through which a rivet is passed, and riveted on each side.

The wheel should always, as in this plate, be placed above the tail water, and not, as in the older forms of wheels (fig. 112), be carried down to the level of the tail-race floor; and the breast of wood, iron, or stone, but usually the latter, which is of so much importance for low falls in retaining the water on the wheel, should break off about ten inches from the extremity of a vertical diameter of the wheel. In fact, the benefits of this form of breast and tail-race are so great they should be strictly carried out where it is desirable to make effective use of the fall.

In high-breast wheels of twenty-five feet in diameter, and upwards, the breast is not required, as the buckets having narrower openings, and their lips extended nearer to the back of the following buckets, retain the water longer on the wheel. In this case the loss from spilling constitutes too small a percentage of the power to compensate for the expense of a lofty and close-fitting breast. In some cases the breasts have been composed of iron and wood, but in the best constructed they are of masonry, and allow little or no space between them and the wheels. It is, however, necessary to be cautious that extraneous matters do not in that case gain admission to the buckets, as by jamming between the buckets and the curb they might cause disaster.

The preceding statements, so far as relates to the method of ventilation, have been principally confined to the form of bucket and description of water wheel suitable for low falls. It will now be necessary to describe the best form of breast wheels for high falls, or falls of from one half to three-fourths of the diameter of the wheel.

High-breast ventilated Wheel.—A water wheel of this kind, constructed for T. Ainsworth, Esq., of Cleator, near Whitehaven, is represented in Plate IV. It is twenty feet in diameter, twenty-two feet wide inside the bucket, and twenty-two inches deep on the shroud. It has a close riveted sole, composed of No. 10 wire gauge iron plate, and the buckets are ventilated from one to the other, as shown on a larger scale in fig. 3. The fall is seventeen feet, and the water is discharged upon the wheel by a circular shuttle, A, which is raised and lowered by a governor as circumstances require. By this arrangement the whole height of fall is rendered available, and the water in dry seasons may be drawn off three or four feet, in order to afford time for the dam to fill in the periods during which the mill is stopped.

The power is taken from each side by two pinions working into the internal spur segments B B, and these again give motion to shafts and wheels at C C, which communicate with the machinery of two different mills, at some distance from each other.

Arrangement of Gearing.—The position of the pinion, or the point where it gears into the spur segments on the water wheel, whether internal or external, is of importance in every water wheel, but pre-eminently so in those constructed on the suspension principle, which are indifferently prepared to resist the torsive strain to which they would be subjected if the power were taken from the unloaded arc of the wheel. Water wheels of this construction, with malleable iron rods only two inches in diameter for their support, could not resist the strain, but would twist round upon the axle, and destroy the wheel.

It is necessary, therefore, in every case, to take the power from the loaded side of the wheel, as near the circumference as possible, in order to throw the weight of the water directly upon the pinion without transmitting it through a larger arc of the wheel than is absolutely necessary. For this purpose the spur pinion should be below the centre of gravity of the water on the wheel, and therefore more or less below the extremity of the horizontal diameter.

In the old water wheels, where the power was generally taken from the axle, the whole of the force passed through the arms to the point, and afterwards by a pit-wheel by some multiplier of speed to the machinery of the mill. In the improved wheels this is no longer the case: the arms, braces, and axle have only to sustain the weight of the wheel, and to keep it in shape, and the power being taken from the circumference, considerable complexity is avoided, and the requisite speed far more easily obtained.

Speed of Water Wheels.—I have usually made breast wheels for high and low falls, with a velocity between 4 and 6 feet per second at the periphery, and between these limits water wheels may be worked with economy. But for a minimum velocity I have taken 3 feet 6 inches per second, for falls of from 40 to 45 feet, and for a maximum velocity, 7 feet per second, for falls of 5 or 6 feet. The higher velocities, namely, from 5 to 6 feet per second, are now very generally adopted for the best constructed wheels, not indeed on the score of economy in the expenditure of water, but for the purpose of obtaining more easily the requisite speed under the variable conditions of supply. In this climate, where the atmosphere is so much charged with moisture, the rivers, for eight months in the year, generally afford an ample supply of water. It is for this reason that an increased velocity is given to the wheel, in order to increase the power in average conditions of supply, so as to work off the surplus rather than adapt the wheel to the minimum expenditure. It would, however, be advantageous to increase the capacity of the wheel, and work at a velocity of four feet, or at most four feet six inches per second.

Area of opening of Bucket.—The width of the opening of the bucket varies according to the point at which the water is laid on. I have made them with openings as low as 4 inches wide and as much as 20 inches, the first being for very high breast and the latter for undershot wheels, but ordinarily the width is from 5½ to 8 inches for high breast and from 9 to 12 inches for low breast wheels. In this matter the millwright must exercise his own judgment, taking into account, 1st, the quantity of water to be delivered upon the wheel; 2nd, the position on the circumference at which the water is to be delivered, a wider opening being necessary for low-breast than for high; and 3rd, he must consider whether the circumstances of

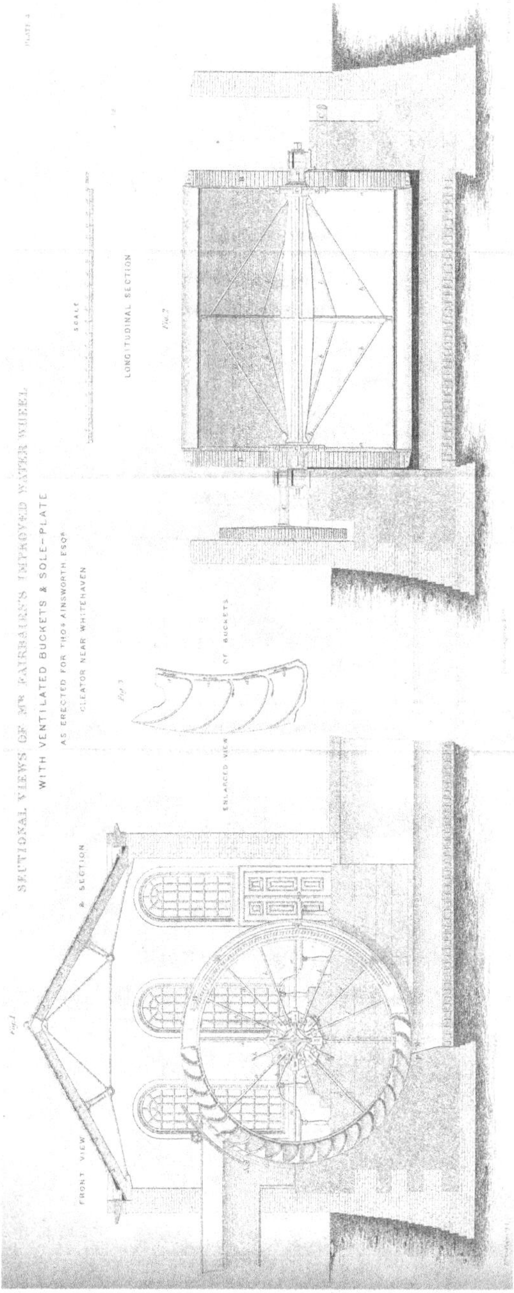

The material originally positioned here is too large for reproduction in this reissue. A PDF can be downloaded from the web address given on page iv of this book, by clicking on 'Resources Available'.

the case in any degree limit the width of the wheel. The width of the opening must be measured perpendicularly to the direction in which the water enters the wheel; thus in fig. 104, x x is the width of opening.

For high falls, the best proportion of the *area* of opening of the bucket, that is, the width multiplied by the length between the shrouds, is found to be such that 5 square feet of sectional area of opening is allowed for 25 cubic feet capacity in the bucket. But in breast wheels which receive the water at a height of not more than 10 degrees above the horizontal diameter, 8 square feet should be allowed for the same capacity. With these proportions the depth of the shrouding is assumed to be about 2 or 2½ times the width of the opening.

The distance of the buckets apart, measured upon the external periphery of the wheel, I have been accustomed to make from 1 foot to 1 foot 6 inches, low breast being somewhat further apart in general than high breast. This proportion fixes the number of buckets in the wheel according to the following table:—

				No. of buckets.		
For wheels	10 feet	diameter,	from	20	to	30.
,,	20	,,	,,	40	,,	60.
,,	30	,,	,,	60	,,	90.
,,	40	,,	,,	88	,,	120.
,,	50	,,	,,	120	,,	150.
,,	60	,,	,,	130	,,	180.

In setting out the curve of the water wheel bucket in breast wheels, a line a b may be drawn cutting the external periphery of the shroud at the point and in the direction in which it is intended that the water shall strike the wheel after passing the guide plates of the pen-trough sluice. If we then measure a distance c equal to the distance of the buckets apart, and from the centre e draw the radius d c, the line a b will be nearly the direction of the lip of the bucket and c d the direction of its start, and the curve must be drawn connecting these lines according to the judgment of the millwright, making some allowance for the velocity of the wheel.

Fig. 118.

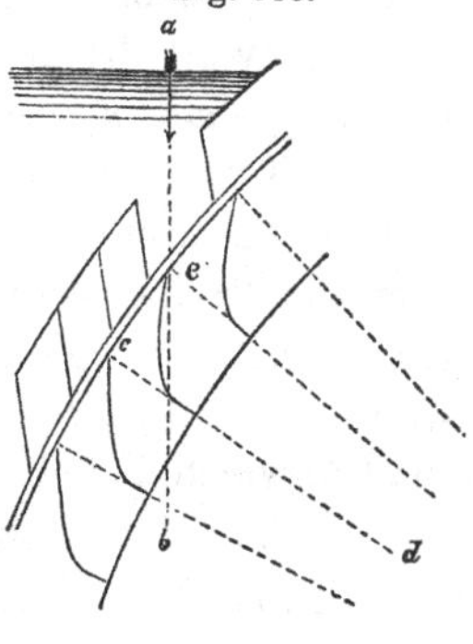

The Shuttle.—The shuttle of these wheels requires a slight notice. The front of the pentrough is of cast-iron, in the form of an arc closely fitting the periphery of the wheel, with an opening extending from side to side for the passage of the water to the buckets. This opening is made of such a breadth and is placed in such a position that when the water in the pentrough is highest it will flow upon the wheel near the top, and when the water is lowest it will still be able to enter the buckets near the bottom. This opening is then fitted with inclined guide plates, arranged so as to prevent the water in entering striking against the sole plate or the back of the succeeding bucket. Over the guide plates is a door, or closely fitting sluice, which slides up or down, according to the height of the water in the pentrough, so as to admit a thin sheet of water flowing over its upper edge through the guide plates into the buckets of the wheel. By this arrangement it will be seen that the water is always drawn off at its highest level and the fall economised to the utmost extent. Racks are fitted to the back of the sluice with pinions, by which its position is altered, and the quantity of water flowing on the wheel adjusted.

In the Catrine wheel, Plates I. and II., the pentrough consists of cast-iron plates bolted together and resting on beams supported on one side by the wall of the wheel-house and on the other on columns.

Figs. 119 and 120 represent the water wheel governor, a very ingenious arrangement, similar in principle to that of the steam-engine, but adapted in its details to a different purpose. It consists of two heavy balls which in revolving take a position further apart or nearer together, according to the velocity at which they are driven. These balls are swung upon the vertical revolving shaft *s s* supported in the strong cast-iron framing A A A. Two cast-iron brackets B B on either side of the frame, and bolted to it, support between them a bridge C C, passing over the driving shaft and clutch box, on which the shaft *s s* rests in a foot step. This vertical shaft is driven by the bevel wheels F and G, the former of which is keyed on the driving shaft *b*, which is hollow, to allow the shaft *a a* connected with the gearing of the sluice to pass through it. A third bevel wheel H, is also placed on a hollow shaft, and is driven by the

Fig. 119.

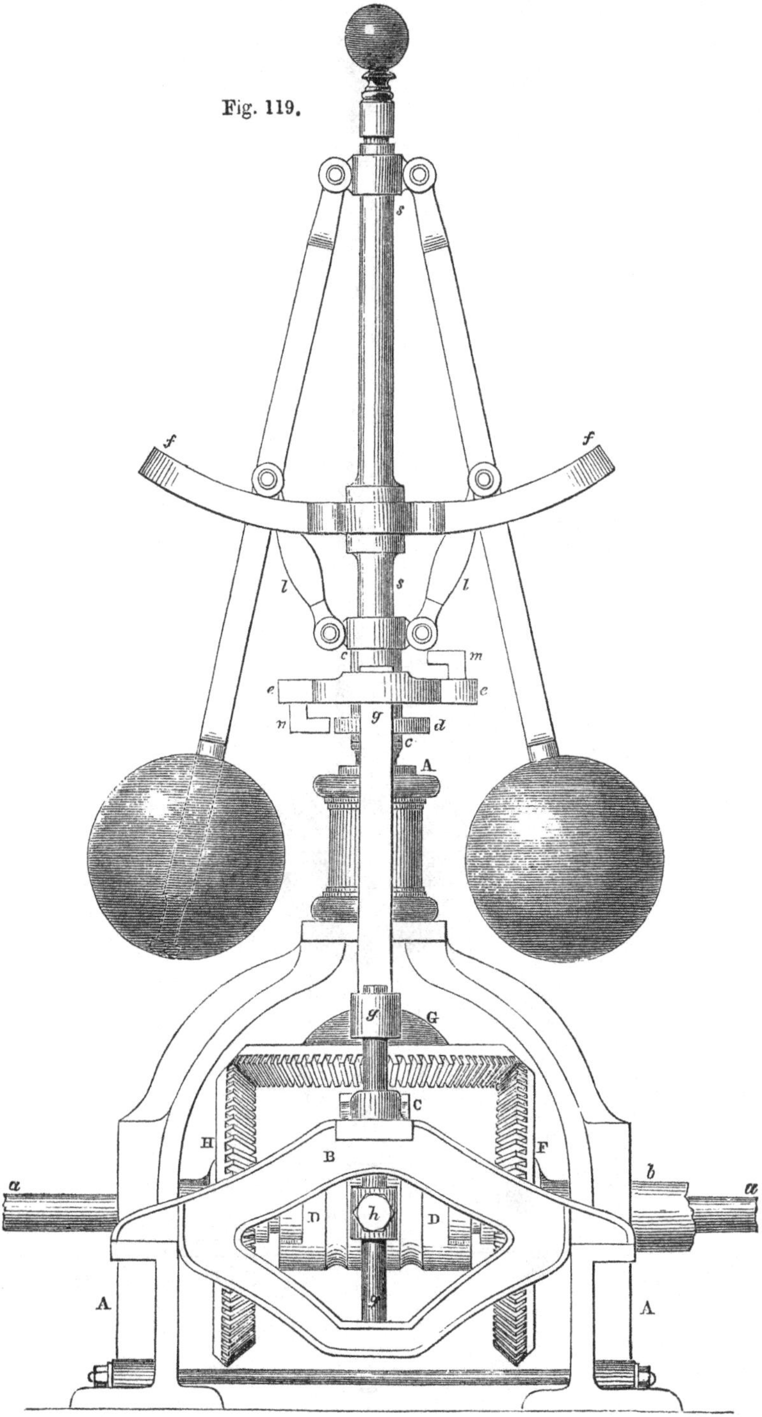

Fig. 120.

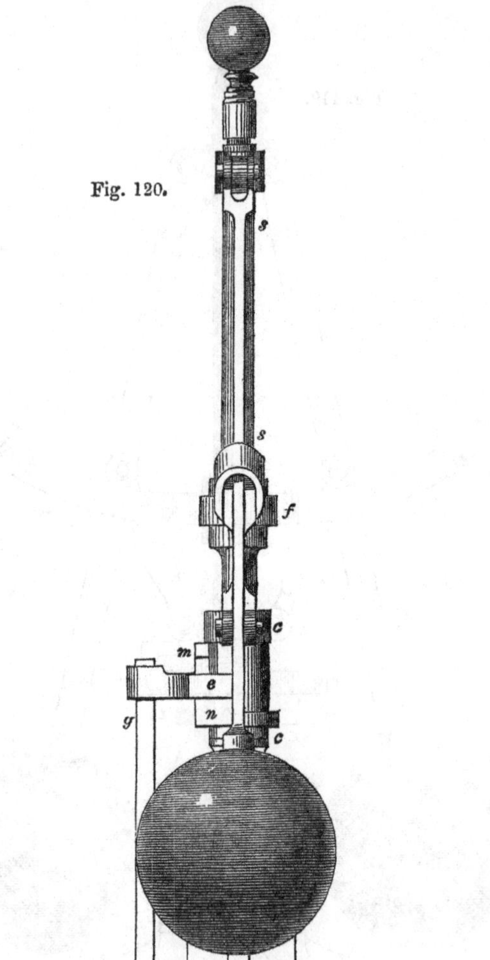

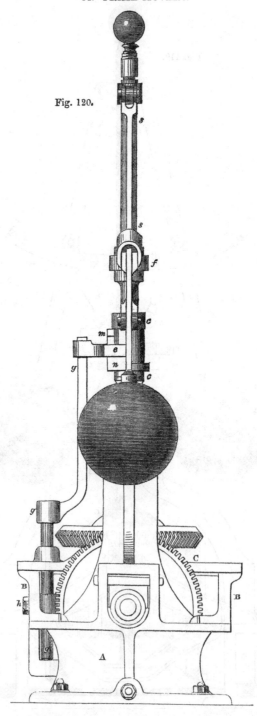

bevel wheel G, revolving of course in an opposite direction to the first wheel F.

The governor balls with the radial arms and slotted arcs *f f* are of the construction usual in steam-engines, but the links *l l* carry a brass slide *c c*, so that, as the governor balls diverge, this slide is drawn up along the vertical shaft, and as they approach it falls. On the slide is fixed the eccentric cam *d*, shown also in fig. 122 as seen in plan. This cam of course revolves simultaneously with the slide, balls, and vertical shaft *s s*. Attached to the bracket B on one side of the framing is a bent lever *g g g* carrying at its upper extremity a fork *e e*, and near the bottom a similar fork placed vertically, *h*, fig. 121. The upper

Fig. 121.

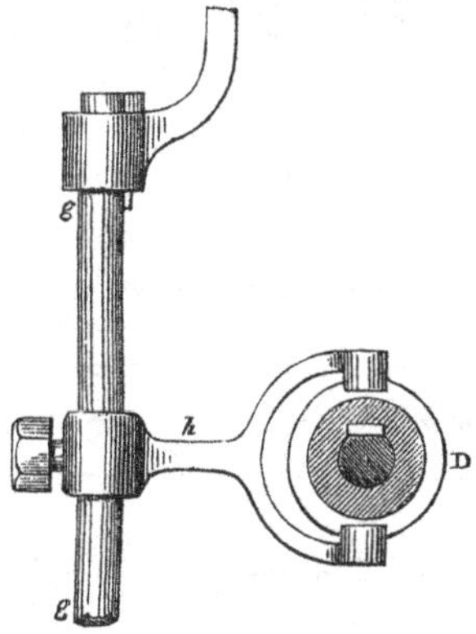

Fig. 122.

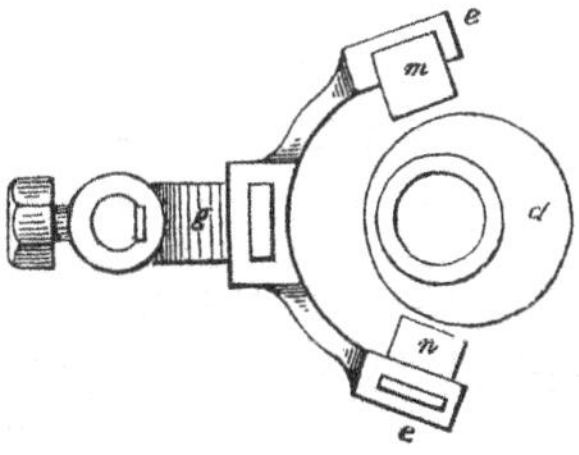

fork is moved by the revolving eccentric *d*, the lower fork moves a clutch box which slides backwards and forwards on the shaft *a a*, and engaging alternately with the bevel wheels H and F. When the motion of the wheel becomes too slow the balls fall and bring the cam *d* in contact with a knee of iron *n* in the upper fork *e e;* this causes the clutch D D to be thrown into gear with the bevel wheel H, and the clutch being keyed so as to slide on the shaft *a a*, causes that also to revolve and the sluice or shuttle to be lowered. On the contrary, when the motion of the wheel is too rapid, the balls diverge, the cam *d* is raised and strikes the upper knee *m*; the clutch is then thrown into gear with F; the shaft *a a* revolves in the opposite direction and causes the shuttle to be raised. At other

TABLE OF PROPORTIONS OF WATER WHEELS.

Diameter of shrouds. Ft. in.	Fall. Ft. in.	Depth of shrouds in inches.	No. of buckets.	Opening in buckets in inches.	Speed of periphery per second. Ft. ins.	Diameter of segments. Ft. in.	No. of Cogs.	Pitch in inches.	Breadth in inches.	Remarks.
60 4	56 4	14	144	3¾	4 0	59 11¾	912	2½	10	External spur.
50 0	48 0	15	120	6	4 1	48 6	560	3¼	15	
46 0	42 0	10	144	4½	4 0	44 6	560	3	10	
46 0	41 10	13	120	4½	4 1	44 7	480	3½	12	
40 0	36 0	18	80	6½	4 2	37 9	432	3¼	14	
39 9	35 4	12	112	5	4 0	39 4	592	2½	8	External spur.
40 0	36 2	10	112	4¾	4 3	38 5¼	480	3	9	
40 0	36 0	18	80	6½	4 2	37 9	432	3¼	14	
36 0	32 0	18	80	7	4 2	33 8½	384	3¼	15	
36 0	31 8	16	80	7	4 4	33 6¼	384	3¼	14	
32 10	28 3	13	96	5½	4 4	30 7½	384	3	9	
30 0	27 0	12	96	5½	4 4	28 9	360	3	10	
30 0	24 0	18	72	7	4 6	28 2	324	3¼	14	
30 6	27 0	15	84	6	4 4	29 0½	364	3	12	
28 6	23 0	18	60	7½	4 6	26 2	300	3¼	13	
28 0	24 9	22	60	7½	4 5	25 1	216	4⅜	14	
26 0	22 8	10	70	5	4 6					
26 0	24 0	18	60	7	4 2	23 8	270	3¼	12	
26 3	22 0	14	70	6	4 4	24 8	310	3	10	
24 0	20 0	14	70	6	4 6	22 4	280	3	10	
24 0	21 0	14	60	6	4 6	22 0	276	3	10	
24 0	20 2	20	48	8½	4 4	22 0	252	3¼	14	
22 9	19 6	8	80	3¾	4 6	21 6½	390	2⅛	6	
22 6	18 10	14	60	6	4 0					
22 3	20 0	14	70	6	4 4					
22 0	20 0	10	70	4¼	4 6	21 10	330	2½	6½	External spur.
22 0	18 3	19	48	7	4 8					
22 0	19 6	19	48	8	4 10	20 3	232	3¼	12	
21 0	17 5	16	50	7	4 7	20 0	250	3	9	External spur.
20 0	17 0	20	48	7½	4 6					

20 0	16 0	22	40	8	4 4	17 10½	224	3	12	
20 0	17 6	14	56	5½	4 6	18 6¼	232	3	9	
19 9	17 0	10	60	4½	4 7	18 6½	280	2½	8	
19 0	16 4	12	48	5	4 10	17 7⅜	264	2½	10	
19 0	16 0	22	42	10	5 6	16 1⅞	174	3½	13	
18 0¼	15 0	15	48	6	4 9					
18 0	14 0	20	32	13	5 4					Ventilated.
18 0	15 0	20	32	11½	5 10	16 0	200	3	14	Ventilated.
18 0	15 9	12	48	5½	4 3	16 6	248	2½	8	
18 0	13 4	22½	32	21	6 0					Ventilated.
18 8	15 0	20	40	15½	5 6	16 9	192	3¼	14	Ventilated.
16 6	13 0	24	36	12½	5 6	12 3	132	3½	10	Ventilated.
16 0	10 6	18	40	15	5 6	15 4¼	192	3	12	Floats, ventilated, external spur.
16 0	12 8	15	40	9	4 10					
16 0	14 0	16	40	7¼	5 0	14 0¾	176 •	3	12	
16 0	12 0	24	32	16½	5 8					Ventilated.
16 0	11 4	21	32	17	6 0	14 0¾	176	3	9	Ventilated.
16 0	12 8	28	32	17	6 4					Ventilated.
16 0	11 9	24	32	15	5 10					
15 6	12 0	20	40	10½	5 8					Ventilated.
15 0	12 0	16	32	10½	5 8	14 6	168	3¼	10	
15 0	13 2	12	40	6	4 8					
15 0	13 0	18	36	7	5 3	13 10	174	3	10	Ventilated.
14 0	10 0	15	36	9	5 6					
14 0	8 6	15	36	9¼	5 10					Ventilated.
14 0	11 8	24	36	4¼	5 2					Ventilated.
12 6	10 0	10	36	4½	5 0					
12 1	8 3	12	30	6	5 1	11 6	174	2½	7	
12 0	8 0	15	36	6¼	5 0	10 10	162	2½	7	
12 0	10 4	8	36	4	4 8	11 3½	282	1½	4½	
10 0	8 0	8	36	3½	4 6					
7 0	5 0	6	32	2⅛	5 0	6 5½	248	1	2½	

times, when the motion does not require adjustment, the clutch is disengaged from both wheels and the whole of the winding apparatus is stationary.

This arrangement of governor is exceedingly compact and effective and a great improvement on the original condition in which I first found it, with rollers and reversing pulleys. It is free from the objection to which those governors are open which directly bring the sluice gearing into operation and retain it so by their momentum.

As examples of the speed at which this part of the machinery is worked, I subjoin a few examples that are working successfully:—

Governor shaft, . . 36 revolutions per minute.

Rack shaft, from 0·0314 to 0·058 revolutions per min.

There is usually a worm on the shaft *a a*, working into a wheel on a cross shaft; on the cross shaft a second worm working into a wheel on the rack shaft; and a small pinion 8 inches in diameter on the rack shaft gears into the rack upon the sluice. This rack should be jointed to the sluice at the middle, and should be of such a length that the rack shaft and pinion can be placed out of water above the pentrough. But the details of the gearing and shafting by which the motion of the governor is transmitted to the sluice vary with the position of the governor and the circumstances of each particular locality, and they must therefore be left to the millwright's own judgment. Only it is important to observe that the motion of the sluice should in every case be slow, as in the above examples, or the acceleration or retardation in the supply of water will cause an irregular motion first faster and then slower in the wheel, conditions inadmissible where machinery is employed.

In designing a water wheel the first important consideration is the height of the fall; this taken in conjunction with the intended outlay will fix the diameter of the wheel. We must next determine the form of bucket as already detailed. Then the quantity of water per second in cubic feet must be ascertained, and this will determine the necessary capacity of the bucket and the consequent breadth of the wheel. Here we have to consider also, 1st, that the bucket is not to be more than $\frac{1}{3}$ or $\frac{1}{2}$ filled; and, 2nd, the rate of revolution of the wheel which determines the number of buckets passing the shuttle per second. (p. 137.)

Suppose a wheel, having 5 feet peripheral velocity per second, supplied with 3000 cubic feet of water per minute, and the breadth of which has to be determined so that the buckets shall be only one-half filled:—

Let depth of shroud = . . .	14 inches.
distance between buckets = . .	14 inches.
section of water in bucket when full, at the pentrough = . . .	144 sq. ins.

Here five buckets pass the sluice per second, and each must contain $\frac{3000}{5 \times 60} = 10$ cubic feet of water per second; but they are to be only one-half filled when containing this quantity of water, hence their capacity must be 20 cubic feet. Their sectional area is 1 square foot, and hence 20 feet is the breadth necessary for the wheel.

The table on pp. 144, 145, of the proportions of water wheels which I have constructed, may afford aid to the engineer and millwright designing wheels, in their adaptation to different heights of falls, quantity of water, &c.

CHAP. IV.

ON THE UNDERSHOT WATER WHEEL.

BEFORE the introduction of iron, undershot water wheels were frequently employed, and were in almost every instance constructed with straight radial floats, as in the annexed sketch,

Fig. 123.

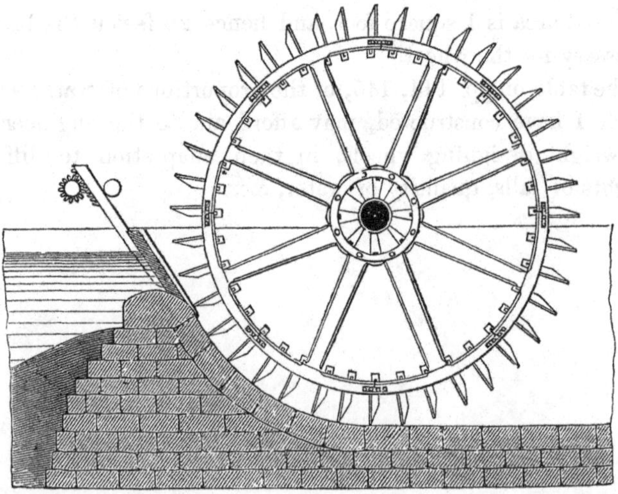

the water being discharged against the float-boards, as it rushed with considerable velocity underneath the shuttle. This was the invariable practice down to Smeaton's time even, the principle being to employ the impulse of the fluid stream, and not its gravity or weight. Indeed, there appeared to be an impression that this was the more effective and economical mode of application, and probably arose out of the circumstances of the original employment of water as a moving power. The earliest wheels of which we read are undershot wheels placed between two boats in a flowing stream, and driven by its impulse, and in Smeaton's own time the works for the supply of water to London obtained their power from some magnificent examples of precisely similar wheels, placed in the tidal stream rushing between the clumsy piers of the old London Bridge. In the old time it was no doubt an advantage to have the prime mover working at a considerable velocity, and an overshot wheel will not do this effectively. Hence wheels were sometimes built of the form shown in fig. 124, the water being carried down from the top of the fall so as to strike the radial floats of the wheel at a very high velocity. Such a wheel is described in Smeaton's Reports.

Fig. 124.

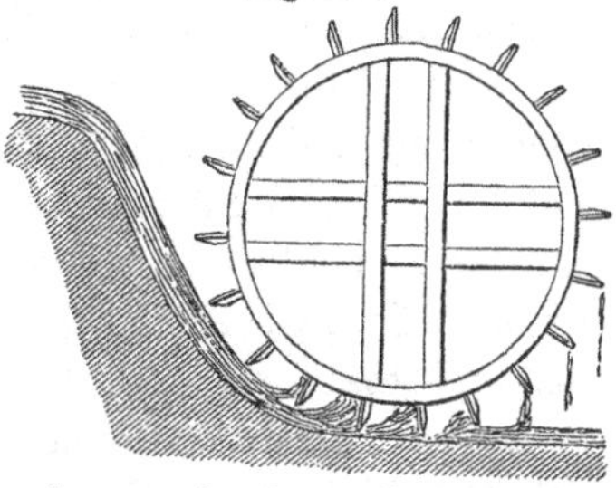

The earliest great advance in the perfecting of the water wheel was effected mainly by Smeaton, and we owe to him the first experimental inquiries on the effect and proper velocity and proportions of water wheels. In all the various applications of water, experimental researches have hitherto been the principal means of advance, and in no department has more labour and talent been expended in such inquiries; the result is, that our hydraulic machinery of the present day is as perfect, and yields as high a proportion of the power to the actual fall of water, as we can ever hope to obtain.

In my own practice I have been accustomed to employ water even for very low falls, solely by gravity, using the arrangement already described, as a low breast wheel, when treating of ventilation, and which is shown in detail in Plate III. This wheel is 16 feet in diameter, 17 feet 6 inches between the shrouds, and is adapted to a fall varying from 5 to 8 feet, according to the condition of the river. The water flows into the wheel at its highest level, over a sliding sluice of precisely the same construction as in high breast wheels; it is retained in the buckets to the bottom of the fall, by the cast-iron and stone breast fitting accurately to the edge of the buckets. The advantages of this construction are manifest, as the water expends its full force on the wheel from the very top of the fall, the buckets being well ventilated, and having a curvature adapted to the position in which they receive the water. By these means, a greatly increased duty is obtained as compared with the wheels with radial floats acted upon by impulse or gravity, or by both. Besides, with this form of wheel, the spider, or suspension principle of construction, may be adopted, and the power taken off at once from an internal segmental spur-wheel, placed on one of the shrouds, and a high velocity at once obtained, independently of multiplying gear. The advantages of this form of construction in iron wheels are very great, and, when combined with an economical application of the water, they form a machine probably as effective as any which can be employed for falls of not less than 5 feet.

Radial float wheels, however, constructed of wood are still in use, and the most important directions in respect to these appear to be to make the depth of the floats large, as compared with the thickness of the lamina of water which strikes them; to place the sluice as close as practicable to the floats; to contract somewhat the aperture of the sluice, and to expand the tail-race immediately beyond the vertical plane passing through the axis, to allow the water escaping from the floats to diffuse itself in the tail-race, and pass freely away. These directions, with the following practical formula for fixing the diameter of the wheel, we have from the dissertation on water wheels in the Engineer and Machinist's Assistant.

Let u = the velocity of the extremity of the floats; N the number of turns desired per minute; h = fall in feet. Assume $u = 2{\cdot}4 \sqrt{h}$ for a maximum effect, then the diameter expressed in terms of the velocity and height of fall will be $19{\cdot}1 \times \frac{2{\cdot}4 \sqrt{h}}{N} = \frac{46}{N} \sqrt{h}$ nearly. Thus supposing the height of fall $= h = 4$ feet; number of turns required per minute $= N = 8$; then the diameter $= \frac{46}{8} \sqrt{4} = 11\frac{1}{2}$ feet nearly. Twelve to twenty-five feet is the usual range of diameter for undershot wheels, and the same writer considers 12 to 16 feet to be the most effective; in my own practice, I have found from 14 to 18 feet perform the best duty. Feathering, or inclining the floats, does not appear to increase the useful effect.

The number of floats is usually equal to $\frac{4}{3}d + 12$, where d is the diameter in feet. The thickness of the vein of fluid striking the floats may be from 6 to 9 inches, and the depth of the floats from 18 inches to 2 feet.

M. Poncelet, one of the first authorities on Hydraulic Machines, and the first writer on Turbines, has contrived a very important modification of the undershot wheel, which has been used on the Continent with very good effect. A series of experiments led him to the conclusion that the floats should be curved instead of plane, and he deduced that for these wheels the velocity which gives a maximum effect was equal to 0·55 the velocity of the current, whilst it may vary from 0·5 to 0·6. He found the dynamic effect to vary from 50 to 60 per cent. of that of the water, being better for small falls with large openings at the bottom of the flood gate, and less for deep falls with small openings.

For describing the curve of Poncelet's floats, let $c\ c$ be the external circumference, and $a\ r$ the radius of the wheel; take $a\ b = \frac{1}{3}$ to $\frac{1}{4}$ the fall, and draw the inner circumference of the shrouding; let the water first strike the bucket at the point a and in the direction $d\ a$, draw $a\ e$ perpendicular to $d\ a$, so that the angle $e\ a\ r$ will be from 24° to 28°. Take on $a\ e$, $f\ g = \frac{1}{6}\ a\ f$, and from centre g, with radius $g\ a$, describe the curve of the float.

Fig. 125.

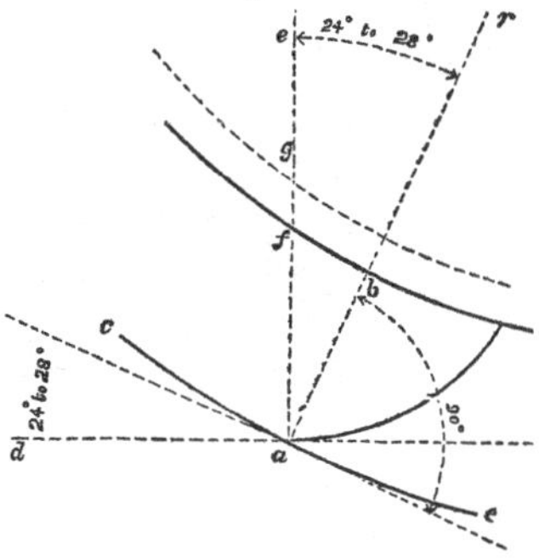

Fig. 125 represents a good example of Poncelet's wheel. The width of the opening should be contracted somewhat towards the wheel so as to assume the form of the fluid vein, and the bottom may be at first inclined $\frac{1}{10}$ to $\frac{1}{15}$, to give the water a greater impetus on the wheel, but over a breadth of 18 or 20 inches at the extremity it should be made to a curve very accurately fitting the periphery of the wheel.

So also the tail-race may be expanded in width and depth to keep the wheel clear of backwater. The buckets are made of wrought iron of the requisite curve, riveted to the shrouds on each side, and the sole plate is altogether dispensed with; as no resistance is opposed by the air, the buckets are made more numerous than in breast or undershot wheels, and as the wheel

Fig. 126.

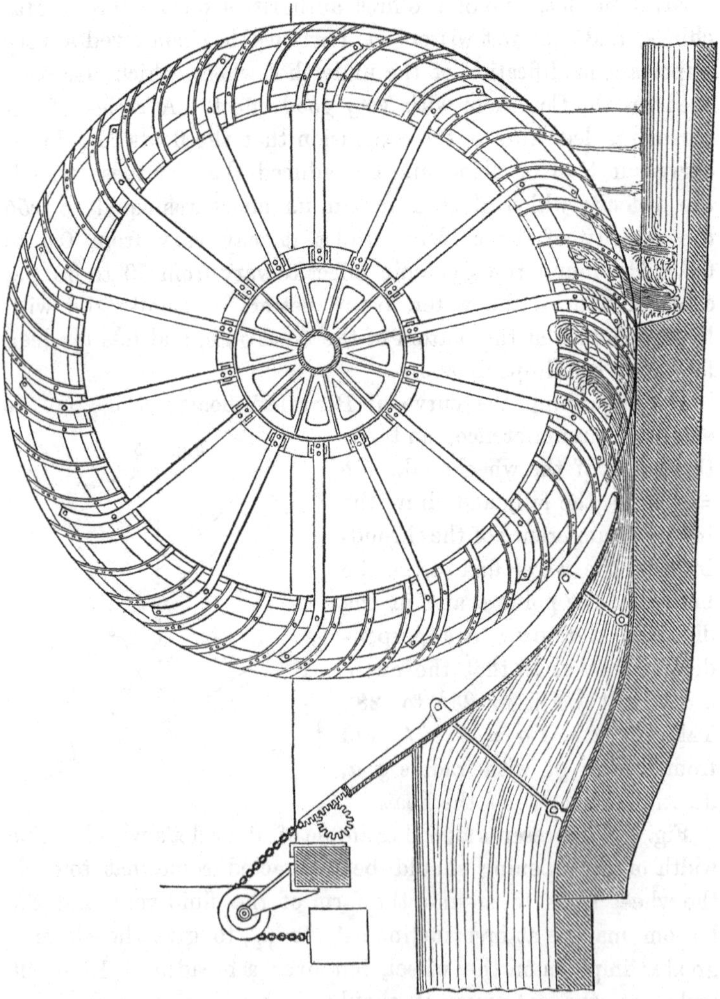

carries no weight of water, it may be made comparatively light. For the number of buckets for wheels of from 10 to 20 feet in diameter, we may take

$$n = \frac{8}{5}d + 16$$

Thus for a wheel 15 feet in diameter,

$$n = \frac{8 \times 15}{5} + 16 = 40$$

The wheel shown in the figure is 16 feet 8 inches in diameter, and 30 feet wide, and is driven by a fall 6 feet 6 inches high, yielding 20,000 cubic feet per minute. With a circumferential velocity of 11 or 12 feet per second, it afforded 140 horses' power.

This wheel gives a useful effect of 50 to 60 per cent. of the water power employed when well constructed, and may be used with advantage for falls not greater than about 6 feet. Above this the low breast wheel is certainly more advantageous and costs less.

Poncelet made some experiments on wheels of this class, with the friction break. The wheel was 11 feet diameter, 28 inches wide, and with 30 floats. He found the efficiency equal to 52 per cent. when the ratio of the velocity of the wheel to the water was 0·52. Morin has also experimented on these wheels, and for falls of from 3 to 4½ feet, with sluice openings of 6, 8, 10, and 11 inches, he found the efficiency 52, 57, 60, and 62 per cent. respectively.*

* In a conversation with General Poncelet on this subject I found that the wheel which bears his name gives a duty of nearly 60 per cent. of the water employed. This is about the same as my own wheel with ventilated buckets for low falls, where the sole is entirely dispensed with. There is, however, this difference, namely, that in the Poncelet wheel the water is discharged upon the floats from *under* the sluice, whereas, in that of the ventilated wheel, it is discharged into buckets *over* the sluice from the upper surface of the fall.

CHAP. V.

ON TURBINES.

It will be impossible in the present work to enter into details on the theory and construction of the immense variety of prime-movers known under the name of turbines, the development of the principles of which we owe chiefly to continental mathematicians. Two varieties of horizontal wheels or turbines have long been employed on the Continent, which although ill-devised and ineffective, yet presented evident advantages in their small size, cheapness, and simplicity of construction. These are known in France as *roues à cuves* and *rouets volants*, the former being a small wheel revolving on a vertical axis, and having inclined curved vanes or buckets arranged radially. It is placed in a pit so that the water passing vertically through it should act by pressure and reaction on the buckets. The *rouet volant* differs from this in having the water applied to the wheel at a small part only of the periphery, so as to drive the wheel by impulse. These wheels of from 3 to 5 feet in diameter with nine to twelve buckets are usually made of cast-iron, and fixed upon a lever foot bridge, so that they can be slightly raised or depressed. The running millstone is fixed on the upper extremity of the vertical axis, so as to obviate the use of any gearing or belting. In regard to efficiency, the *roues à cuves* yield about 27 per cent. and the *rouets volants* about 30 to 40 per cent. of the water used.

General Poncelet was the first to demonstrate the principle and superior advantages of the turbine, and in 1827 M. Fourneyron recalled public attention in France very forcibly to the construction of the horizontal wheels by a turbine very happily conceived and executed. For this invention he received in 1833 a prize of 6000 francs; and the principles of his machine have been investigated, and its superiority proved, by the ablest continental experimenters on hydraulics. In its present form it is equal in efficiency to the best hydraulic machines, and in many circumstances is very advantageously employed. Since then the manufacture of these turbines in countries where water power is much depended upon has assumed considerable importance, and very numerous modifications of its form and construction have been adopted.

1. *Turbines in which the water passes vertically through the wheel.*

Wheels of this class are composed of two annular cylinders, the upper fixed and the lower revolving on a vertical axis. The upper is fitted with guides to direct the water most effectively against similar curved vanes or buckets, turned in the opposite direction, in the lower wheel. The water passes from the reservoir or cistern placed over the upper cylinder, vertically downwards, acting on the revolving wheel by pressure as it glides over the surface of the vanes.

Fig. 127.

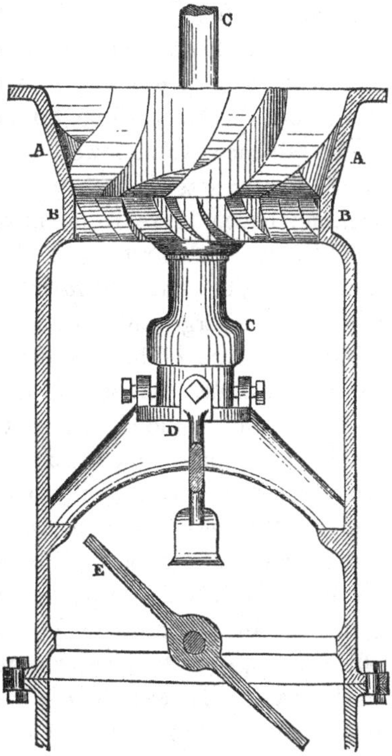

Burdin about 1826 invented a turbine of this description (*turbine à evacuation alternative*), the efficiency of which was as much as 67 per cent. of the water power expended.

Fig. 127 represents Feu Jonval's turbine (known also as the Koechlin turbine). The fixed wheel is shown at A A, the revolving wheel at B B. The wheels consist of cast-iron rims having wrought-iron guides grooved and riveted to them. The running wheel is keyed on the shaft C C, which is supported on a step D, firmly fixed by screws on the cast-iron bridge attached to the cylinder forming the tail-race. The regulation of the water is effected partly by a valve E resembling the throttle-valve of a steam-engine and placed beneath the wheel, or in some cases by a sluice at the opening of the conduit into the tail-race. This method, when much variation of power is required, reduces the efficiency of the wheel, but it has the merit of great simplicity and facility of construction. In the construction represented the vane carries outside the cylinder in which it is placed a wheel, acted on by a worm from a hand-wheel placed at any convenient point above the upper cistern. There are also employed movable divisions by which part of the inner periphery of the revolving wheel is enclosed, and the water passes through a narrower annular aperture on the external periphery. This arrangement is said to have operated effectively in America, so that a wheel giving 60 H. P. in wet seasons can work at 40 H. P. in dry seasons, without losing more than 15 or 16 per cent. of its efficiency.

These wheels are placed in an air-tight cylinder, for low falls at a depth of 4 to 6 feet below the surface, and for high falls at a distance not exceeding 30 feet above the level of the water in the tail-race, when lowest; so that in the upper part of the fall the water acts by pressure, in the part below the wheel by suction; hence, there is no inconvenience from backwater beyond the inevitable reduction of fall, and the waste water may, if necessary, be conveyed in an air-tight pipe to any convenient point of discharge, only taking care that its mouth be under water. In case of break down the wheel is very easily rendered accessible. These wheels are said to yield 75 per cent. of the power expended on falls above 12 feet.

Fig. 128 represents part of a similar turbine by M. Fromont, which received the Council Medal at the Great Exhibition of 1851. It differs from the last in the method of regulating the water, and is known as M. Fontaine Baron's turbine. A number of sluices *s s* are suspended in the fixed wheel by wrought-iron rods, and are raised or lowered simultaneously by means of wheel-work, so as to open or contract the orifices for the passage of the water. In awarding a medal to this turbine, the jury made the following remarks on its merits:—" 1st. It occupies a small space; 2nd. Turning very rapidly, it may, when used for grinding flour, be made to communicate the motion directly to the millstones; 3rd. It works equally well under great and

Fig. 128.

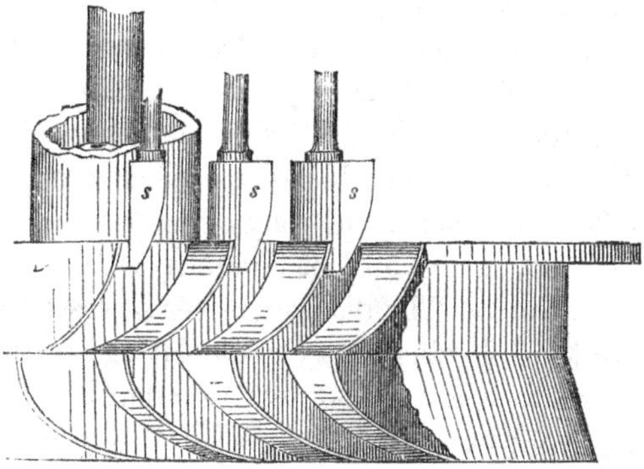

small falls of water; 4th. It yields, when properly constructed, and with the supply of water for which it was constructed, a useful effect of 68 to 70 per cent., being an efficiency as high as any other hydraulic machine; 5th. The same wheel may be made to work at very different velocities, without materially altering its useful effect."—*Reports of Juries.*

Fig 129.

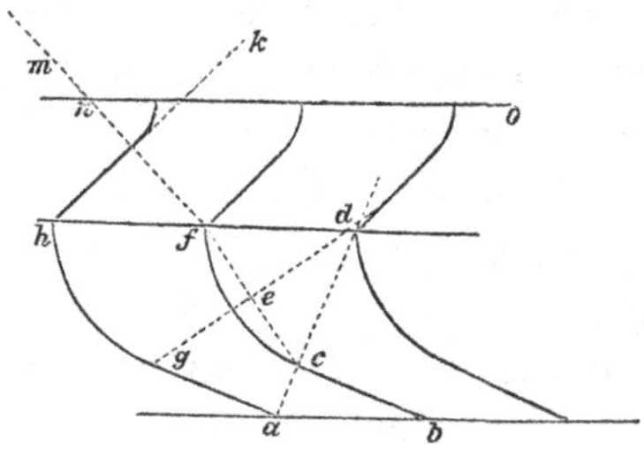

In designing a wheel of this description, we must take a distance $a\ b$ equal to the distance between the floats, or $\frac{6{\cdot}29 \times \text{radius}}{\text{No. of vanes}}$. Take the angle $a\ b\ c = 15°$ to $20°$, and draw $a c$ perpendicular to $c\ b$. Lay off $d\ c\ f$ equal to $\frac{\delta+\beta}{2}$ where δ is the angle $a\ b\ c$, and β is taken arbitrarily equal to 100° to 110°. Bisect $c\ f$ in e, and through e draw $g\ d$ perpendicular to $c\ f$, and cutting $c\ d$ in d. From d with radius $d\ c$, or $d\ f$, draw the arc to which $c\ b$ will be a tangent. For the guides, take the angle $f\ h\ k = a$, so that

$$\text{Cot}\ a = \text{Cot}\ \beta + \frac{1}{\sin \delta}$$

draw $f\ m$ perpendicular to $h\ k$, cutting the top of the guide wheel $n\,o$ in n; from n draw arc, touching $h\,k$. These directions are from Weisbach.

In America the Koechlin turbine has been experimented upon by the Franklin Institute, with the following results:—The turbine experimented upon is intended to afford 7 horse power under a fall of 10 feet. It is $21\frac{1}{4}$ inches in diameter, $3\frac{1}{2}$ inches deep, and intended to make 190 revolutions per minute, giving $63\frac{1}{3}$ revolutions of a horizontal shaft to which it is geared 3 to 1. To this shaft was attached a Prony dynamometer, whose lever was 7·96 feet long, giving 50 feet circumference.

Experiment No. 1.—The discharge over a waste board in the tail-race gave the following data for calculating the discharge:—L = width of waste board = 3·83 feet, h = depth of water on it, 0·74. Then $Q = \cdot 383 \times 3\cdot 83 \times \cdot 74 \sqrt{64 \times \cdot 74} = 7\cdot 468$ cubic feet per second. Hence the theoretical power $= 7\cdot 468 \times 62\cdot 5 \times 9\cdot 34 \times 60 = 261{,}537$ foot-pounds per minute, $= 7\cdot 92$ horses power.

It was found that at 63 revolutions per minute of the horizontal shaft 63 lbs. balanced the lever. Hence the power developed by the wheel was $63 \times 63 \times 50 = 198{,}450$ lbs. $= 6\cdot 014$ horses power.

Experiment 2.—The gates from the head race were so far closed as to reduce the head one foot, and maintain it at that level during the experiment. The depth of water on the waste board was $8\frac{1}{8}$ inches, and the fall 8·41 feet. $\therefore Q = 0\cdot 39 \times 3\cdot 83 \times \cdot 677 \sqrt{64 \times \cdot 677} = 6\cdot 66$ cubic feet per second. Hence theoretical power $= 6\cdot 66 \times 62\cdot 5 \times 8\cdot 41 \times 60 = 210{,}000$ foot-pounds per minute $= 6\cdot 36$ horses power.

It was found that 63 lbs. balanced the lever at 49 revolutions per minute of the shaft. Hence the power developed by the wheel was $49 \times 63 \times 50 = 164{,}350$ lbs. $= 4\cdot 98$ horses power.

The coefficients are then for No. 1, $\dfrac{6\cdot 014}{7\cdot 92} = 0\cdot 76$.

" " " " 2, $\dfrac{4\cdot 98}{6\cdot 66} = 0\cdot 78$.

And making allowance for leakage round the waste board, the experimenters conclude that the wheel yielded 75 per cent. of the power expended.

Another experiment on a 60-horse power turbine gave the following results:—

$$\left.\begin{array}{l}\text{Effective power } 56{\cdot}30\\ \text{Theoretical power } 63{\cdot}92\end{array}\right\} 0{\cdot}88.$$

Perhaps this very large coefficient is not quite reliable.

2. *Turbines in which the water flows horizontally and outwards.*

In turbines of this class the revolving wheel is placed outside of the fixed wheel, so that the water directed by guide plates on the inner wheel strikes the curved vanes of the outer wheel, and forces them round by pressure and reaction. The water is regulated by a cylindrical sluice fitting between the fixed and movable wheels.

M. Fourneyron's turbine is the chief example of this class. Its advantages, as stated in M. Poncelet's Report to the Academy of Sciences at Paris, are the high velocity at which it may be worked without reducing its useful effect, its small size, and lastly its capability of working equally well under backwater. From the experiments of M. Morin, the coefficient of useful effect appears to range from 0·60 to 0·80. On the other hand it has to the full the defects of this class of machines, requiring the utmost nicety of design and execution, and being very susceptible to injury, from small bodies carried into it by the water. It requires for its successful application both a large acquaintance with the principles of its construction and a considerable experience of its use: hence it will be unnecessary to do more in this place than select for illustration one of the most successful instances of their application.

Fig. 130 represents a vertical section, and fig. 131 a plan, of the celebrated turbine erected under M. Fourneyron's direction at St. Blazien for a fall of 354 feet. This small wheel, of only about 26 inches diameter, is employed in driving the machinery of a spinning factory of 8000 throstle-spindles, with the necessary preparing apparatus. In comparison with the work it has to perform, it is therefore of a size altogether unique.

The wheel consists of a cast-iron concave plate *t t*, keyed on the main axis *a a*; on this is fixed the annular wheel *s s*, consisting of an upper and lower plate of wrought iron, in which are fixed the 36 curved diaphrams seen in the plan, fig. 131. Opposite each of these curved plates on the outer revolving wheel, there is a similar guide on the inner fixed wheel *v v*, which are carried on a massive cast-iron plate attached to the

Fig. 130.

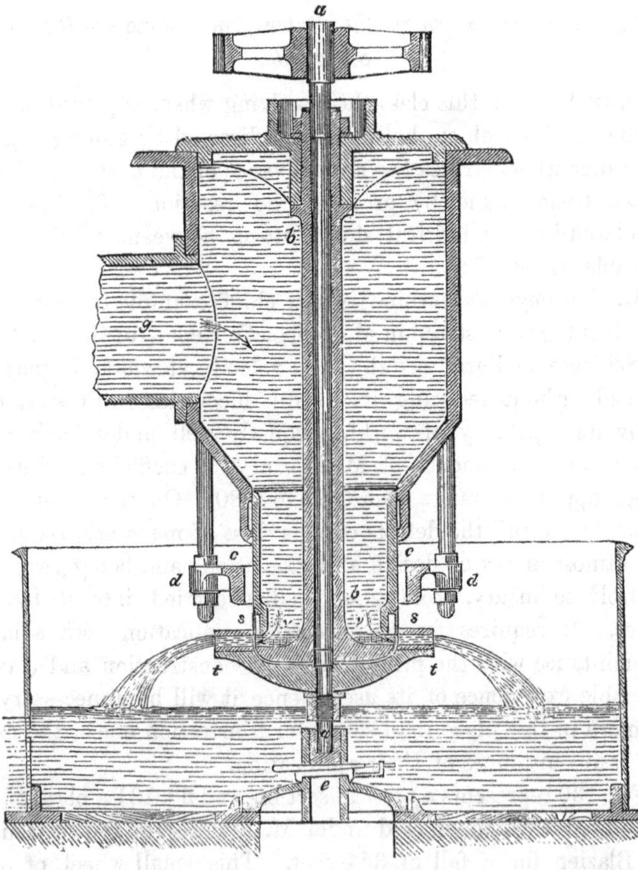

hollow tube *b b*, in which is placed the main axis. This plate not only sustains the guide plates, but takes off from the main axis the weight of the water, and thus reduces the friction on the foot step. The cylinder *c c* slides up and down in the larger water cistern, and forms a circular sluice between the revolving and fixed wheel, by which, within certain limits, the discharge of water and velocity of the turbine can be regulated. This sluice is raised or lowered by 4 rods, *d d*, which are screwed above into the eyes of 4 pinions (not shown). These pinions all gear into one larger wheel, and in this way the 4 rods may be raised or lowered simultaneously. The supply of water is brought to the cistern by a pipe *g* of 16½ inches

Fig. 131.

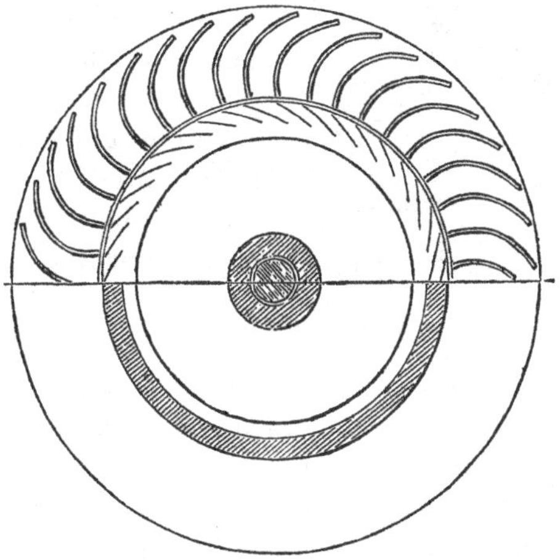

diameter, and 1200 feet in length. The spindle works on a steel pivot in a footstep adjusted by gibs and cotters *e*. This turbine makes from 2200 to 2300 revolutions per minute.

Another form of turbine, in which the flow is horizontally outwards, has been made to some extent in this country by Messrs. Whitelaw and Stirrat, and is sometimes called the Scotch turbine or reaction wheel. It is precisely on the principle of Barker's mill, and works by reaction. The principal improvement effected by Mr. Whitelaw is the form of the arms, which are curved in an archimedean spiral. Fig. 132 shows the method of striking these curves, the centre line of the arm being first drawn

Fig. 132.

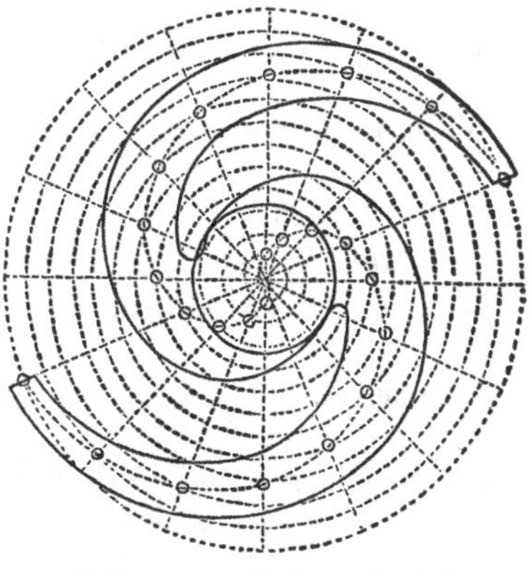

and half the breadth set off on each side of it, so that the capacity of the arm increases from the extremity towards the centre in the inverse ratio of the velocity at each point. Fig. 133 shows the arrangement of this wheel, the water being brought in a pipe A, curved at bottom, so as to enter the reaction wheel on the under side. The wheel is firmly stayed to prevent its rising from the upward pressure of the water, and carries directly the vertical first motion shaft. The most effective velocity at the extremity of the arms is said to be nearly

Fig. 133.

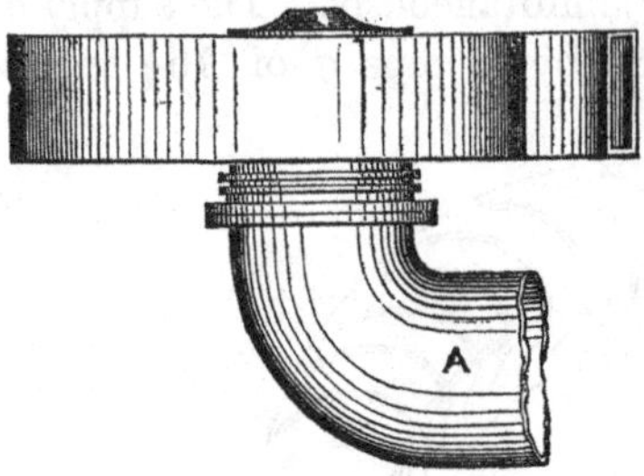

$$v = \sqrt{2gh} \ldots (1),$$

where v = velocity in feet per second, h = head of water in feet, and g is the accelerating force of gravity = 32·19.

The following are the rules which Mr. Whitelaw gives for proportioning this machine:—

Let Q be the number of cubic feet of water supplied per minute.

H, the height of the fall or head of water.

E, the useful work in units of horse-power.

$$\therefore \text{E} = \frac{\text{Q H}}{696{\cdot}73} \ldots (1).$$

Then for two properly formed jets:—

Width of each discharging orifice $= w_1 = \sqrt{\dfrac{135\,\text{E}}{1000\,\text{H}\sqrt{\text{H}}}}$.

Width of each arm of machine $= 4\,w_1 = w_2$.
Diameter of machine $= \text{D} = 50\,w_1$.
Diameter of central opening $= 10\,w_1$.

Number of revolutions in a minute $= \dfrac{149{\cdot}4338\sqrt{\text{H}}}{\text{D}}$.

In experiments with models the wheel is said to have realised from 74 to 77·8 per cent. of the available power.

3. *Turbines in which the water flows horizontally inwards; vortex wheels.*

We owe the invention of this class of turbines to one of my own pupils, Mr. James Thomson, C. E. of Belfast, and probably no turbines are more efficient or capable of more general application to every variety of fall than the vortex wheels which he has constructed. For this reason, and also because from their recent introduction they are less known than the varieties which have been longer in use, we shall illustrate them rather more fully with the aid of working drawings, supplied by Messrs. Williamson and Brothers of Kendal, who we believe have at present erected all which are employed in this country.

The peculiarity of these vortex wheels consists in the arrangement of the fixed guide blades on the outside of a circular chamber in which is placed the revolving wheel, so that the water flowing inwards strikes the curved plates of the revolving wheel tangentially, and leaves the wheel at the centre at a minimum velocity; the whirlpool created in the wheel chamber giving to this description of turbine its designation of vortex wheel.

Fig. 134 shows the general form of the guides and passages of a vortex wheel; *a, a* are the fixed guides, four in number, which direct the water tangentially into the passages of the wheel *b b;* after having done its work in these, the water leaves the wheel at the open passage at the centre *c; s* is the vertical shaft carrying the wheel and communicating its motion to the mill. The chamber in which the guide blades *a, a* are fixed, forms part of the supply chamber, and the supply of water to the wheel may be regulated by altering the position of the guide blades, and thus diminishing or increasing the area of opening between them. For this purpose the guide blades are fixed on

Fig. 134.

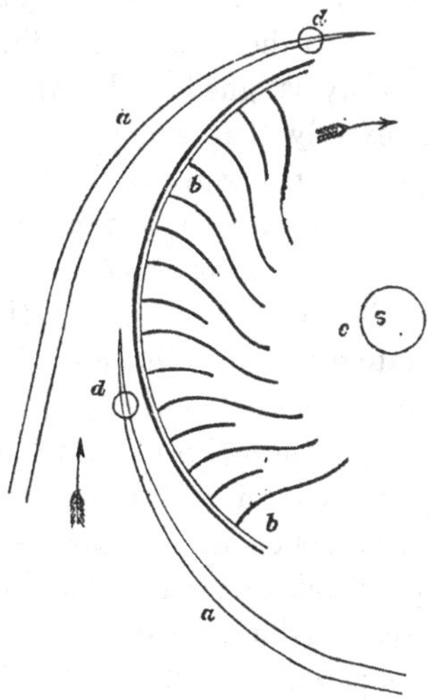

gudgeons *d*, *d*, near their extremities, and are connected by levers and links so that they may be shifted simultaneously by a spindle. The inner radius of the wheel is usually half the external radius, and the obliquity of the inner ends of the vanes 30° to 45°.

The general principles of these turbines Mr. Thomson thus explained at the meeting of the British Association in 1852:—"The velocity of the circumference is made the same as that of the entering water, and thus there is no impact between the water and the wheel; but, on the contrary, the water enters the radiating conduits of the wheel gently, that is to say, with scarcely any motion in relation to their mouths. In order to attain the equalisation of these velocities, it is necessary that the circumference of the wheel should move with the velocity which a heavy body would attain in falling through a vertical space equal to half the vertical fall of water, or, in other words, with a velocity due to half the fall, and that the orifices through which the water is injected into the wheel chamber should be conjointly of such area, that when all the water required is flowing through them it may also have a velocity due to half the fall. Thus one half only of the fall is employed in producing velocity in the water, and therefore the other half still remains acting on the water in the wheel chamber at the circumference of the wheel in the condition of fluid pressure. Now, with the velocity already assigned to the wheel, it is found that this fluid pressure is exactly that which is requisite to overcome the centrifugal force of the water in the wheel, and to bring the water to a state of rest at its exit, the mechanical work due to both halves of the fall being transferred to the wheel during the combined action of the moving water and the moving wheel. In the foregoing statements, the effects of fluid friction, and of some other modifying influences, are, for simplicity, left out of consideration; but in the practical application of the principles, the skill and judgment of the designer must be exercised in taking all such elements as far as possible into account. To aid in this, some practical rules, to which the author (Mr. Thomson) as yet closely adheres, were made out by him previously to the date of his patent. These are to be found in the specification of the patent, published in the *Mechanics' Magazine* for January 18 and January 25, 1851."

Fig. 135.

Elevation.

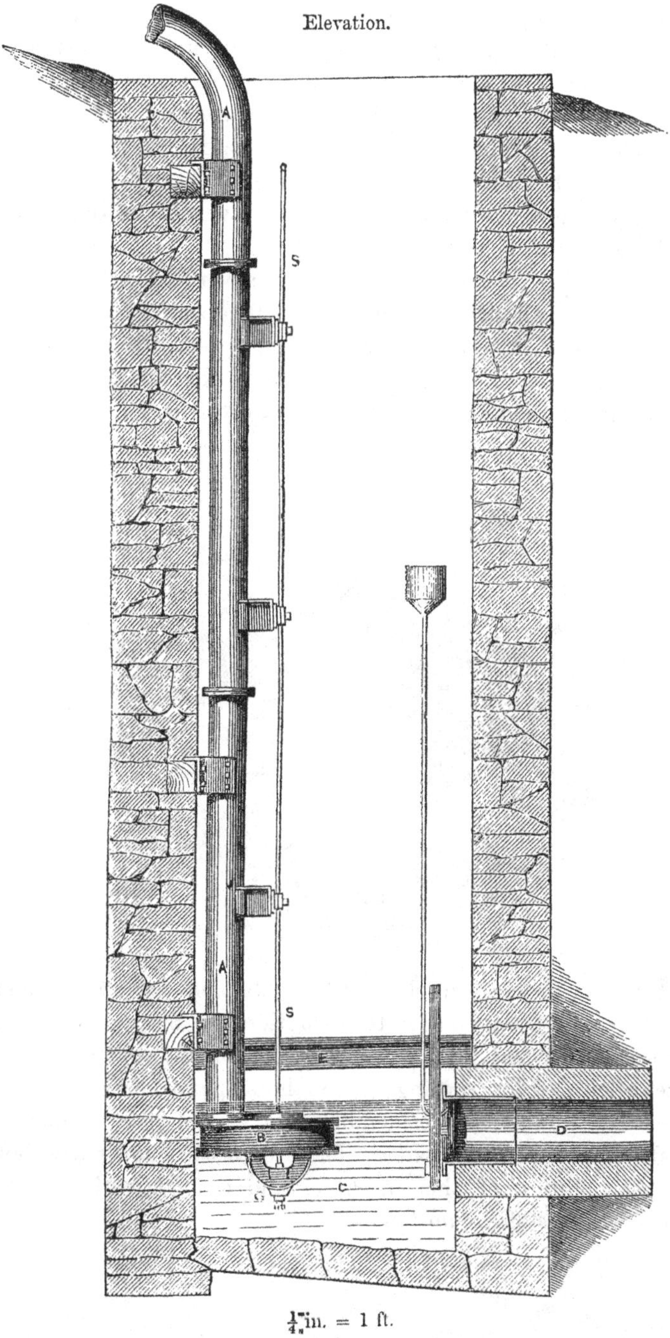

$\frac{1}{4}$ in. = 1 ft.

Mr. Thomson claims for his wheel the peculiar advantages (1.) that the injection passages are large and well formed. (2.) That it permits the employment of a most advantageous mode of regulating the power, by contracting the areas of the injection passages, without reducing the efficiency of the machine. (3.) That the maximum velocity of the water in the wheel does not exceed that due to *half* the fall. (4.) That the centrifugal action of the water tends to regulate the velocity of the wheels under a varying load.

In his paper, Mr. Thomson describes a vortex for a fall of 37 feet, and for an average supply of 540 cubic feet per minute, yielding 28 effective horse-power. The speed, 355 revolutions per minute; diameter, $22\frac{5}{8}$ inches; and extreme diameter of case, 4 feet 8 inches; also a low-pressure vortex for a fall of 7

Fig. 136.

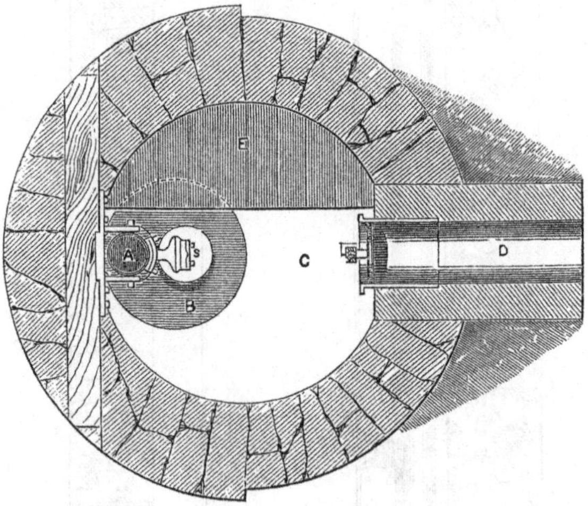

feet, for an average supply of 2,460 cubic feet per minute, and yielding 24 horse-power, at 48 revolutions per minute. Another he has constructed for a fall of 100 feet, and a fourth of large size, calculated for working at 150 horse-power, on a fall of 14 feet, and through a considerable part of the year submerged under 7 feet of back-water. These data will sufficiently show the capabilities of this machine, and its adaptation under great varieties of circumstances.

Figs. 135 and 136 exhibit an elevation and plan of a high-pressure vortex wheel, constructed by Messrs. Williamson and Brothers of Kendal. It is of 5 horse-power, on a 30 feet fall, and consumes 118 cubic feet per minute. The water is conveyed to the wheel in the 9-inch pipe A A, at a velocity of 4·4 feet per second. B is the supply chamber, or wheel case, fixed on masonry in the tail race C, from which the water passes away by the tunnel D. In the drawing the tunnel is shown closed, as is occasionally necessary, for access to the wheel or other purposes. E is a platform just above the ordinary level of the water; S S is the first motion shaft, to which the wheel is attached, and which is supported on the footstep at G, and by pedestals attached to the supply pipe A A.

Fig. 137 shows the wheel case in section. H H is the supply chamber or guide blade chamber, cast in parts and bolted

Fig. 137.

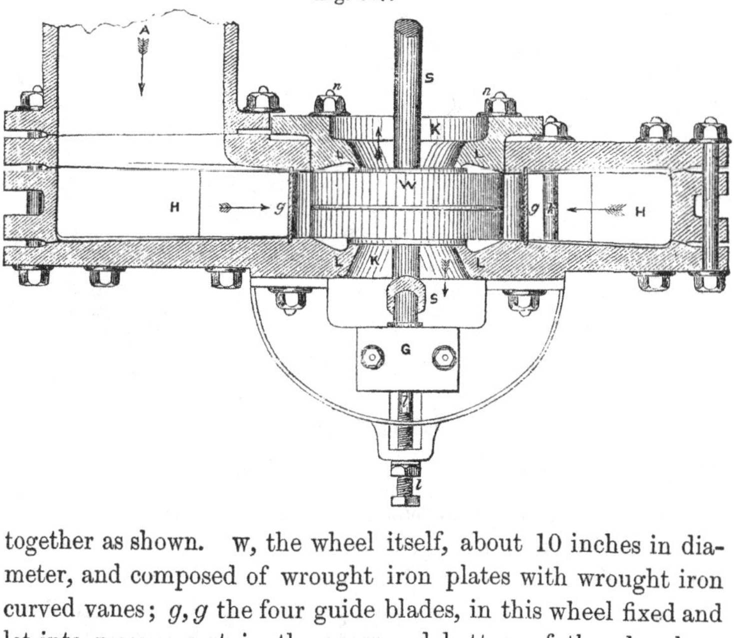

together as shown. W, the wheel itself, about 10 inches in diameter, and composed of wrought iron plates with wrought iron curved vanes; *g*, *g* the four guide blades, in this wheel fixed and let into grooves cast in the cover and bottom of the chamber; K, K, four bolts tying the cover and bottom of the supply chamber together to strengthen it against pressure. A the supply pipe as before, and K, K the openings in the centre of the wheel for the escape of the waste water after it has done its work on the wheel. The joint between the wheel and its case is made by means of the accurately fitting annular parts L, L, adjusted for the wheel to run without friction by bolts *n*, *n* in the upper piece. S S is the first motion shaft resting on a lignum vitæ pivot firmly fixed in the foot bridge G, which is bolted on below the supply chamber, the height of the pivot as it wears being adjusted by the screw *l l*. The pivot is lubricated by the water

Fig. 138.

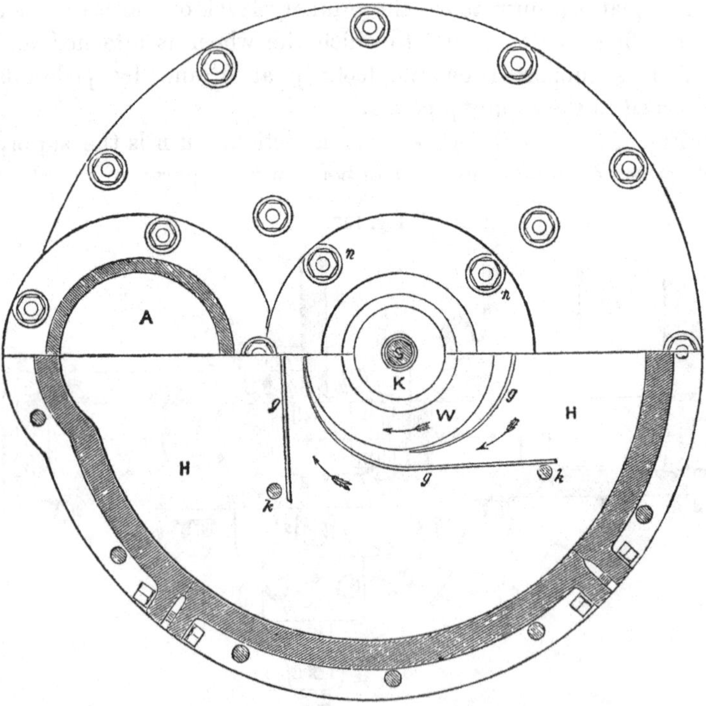

in which it works spread over it by a radial groove. In other cases Mr. Thomson makes the shaft to terminate in an inverted cup containing a concave brass disc working on a fixed steel pivot, with a radial groove for spreading the water. He does not consider the lubrication with oil so essential as other engineers insist, and believes the cases in which turbine pivots have been rapidly destroyed to be attributed to the absence of a proper provision for the escape of the air between the rubbing surfaces. Fig. 138 represents a half-plan and half-horizontal section of the same wheel, the same letters of reference being used as in fig. 137.

Fig. 139 exhibits the arrangement, in sectional elevation, of a low-pressure vortex wheel with its pentrough and tail race. This wheel is of 34 horse-power, with an effective fall of 14 feet 3 inches, and a supply of 1680 cubic feet per minute. The

Fig. 139.

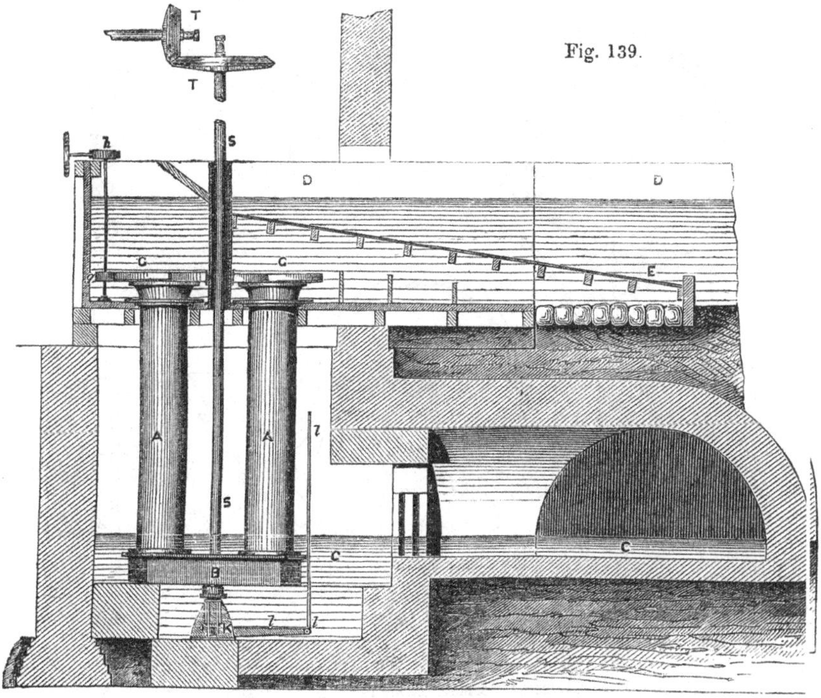

wheel is 52 inches in diameter, and makes 94 revolutions per minute. A, A are the four supply pipes, two feet in diameter, so that the water in them has a velocity of about 2·3 feet per second. B is the square supply chamber, C C the tail race, and D D the conduit and pentrough. The water as it arrives passes through a perforated metal strainer E E, to prevent the choking of the narrow passages of the wheel by floating leaves, &c. Over the four supply pipes A, A, is fixed a circular cast iron plate G G, with four holes corresponding in one position with the trumpet mouths of the supply pipes. On the edge of this plate is a rack into which the pinion *g* gears, so that by moving the worm and wheel *h* the sluice plate G G may be revolved and the entrance for the admission of the water to the wheel more or less closed or opened. This is an effective and inexpensive means of regulating the power of the wheel, where the supply of water is abundant and it is not necessary to economise its

Fig. 140.

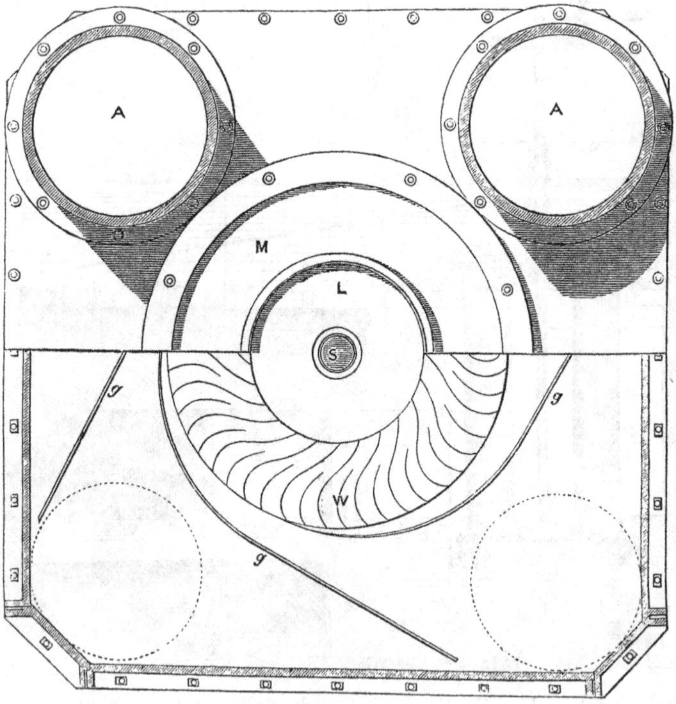

Scale ½ in. = 1 ft.

expenditure to the utmost extent. S S is the first motion shaft, and T, T the bevel wheels by which it gives off the power of the wheel to the mill. K is the footbridge carrying a step which can be raised by the lever *l l* as it wears away. Fig. 140 exhibits a half-plan and half-section of the wheel and supply chamber. A, A, as before, supply pipes, W, the wheel itself, *g*, *g*, *g*, *fixed* guide blades, the regulation of the wheel being effected by the sluice as before described. S, first motion shaft, L central opening for the escape of the water, M wheel cover, forming at its inner periphery a close and accurate joint with the revolving wheel.

Another plan which has been adopted with these wheels for regulating the speed, when they are applied to high falls, is to bring the supply pipe when near the wheel into a horizontal direction, fitting to it an ordinary sluice such as is used in high-pressure mains. A ten horse-power turbine, on a fall of 80 feet, has been erected on this principle by Messrs. Williamson and Brothers in Yorkshire. The wheel is only 13 inches in diameter, and consumes 44 cubic feet of water per minute, which is brought a distance of 340 yards, in a 9-inch pipe, the pen-trough and strainer being placed at the upper end, and the sluice at the bottom, close to the wheel.

But beyond question the most economical arrangement for regulating the expenditure of water, although somewhat more complicated in its details, is the adjustment of the guide blades themselves in the manner already alluded to. Fig. 141 shows a plan of a turbine, arranged with *movable* guide blades. A, A, two supply pipes, B, wheel cover; C, C, C, C, bell cranks connected together by links. The whole of these bell cranks are worked by a vertical spindle, D, and worm and wheel in the mill; they carry in the supply chamber links shown by the dotted lines, by which the guide blades *g*, *g*, *g*, *g*, movable on centres at *h*, *h*, *h*, *h*, can be opened or closed.

Fig. 141.

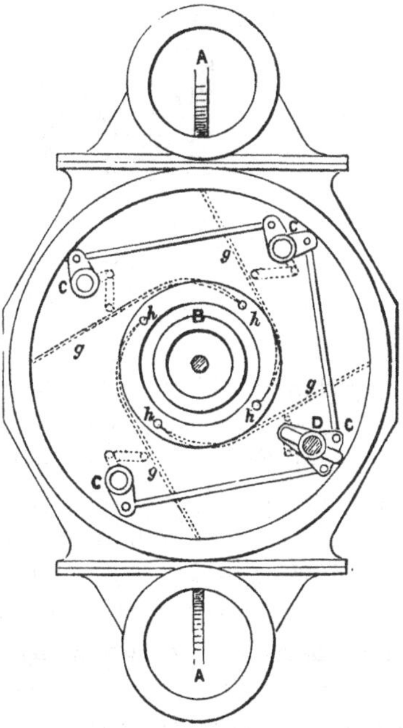

These turbines yield 75 per cent. of the power expended, and are therefore as efficient as the best water wheels or turbines. They work equally well under backwater, and if it be necessary they can be placed at any height less than 30 feet above the water in the tail race, the lower part of the fall being made to do its work on the wheel by suction in pipes descending from the central discharge orifice, and terminating in the water of the tail race.

In America, turbines of various kinds have come into extensive use, and some erected there are of unprecedentedly large size. The better forms have been copied in their main features, from European machines already described, with some variation in the constructive details. Thus Mr. Boyden has introduced a diffuser, or annular mouth-piece, round the outer or revolving wheel of the Fourneyron turbine, instead of permitting the water to escape into the free space of the tail water. He has also, to avoid the difficulties arising from the rapid wear of the

Fig. 142.

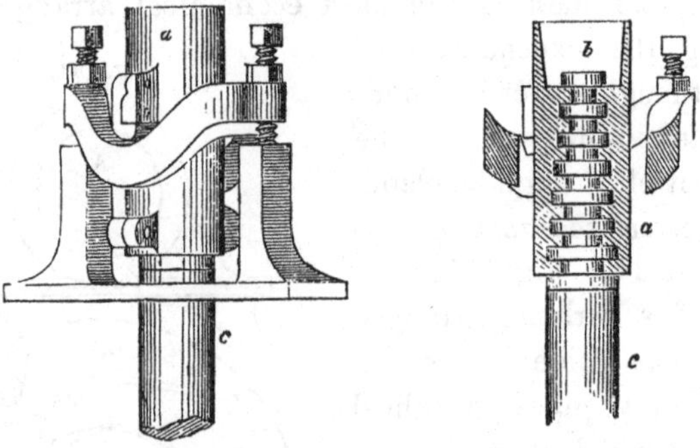

Elevation. Section.

foot-step working under water in large turbines, suspended them from above, instead of supporting them below. This he accomplishes by the peculiar form of bearing shown in fig. 142. The top of the main vertical shaft of the turbine *c*, is cut so as to form a series of bearing surfaces; these fit into corresponding grooves in the metal of the suspension box *a*, which is supported as shown in the elevation by gimbals. The height of the shaft can be accurately adjusted by screws, so that the weight of the turbine may rest on the collars in the suspension shaft and the lower bearing beneath the water serve merely to retain the shaft in its place. By lining the suspension box *a* with a soft metal, principally tin, melted and poured in with the necks in place, sufficient accuracy can be attained to prevent any undue strain on particular collars. This form of bearing is said to have been successfully employed, and to obviate the difficulties of oiling beneath the water.

Efficiency of Turbines.—It may be useful to revert here for a moment to the experiments which have been made upon different forms of turbine to ascertain their relative efficiency. In all these machines, the useful work rendered, is less than the entire force of the fall of water which acts upon them by the loss of work expended in overcoming the friction and inertia of the machine, together with the loss from the *vis viva* expended in shocks and impact and passing away in the water of the tail race, and from other causes in special cases. The fraction which expresses the ratio of the total work expended by the water to the useful work returned by the machine is the *efficiency* of the machine. Commonly we express this ratio in a percentage, taking the work of the fall as 100, and calling the work accomplished useful effect or return.

For the turbines of Fontaine and Jonval, in which the flow is vertical, a return of 70 to 72 per cent. was obtained by M. Morin; 67 per cent. by MM. Alcau and Grouvelle; 74·5 per cent. by MM. Hulze, Borneman, and Bruckman.

The turbine of Fourneyron yields, according to M. Morin; 74 per cent.; but 64½ according to MM. Redtenbacher and Marozeau; M. Fourneyron has obtained results varying from 65 to 80 per cent. according to the fall and immersion of the turbine. The turbine of St. Blazier is said to yield from 70 to 75 per cent.

The turbine of Poncelet, in which the water is laid on tangentially, yields from 65 to 75 per cent.; according to M. Hulze, 70 per cent.

The turbine of Cadiat, with an outward flow like that of Fourneyron, but regulated by an exterior circular sluice, gave 65 per cent. to M. Redtenbacher.

The reaction wheel of Whitelaw and Stirrat has yielded in experiments with models 70 to 78 per cent.

Mr. Thomson's vortex wheel yields according to his experiments 75 per cent.

All these returns appear to approximate closely to the duty performed by water wheels; probably not so high as that given by a well-constructed iron water wheel, but the difference is inconsiderable. Smeaton's experiments gave, on his overshot wheels, as much as 76 per cent., and the results obtained from experiments on the breast wheel, with ventilated buckets on a large scale, gave nearly 78 per cent. of the actual power of the water employed.

Certain advantages, it must be admitted, are obtained by the turbine in certain localities under certain conditions; but it is very doubtful whether they are equal, either on the score of expense or ultimate efficiency, to well-constructed water wheels. In some situations favourable for their reception they are doubtless preferable in effecting a reduction of the original cost, but taking to account the conveyance of the water in pipes and other charges, it will be found as a general rule that the difference is not considerable, and that a well-constructed water wheel of 50 years' duration is an effective and excellent substitute for the turbine.

4. *Water Pressure Engines.*

In the water pressure engine the power obtained from the *pressure* of a column of water is employed in generating a reciprocating instead of a rotatory motion. Engines of this description have long been employed in the mining districts of the Continent, but in England their use appears to date from 1765, when a single-acting water pressure engine was erected for draining a mine in Northumberland by Mr. Westgarth.

For the most successful application of these engines, as regards efficiency, it is necessary that the motion of the water should be slow, and as far as possible without shock. Three to six strokes per minute, or a velocity for the piston of one foot per second, is about the ordinary speed. The stroke also should be long, and therefore "the most advantageous use to which a water pressure engine can be put is the pumping of water, to which slow motion and a long stroke are well adapted, because they are favourable to efficiency, not only in the engine but in the pump which it works."—*Rankine.*

The valves now usually employed in these engines are solid pistons working in the supply pipe, with leather or metal packings. Figs. 143 and 144 showing the valves for a single-acting engine will sufficiently indicate the principle. A A is the supply pipe, B B the entrance to the cylinder, and C C the eduction pipe. When the cylinder is being filled, fig. 144, the valve D is below the entrance and closes the eduction pipe. When, however, the cylinder is emptying, fig. 143, the valve is raised and then closes the supply pipe. Deep notches are cut in these valves in order that they may very gradually open and close the passages to prevent shock.

Fig. 143.

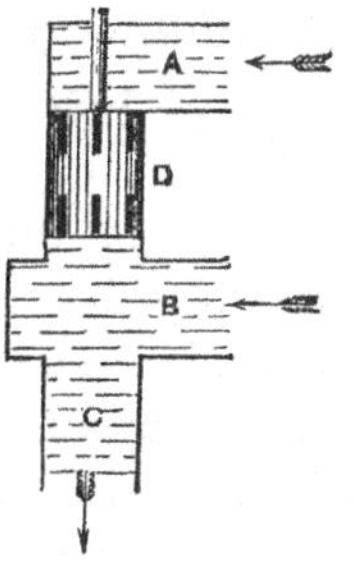

Fig. 144.

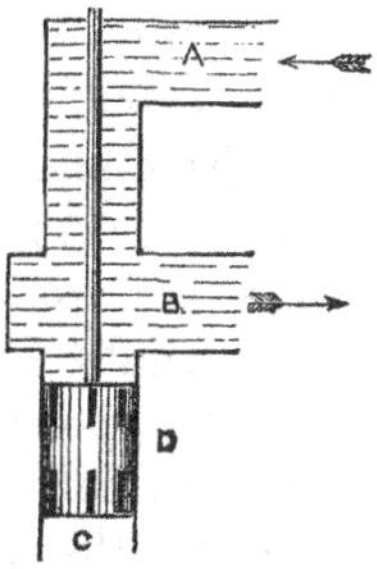

These valves are usually worked by a small water pressure engine, acting in the reverse direction to the general engine, and worked from it by tappets. Fig. 145 shows such an arrangement, from the single-acting engine of M. Junker.

Fig. 145.

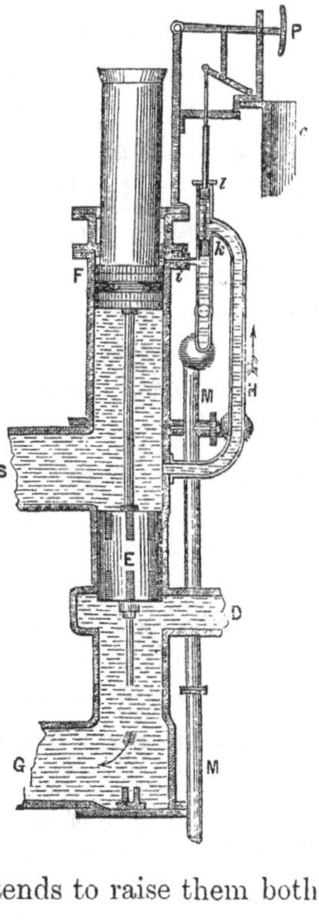

In this drawing *c* represents the upper edge of the main cylinder, S the supply pipe, D the port connecting the main cylinder with the valve chest, G the discharge pipe: E is the valve, which when above D, as in fig. 145, permits the water to escape from the cylinder, and when below D, closes the discharge pipe and opens a passage from the supply pipe. The area of the valve E, is made less than that of the piston F, with which it is connected by a rigid rod. Hence the pressure of the water between E and F tends to raise them both.

The upper side of F is provided with a trunk working in a stuffing box in the top of the valve cylinder. The use of this is to diminish the effective area of the upper side of the piston F, so that it shall not be more than is requisite to enable the water when admitted through the port *i* to overcome the upward tendency of the piston together with the friction of the piston and valve.

H is the supply pipe, and M the discharge pipe of the auxiliary engine for working the valves; K is the valve of this engine which regulates the admission and discharge of the water through the port *i*, precisely in the same manner as the valve E regulates the admission and discharge from the main cylinder. *l* is a plunger of the same size as K, that the pressure between them may be equalised and not tend to move K upwards or downwards. The rod to which *k* and *l* are fixed is connected by means of a train of levers and link work with a lever carrying the crutch P. This is alternately raised and depressed by a tappet rod carried by the piston in the main cylinder *c*.

Suppose now the piston valve E is raised, and the water discharging from the main cylinder, as shown in fig. 145. When the main piston approaches the bottom of its stroke, the upper tappet strikes the lower hook on P and depresses it, along with the auxiliary valve *k*. This admits water from S through H and *i* to the upper side of the counter piston F, so as to depress it along with the valve E. The valve E then closes the discharge pipe, and admits water from S to the main cylinder; the piston rises, and near the termination of its stroke strikes the upper hook on P, and raises the auxiliary valve *k*. This allows the water to discharge from the upper side of F, and then the surplus pressure on its lower side lifts it with E, and the operation is repeated.*

Fig. 146 exhibits an elevation of a single-acting water pressure engine, which I erected some years since in Derbyshire for the purpose of raising water from the Alport lead mines. It does not widely differ in its action from that of M. Junker just described. C is the main cylinder, and P its piston or plunger.

* The description of this valve is abridged from Mr. Rankine's and Prof. Weisbach's Treatises.

S the supply pipe, and D the discharge pipe, connected with the valve apparatus E. F is the cataract or auxiliary engine for

Fig. 146.

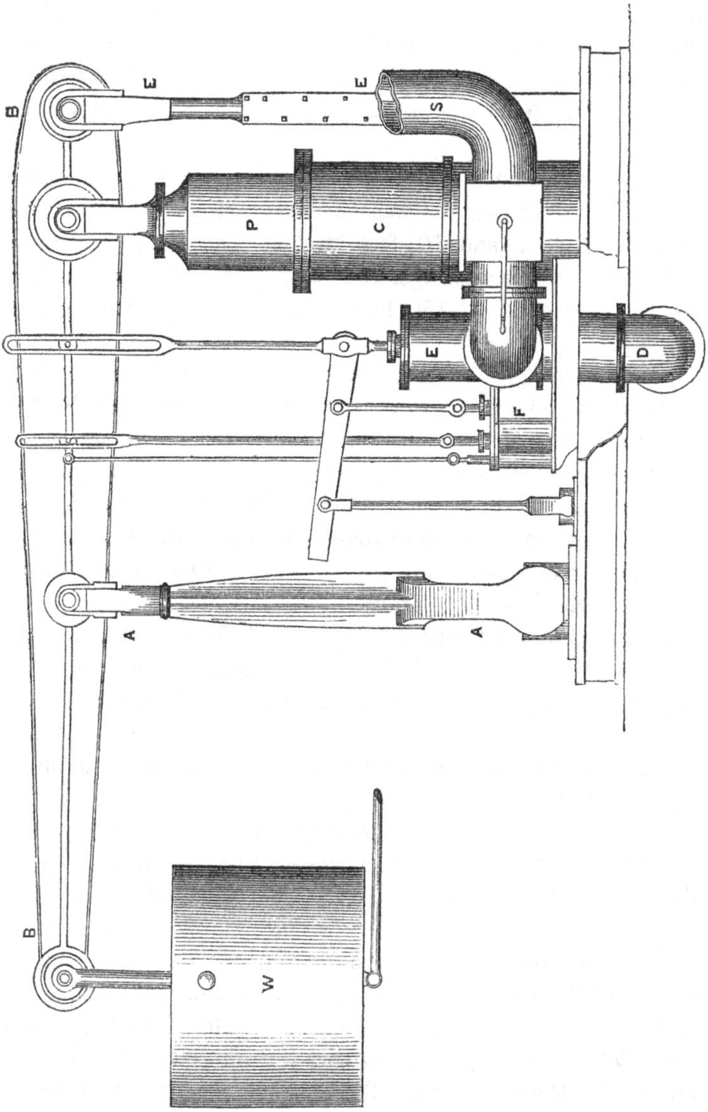

working the valves. The piston P is connected with the sway beam B B, which at its other extremity is attached to the oscillating connecting rod A A, which is fixed on a pivot or joint at its lower extremity. By this arrangement the piston is permitted to rise vertically, and the spear rod of the pumps E is also nearly vertical in its movement. A heavy balance weight W is attached at the opposite end of the sway beam to balance the pump rods at the other, so that the piston should fall in the cylinder C at an appropriate velocity, and without shock.

Mr. Joseph Glynn erected a similar engine at the same mines in 1842. This engine was of larger size, namely with a 50-inch cylinder, and 10 feet stroke. The head of water is 132 feet, and lifts a plunger rod 42 inches in diameter, affording a power of about 150 horses when working at its greatest velocity.

Hydraulic engines of this description are not the most effective even for pumping water, as the motion is exceedingly slow, and the friction of the water and the organic parts of the engine absorbs a considerable amount of the power employed. To remedy this evil it is found desirable in some cases, wherever the fall is not too high, to introduce the water-wheel with cranks and spear rods, communicating a reciprocating motion to the pumps in the shaft of the mine.

In mountainous countries, where high falls descending from great elevations are found, the reciprocating engine is probably the best application for draining purposes, as the motion is conveyed direct from the main cylinder to the pumps, and that probably at the smallest outlay of capital, when a supply of water is at hand.

It is otherwise when large supplies of water on low falls are present. Then the water-wheel, with its machinery, is the most effective and the most economical application of the power.

The recent introduction of the turbine may, however, effect a change in this class of machinery, as it is admirably adapted to high falls, and may be advantageously employed at a moderate cost. The great objection to its use in this form is the great velocity it attains on high falls, and the consequent reduction which would be requisite to work pumps at 10 to 12 strokes per minute, when the machine itself is moving at the rate of 400 to

500 revolutions per minute. This appears to be the only drawback, and it is not improbable that the simple cylinder here described may, under certain conditions, be best adapted to meet all the requirements of raising water from deep mines with the aid of convenient streams on high falls.

CHAP. VI.

ON THE PROPERTIES OF STEAM.

Before considering the application of the steam-engine as a prime-mover, it may be interesting to know something of the properties of steam by which it is moved, in regard to pressure, temperature, and density, as ascertained by various philosophers since the days of Newcomen and Watt. Of late years a great change has gradually taken place in the system of working the steam-engine. At the time of the introduction of the double acting engine of Watt, the makers of engines never dreamed of employing steam at a greater pressure than 10 lbs. on the square inch, and up to 1840 that was the maximum pressure at which steam-engines were worked, with the exception of a few constructed on Wolf's principle of double cylinders, where the steam is first admitted to the piston of the smaller cylinder at a pressure of 30 to 40 lbs. per square inch, and after having performed its office there, is allowed to expand into the second cylinder of three or four times greater capacity, and thus to unite its force with that of the small cylinder, as it moved from one extremity of the stroke to the other. To work this description of engine with high-pressure steam, it was necessary to proportion the strength of the parts of the engine as well as the boiler to a much greater extent of pressure than in the double-acting engine of Watt. Hence it was soon found that the waggon form for the latter, as employed by Watt, was not calculated to resist a pressure exceeding 10 or 12 lbs. per square inch without the introduction of numerous wrought-iron stays to retain it in form. To raise steam for the compound engine such a boiler was wholly inadequate, and a series of small boilers, with hemispherical ends, were introduced in its stead wherever steam of high-pressure was required.

The single pumping engines of Watt, and the compound engines of Wolf, employed at the mines in Cornwall, gave, however, extraordinary results as regards the work accomplished for the quantity of coal consumed, which was less than half the quantity used in the rotative engines employed in mills. It was also asserted that the double cylinder engine in use on the Continent (but chiefly made in this country) was performing a more satisfactory duty than could possibly be attained by the single cylinder low-pressure engine.

These assertions, often repeated, and the returns of Cornish engines, published from year to year, led to a close inquiry into the subject, first in my own works at Manchester, and subsequently before the British Association for the Advancement of Science, where the whole question was ably discussed, and ultimately led to a better system of working in factory engines, with a saving of one-half the fuel formerly consumed in effecting the same quantity of work. In these investigations it was found that the compound engine had no advantage over the single cylinder engine, as constructed by Watt, when worked at the same pressure of steam and the same rate of expansion; that is, a single cylinder engine, with properly constructed valves, having the power of cutting off the steam at any point of the stroke, is quite as effective, and more simple in construction, than the double cylinder engine. It is true, that at first the double cylinder engine had an advantage over the single cylinder engine in its greater uniformity of motion, but this is no longer the case, as an increase of the velocity of the piston from 240 to 320 and 360 feet per minute effectually remedies that evil, and increases the power of the engine in the ratio of the increase of speed.

Thus it will be seen that a great change has come over the system of employing steam; the pressure is quadrupled in factory engines, and more than doubled in marine engines. Every engine of recent construction is provided with boilers of great resisting powers, and on an average cuts off the steam in the cylinder at one-fourth, and at other times one-fifth or one-sixth of the stroke, the steam acting by expansion alone during the remaining three-fourths, four-fifths, or five-sixths, as the case may be. This system is found to be of great value, as the quantity of fuel consumed does about double the amount of work which could be got out of it on the low-pressure principle.

The important results already obtained by a judicious system of working steam expansively has given a powerful stimulus to the extension of our commerce and manufactures, and the question naturally arises, whether or no we have attained the full benefit from the introduction of the methods of working now employed, or whether we may not reap a still greater advantage from progressing in the same direction and using steam of higher pressure, expanded to still greater lengths than has yet been attained in our present practice. This is a question which remains for solution, and it appears most desirable that we should ascertain by direct experiments to what extent of pressure and expansive action we may safely venture with perfect security to the boilers and the working parts of the engines. Assuming for a moment that an increased pressure, accompanied by increased expansion, would in the same proportion increase the economy of working, we have then to consider the capabilities of our vessels for resisting those pressures. And lastly, the observation of the action of steam in expanding has led many to expect still further advantage from the use of superheated or gaseous steam. To make sure progress in either of the directions here indicated two things are necessary: we must cultivate a more intimate acquaintance with the resisting powers of materials, and the strength of vessels of different forms before we can assure ourselves of success; and we must attain increased and increasing knowledge of the properties of the agent we employ under the various conditions of expansion and superheating. In regard to the first of these requisites a steady progress has been made, and experimental inquiries have been extensively carried on in regard to the resisting powers of vessels and the causes of their failure, and the difficulty of constructing boilers to resist very high pressures has been greatly diminished. Our knowledge of steam has also rapidly increased, and many of the necessary questions relating to its properties have been for ever set at rest by the recent and classical labours of Regnault, carried on at the instance and with the assistance of the French Government. The questions of the density and law of expansion of steam, however, still require solution.

They are being investigated, from a theoretical point of view, with considerable success, by Mr. Rankine of Glasgow. The experimental inquiry I have undertaken in conjunction with my friend Mr. Tate, and a part of the results, comprising experiments up to a pressure of 60 lbs. per square inch, will appear in the Transactions of the Royal Society. We are now preparing to enter on the more arduous and dangerous task of ascertaining the density, volume, &c. at much higher pressures. The accumulation of facts on this subject, bearing directly upon the application of steam, cannot be otherwise than acceptable to the general reader, and I shall, therefore, without further preface, insert such an abstract as bears directly on the subject under consideration.

General Laws of Vaporisation.

When a liquid is heated in any vessel, its temperature progressively rises up to a certain point, at which it becomes perfectly stationary. At that point the heat continuously absorbed becomes *latent*, or is no longer registered by the thermometer; ebullition commences, and vapour, of a bulk enormously greater than that of the liquid from which it is formed, rises in bubbles and fills the vessel. In this condition the temperature of the liquid is perfectly constant; no urging of the fire will cause it to rise; the heat, absorbed continuously, expands itself in effecting that change in the state of aggregation of the liquid which we know as vaporisation.

This remarkable constancy in the temperature of liquids undergoing vaporisation in open vessels has long been known and applied to the graduation of thermometers. The point at which a liquid boils in an open vessel is called its boiling point. The following table gives the boiling points of some of the more important liquids:—

	Boiling Point. Fahr.	Authority.
Water	212°·0	
Ether	94·8	Kopp.
Alcohol	173·1	Pierre.
Sulphuric Acid	640·0	Marignac.
Mercury	662·0	Regnault.

We have said that the boiling point of a liquid is constant

when in an *open vessel*, that is, when subject to the atmospheric pressure. If we change the pressure the temperature of ebullition changes also. Thus, if we place a vessel of hot but not boiling water under the receiver of an air-pump, and rapidly exhaust the air, the liquid will after some time begin to boil, and we may notice that the lower its temperature the more perfect must we make the vacuum before ebullition commences. Or again, if water be subject to pressure greater than that of the atmosphere, its temperature must be raised higher than 212° before it will boil. Experiment, therefore, shows that the boiling point, constant at the same pressure, varies at different pressures, rising higher as the pressure increases, and *vice versâ*.

Strictly speaking, the pressure of the atmosphere is not always the same; it varies within narrow limits from day to day; it decreases as we ascend higher into it, and hence there will be a small but corresponding variation in the boiling point at different times and places. This last fact has afforded the means of measuring the altitude of mountains, by determining the difference of the boiling point at their base and their summit. Measuring the atmospheric pressure by the column it supports in the barometer, we may draw up the following table of the relation of the boiling point to the height of the barometer column and the altitude of the observer, assuming that the barometer stands at 29·922 inches, and water boils at 212° Fahr. at the level of the sea.

TABLE I.—EXHIBITING THE INFLUENCE OF CHANGES OF ATMOSPHERIC PRESSURE ON THE BOILING POINT OF WATER, AND THE BOILING POINT AT DIFFERENT ALTITUDES.

Height of Barometer in inches.	Altitude in feet.	Boiling Point of Water. Fahr.	Height of Barometer in inches.	Altitude in feet.	Boiling Point of Water. Fahr.
29·922	0	212°·0	25·888	3,926	204°·9
29·396	462	211·1	25·468	4,460	204·0
28·774	933	210·0	25·014	5,000	203·0
28·559	1,411	209·3	24·046	6,111	201·2
27·846	1,897	208·5	23·454	7,263	200·0
27·348	2,392	207·6	18·992	13,700	190·0
26·852	2,895	206·7	15·135	18,000	180·0
26·372	3,407	205·8	12·145	26,000	170·0

But, besides pressure, certain other circumstances exercise a slight but sensible influence on the boiling point. In a glass vessel the boiling point of water is about two degrees higher than in a metal one, owing apparently to some adhesion between the glass and the liquid. Dr. Miller states that if the glass be varnished with shellac the temperature of the water may be raised to 221° in the open air, when a sudden burst of steam will take place, during which the temperature falls to 212°. From a similar cause the presence of salts in solution raises the boiling point in some cases considerably. A saturated solution of common salt boils at 227° Fahr., and a saturated solution of chloride of calcium, which has an enormous affinity for water, does not boil at a less temperature than 355° Fahr.

There is yet one other remarkable condition of evaporation which should be noticed here. If water be dropped upon a clean metallic surface heated sufficiently high, instead of entering into ebullition it assumes a globular form, and rolls about very slowly and quietly evaporates. This condition, known as the spheroidal state, has been investigated by Mr. Boutigny. He finds that the temperature of the liquid globule never rises so high as its boiling point, being indeed usually 5° to 10° below it; that the temperature of the plate necessary to cause the spheroidal state varies with different liquids, and depends in part on the conducting power of the plate; and he considers the temperature of the spheroid to be constant, being for water 205°·7; for alcohol 167°·9, and for ether 93°·6.

If, whilst the spheroid is rolling upon the metal plate, the temperature of the plate is allowed to fall below a certain temperature (340° for water), the spheroid breaks, and is suddenly dispersed in vapour.

The temperature of the vapour rising from a liquid is necessarily identical with that of the liquid from which it rises, except in those cases in which the boiling point has been affected by adhesion, when the vapour at once adjusts itself to the normal temperature at that pressure.

So long as vapour is in contact with the liquid from which it has been formed its temperature continues the same as that of the liquid, for if it be heated it takes up fresh liquid, and the temperature falls from the absorption of the heat rendered latent, until the normal temperature of the boiling point is regained.

The Vaporisation of Water and the Formation of Steam.

The temperature at which water boils is therefore constant at each pressure, and in consequence the temperature of the steam itself, when in contact with water, is constant at each pressure. The relation between the temperature and pressure of steam has been ascertained by experiment.

When in contact with the water producing it, steam is at the maximum density consistent with that temperature and pressure, and is then called *saturated* steam, or vaporous steam, and its temperature is called the maximum temperature of saturation at the given pressure. Usually when the *pressure of steam* is spoken of, the pressure of saturated steam is intended.

When isolated from the water producing it and heated, the steam expands, and decreases in density if the pressure be constant, or if the volume be constant it increases in pressure; it is then called variously anhydrous, gaseous, or superheated steam. The rate of expansion of superheated steam must be determined by experiment.

By the *density* of steam we mean the relative weight of a unit of volume. The *specific volume* of the steam is the reciprocal of the density, or the ratio of the volume of the steam to that of the volume of water which produced it. The density of saturated steam is constant at each temperature, and must be determined by experiment.

The *latent heat* of evaporation of steam is the quantity of heat which disappears in effecting the conversion of the water into vapour, or which reappears in the condensation of the steam. The latent heat of evaporation added to the sensible heat, or heat required to raise the temperature of the water up to the temperature of ebullition, is called the total heat of the steam.

The Relation between the Pressure and Temperature of Saturated Steam.

Probably the earliest experiments on this subject were made

by Watt *, who tells us that when inventing the separate condensation he made some trials (in 1774) from which he constructed a curve, of which the ordinates represented the pressures, and the abscissæ the temperatures of the steam, and thus enabled him to calculate the one from the other at sight, with sufficient accuracy for his purposes. Watt first surmised that the elastic force or pressure of the steam increased in a geometric progression for temperatures increasing in an arithmetical progression.

Fig. 147.

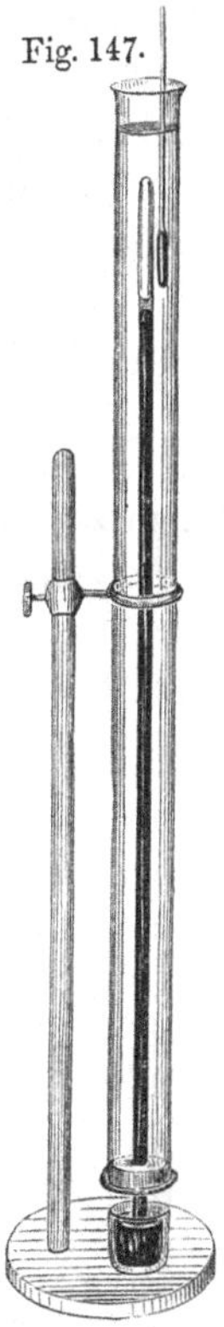

Robison † made experiments upon the same subject at elevated temperatures, ascertaining the temperature at which the steam began to blow off from a safety valve loaded with weights, a proceeding susceptible of little accuracy. Dalton ‡, however, was more successful in devising an accurate method. He employed a barometer carefully purged of air, into which he introduced a small quantity of water. The barometer was surrounded by an outer water bath, by which the vapour in its chamber was heated to various temperatures. The mercury in the barometer tube adjusted itself so as to be in equilibrium at each temperature between the pressure of the atmosphere on the outside and the pressure of the vapour within, and the column fell as the temperature rose to an extent which is an exact measure of the pressure of the vapour within. The difference of height of an ordinary barometer and the barometer containing the water gives directly the pressure of the steam, so that by a series of careful measurements of a humid barometer and ordinary dry barometer, the pressures corresponding to various temperatures may be observed.

This method, under various modifications, has been frequently employed, both for water and other liquids, at pressures which

* Muirhead's Life of Watt, p. 76.

† Mechanical Philosophy, vol. ii. p. 23.

‡ Memoirs of the Manchester Literary and Philosophical Society, vol. xv. p. 409. New Series, vol. v. p. 553.

are less than that of the atmosphere. The chief difficulty is to maintain the liquid in the bath by which the barometer is heated at a uniform temperature, and to prevent it from dividing into strata unequally heated. To obviate this, the temperature may be observed at various depths and the arithmetical mean taken, or the length of the barometer may be decreased as the temperature rises. Or a barometer with two limbs may be employed, as in the researches of Dr. Ure, or, lastly, the varying temperature of the atmosphere may be substituted for that of the liquid bath, as in the experiments of Kaemtz *, intended to supply data for meteorological purposes, which extended over a period of two years, and ranged from $-15°$ to $+80°$ Fahr.

Dr. Ure's † modification enables the experiments to be carried to pressures higher than that of the atmosphere. The space in the barometer tube occupied by the vapour need never be large, and the increase of elastic force is measured by the quantity of mercury which must be added to a second limb of the barometer in order to maintain the quicksilver in the first at a constant level. Thus, in fig. 148, *a b c* is the bent barometer tube for experiments above the atmospheric pressure, the shorter limb being enclosed in a glass vessel, which can be filled with oil and heated progressively to any required temperature. Fine rings of platinum wire are firmly fixed round the tubes at the level *d d*, and as the temperature rises the mercury in the limb in the bath is maintained at this level by adding mercury in the other limb, when the column *d e*, supported by the steam, measures its elastic force.

Fig. 148.

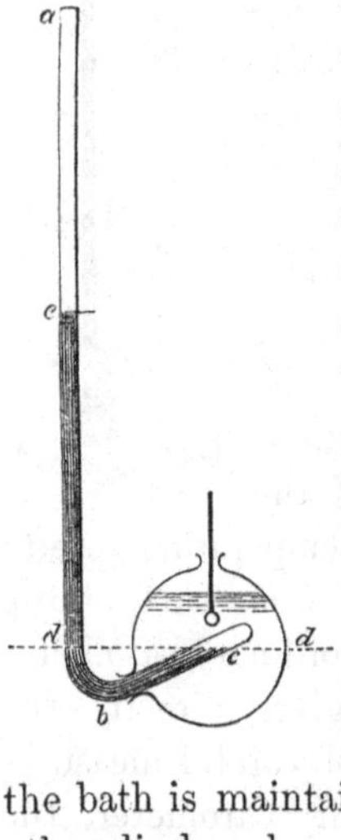

Dalton, whose experiments were, on the whole, accurate, inferred from the results which he obtained with water and alcohol, that the tension of all vapours was equal at temperatures

* Traité de Météorologie, vol. i. p. 290.

† Phil. Trans., 1818, p. 338.

equally distant from their boiling-points under atmospheric pressure. This law, which has since borne his name, has not been confirmed by experiments on a larger number of liquids. For many liquids, however, it is nearly true, at small distances above the boiling point. Thus:—

Degrees from the boiling point. Fahr.	Elasticity in inches of Mercury. Water.	Alcohol.	Ether.
+ 30	52·90	56·60	50·90
+ 20	44·06	46·30	42·64
+ 10	36·47	37·00	35·20
0	30·00	30·00	30·00
− 10	24·50	24·20	24·70

In the above table the boiling-point of water is 212°, of alcohol 173°, and of ether 104°.

The following table gives a few of Dalton's results for comparison with those of other experimenters which will be given presently.

TABLE II.—ELASTIC FORCE OF THE VAPOUR OF WATER, ACCORDING TO DALTON.

Temperature. Fahr.	Elastic Force of Vapour. In inches Mercury.	In lbs. per square inch.
0	·066	·033
10	·090	·045
20	·129	·064
40	·263	·131
60	·524	·262
80	1·000	·500
100	1·86	·93
150	7·42	3·71
212	30·00	15·00

In 1823 the French Government, then legislating on the subject of steam, and requiring some further knowledge of its properties, intrusted to the French Academy the conduct of some important experiments on this subject. The Academy appointed a Commission, consisting of MM. Prony, Arago, Girard, and Dulong, to investigate the subject, and their report was published in the Memoirs of the Academy for 1831.* The experiments detailed in this Report were made chiefly by MM. Dulong and Arago, by a new method, and with all the care

* Mémoires de l'Institut, tom. x. p. 194, and Annales de Chimie et de Physique, tom. xliii. p. 74.

and accuracy which was possible in the state of science at the time. They were also on a scale which is only possible where private effort is seconded by the munificence of the Government.

Their apparatus consisted of, 1st, a boiler to generate the steam, 2nd, a manometer to measure the pressure.

The pressure, which extended to twenty-four atmospheres, was in fact measured by the column of mercury it would support in an open glass tube, but as the length of tube necessary for this purpose rendered it very inconvenient, they employed an intermediate measurer, consisting of a closed air manometer, graduated by experiment with the open mercury column. At the centre of the tower of the ancient church of St. Geneviève they erected a firmly supported wooden column, to which they attached the glass tubes containing the mercury column. These tubes, thirteen in number, were each 6½ feet in length, so that the mercury column for the graduation of the manometer could be as much as 86 feet in height, corresponding to a pressure of thirty atmospheres, or 450 lbs. per square inch. This column was adjusted precisely vertical, and communicated with a cistern containing 100 lbs. of mercury. The manometer, which consisted of a carefully dried glass tube, closed at the upper extremity, and 67 inches long, communicated with the same cistern, and was maintained at a uniform temperature by a stream of water circulating round it. The height of its mercury was read by means of a vernier, similar to that of a standard barometer. It is easy to see how, by means of a force pump, the pressure in the cistern of mercury could be increased at pleasure, and how the pressure could be registered by reading off simultaneously the height of the mercury in the open tube and its corresponding level in the manometer. When the value of the divisions of the manometer had been thus determined up to twenty-seven atmospheres, it became an instrument for measuring pressure of as great accuracy and delicacy as could be desired.

The boiler for generating the steam was of a capacity of 17·6 gallons, to ensure a uniform temperature, and communicated with the manometer by a tube filled with water, cooled by a refrigeratory apparatus. The temperature was measured by means of mercurial thermometers, placed in thin metal tubes, containing mercury, to protect them from pressure.

The boiler being charged, and a convenient quantity of fuel introduced into the furnace, the temperature was allowed to rise until it nearly attained a maximum. A series of readings were then taken simultaneously from the manometer and four thermometers, until the temperature passed its maximum, and began sensibly to decrease. The readings at the maximum were alone retained for calculation. Fresh fuel was then added, and a second experiment obtained.

The method, carried out with the skill for which MM. Arago and Dulong have earned so high a reputation, possesses most of the essentials of complete accuracy. Its chief defect, as M. Regnault has pointed out, lies in this, that when the pressure and temperature are changing, however slowly, it is impossible to be absolutely certain that the thermometers have followed that change with the necessary rapidity, and that they do really register the temperature at the time the observation is made. There is in these experiments one other source of possible error, namely, the use of the mercurial thermometer, which, in the higher parts of its scale, does not possess the accuracy necessary in experiments of this nature. Be this as it may, these experiments are of high value and permanent importance. The results obtained in thirty experiments are given in Table IV. on next page.

Next to the experiments of the French Academy, the most important experiments on the relation of temperature and pressure of steam were those of the Franklin Institute in America. They differed considerably from those of the French physicists, and are probably less reliable. The following table gives an abstract of the results:—

TABLE III.—ELASTIC FORCE OF STEAM FROM THE EXPERIMENTS OF THE FRANKLIN INSTITUTE.

Pressure in Atmospheres.	Temperature in degrees Fahr.	Pressure in Atmospheres.	Temperature in degrees Fahr.	Pressure in Atmospheres.	Temperature in degrees Fahr.
1	212°	4½	298°·5	8	336°
1½	235	5	304·5	8½	340½
2	250	5½	310	9	345
2½	264	6	315·5	9½	349
3	275	6½	321	10	352½
3½	284	7	326		
4	291·5	7½	331		

TABLE IV.—RESULTS OF MM. ARAGO AND DULONG'S EXPERIMENTS ON THE RELATION OF PRESSURE AND TEMPERATURE OF SATURATED STEAM.

No. of Experiment.	Temperature observed. Cent.		Mean Temperature reduced to Fahr. scale.	Elastic force of Steam in Metres of Mercury, at 0° C.	Elastic force in inches of Mercury.	Elastic force in Atmospheres of 29·922 inches.
	Small Thermometer.	Large Thermometer.				
	°	°	°			
1	122·97	123·7	253·99	1·62916	64·141	2·14
2	132·58	132·82	270·86	2·1767	85·698	2·87
3	132·64	133·3	271·34	2·1816	85·891	2·88
4	137·70	138·3	280·40	2·5386	99·947	3·348
5	149·54	149·7	301·31	3·4759	136·85	4·584
6	151·87	151·9	305·38	3·6868	145·15	4·86
7	153·64	153·7	308·60	3·881	152·80	5·12
8	163·00	163·4	315·76	4·9383	194·42	6·51
9	168·40	168·5	335·21	5·6054	220·69	7·391
10	169·57	169·4	337·06	5·7737	227·31	7·613
11	171·88	172·34	341·79	6·151	242·17	8·114
12	180·71	180·7	357·26	7·5001	295·29	9·893
13	183·70	183·7	362·66	8·0352	316·35	10·60
14	186·80	187·1	366·51	8·6995	342·51	11·48
15	188·30	188·5	371·12	8·840	348·04	11·66
16	193·70	193·7	380·66	9·9989	393·66	13·19
17	198·55	198·5	389·33	11·019	432·83	14·53
18	202·00	201·75	395·36	11·862	467·02	15·65
19	203·40	204·17	398·80	12·2903	483·88	16·21
20	206·17	206·10	403·03	12·9872	511·32	17·13
21	206·40	206·8	403·88	13·061	514·22	17·23
22	207·09	207·4	405·03	13·1276	516·84	17·30
23	208·45	208·9	407·62	13·6843	538·76	18·05
24	209·10	209·13	408·39	13·769	542·10	18·16
25	210·47	210·5	410·86	14·0634	552·41	18·55
26	215·07	215·3	419·32	15·4995	610·23	20·44
27	217·23	217·5	423·25	16·1528	635·95	21·31
28	218·3	218·4	425·03	16·3816	644·96	21·60
29	220·4	220·8	429·08	17·1826	676·50	22·66
30	223·88	224·15	435·22	18·1894	716·13	23·994

The experiments of Arago and Dulong give a temperature of 358°·88 Fahr. for a pressure of ten atmospheres, or 6°·38 higher than that of the American Institute. This notable difference, too great to be merely accidental, Regnault, whose experience in the matter entitles him to speak with certainty, attributes to the use of mercurial thermometers, which, although agreeing perfectly between 32° and 212°, often present at elevated temperatures a difference of many degrees. Regnault's own experiments give 356·54 on the air-thermometer as the temperature at the pressure of ten atmospheres, which is 2°·34 lower than the French Academy, and 4° higher than the Franklin Institute. As at this temperature the mercurial-thermometer gives higher indications than the air-thermometer, MM. Arago and Dulong's experiments appear the most reliable.

The uncertainty arising from the discordance of the numerical results of the different physicists who had studied this question, and especially the difference above noted, called for a new investigation.

The experiments of M. Regnault on the reliability of the various instruments employed in measuring temperature, led him to the conclusion that at elevated temperatures the indications of different mercurial-thermometers were too variable to be trusted, unless they were made of the same description of glass, and that even in that case they require reduction to the absolute temperature of the air-thermometer. The following table gives some of the results obtained :—

Temperature by the Air Thermometer.	Temperatures by Mercurial Thermometers.			
	Crystal of Choisi-le-Roi.	Ordinary Glass.	Green Glass.	Swedish Glass.
100°	100°	100°	100°	100°
130	130·20	129·91	130·14	130·07
150	150·40	149·80	150·30	150·15
180	180·80	179·63	180·60	180·33
200	201·25	199·70	200·80	200·50
250	253·00	250·05	251·85	251·44
300	305·72	301·08		
350	360·50	354·00		

The above numbers are in centigrade degrees. They show for the thermometer of ordinary glass, when its indications are reduced to Fahrenheit's scale, the following divergences from the true temperature shown by the air thermometer : —

Temperature by Air Thermometer. Fahr.°	Temperature by Mercurial Thermometer. Fahr.°
302	301·64
392	391·46
482	482·09
572	573·94
662	669·20

To M. Regnault the French Government committed, on the proposition of M. Legrand, the task of carrying on a series of experiments to determine, with the greatest precision, the principal laws and numerical data which enter into the calculation of the duty of steam-engines, and supplied the funds for fulfilling its intentions on a scale such as would have been impossible in any private enterprise. The papers, which were the result of this munificence, are amongst the most important in the recent history of science. They cover a large ground, and possess a precision and completeness before scarcely ever attained in researches of this kind.

In relation to the steam-engine, the most important questions which M. Regnault has set himself to solve are—1st. The elastic force of the vapour of water, both at pressures lower than that of the atmosphere, and at high pressures, up to 400 lbs. per square inch. 2nd. The latent heat of the vapour of water through a similar range of temperature and pressure. 3rd. The specific heat of liquid water. The laws of the density and expansion of steam, it will be observed, Regnault did not touch, but, on the subjects above named, his researches are not likely to be superseded in accuracy or extent.

To ascertain the relation of temperature and volume of steam at low temperatures, Regnault adopted the plan of employing two barometers, placed side by side, under precisely similar circumstances, into one of which was introduced a portion of water perfectly freed from air. The upper part of these barometers was surrounded by a large bath of water, maintained by agitation at a constant temperature. The difference of level of the mercury in the humid and dry barometers gave directly the elasticity of the steam at the temperature of the bath.

Fig. 149, shows one of the forms of apparatus employed. The two barometers, *e g*, *o k*, were plunged in the same cistern *v*, and maintained vertical against a firm board. In the form shown, the moist barometer, *e g*, communicated with a glass globe A, of a capacity of 80 cubic inches, and exhausted of air by means of an air-pump, after which the tube *l* was hermetically sealed The tension of the air remaining was accurately ascertained, and did not exceed 1 to 2 millimetres. The bath of galvanised iron, V V, was of a capacity of about 2746 cubic inches; over a rectangular opening opposite the barometers the plate of glass, E G, was fixed, and through this the readings were taken, after the error arising from refraction had been determined. By means of a lamp placed underneath, a constant temperature could be maintained in this bath as long as was necessary to take a series of readings. When these were complete, water was withdrawn from the bath and replaced by boiling water, then, when a constant temperature was again arrived at, a new series of readings could be obtained.

These methods answered with perfect accuracy up to about 150° Fahr., above this the tendency of the water to separate into strata of unequal temperature began to manifest itself so as to

Fig. 149.

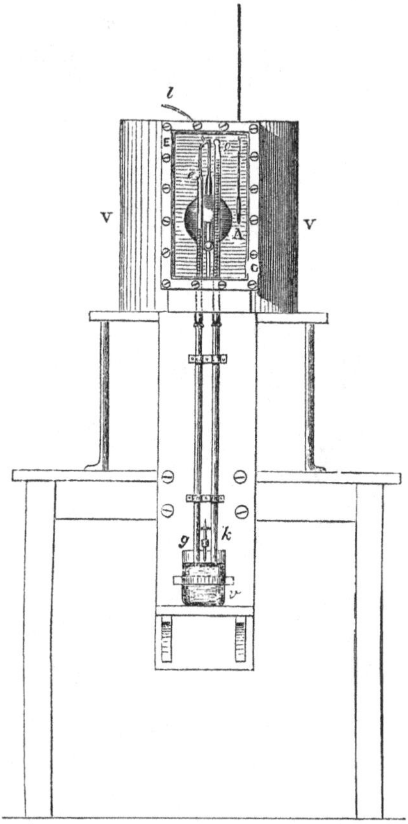

introduce errors into the experiments. Regnault, therefore, had recourse to the plan of observing the temperature at which water boils at determined pressures. This was the proceeding adopted by Arago and Dulong, but with a new precaution.

Those physicists were, by the method they adopted, compelled to regulate their experiments by the condition of their fire, so that it was impossible to maintain a constant temperature for any great length of time. By adding to their apparatus a large vessel containing air and acting as an artificial atmosphere, together with some large air-pumps, by which the pressure on the water could be predetermined and maintained perfectly constant, Regnault, in fact, obtained means for regulating his experiments altogether independently of the furnace, for the temperature of ebullition, we have already seen, is perfectly constant under a constant pressure. The conditions were identically those of water boiling in air.

Fig. 150 shows the larger of the two forms of apparatus em-

Fig. 150.

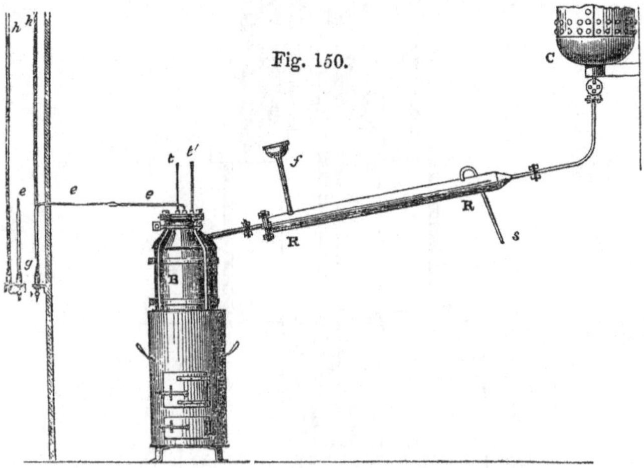

ployed in the experiments on the elasticity of steam at high temperatures.

It consists of a boiler, condensing tube, artificial atmosphere, mercurial manometer, and an air-pump. The boiler, B, is of red copper, of 13·7 inches diameter and 123 pints capacity. The cover carries two tubes, in which were placed mercurial thermometers protected from pressure, and a third for an air thermometer. The boiler was strengthened by iron rings bolted round it. The refrigerator, R R, of a copper tube 5 feet long, communicating with the boiler, surrounded by a larger tube, was arranged so that a continuous stream of cold water flowed into the funnel *f*, and away by the siphon, *s*. The reservoir of air forming the artificial atmosphere for maintaining a constant pressure was formed of a cylinder, C, of 62 gallons capacity, and riveted and brazed so as to be perfectly air-tight. The manometer for ascertaining the pressure consisted of an open mercurial column in every instance, than which no more perfect instrument, or one more free from corrections depending on theoretical calculations, could be devised. The indications are of equal sensitiveness at all pressures, instead of continually decreasing in value, as in the ordinary compressed air-gauge employed by Arago and Dulong.

This manometer is not shown in the sketch, but it consisted of a cistern of mercury with a pipe attached having four openings; one of these was usually closed; the other three were for the attachment of glass tubes to contain the mercury columns in the various experiments in which this apparatus was employed. The column open to the atmosphere consisted of ten to twenty-two glass tubes, nearly 10 feet long and $\frac{4}{10}$ inch bore. These were carefully connected together to form a column perfectly vertical, and from 40 to 80 or more feet in height, as required, being supported against a vertical wall. Up to fifteen atmospheres the levels were taken by two cathetometers; for higher pressures the glass tube itself was graduated into millimetres.

The air-pump for maintaining the artificial atmosphere in the large cylinder, C, consisted of three single-acting cylinders, each discharging 42 cubic inches per stroke.

To measure the temperature two mercurial thermometers, *t*, *t*, perfectly accordant, and an air thermometer, were employed; the latter consisting of a thin glass cylinder of about 1·17 inches diameter, and 11·7 inches long. This communicated by the capillary tube, *e e*, with the manometer, *g h*. The capacity of the air thermometer, its temperature, and the pressure were therefore known, and with the corrections which M. Regnault applied, he considers that it indicated sensibly $\frac{1}{25}$ of a degree.

It will be evident how, with this apparatus, a continuous and energetic ebullition was maintained in the boiler, B, under any pressure at which the observer wished to determine the temperature of the steam. Condensation went on at the same time with a corresponding rapidity in the cooling apparatus, R R; the pressure being maintained constant by the air-pump, the thermometers would in time become identical in temperature with the steam in the boiler B; simultaneous observations at this period of the thermometers and the manometer gave the relation of temperature and pressure which was required.

We have now described all the principal methods by which it has been sought to determine, experimentally, the relation of the pressure and temperature of saturated steam. It is necessary that we should next consider how they may be expressed in a formula suitable for calculation.

The law to which Watt was led, and which is usually known as Dalton's, from the care with which he verified it, so far as his experiments went, is that which, in general terms, most nearly expresses this relation; it is, that the elastic force of vapours increases in a geometrical progression, for a series of temperatures increasing in an arithmetical progression, and many of the formulæ which have been constructed to express the results of experiments, have been based upon it. Strictly speaking, it is, however, only an approximate expression of the true law.

One of the earliest and best of the formulæ which have been proposed, is that first applied to steam by M. Prony, of the form

$$\mathrm{F}=a\alpha^t+b\,\beta^t+c\gamma^t+\ \dots\ (1)$$

where F is the elastic force, and t the temperature. The other quantities are constants derived from experiment. This formula is accurate, but requires a large amount of calculation.

Dr. Young proposed the formula

$$\mathrm{F}=(a+bt)^m\ \dots\ (2)$$

which has been the basis of several formulæ employed for interpolation by physicists. Thus MM. Arago and Dulong give from their own experiments the following constants:—

$$e=(1+0{\cdot}7153\ \mathrm{T})^5\ \dots\ (3)$$

where e expresses the elasticity in atmospheres of 29·922 inches of mercury; T the temperature in centigrade degrees reckoned from 100°, positive above that point and negative below, taking for unity an interval of 100°. For pressures greater than one atmosphere this formula is satisfactory, but it deviates greatly from experiment at lower pressures. At the same time it has a great simplicity.

The formula proposed by the Franklin Institute was of the same kind, the constants being

$$e = (\cdot 00333\ \text{T} + 1)^6 \ \dots\ (4)$$

where e is the pressure in atmospheres of 30 inches of mercury, and T the excess of the temperature above 212° in Fahrenheit. This formula also does not apply below atmospheric pressure with accuracy. For calculation we may write it

$$\log e = 6 \log \{\cdot 00333\ (t - 212^\circ) + 1\} \ \dots\ (5)$$

and for calculating the temperature from the pressure,

$$= \frac{\sqrt[6]{e} - 1}{\cdot 00333} + t\ 212 \ \dots\ (6)$$

Another form of expression was given by M. Biot in 1844, viz.

$$\log \text{F} = a + b\,a^t + c\,\beta^t \ \dots\ (7)$$

The constants for this formula M. Regnault has calculated from the following values, obtained from the graphic curve, which represented his experiments:—

$t_0 = 0°$	$\text{F}_0 = 4\cdot 60$ mm.
$t_1 = 25$	$\text{F}_1 = 23\cdot 55$
$t_2 = 50$	$\text{F}_2 = 91\cdot 98$
$t_3 = 75$	$\text{F}_3 = 288\cdot 50$
$t_4 = 100$	$\text{F}_4 = 760\cdot 00$

whence he deduced

$$\log a_1 = 0\cdot 006865036$$
$$\log \beta_1 = \bar{1}\cdot 9967249$$
$$\log b = \bar{2}\cdot 1340339$$
$$\log c = 0\cdot 6116485$$
$$a = +4\cdot 7384380$$

where the third term $c\beta_1^t$ is negative. Between 0° and 100° centigrade, this expresses M. Regnault's results with very great exactness. Above this temperature, it gives results which become sensibly different from those of experiment.

For temperatures between 100° and 230° centigrade, M. Regnault obtained the following values for the constants in M. Biot's formula:—

	For the Air Thermometer.	For the Mercurial Thermometer.
$\log \alpha_1$	$= 1{\cdot}997412127$	$= 1{\cdot}997443007$
$\log \beta_1$	$= 0{\cdot}007590697$	$= 0{\cdot}01182377$
$\log b$	$= 0{\cdot}4121470$	$= 0{\cdot}4163766$
$\log c$	$= 3{\cdot}7448901$	$= \bar{4}{\cdot}9731198$
a	$= 5{\cdot}4583895$	$= 5{\cdot}4882878$

Where in the formula the second term is negative, and

$$\log F = a - b\,\alpha_1{}^x + c\,\beta_1{}^x \;\ldots\; (8)$$
$$x = t^\circ - 100^\circ$$

which gives the relation of temperature in centigrade degrees and pressure in millimetres. The mercurial thermometer was constructed of crystal of Choisi-le-Roi.

Mr. Rankine, in 1849, urged some theoretical objections to the formulæ employed by M. Regnault, and proposed the following: —

$$\log p = A - \frac{B}{\tau} - \frac{C}{\tau^2} \;\ldots\; (9)$$

$$\tau = 1 \div \left\{ \sqrt{\left(\frac{A - \log p}{C} + \frac{B^2}{4\,C^2} \right)} - \frac{B}{2\,C} \right\} \;\ldots\; (10)$$

when $\tau = T + 461^\circ{\cdot}2$ Fahr.

The constants for this formula are —

$$\left.\begin{array}{lcl} A & = & 8{\cdot}2591 \\ \log B & = & 3{\cdot}43642 \\ \log C & = & 5{\cdot}59873 \\ \dfrac{B}{2\,C} & = & 0{\cdot}003441 \\ \dfrac{B^2}{4\,C^2} & = & 0{\cdot}00001184 \end{array}\right\} \text{giving } p \text{ in lbs. per square foot.}$$

$A = 6{\cdot}4095$ giving p in inches of mercury.
$A = 6{\cdot}1007$ giving p in lbs. per square inch.

For accuracy this formula leaves little to be desired, but it requires considerable calculation, especially for finding the temperature from the pressure. Where so great accuracy is not required, the following simple formula gives results that may be relied upon for practical purposes over a large range of the scale: —

$$\log p = \frac{5(\text{T}-212)}{\text{T}+367} \dots (11)$$

$$\text{T} = \frac{2895}{5-\log p} - 367 \dots (12)$$

which gives the pressure p in atmospheres of 29·922 inches of mercury or 14·7 lbs. per square inch.

Example 1. — For instance, let T be given = 230° Fahr.,

$$\text{then } \log p = \frac{5 \times 18}{597} = 0{\cdot}15075 = \log 1{\cdot}415.$$

At 230° Fahr. therefore the pressure is 1·415 atmosphere = 1·415 × 29·922 = 42·339 inches of mercury = 1·415 × 14·7 = 20·8 lbs. per square inch, or 20·8 − 14·7 = 6·1 lbs. above the atmospheric pressure.

Example 2. — Again, let p = ·691 atmosphere

$$= \frac{2895}{5-\bar{1}{\cdot}8395} - 367 =$$

$$\frac{2895}{5{\cdot}1605} - 367 = 193^\circ{\cdot}99 \text{ Fahr.}$$

Example 3. — We may also calculate the case given in Example (1) by Mr. Rankine's formula; here $\tau = \text{T} + 461{\cdot}2 = 230^\circ + 461^\circ{\cdot}2 = 691^\circ{\cdot}2$.

$$\begin{array}{rcl} \log \text{B} &=& 3{\cdot}43642 \\ -\log \tau &=& \underline{2{\cdot}83960} = 0{\cdot}59682 = \log\ 3{\cdot}9521 \end{array}$$

$$\begin{array}{rcl} \log \text{C} &=& 5{\cdot}59873 \\ -2\log \tau &=& \underline{5{\cdot}67921} = \bar{1}{\cdot}91952 = \log.\ 0{\cdot}8308 \\ && \qquad\qquad\qquad\qquad \overline{4{\cdot}7829} \end{array}$$

For lbs. per square inch 6·1007 − 4·7829 = 1·3178 = log 20·79 lbs.

For inches of mercury 6·4095 − 4·7829 = 1·6266 = log 42·33 inches.

These results are almost identical with those given by the preceding formula.

The following table may serve as a guide in the use of these formulæ, showing how far they are accurate and within what limits on the scale they may be used with safety: —

Temperature Fahr.	Pressure of Steam.				
	Regnault's Tables.	Tate's Formula. (11.)		Rankine's formula. (9.)	
	Inches.	Inches.	Error.	Inches.	Error.
− 25·6	·0126	·0099	− ·0027	·01113	− ·0015
+ 32·0	·1811	·1661	− ·0150	·1734	− ·0077
69·8	·7265	·7051	− ·0214	·7200	− ·0065
100·4	1·9410	1·915	− ·0260	1·936	− ·0050
150·8	7·6791	7·674	− ·0051	7·695	+ ·0159
212·0	29·9218	29·922	0	29·922	0
257·0	68·658	68·640	− ·018	68·65	− ·008
302·0	140·995	140·81	− ·185	140·87	− ·125
347·0	264·471	263·86	− ·611	264·20	− ·271
392·0	460·204	458·98	− 1·224	459·90	− ·304
437·0	751·866	750·31	− 1·556	751·98	+ ·12

The errors of the numbers given by the formulæ are placed beside them for comparison.

Table V.—Of the Pressure and Corresponding Temperature of Saturated Steam, obtained from the Tables of M. Regnault by Interpolation and Reduction to English Measures.

Pressure in lbs. per square inch.	Temperature in degrees Fahr.	Rise of Temperature for 1 lb. Pressure.	Pressure in lbs. per square inch.	Temperature in degrees Fahr.	Rise of Temperature for 1 lb. Pressure.	Pressure in lbs. per square inch.	Temperature in degrees Fahr.	Rise of Temperature for 1 lb. Pressure.
1	101·98	24·28	31	252·09	1·85	70	302·71	0·93
2	126·26	15·35	32	253·94	1·76	75	307·38	0·89
3	141·61	11·47	33	255·70	1·77	80	311·83	0·85
4	153·08	9·25	34	257·47	1·68	85	316·00	0·81
5	162·33	7·79	35	259·15	1·68	90	320·03	0·77
6	170·12	6·78	36	260·83	1·61	95	323·87	0·74
7	176·90	6·00	37	262·44	1·60	100	327·56	0·71
8	182·90	5·41	38	264·04	1·54	105	331·10	0·68
9	188·31	4·92	39	265·58	1·53	110	334·51	0·66
10	193·23	4·54	40	267·12	1·49	115	337·84	0·63
11	197·77	4·19	41	268·60	1·47	120	340·99	0·61
12	201·96	3·92	42	270·07	1·43	125	344·06	0·59
13	205·88	3·67	43	271·50	1·41	130	347·05	0·57
14	209·55		44	272·91	1·39	135	349·93	0·57
14·7	212·00	3·47	45	274·30	1·35	140	352·76	0·56
15	213·02	3·27	46	275·65	1·34	145	355·6	0·55
16	216·29	3·14	47	276·99	1·31	150	358·3	0·51
17	219·42	2·96	48	278·30	1·29	160	363·4	0·48
18	222·37	2·82	49	279·59	1·26	170	368·2	0·47
19	225·19	2·72	50	280·85	1·25	180	372·9	0·46
20	227·91	2·63	51	282·60	1·22	190	377·5	0·43
21	230·54	2·54	52	283·32	1·21	200	381·8	0·42
22	233·08	2·35	53	284·53	1·20	210	386·0	0·39
23	235·43	2·32	54	285·73	1·17	220	389·9	0·39
24	237·75	2·25	55	286·90	1·15	230	393·8	0·37
25	240·00	2·16	56	288·05	1·14	240	397·5	0·36
26	242·16	2·10	57	289·19	1·12	250	401·1	0·34
27	244·26	2·04	58	290·31	1·11	260	404·5	0·34
28	246·32	1·98	59	291·42	1·09	270	407·9	0·33
29	248·30	1·93	60	292·51	1·05	280	411·2	0·32
30	250·23	1·86	65	297·77	1·01	290	414·4	0·31
31	252·09		70	302·71		300	417·5	

On the Relation of Temperature and Density of Saturated Steam.

Notwithstanding the very numerous experimental researches on the relation of pressure and temperature of steam, the relation of temperature and density, which is equally important in the calculations of the steam-engine, has, till recently, been investigated by theoretical investigations alone. By the method of Dumas it was found that, in becoming vapour, a cubic unit of water expanded to 1669 cubic units of steam, and from this single datum the density and volume at all other temperatures has been calculated, on the assumption that steam follows the same laws of expansion and contraction, under the influence of temperature and pressure, as a perfect gas.

The gaseous laws, or the laws of the relation of volume, pressure, and temperature of a perfect gas may be enumerated as follows:—

1. Mariotte's or Boyle's law; the pressure or elasticity is inversely as the volume when the temperature remains the same. That is, if a volume of gas of 10 cubic feet volume, under a pressure of 15 lbs. per square inch, be subjected to a pressure of 30 lbs. per square inch, the volume will be diminished to 5 cubic feet; or, on the other hand, if the pressure be decreased to $7\frac{1}{2}$ lbs. per square inch, the volume will increase to 20 cubic feet. Expressed in a formula, putting P for the pressure when the volume is V, P_1 the pressure when the volume is V_1,—

$$\frac{P}{P_1} = \frac{V_1}{V} \ldots (13).$$

2. Gay-Lussac's or Dalton's law; the expansion of a given weight of an elastic fluid under a constant pressure is $\frac{1}{459}$th part of its volume at 0° Fahr. for every degree of increase of temperature. Expressed in a formula this law is,—

$$\frac{V}{V_1} = \frac{459 + t}{459 + t_1} \ldots (14).$$

Hence, also, if the volume be constant,

$$\frac{P}{P_1} = \frac{459 + t}{459 + t_1} \ldots (15)$$

and combining the two formulæ

$$\frac{V \times P}{V_1 \times P_1} = \frac{459 + t}{459 + t_1} \ldots (16)$$

that is, the product of the volume and pressure at one temperature, is to that product at another temperature, as the temperature in the first case to the temperature in the second, the temperatures being counted from the absolute zero, or a temperature of $-459°$ Fahr.

Now we have seen, that it has been determined experimentally for steam that when $t_1 = 212°$ Fahr., $P_1 = 14{\cdot}7$ and $V_1 = 1669$, and if we assume that steam is strictly gaseous, these data suffice for calculating the volume or density of the same weight of steam at any other temperature and pressure; substituting in (16) we get

$$V = 1669 \times 14{\cdot}7 \times \frac{459 + t}{671 \times P}$$

$$= 36{\cdot}5 \frac{459 + t}{P} \ldots (17).$$

Thus, if we take from the preceding table of the relation of temperature and pressure the corresponding numbers, and substitute them for t and P in the above formula, we shall get the theoretical volume at that temperature and pressure. Thus from Table V. we have $t = 281°$ when $P = 50$ lbs., then

$$V = 36{\cdot}5 \frac{459 + 281}{50} = 540$$

that is, a volume of 1669 cubic feet at 212°, would be reduced to 540 at 281°, and of course the density increased in the inverse ratio.

From this well-known formula all the tables of the density of steam, with one recent exception, have been deduced, on which calculations of the duty of the steam-engine have been founded.

Although experimentalists have for some time questioned the

truth of this theoretical formula, yet, up to a recent time, no reliable direct experiments had been made to test its truth. Yet a few years since, Dr. Joule and Professor Thomson announced, as the result of the application of the dynamical theory of heat, that for temperatures above 212° Fahr. there would prove to be a considerable deviation from the gaseous laws in the case of steam. In 1855, Professor Rankine gave a theoretical formula for the density of steam, confirmatory of Professor Thomson's views.* This formula deduces the volume from the latent heat, and is of the form

$$v - v' = \frac{\text{H}}{\text{L}} \dots (17)$$

where L is the latent heat of evaporation per cubic foot in foot pounds of energy, and H the latent heat of evaporation of one pound of steam in units of energy, and $v - v'$ is the increase of volume of one pound of the fluid in evaporating. As we have as yet not considered the subject of latent heat, we may express Professor Rankine's formula in another form, as giving the volume from the pressure and temperature. It is then

$$\text{v}' = \frac{772\ \{1091{\cdot}7 - {\cdot}7\,(\text{T} - 32)\} \times (\text{T} + 461{\cdot}2)^2}{2{\cdot}3026\ v'\ \text{p}\ \{\text{B}\,(\text{T} + 461{\cdot}2) + 2\ \text{C}\}} + 1.$$

where $\frac{v}{v'}$ is the specific volume of the steam, v' the volume of one pound of water at the temperature T; p the pressure of the steam at T temperature in pounds on the square foot; log B = 3·43642 ; log C = 5·59873.

About the same time Mr. Tate made some experiments with ether, which led him to the conclusion that, at pressures somewhat above the atmospheric, the vapour of this substance does not follow the gaseous laws. These experiments led to a comprehensive series of researches, undertaken by Mr. Tate in conjunction with myself, to ascertain the density of saturated steam at all pressures, by a new and original method.

The general features of our method of ascertaining the density of steam, consist in vaporising a known weight of water in a

* Proc. Roy. Soc. Edinb. 1855. These views have been further developed by Mr. Rankine in his Manual of the Steam Engine and other Prime Movers, in which are given full tables of the density of steam, agreeing well with the experimental results about to be detailed.

large glass globe—with a stem—of known capacity and devoid of air, and observing the exact temperature at which the whole of the water is just vaporised. Then, knowing the weight, volume, and temperature of the steam, its specific gravity may be calculated. In order to pursue this method with safety and with the requisite amount of accuracy, the following peculiarities of construction of the apparatus were adopted.

First, in order to secure the thin globe from bursting, and at the same time to have it uniformly heated, it is placed in a strong closed copper steam bath, having a thermometer and pressure gauge attached, and a strong glass tube, closed at its exterior extremity, for receiving the stem of the globe. By this arrangement the glass globe is secured from bursting, for whatever may be the elasticity of the steam, the internal pressure in the globe is balanced by the external pressure in the steam bath.

Second, when a given weight of water is vaporised in a closed vessel devoid of air, the steam is said to be in a state of saturation so long as any portion of the liquid remains in the vessel. But after all the water is vaporised, heat being still applied, the steam becomes superheated, or heated beyond the temperature just requisite for vaporising all the water. By way of distinction we call this point the maximum temperature of saturation. Now as we have to find by observation the temperature of the steam exactly at the point when the whole of the water is vaporised, the determination of this with sufficient accuracy and delicacy has hitherto formed the great practical difficulty attending experimental researches on the density of vapours. We have overcome this difficulty by using what may be called a *saturation gauge*, the form of which varies according to circumstances, but the principle on which it is constructed may be illustrated as follows:

Imagine two globes, A, B, fig. 151, connected by a bent tube containing mercury, and immersed in a large bath of liquid to secure uniformity of temperature; suppose these globes devoid of air but containing weighed portions of water, say twenty grains in A and thirty in B. If heat be now applied to the liquid bath so as to increase progressively the temperature of the globes, this weighed portion of water will gradually pass into steam, and the elastic force in each globe will increase in a ratio corresponding with the temperature, but without in the least affecting the uniformity of level of the mercury columns C and D, because the pressure on each side will be the same. But when the whole of the water in globe A has been evaporated, this equality of pressure will no longer exist and the column C will rise. The pressure in B increases in the ratio for saturated steam, whilst that in A increases in the much smaller ratio of superheated steam, and hence the difference of level of the columns. The instant at which the columns begin to rise on one side and fall on the other, is the point at which the whole of the water in A is converted into steam; and the temperature then noted is the maximum temperature of saturation. The following theoretical table gives approximately the rise of the mercury column at several temperatures: —

Fig. 151.

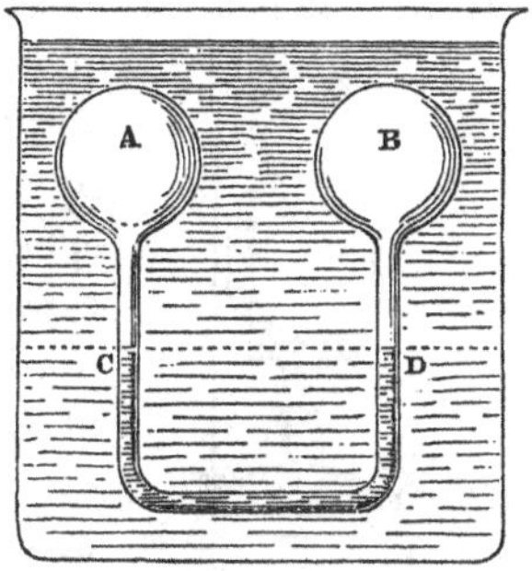

Saturated Steam.		Increments of Pressure for 1° Fahr.		
Pressure.	Temperature.	For expansion.	For vaporisation.	Difference.
At 4 lbs. and	152°	0·012	0·222	0·210
7 ,,	176°	0·022	0·32	0·30
15 ,,	213°	0·044	0·60	0·56
20 ,,	228°	0·060	0·80	0·74
61 ,,	295°	0·160	2·00	1·84
74 ,,	308°	0·200	2·22	2·02

The increments of pressure in this table are measured in inches of mercury. Their difference shows the rise of the mercury column on the side on which expansion from superheating is taking place. That is, the columns would diverge from the level 210 inch at 152° F., 0·56 inch at 213°, 2·02 inch at 308°, and so on.

For reasons which will hereafter be obvious, it was found impossible to determine the instant at which the whole of the water in the globe was vaporised and the columns diverged. The cohesion of the glass to the last particles of water, the foggy condition of the steam, and other causes, rendered it necessary to superheat the steam a few degrees, and then having very carefully determined the difference of level of the columns, to estimate from these data the maximum temperature of saturation.

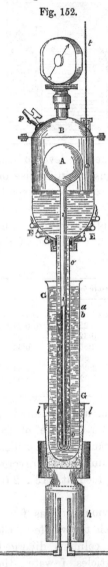

In fig. 152 is shown a sectional elevation of the apparatus employed in these researches for pressures varying from 15 lbs. to 70 lbs. on the square inch, or from one to five atmospheres. A is the glass globe of measured capacity for the reception of the weighed portion of water, drawn out into a stem about 32 inches long. The average size of the globes was $5\frac{1}{4}$ inches diameter or 75 cubic inches capacity; the stems were $\frac{3}{8}$ to $\frac{7}{16}$ inch bore. B B is the copper boiler, or steam bath in which the globe was heated uniformly throughout. The copper bath is prolonged by a strong glass tube, *o o*, $1\frac{1}{4}$ inches in diameter, and closed at the bottom; this tube is fixed to the boiler by a stuffing box, its upper part being trumpet-mouthed to prevent its being forced out by the pressure. The joint in the stuffing box was made by a ring of vulcanised india-rubber, which at the temperatures required in this series of experiments, answered its purpose perfectly. To heat this outer glass tube, which was peculiarly liable to explode, and, in fact, on two occasions did so, an outer oil bath, G G, was used, made of blown glass, twenty inches long, and resting in a sand bath, *l l*. This bath was supported on a tripod. The copper bath was heated by a coil of gas jets, E E; and the oil bath by a large wire gauge lamp, *h*, protected from draughts by a muffle, K K. The temperature thus obtained and distributed uniformly throughout the glass tube and steam bath by convection, was measured by a thermometer in the

Fig. 152.

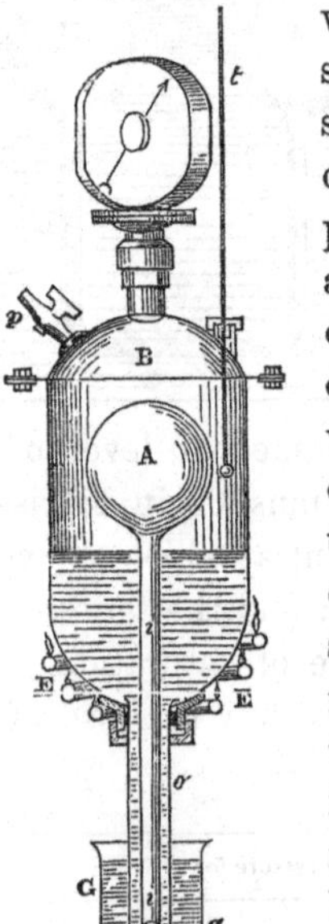

oil bath, and another *t*, exposed naked on the steam bath, and fixed in a stuffing box. Opposite the thermometer is a stopcock *p*, and on the top of the boiler a pressure gauge, for roughly indicating the pressure in the boiler. The copper boiler replaced the globe B in the diagram, fig. 151. The two mercury columns, the outer in the tube *o o*, and the inner in the stem of the globe *i i*, separate the vapour and water in the steam bath from that in the globe, and form the saturation gauge to which reference has been made. So long as the steam in the globe A remains in a state of saturation, the inner column remains stationary at a point a little above the level of the outer column, so as to balance the column of water in the steam bath B B. But when in raising the temperature the whole of the water in A is evaporated, and the steam begins to superheat, then the pressure of the steam in A no longer balances that of the steam in B, and the columns diverge: the difference of level forming a measure of the expansion of the steam. It was found in practice a matter of the utmost importance that the observer should not, in these experiments, trust to the unaided eye to determine the point at which the columns began to diverge, but that a careful series of measurements of the difference of level of the columns should be made, not only near the saturation point, but also at various temperatures of superheating; thus affording data for determining the law of expansion near the saturation point, and for estimating the maximum temperature of saturation from a point at which the error from the cohesion between the water and the glass, and the error from the retention of portions of water in the steam itself, might both be eliminated. It was also found advisable to take these readings of the levels of the columns, rather in a descending than in an ascending series of temperatures.

To read the column levels with rapidity and facility, seeing that they could not be approached within six or eight inches, a simple form of cathetometer was devised, sufficiently accurate for the purpose. It consisted of a telescope with cross wires sliding on a vertical graduated iron stem, and carrying a vernier for reading off the levels to the one-hundredth of an inch.

Fig. 153.

Fig. 154.

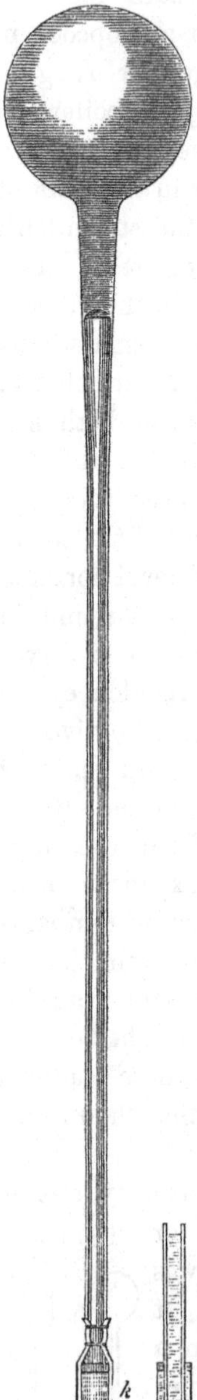

The steps in the process for determining the specific gravity of steam by this apparatus were as follows: —

A glass globule of a size to contain, as nearly as might be, the required quantity of water for vaporisation, was selected from a series (fig. 153). These globules had open stems, and after being filled with water were immersed hot in a cup of mercury, so that, in cooling, the mercury should rise into and fill the capillary stem. The weight of the water introduced was easily ascertained by deducting from the weight after filling, the weight of the dry cup, globule, and mercury. In this state the cup of mercury was transferred, and the globule passed into the large globe, in which a Torricellian vacuum had been previously formed.

To form the Torricellian vacuum, the globe, dried and filled with warm mercury, was heated on a sand bath until the mercury boiled; the stem was then filled with dry mercury, and the globe inverted, with its stem inserted in a basin of mercury. The globule was then introduced into the stem, and allowed to ascend into the globe. In order to transfer the globe from the basin to its place in the steam bath, a cup *k*, fig. 154, filled with mercury was suspended from the stem by an indian-rubber strap, a platinum wire being inserted between the cup and globe stem to ensure free passage for the mercury. The cover of the boiler B B being then taken off, and the outer tube *o o* dried and partially filled with dry mercury, the globe was raised and inserted into its place, resting on a tripod in the boiler. The cover was then fixed with a flax and red lead joint, and the cock *p* connected with an air pump. Exhaustion was effected, so that the columns in the globe stem and outer tube stood nearly level; the air pump was then removed, and a portion of water allowed to enter through the cock. The gas lights were then kindled; and until the water attained the boiling point, the columns were maintained level by means of the air pump, to prevent the possible entrance of water into the globe. After boiling for a time the cock p was closed, and the process of vaporisation went on simultaneously in the bath and globe, the temperature being kept sufficiently high in the oil bath G G to maintain the water in the outer tube in a state of ebullition. The temperature of the baths is slowly and uniformly raised, until the temperature of the vapour in the globe is considerably above the maximum temperature of saturation. After having been maintained for a considerable period at this temperature, the levels of the columns were observed; then the temperature being allowed to sink some degrees, the operation was repeated, and the temperature again reduced; and so on until the columns became stationary, indicating saturated steam in the globe as well as on the boiler. A series of readings was taken at each temperature, to make sure that the globe had attained a uniformity of temperature. At the same time the levels of some file marks on the stem were taken, by which the capacity of the globe in each position of the mercury column could be determined. All the elements were thus obtained for calculating the density of the steam.

Let W be put for the weight of distilled water at 39°·1 Fahr., filling the globe to the point at which the mercury columns stood at the maximum temperature of saturation. Let w be the weight of water vaporised; v the specific volume of the steam, or the number of times the volume of steam exceeds the volume of the water from which it is raised, then:—

$$v = \frac{W}{w} \dots (19).$$

By at once superheating the steam in the globe, and then slowly reducing the temperature until the maximum temperature of saturation is reached, we secure the following advantages:— The cohesion of the water to the surface of the glass being overcome, that force, it may be presumed, cannot be regained until the glass again becomes wet, which can only occur on condensation, that is, by the reduction of the temperature below that which corresponds to the maximum temperature of saturation.

Moreover, the observation of the columns at different temperatures of superheating, not only supplies us with data for ascertaining the maximum temperature of saturation, but also for determining the law of expansion of superheated steam near the saturation point.

The following table gives the temperature of saturation deduced from the experiments with the above apparatus from the two highest temperatures of superheating attained in each case. Where the lower of these temperatures is manifestly within the limits of imperfect expansion, the reduction from higher temperature only has been retained.

TABLE VI.—RESULTS OF EXPERIMENTS ON THE DENSITY OF STEAM AT PRESSURES OF FROM 15 TO 70 LBS. PER SQUARE INCH.

Number of Exper.	Maximum Temperature of Saturation, Fahr.	Pressure of Steam in Inches of Mercury.	Specific Volume of the Steam.
1	242·89° 242·92 } 242·90	53·60 53·63 } 53·61	943·1
2	244·90 244·74 } 244·82	55·60 55·44 } 55·52	908·0
3	245·42 245·02 } 245·22	56·08 55·70 } 55·89	892·5
4	255·37 255·62 } 255·50	66·70 66·97 } 66·84	759·4
5	263·20 263·09 } 263·14	76·26 76·13 } 76·20	649·2
6	267·35 267·08 } 267·21	81·71 81·36 } 81·53	635·3
7	269·24 269·16 } 269·20	84·36 84·25 } 84·20	605·7
8	274·76	92·23	584·4
9	273·30	90·08	543·2
10	279·42	99·68	515·0
11	282·55 282·61 } 282·58	104·48 104·60 } 104·54	497·2
12	287·49 287·00 } 287·25	112·82 112·75 } 112·78	458·3
13	292·53	122·25	433·1
14	288·25	114·25	449·6

A similar series of experiments was obtained at pressures less than 15 lbs. per square inch, but in this case the saturation gauge was abandoned. The stem of the globe was immersed at bottom into a cistern of mercury open to the atmosphere; in other respects the method of the experiment was precisely the same. The water was introduced, the globe heated; and as vaporisation went on, the mercury column descended in proportion to the increase of the elasticity of the vapour. Simultaneous readings of a barometer were taken; and by deducting from the height of the mercurial column in the barometer, the height of that in the globe stem. So long as the vapour in the globe was in a condition of saturation, its elasticity thus found corresponded with that in M. Regnault's tables. When it became superheated, the ratio of increase of elasticity was very greatly reduced, and the column became almost stationary. The superheating was carried in these experiments to twenty or thirty degrees above the saturation point. The principle of the experiments was therefore entirely unchanged, the only alteration being that the elasticity of the saturated steam was obtained from previous experiments, and that of superheated steam observed, and the difference of level of saturated and superheated steam obtained by subtracting the one from the other, instead of being directly observed.

The following table gives the results obtained in this series of experiments reduced on the same principle as the last:—

Table VII.—The Results of Experiments on the Density of Steam at Pressures below that of the Atmosphere.

Number of Exper.	Maximum Temperature of Saturation, Fahr.	Pressure of Steam in Inches of Mercury.	Specific Volume of the Steam.
1	136°·85 136·70 } 136·77	5·36 5·34 } 5·35	8275·3
2	155·38 155·28 } 155·33	8·64 8·61 } 8·62	5333·5
3	159·35 159·35 159·40 } 159·36	9·45 9·45 9·46 } 9·45	4920·2
4	170·88 170·96 } 170·92	12·46 12·48 } 12·47	3722·6
5	171·52 171·44 } 171·48	12·63 12·60 } 12·61	3715·1
6	174·92	13·62	3438·1
7	182·26 182·34 } 182·30	16·01 16·02 } 16·01	3051·0
8	188·30	18·36	2623·4
9	198·78	22·88	2149·5

These results show that the density of saturated steam at all temperatures above as well as below 212°, is invariably greater than that derived by calculation from the gaseous laws.

As we propose extending these experiments to higher pressures, it is premature to venture on any elaborate generalisation of the results we have attained. The following formulæ, however, express with much exactness the relation between temperature and volume, and between pressure and volume, as indicated by our experiments.

Let v be the specific volume of saturated steam, at the pressure P, measured by a column of mercury in inches; then

$$v = 25{\cdot}62 + \frac{49513}{P - 72} \quad \ldots \; (20)$$

$$P = \frac{49513}{v - 25{\cdot}62} - 0{\cdot}72 \quad \ldots \; (21)$$

The following numbers show the agreement of these formulæ with the experimental results: —

Temperature Fahr.	Specific Volume		Proportional Error of Formula.
	By Experiment.	By Formula.	
136°·77	8275·3	8183	$-\frac{1}{90}$
155·33	5333·5	5326	$-\frac{1}{762}$
159·36	4920·2	4900	$-\frac{1}{246}$
170·92	3722·6	3766	$+\frac{1}{87}$
171·48	3715·1	3740	$+\frac{1}{149}$
174·92	3438·1	3478	$+\frac{1}{86}$
182·30	3051·0	2985	$-\frac{1}{46}$
188·30	2623·4	2620	$+\frac{1}{874}$
198·78	2149·5	2124	$-\frac{1}{90}$
242·90	943·1	937	$-\frac{1}{157}$
244·82	908·0	906	$-\frac{1}{454}$
245·22	892·5	900	$+\frac{1}{111}$
255·50	759·4	758	$-\frac{1}{759}$
263·14	649·2	669	$+\frac{1}{32}$
267·21	635·3	628	$-\frac{1}{91}$
269·20	605·7	608	$+\frac{1}{304}$
274·76	584·4	562	$-\frac{1}{26}$
273·30	543·2	545	$+\frac{1}{271}$
279·42	513·0	519	$+\frac{1}{128}$
282·58	497·2	496	$-\frac{1}{497}$
287·25	458·3	461	$+\frac{1}{152}$
292·53	433·1	428	$-\frac{1}{86}$
288·25	449·6	456	$+\frac{1}{75}$

We have also computed the following table from the experimental formula, which exhibits at a glance the pressure, volume, and weight of saturated steam, and will enable the reader to ascertain the necessary data for calculations at all pressures from 1 to 250 lbs. per square inch :—

General Table (VIII.) of the Relation of Pressure, Volume, and Weight of Saturated Steam deduced from experimental Data.

Pressure. In lbs. per sq. inch.	Pressure. In Inches of Mercury.	Specific Volume.	Decrease of specific Volume per lb. Pressure.	Weight of a Cubic Foot of Steam.
1	2·0361	17990·6	7633·0	·00347
2	4·0722	10357·6	3081·0	·00602
3	6·1083	7276·6	1666·0	·00858
4	8·1444	5610·6	1042·5	·01112
5	10·1805	4568·1	715·5	·01258
6	12·217	3852·6	521·0	·01620
7	14·253	3331·6	395·0	·01874
8	16·289	2936·6	311·2	·02126
9	18·325	2625·4	251·1	·02377
10	20·361	2374·3	206·9	·02630
11	22·397	2167·4	173·4	·02880
12	24·433	1994·0	147·3	·03131
13	26·469	1846·7	126·9	·03380
14	28·505	1719·8		·03630
Atmospheric 14·7 Pressure.	29·922	1641·5	110·4	·03803
15	30·541	1609·4	96·8	·03878
16	32·577	1512·6	85·7	·04127
17	34·614	1426·9	76·3	·04375
18	36·650	1350·6	68·5	·04622
19	38·686	1282·1	62·1	·04869
20	40·722	1220·0	55·6	·05117
21	42·758	1164·4	50·9	·05361
22	44·794	1113·5	46·6	·05606
23	46·830	1066·9	42·8	·05851
24	48·866	1024·1	39·3	·06096
25	50·902	984·8	36·4	·06339
26	52·938	948·4	33·8	·06582
27	54·975	914·6	31·4	·06826
28	57·011	883·2	29·2	·07068
29	59·047	854·0	27·2	·07310
30	61·083	826·8	25·6	·07550
31	63·119	801·2	24·0	·07792
32	65·155	777·2	22·5	·08032
33	67·191	754·7	21·2	·08272
34	69·227	733·5	18·9	·08510
35	71·264	713·4	17·9	·08751
36	73·299	694·5	16·9	·08988
37	75·336	676·6	16·1	·09227
38	77·372	659·7	15·4	·09463
39	79·408	643·6	14·8	·09700
40	81·444	628·2	14·1	·09937
41	83·480	613·4	13·4	·10040
42	85·516	599·3	12·1	·10416
43	87·552	585·9	12·0	·10654
44	89·588	573·8	11·4	·10880
45	91·625	561·8	10·9	·11111
46	93·607	550·4	10·5	·11342
47	95·697	539·5	10·4	·11571
48	97·733	529·0	10·1	·11801
49	99·769	518·6	9·4	·12037
50	101·805	508·5		·12276

Pressure. In lbs. per sq. inch.	Pressure. In Inches of Mercury.	Specific Volume.	Decrease of specific Volume per lb. Pressure.	Weight of a Cubic Foot of Steam.
51	103·841	499·1	9·0	·12508
52	105·877	490·1	8·7	·12737
53	107·913	481·4	8·5	·12967
54	109·949	472·9	8·2	·13200
55	111·985	464·7	7·7	·13434
56	114·022	457·0	7·4	·13660
57	116·058	449·6	7·2	·13884
58	118·094	442·4	7·1	·14111
59	120·130	435·3	6·8	·14341
60	122·166	428·5	6·5	·14568
61	124·202	422·0	6·4	·14793
62	126·238	415·6	6·2	·15021
63	128·274	409·4	5·9	·15248
64	130·310	403·5	5·8	·15471
65	132·346	397·7	5·6	·15697
66	134·383	392·1	5·5	·15921
67	136·419	386·6	5·3	·16147
68	138·455	381·3	5·2	·16372
69	140·491	376·1	4·9	·16598
70	142·527	371·2	4·8	·16817
71	144·563	366·4	4·7	·17038
72	146·599	361·7	4·6	·17259
73	148·635	357·1	4·5	·17481
74	150·671	352·6	4·3	·17704
75	152·708	348·3	4·2	·17923
76	154·744	344·1	4·1	·18142
77	156·780	340·0	4·0	·18360
78	158·816	336·0	3·9	·18579
79	160·852	332·1	3·8	·18797
80	162·888	328·3	3·7	·19015
81	164·924	324·6	3·7	·19232
82	166·960	320·9	3·6	·19454
83	168·996	317·3	3·5	·19674
84	171·032	313·9	3·4	·19887
85	173·069	310·5	3·3	·20105
86	175·105	307·2	3·2	·20321
87	177·141	304·0	3·2	·20535
88	179·177	300·8	3·1	·20753
89	181·213	297·7	3·0	·20970
90	183·249	294·7	2·9	·21183
91	185·285	291·8	2·9	·21393
92	187·321	288·9	2·8	·21608
93	189·357	286·1	2·8	·21819
94	191·393	283·3	2·7	·22045
95	193·429	280·6	2·6	·22247
96	195·466	278·0	2·6	·22455
97	197·502	275·4	2·6	·22667
98	199·538	272·8	2·5	·22883
99	201·574	270·3	2·4	·23095
100	203·610	267·9	2·19	·23302
110	223·971	246·0	1·84	·25376
120	244·332	227·6	1·54	·27428
130	264·693	212·2	1·33	·29419
140	285·054	198·9	1·20	·31386
150	305·415	186·9	0·96	·33400

Pressure. In lbs. per sq. inch.	Pressure. In Inches of Mercury.	Specific Volume.	Decrease of specific Volume per lb. Pressure.	Weight of a Cubic Foot of Steam.
160	325·776	177·3	0·90	·35209
170	346·137	168·3	0·78	·37092
180	366·498	160·5	0·72	·38895
190	386·859	153·3	0·64	·40722
200	407·220	146·9	0·57	·42496
210	427·581	141·2	0·53	·44211
220	447·942	135·9	0·47	·45935
230	468·303	131·2	0·44	·47581
240	488·664	126·8	0·41	·49232
250	509·025	122·7		·50877

In the above table, which is probably the first calculated from direct experimental data, the third column is calculated by means of formula (20), and the last by dividing the weight of a cubic foot of water, at the temperature of 39°·1 Fahr., by the specific volume of the steam, that is, by the volume to which a cubic foot of water expands when converted into steam.

It will be interesting to compare the numbers given by the above table with those which are obtained in an entirely independent manner from Mr. Rankine's formula in which the volume is deduced from the latent heat. The following numbers show a near agreement in the results. With these has been placed at the same time a column giving the results deduced from the gaseous laws as they have hitherto been generally received.

Comparison of the Values of the specific Volume of Saturated Steam, from the Formulæ of Mr. Fairbairn and Mr. Tate, Mr. Rankine, and from the Gaseous Laws.

Temperature, Fahr.	Pressure in Inches of Mercury.	Specific Volume. Fairbairn and Tate.	Specific Volume. Rankine.	Specific Volume. Gaseous Laws.
104°	2·1618	17207	$19520+\frac{1}{7}$	$19390+\frac{1}{7}$
140	5·8578	7553	$7620+\frac{1}{116}$	$7611+\frac{1}{130}$
176	13·9621	3397	$3367-\frac{1}{113}$	$3385-\frac{1}{283}$
212	29·9218	1641·5	$1645+\frac{1}{410}$	$1700+\frac{1}{28}$
248	58·7116	858·7	$874+\frac{1}{37}$	$941{\cdot}8+\frac{1}{10}$
284	106·9930	485·3	$498+\frac{1}{37}$	$516{\cdot}8+\frac{1}{15}$
320	183·1342	294·9	$301+\frac{1}{50}$	$316{\cdot}6+\frac{1}{13}$
356	297·1013	191·9	$191-\frac{1}{213}$	$204{\cdot}1+\frac{1}{15}$

The fractions in the two last columns indicate the proportional deviation from the experimental formula. Within the limits of the experiments below the atmospheric pressure, that is, between 136° and 212°, both Mr. Rankine's and the gaseous results agree closely with our experiments. Above 212° both Mr. Rankine's and our own results deviate considerably from those obtained on the assumption of the gaseous laws, whilst at the same time the two former approximate nearly in value.

Fig. 155, represents graphically the relations which have been described in this section, the spheres representing the volume assumed by the same weight of steam at the respective pressures shown on the figure. It may also serve to show how the density increases and the volume decreases with the pressure according to the law determined by our experiments.

On the Latent Heat of Steam at different Pressures.

It has already been explained that in all changes in the state of aggregation of bodies heat becomes latent or sensible. If a body passes from the solid to the liquid, or from the liquid to the gaseous state, heat becomes latent; in the inverse process an equal amount of heat becomes sensible.

Black determined the amount of increase of heat in the water surrounding the worm of a still by the condensation of a weighed portion of steam, and found that the condensation of one pound of steam raised the temperature of an equal quantity of water 954° Fahr., an estimate which has since proved too low.

Watt investigated this subject in relation to the action of his condenser, and from his experiments concluded "*that the quantity of heat necessary to convert one pound of water at 32° into steam at any pressure is constant.*" That is, that the latent heat of steam decreases as the pressure rises, by as much as the sensible heat increases, the total heat being constant. If we take the total units of heat of steam at 212° to be 1146·6, and if λ be put for the total units of heat, and λ^1 for the latent heat at any other temperature T, then the law of Watt will be expressed by the formula —

$$\lambda = \lambda^1 + T = 1146{\cdot}6. \quad . \quad . \quad (22).$$

Fig. 155.

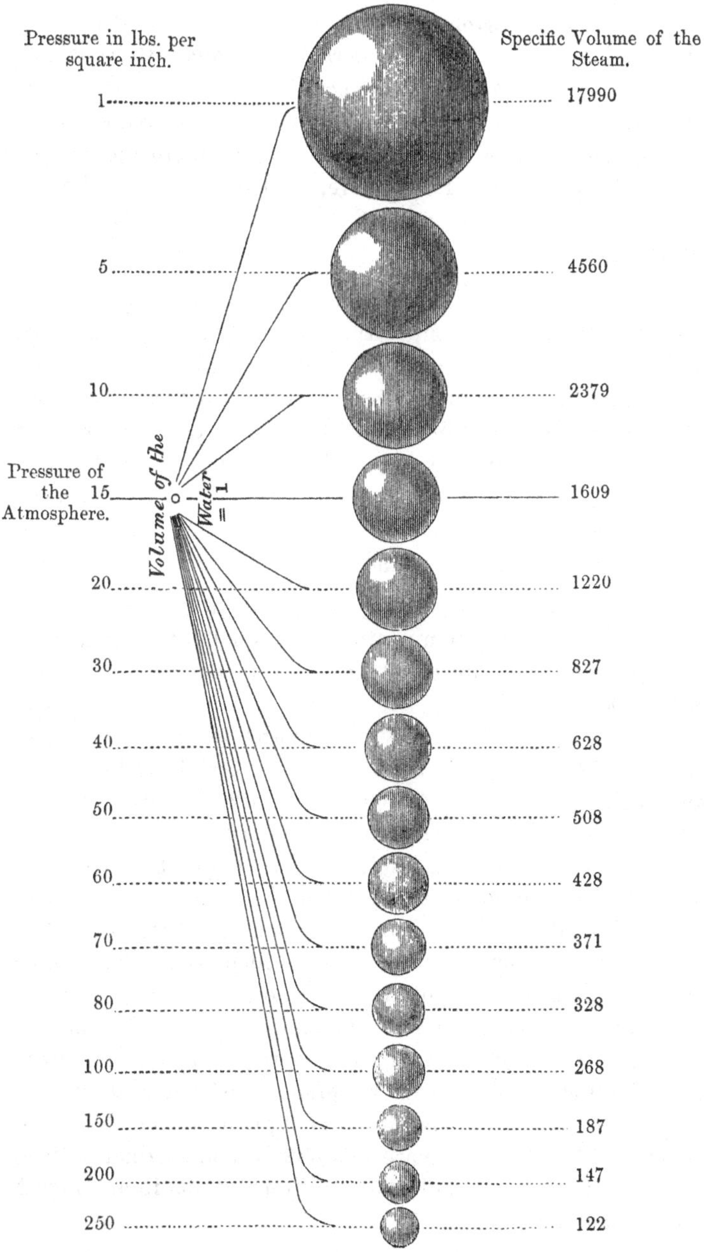
Pressure in lbs. per square inch.
Specific Volume of the Steam.
1
17990
5
4560
10
2379
Pressure of the Atmosphere. 15
1609
Volume of the Water = 1
20
1220
30
827
40
628
50
508
60
428
70
371
80
328
100
268
150
187
200
147
250
122

Subsequently to Watt, in 1803, Southern made some experiments on the same subject, from which he deduced a different law, which has since borne his name. He concluded "*that the latent heat of evaporation is constant at all pressures*," and that the total quantity of heat increases as the pressure rises, by as much as the sensible temperature rises. Taking the same numbers as before, we have the latent heat at 212° = 1146°·6 − 180° = 966·6; then Southern's law is expressed in the formula:—

$$\lambda = 966{\cdot}6 + (T - 32°) \quad . \quad . \quad . \quad (23).$$

Watt's law, from its simplicity, has generally been employed in calculations on the duty of the steam-engine; it has been reserved for Mr. Regnault to ascertain, with the same accuracy and care as he determined the relation of temperature and pressure of steam, the true law of the relation of the latent heat and the pressure. The law he has discovered is expressed in its simplest form in the formula—

$$\lambda = 1082 + \ 305\,T, \quad . \quad . \quad (24)$$

which shows that the total heat, λ, incorporated in a pound of saturated steam at the temperature T, is equal to the latent heat of evaporation of steam at 32° (nearly), increased by the product 0·305 T.

At 212° the total heat by all three formulæ will be the same, namely, 1146·6 units; but at 300° the total heat by Watt's law would still be only 1146·6; by Southern's it would be 1194·6; and by Regnault's it would be 1173·5.

The quantities of heat in the above paragraphs have been measured by "units of heat:" it will be necessary to explain what is intended by the phrase. The unit of heat, or *British thermal unit,* is the quantity of heat which would have to be added to one pound of pure liquid water, at or near its point of maximum density, to raise its temperature 1° Fahrenheit. The French thermal unit is the quantity of heat necessary to raise the temperature of 1 kilogramme of water at or near its point of maximum density 1° centigrade. It will be convenient to state that there are 3·96832 British thermal units in a French thermal unit, and 0·251996 French units in an English unit. (Rankine.)

The apparatus employed by Regnault to determine the latent heat of steam consisted of a boiler, condenser, artificial atmosphere and two precisely similar calorimeters. The boiler, made immensely strong, was of a capacity of 66 gallons, and was filled when required with newly-distilled water. From this a copper tube conveyed the steam to the calorimeters and condenser. This tube was surrounded by a dense stratum of steam in a jacket, which communicated with the same boiler, and terminated in a peculiarly formed cock, by which the steam could be sent to the calorimeters or condenser, as necessary. The condenser, a vessel of 13 gallons' capacity, kept cool by a stream of water, was employed merely to cause a continuous flow of steam through the apparatus, lest any portion should be cooled before entering the calorimeters. The air receiver, or artificial atmosphere, communicating with air pumps, was of the same character, and employed for the same purpose as in the experiments upon the relation of pressure and temperature, viz. to regulate the temperature of the steam in the boiler, by maintaining perfectly uniform the pressure at which ebullition takes place. The calorimeters for measuring the heat disengaged were the most essential part of this apparatus, and consisted of two red copper cylinders, with thin metal covers. The worm consisted of a first bulb A, of red copper, ·078 inch in thickness, into which the steam to be condensed passed directly; the condensed water and steam thence passed into a second bulb B, with a cock *r* placed outside the calorimeter; the same bulb B, had an upper tubulure, by

Fig. 156.

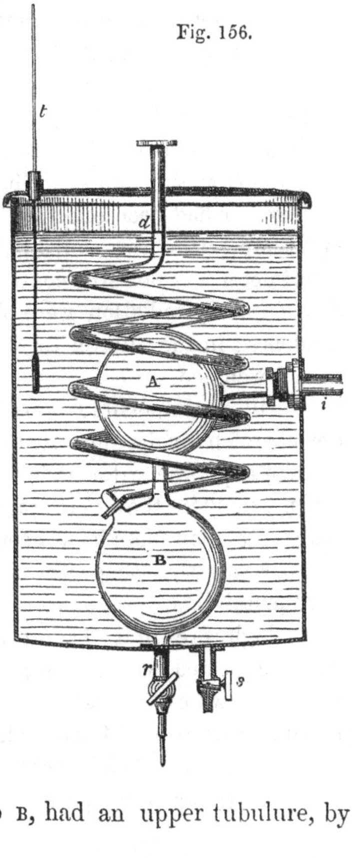

which it communicated with a copper worm. An agitator, or fan, of two discs of fluted copper, served to blend together the strata of water in the calorimeter during the experiment. The same volume of water was introduced into the calorimeters at every experiment, being measured in a gauging vessel. The mercurial manometer and forcing pumps were identical with those employed in the experiments on the relation of pressure and temperature. When complete, the apparatus was tested with a pressure of air of ten atmospheres, and every leak perfectly closed.

For experiments at the atmospheric pressure, every part of the apparatus exposed to condensation was covered with flannel. The distributing cock was placed so that no steam reached the calorimeters, and distillation was commenced and carried on for about an hour, to secure perfect uniformity of temperature throughout, and to expel the air. Cold water was then introduced into the calorimeters, and a portion of the steam sent through them, after a previous experiment for five minutes had been made as to the amount of heat communicated to the water by conduction at the joints. When sufficient condensation in the calorimeter had been effected, the cock was closed, and the time and temperature were noted; the agitation of the water in the calorimeter was however continued, and observations of the rate of cooling by radiation were obtained. The quantity of water condensed in the calorimeter was allowed to flow out at the cock *r*, and weighed.

It will be impossible here to enter in detail into all the devices to obtain data for calculating the corrections to be applied in each experiment for radiation, conduction, &c; nor the formulæ by which they were calculated. It is certain, however, that these allowances have been made with very great accuracy; in every case the theoretical formula has been checked by experimental data.

At high pressures the air pumps and artificial atmosphere were connected with the apparatus; in other respects the experiments were identical, and in this way results were obtained up to a pressure of 14 atmospheres, or 205 lbs. per square inch.

For pressures below that of the atmosphere the forcing pumps were replaced by the ordinary exhausting air pump, communicating with the reservoir of air. In this way experiments were obtained at pressures varying from 0·22 atmosphere to 0·64 atmosphere.

The most accurate formula for the latent heat of evaporation is Mr. Rankine's: —

$$l = 1091{\cdot}7 - 0{\cdot}695\,(\text{T} - 32^\circ) - 0{\cdot}000000103\,(\text{T} - 39^\circ{\cdot}1)^3$$

but for practical purposes the last factor may be omitted. In this way the following table of latent and total heat has been computed. The latent heat being calculated for a pound weight of steam, the units of heat required to raise the water from 32° to the boiling point are very nearly T − 32°.

TABLE IX.—THE LATENT AND TOTAL HEAT OF STEAM FROM 1 LB. TO 150 LBS. PER SQUARE INCH.

Pressure in lbs. per sq. inch.	Corresponding Temperature, Fahr.	Latent Heat of Evaporation in British thermal Units.	Total Heat from 32° Fahr. in British thermal Units.	Increment of total Heat per lb. Pressure.
1	101·98	1043·1	1113·1	
2	126·26	1026·1	1120·4	7·3
3	141·61	1015·4	1125·0	4·6
4	153·08	1007·4	1128·5	3·5
5	162·33	1006·3	1131·2	2·7
6	170·12	995·5	1133·6	2·4
7	176·90	990·8	1135·7	2·1
8	182·90	986·6	1137·6	1·9
9	188·31	982·8	1139·1	1·5
10	193·23	979·4	1140·6	1·5
11	197·77	976·1	1142·0	1·4
12	201·96	973·2	1143·2	1·3
13	205·88	970·5	1144·4	1·2
14	209·55	967·8	1145·4	1·0
14·7	212·00	966·1	1146·1	1·0
15	213·02	965·4	1146·4	
16	216·29	963·1	1147·3	·9
17	219·42	960·8	1148·2	·9
18	222·37	958·8	1149·1	·9
19	225·19	956·9	1150·1	·8
20	227·91	955·0	1150·8	·7
21	230·54	953·0	1151·5	·7
22	233·08	951·3	1152·3	·7
23	235·43	949·6	1153·0	·7
24	237·75	948·0	1153·7	·7
25	240·00	946·4	1154·4	·7
26	242·16	944·9	1155·1	·7
27	244·26	943·4	1155·7	·6
28	246·32	942·0	1156·3	·6
29	248·30	940·5	1156·9	·6
30	250·23	939·1	1157·5	·6
31	252·08	937·8	1158·0	·5
32	253·94	937·4	1158·3	·5
33	255·70	935·2	1159·0	·5

Pressure in lbs. per sq. inch.	Corresponding Temperature, Fahr.	Latent Heat of Evaporation in British thermal Units.	Total Heat from 32° Fahr. in British thermal Units.	Increment of total Heat per lb. Pressure.
34	257·47	934·0	1159·5	·5
35	259·65	932·8	1160·0	·5
36	260·83	931·6	1160·4	·5
37	262·43	930·4	1160·8	·5
38	264·04	929·3	1161·3	·4
39	265·57	928·2	1161·8	·4
40	267·11	927·1	1162·2	·4
41	268·59	926·1	1162·7	·4
42	270·07	925·1	1163·2	·4
43	271·49	924·0	1163·6	·4
44	272·91	923·0	1163·9	·4
45	274·28	922·0	1164·3	·4
46	275·65	921·1	1164·7	·35
47	276·97	920·1	1165·1	·35
48	278·30	919·2	1165·5	·3
49	279·57	918·2	1165·8	·3
50	280·85	917·3	1166·1	·3
52	283·32	915·6	1166·9	·3
54	285·73	913·9	1167·6	·3
56	288·05	912·2	1168·2	·3
58	290·31	910·6	1168·9	·3
60	292·51	909·1	1169·6	·3
65	297·77	905·3	1171·1	·3
70	302·71	901·8	1172·5	·3
75	307·38	898·5	1173·8	·26
80	311·83	895·1	1175·1	·26
85	316·00	892·2	1176·2	·22
90	320·03	889·2	1177·3	·22
95	323·87	883·9	1178·5	·22
100	327·56	881·3	1179·5	·20
105	331·10	878·9	1180·4	·20
110	334·51	876·5	1181·4	·20
115	337·84	874·2	1182·3	·20
120	340·99	872·0	1183·1	·19
125	344·06	872·0	1184·0	·18
130	347·05	869·9	1184·9	·16
135	349·93	867·8	1185·7	·16
140	352·76	865·7	1186·5	·15
150	358·3	861·7	1188·0	·13
200	381·7	845·0	1194·7	·11
250	401·1	831·2	1200·3	

On the Law of Expansion of Superheated Steam.

When steam is isolated from water and heated, it expands and decreases in density if the pressure be constant, or increases in pressure if the density (volume) be constant. In this state it is said to be *surcharged* or *superheated*, or, perhaps better, *gaseous* steam.

The earliest experiments on superheated steam are perhaps those of Frost in America, which, however, do not appear to be reliable. Mr. Siemens adopted a simple apparatus not long since, with which he sought to determine the rate of expansion of steam isolated from water, and which gave a high rate of expansion, but his conclusions are not at all borne out by my own experiments, and his apparatus does not appear calculated to yield very accurate results.

We are at this time prosecuting some researches on this subject, which are not yet advanced sufficiently for publication. In the meantime some results within a small range of superheating, obtained during the experiments on the density of saturated steam, approximate so closely to what might have been expected to be the law in this case, whilst at the same time they were made with great care, that I believe they are entitled to greater confidence than any previous attempts at the determination of this question.

Mr. Rankine, in the absence of data, has taken as the basis of his calculations on superheated steam, in his recently published "Manual of the Steam Engine," the assumption that superheated steam follows precisely the gaseous laws in its expansion under the influence of heat, that is, that

$$\frac{\mathrm{v}}{\mathrm{v}_1} = \frac{461{\cdot}2 + t}{461{\cdot}2 + t_1}$$

Mr. Siemens' experiments do not at all agree with this assumption; they would give a higher rate of expansion. But, with a certain proviso, my own results accord with it very nearly, and would seem to show that superheated steam expands at the same rate as a perfect gas.

The proviso to which I allude is this, that within a short distance of the temperature of maximum saturation, not exceeding about 20° Fahr., the rate of expansion is variable; close to the saturation point it is much higher than that of a perfect gas, but it rapidly decreases till, at a point at no great distance above the temperature of saturation, it becomes sensibly identical with that of a perfect gas. These results, however, do not extend over a sufficient range of temperature at present for us to deduce the true law, although their entirely independent coincidence with the laws already known to physicists is some guarantee for their accuracy.

By the rate of expansion we mean here the fraction expressing the increment of volume for one degree of temperature Fahrenheit; for air this fraction is:

$$r = \frac{1}{459 + t}$$

where t is the temperature of the gas. Thus at 212° the rate of expansion of a perfect gas is $\frac{1}{671}$, at 300° it is $\frac{1}{759}$, at 400 it is $\frac{1}{859}$; and so on at other temperatures.

Now in our experiments we may deduce the rate of expansion in a similar way, assuming it to be uniform for small increments of temperature; thus in experiment 6, in which the maximum temperature of saturation is 174°·92 Fahr., the coefficient of expansion for the steam between that temperature and 180° Fahr. is $\frac{1}{190}$, or three times that of air; whereas between 180° and 200° the coefficient is very nearly the same as that of air, being $\frac{1}{637}$ when air would be $\frac{1}{639}$; and the same rule is found in every experiment. The mean coefficient at zero of temperature from seven experiments below the atmospheric pressure, and calculated from a point several degrees above that of saturation, is $\frac{1}{438}$, whereas for air it is $\frac{1}{459}$, so that within the range of superheating obtained in these experiments the formula of expansion would be,

$$\frac{V}{V_1} = \frac{438 + t}{438 + t_1}.$$

The experiments seem to indicate that if the superheating had been carried further, the coefficient would have still more closely agreed with that which applies to incondensible gases. The following table gives the results upon which the previous generalisations have been founded, and which seem for the present the most reliable results we possess upon this subject. Before long I hope that we shall be able to lay before the public some direct experiments upon this subject, carried to a high degree of superheating.

The following table gives the value of the coefficient of expansion for superheated steam, taken at different intervals of temperature from the maximum temperature of saturation:

TABLE SHOWING THE COEFFICIENT OF EXPANSION OF SUPERHEATED STEAM.

No. of the Exper.	Maximum temperature of saturation.	Temperatures between which the expansion is taken.		Coefficient of Expansion of Steam.	Coefficient of Expansion of Air.
1	136°·77	140°	170°	$\frac{1}{593}$	$\frac{1}{599}$
2	155 ·33	160	190	$\frac{1}{556}$	$\frac{1}{619}$
3	159 ·36	159 ·36	170·2	$\frac{1}{150}$	$\frac{1}{618}$
		170 ·2	209·9	$\frac{1}{624}$	$\frac{1}{629}$
5	171 ·48	171 ·48	180	$\frac{1}{200}$	$\frac{1}{630}$
		180	200	$\frac{1}{604}$	$\frac{1}{639}$
6	174 ·92	174 ·92	180	$\frac{1}{190}$	$\frac{1}{634}$
		180	200	$\frac{1}{637}$	$\frac{1}{639}$
7	182 ·30	182 ·3	186	$\frac{1}{230}$	$\frac{1}{641}$
		186	209·5	$\frac{1}{630}$	$\frac{1}{645}$
8	188 ·30	191	211	$\frac{1}{604}$	$\frac{1}{650}$
1	242 ·9	243	249	$\frac{1}{517}$	$\frac{1}{702}$
4	255 ·5	257	259	$\frac{1}{392}$	$\frac{1}{716}$
		257	264	$\frac{1}{600}$	$\frac{1}{716}$
6	267 ·21	268	271	$\frac{1}{210}$	$\frac{1}{727}$
		271	279	$\frac{1}{640}$	$\frac{1}{730}$
7	269 ·2	271	273	$\frac{1}{232}$	$\frac{1}{730}$
		273	279	$\frac{1}{551}$	$\frac{1}{733}$
9	279 ·42	283	285	$\frac{1}{298}$	$\frac{1}{742}$
		285	289	$\frac{1}{633}$	$\frac{1}{744}$
13	292 ·53	297	299	$\frac{1}{281}$	$\frac{1}{756}$
		299	302	$\frac{1}{638}$	$\frac{1}{758}$

The limited extent to which the superheating is carried leaves the question of efficiency at higher temperatures unsolved. We believe we are perfectly conversant with the costly machinery that is used for this purpose on board ship and in other places, but until more reliable data have been determined by direct experiment, it would be premature to pronounce by what law the advantages assumed to result from its use are produced. We hope in entering on this inquiry to arrive at conclusions founded on the unerring principles of physical truth.

CHAP. VII.

VARIETIES OF STATIONARY STEAM ENGINES.

The steam engine as an instrument of propulsion is at the present time of such vast importance as to sink into insignificance every other known agent as a motive power. We have already considered the best methods by which the power of water can be utilised, but the whole of the water power in Great Britain falls immeasurably short of that obtained from steam, in every department of useful art. If we were to stop for a moment to compare the amount of steam power employed in industrial operations, with that of wind or water, we should find that the latter were mere fractions in the sum; and looking forward to still further developments in its application, I have taken some pains in the preceding chapter in giving a concise account of the properties of water when converted by the agency of heat into vapour or steam. I have considered these facts of vital importance to a knowledge of its economical employment and application, and I have dwelt longer on the inquiry than I originally intended, in order that I might have an opportunity of rendering accessible the results of experiments on the density of steam, and that the subject might be clearly and distinctly understood before treating of the construction of the steam engine.

It is not my object in this treatise to follow historically the many changes and improvements which have been effected in the steam engine since it left the hands of Watt. Suffice it to observe, that there has been no change in the principles of its action, unless we are to reckon as such the recent employment of gaseous instead of saturated steam. All the other improve-

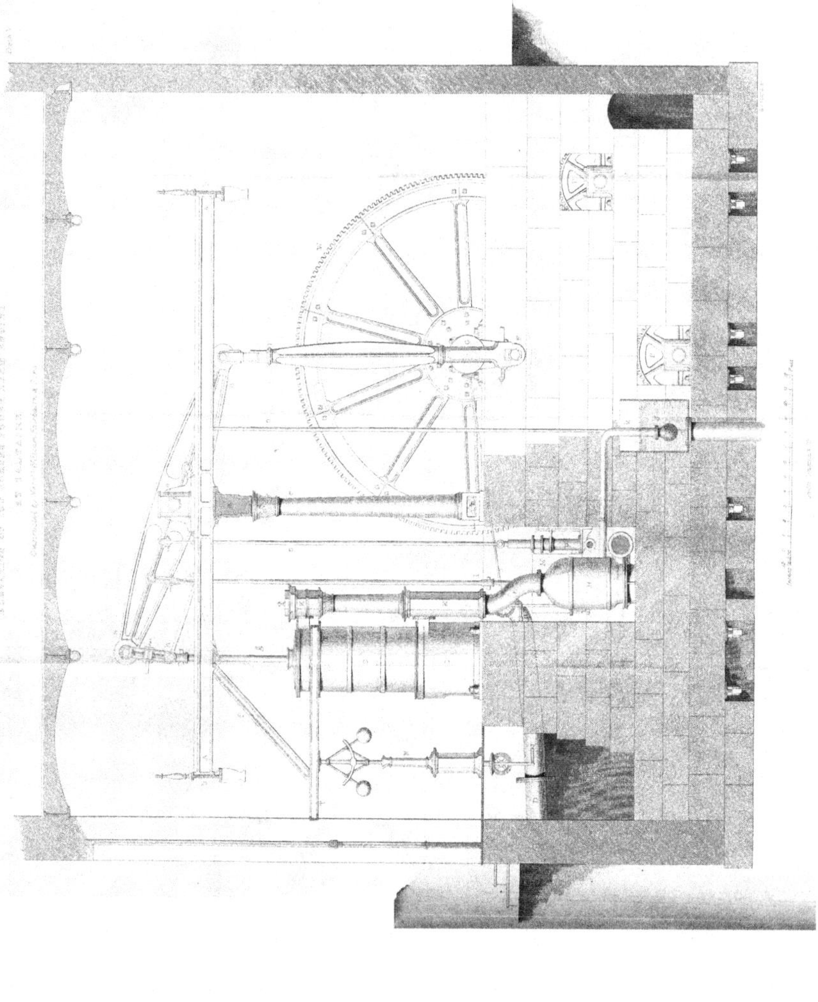

The material originally positioned here is too large for reproduction in this reissue. A PDF can be downloaded from the web address given on page iv of this book, by clicking on 'Resources Available'.

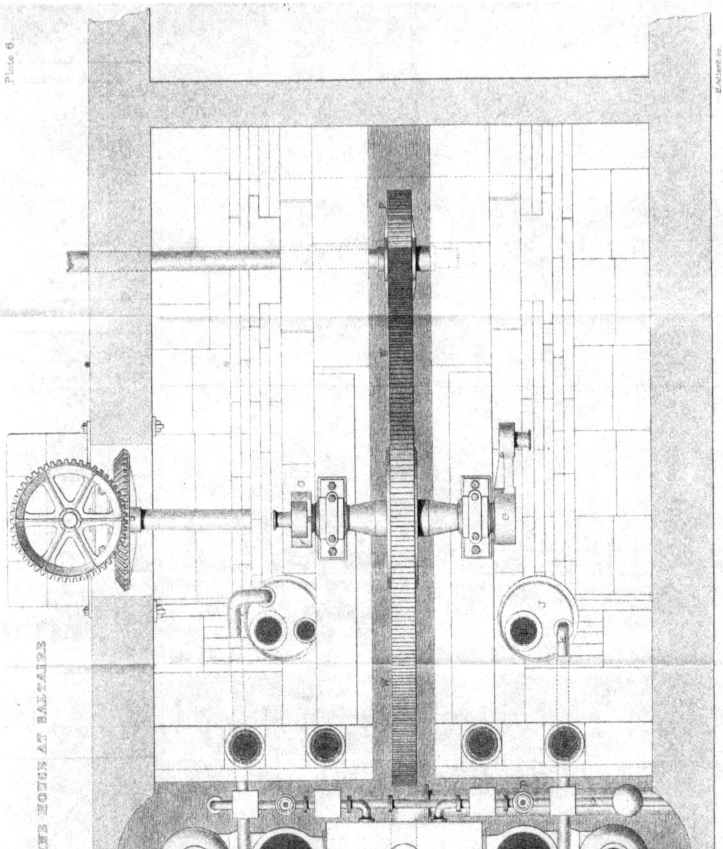

The material originally positioned here is too large for reproduction in this reissue. A PDF can be downloaded from the web address given on page iv of this book, by clicking on 'Resources Available'.

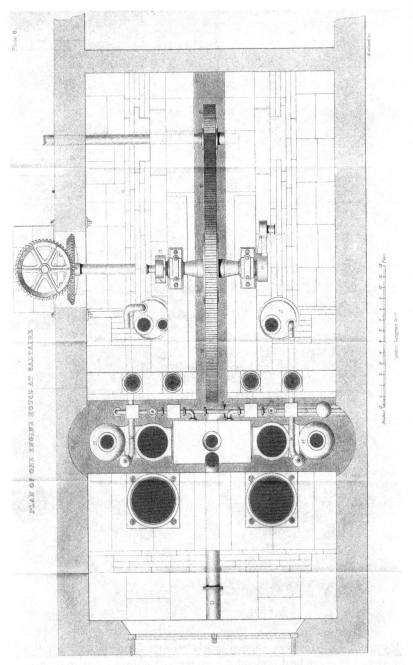

ments, in whatever form they present themselves, are confined to alterations of the organic parts of the engine, but have effected no change in the principle of action.

Taking then the condensing engine in its best and most economical form, I shall endeavour to lay before the reader some examples of the best and most recent construction, adapted for mill purposes in all the conditions of manufacture to which they are applied. In making a selection of those of medium size, I have chosen for illustration those at Saltaire, near Bradford, of 100 nominal horse-power each. Those engines are the best known for mills and factories, and the description of them will include the essential features of every other condensing beam engine.

Stationary Beam Engines.

At Saltaire the engines required to drive the machinery consist of two pairs of condensing beam engines, each engine being of 100 nominal horse-power, or collectively 400 nominal horse-power. They are placed on either side of the principal entrance to the mills, and are supplied with steam from boilers placed underground, at a short distance in front of the mill.

Plate V. contains a side elevation of one of these engines, and Plate VI. a plan of one engine-house with its pair of engines. The general arrangement will be understood when it is noticed that the power generated in the cylinders C, and transmitted through the working beam B B, to the large spur fly-wheel W, 24 feet in diameter, is taken direct from its circumference by the pinions P P, which give it off at the required velocity to the shafting of the mill.

The working beam B B is supported on two massive columns *c* 16 feet high, 14½ inches in least diameter, and 1½ inch thick of metal; these columns are bolted down beneath the whole mass of masonry supporting the engine. The heavy entablature *e* bolted to each column, and to the columns of the adjoining engine, is firmly fixed in the walls of the engine-house on each side, and the spring beams A A over this and at right angles with it are similarly attached to the cross beams *b b*. In this way an exceedingly strong and rigid support is secured for the main centre of the engine, which, resting in its pedestal *a*, has to sustain the principal strain of working. The spaces between the spring beams and the walls, excepting where the main beam vibrates, are filled with ornamental perforated metal plates, forming the beam-room, approached by the staircase *f*, for the purpose of oiling the centres, repairs, &c. The working-beam receives its motion from the piston-rod *g*, through the parallel motion *h h*, and transmits it by the connecting-rod F and crank G to the fly-wheel W.

The steam is brought from the boilers through a prolongation of the tunnel or flue in which the smoke passes to the chimney, and enters the engine-house by the pipe D. Having thence been admitted to the cylinder through the valve chests K K, it repasses after it has completed its work to the condenser H, through the eduction pipe E, in the usual way. The condenser is supplied with cold water from the river Aire, by the pipe *k k*, which communicates with the cold-water cistern J; the injecter through which the water enters the condenser is in these engines 6 inches bore, but the supply of water may be diminished if necessary by the injection gear hereafter described. Beside the condenser is the air-pump for pumping out the water and the air which enters with the water into the condenser, and is worked by the rod *l l* from the beam through a part of the parallel motion. A pump to supply the cold-water cistern is worked by the rod *n*, and another pump is worked by the rod *p p*, by which part of the hot water from the condenser is pumped back again for the supply of the boilers, in proportion as the water in them is decreased by its evaporation into steam. The supply of steam to the engine is regulated by the governor N acting on the throttle valve *q*, and thus the speed of the engine is kept uniform. A shaft *s s*, receiving motion from a bevel wheel on the crank shaft, works the equilibrium valves in the chests K K, as will be described; T T is a flooring or stage by which access is gained to the cylinder covers for oiling and cleaning. The cylinder is 50 inches in internal diameter, and has 7-feet stroke; it stands on the circular cylinder bottom C′, which is firmly bolted to the masonry by the long holding down bolts *r r*.

The length of the engine-house is 50 feet, and its breadth 24 feet. It will be seen that the two engines are combined so as to act in concert upon the same crank shaft and fly-wheel, the cranks being placed at right angles to each other, that when one engine is passing its top and bottom centres, and exerting least power, the other is in mid stroke and exerting its whole power upon the full leverage of the crank. In this way the action of the engines is equalised, and the motion rendered smoother than is possible with an independent engine, whilst, in case of accident to either of the pair, its fellow may be employed alone until the damage is made good.

Plate VII. exhibits a half-elevation and half-section of the valve chests, condensers, air pumps, &c., of a pair of engines, showing the valves and the manner of working them. As before, C C are the cylinders, C^1 C^1 the cylinder bottoms, K K the upper, and K^1 K^1 the lower valve chests, fixed right over the cylinder ports and communicating by the side pipes t t^1. D D is the steam pipe, H H the condensers, L L air pumps, with their valves *v v v*; M the hot-well into which the air-pump lifts the water accumulating in the condenser. This water passes away by the overflow pipe *m*; *p p p* are feed pumps for supplying the boilers, with an air vessel p^1, for equalising the pressure and preventing any sudden shocks in the pipes; *u*, injection cock and injecter, the quantity of water admitted being regulated by the injection cock worked by the hand wheel *f*, through the medium of the small shafts and bell cranks *n n*.

The valves in these engines are of a peculiar construction, being modifications of the double beat or equilibrium valve, invented by Mr. Hornblower, and generally employed in the mining engines in Cornwall, where the high price of coal has led to that rigid economy for which its engineers have long been so justly famous. Most of the appliances for using steam expansively in rotative engines (*i. e.* in mill engines as distinguished from pumping engines), are open to the objections,—1st, of wire-drawing the steam; 2nd, of cutting it off too slowly; and 3rd, of leaving too much space between the cut-off valve and the cylinder, whereby much steam is wasted without producing its due mechanical effect. To remedy these defects I have employed the particular arrangement of valves shown in the plate, which are applicable to all rotative engines working expansively, whether with high or low-pressure steam.

The steam entering the upper steam chest K, through the stop valve *a*, has free access also to the lower steam chest K^1, through the side pipe *t*; whilst the exhaust steam has also clear access to the condenser, through the other side pipe t^1. The steam is admitted to the cylinder from the valve boxes by means of the valves *x* and x^1; and after having completed its work it passes through the exhaust valves *y* and y^1 to the condenser, these valves being opened and shut, alternately, at the right instant by an apparatus yet to be described. Each of the valves consists of two single conical valves *x*, 1 and 2, carefully secured together and accurately fitting their seats; the lower valve is slightly smaller than the upper. The steam is admitted on the upper and lower side of each of these pairs of valves and presses in opposite directions, so that the downward pressure on the upper valve is neutralised by the upward pressure on the lower, excepting that a slight preponderance is given to the former in consequence of the difference of area in the valves, in order to aid in keeping the valves firmly pressed upon their seats when released by the cams. Hence they lift with the greatest ease and expose any required opening for the admission or exit of the steam.

The mode of working these valves is very simple; a shaft *s s* (Plate V.), receives motion from the crank shaft and imparts it by the bevel wheels *b b*, to the horizontal shaft *c c*; this in turn gives motion to the valve spindles *d d*, which pass continuously through bearings in the valve chests and are supported on footsteps on the brackets *e e*. Upon each of these spindles are fixed two discs *g g*, carrying cams upon their upper surfaces, so arranged as to lift and release each valve at the proper instant of time. This is effected by a direct and simple action; the height of the cam corresponds with the lift of the valve, its length with the duration of the lift, and its position on the cam disc, which makes one revolution for every stroke, regulates the instant of time in the course of the stroke at which the valve is opened and shut. The action of the cams is transferred to the valves through the medium of friction pulleys *k k k k*, fixed upon small cross-heads, which are guided in their upward

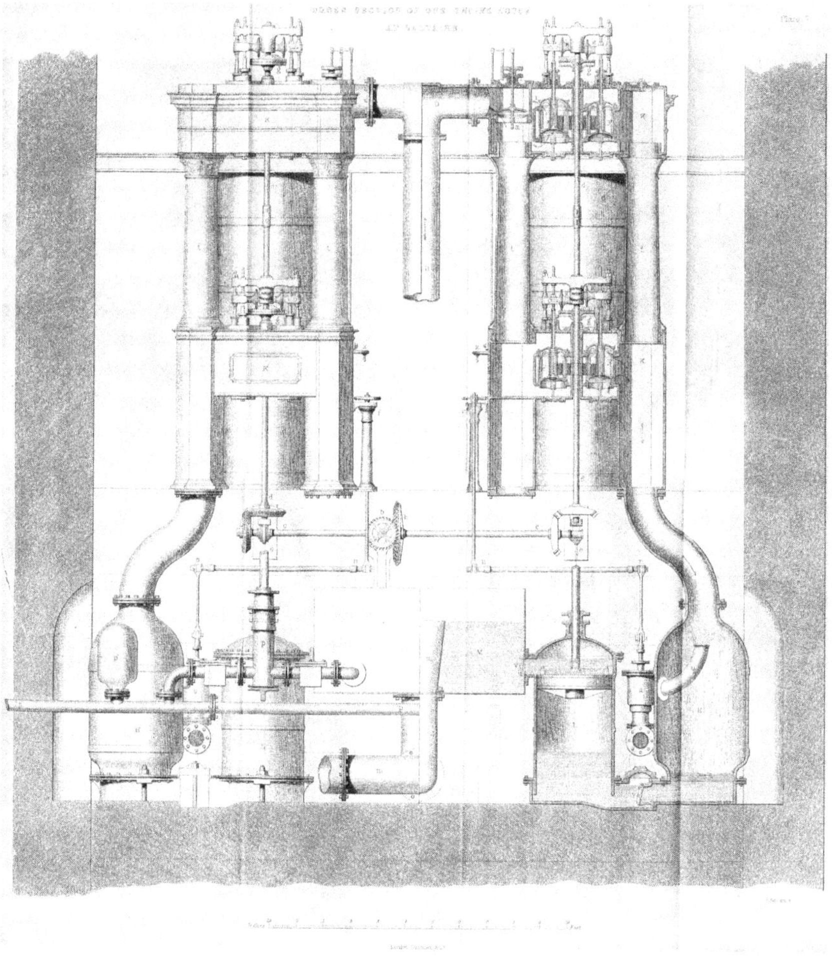

The material originally positioned here is too large for reproduction in this reissue. A PDF can be downloaded from the web address given on page iv of this book, by clicking on 'Resources Available'.

and downward motion by the brass standards in which they work. In the case of the steam valve these pullies are capable of adjustment by sliding them along the cross-head, towards or away from the valve spindle, so as to bring them over different parts of the cam, which is so arranged that the steam may be cut off at $\frac{1}{2}$, $\frac{1}{3}$, $\frac{1}{4}$, or any required portion of the stroke, the remainder being effected by the expansion of the steam.

The exhaust steam requiring a full opening into the condenser, it is desirable to retain the exhaust valve fully open during the whole length of the stroke. By the present arrangement this is effected with a greater degree of certainty than by any other means hitherto proposed. The exhaust valves rise suddenly on the short inclined planes of the cams, and having allowed time for the escape of the steam through a wide passage to the condenser, they fall with equal celerity by their own weight; thus a more complete vacuum is formed under the piston than is perhaps possible to obtain by any other process.

The stop valve *a* is a simple conical valve, worked by a lever and hand wheel *z*, fixed by a bracket to the side of the steam chests, and is chiefly used for shutting off the steam from the engine.

The following diagrams were taken from these engines on May 4th, 1859. The engines were then working at 25 revolutions per minute, and one pair with part of the load off:—

Diameter of cylinder . . .	50 ins.
Area " . . .	1963·50 ins.
Speed of piston . .	350 feet per minute.
Scale of diagrams .	$\frac{1}{16}$ inch per lb. pressure.

Engine A.

From this diagram we get:—

	Lbs. per sq. in.
Mean pressure of steam . . .	= 7·1684
Deduct for friction, &c. . . .	= 2·0000
Effective pressure . . .	= 5·1684

∴ Actual horses-power = 107·63.

INDICATOR DIAGRAMS.

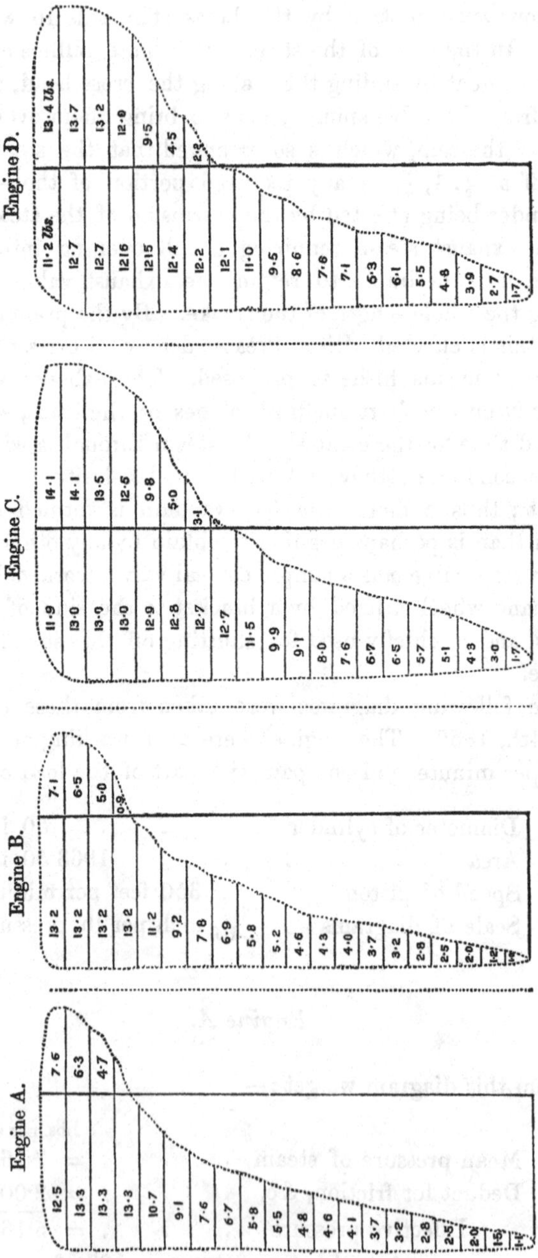

Fig. 157.

Engine B.

From this diagram we get:—

Mean pressure of steam . . . =	7·3646
Deduct for friction, &c. . . . =	2·0000
Effective pressure . . . =	5·3646

∴ Horses-power = 111·46.

Engine C.

From this diagram we get:—

Mean pressure of steam . . . =	13.301
Deduct for friction, &c. . . . =	2·000
Effective pressure . . . =	11·301

∴ Horses-power = 235·34.

Engine D.

From this diagram we get:—

Mean pressure of steam . . . =	12·946
Deduct for friction, &c. . . . =	2·000
Effective pressure . . . =	10·946

∴ Horses-power = 227·95.

Collectively 682·38 horses-power.

With a higher pressure of steam, or a shorter expansion, these engines will work to nearly double the above or 1200 horses-power.

Fig. 158 represents the cylinder, and its base, of the Saltaire engines. The cylinders are 50 inches internal diameter, with 7 feet stroke, of metal $1\frac{1}{2}$ inches thick. The ports are twenty inches wide, by 6 inches deep, so as to give $\frac{1}{16}$ the area of the piston for the admission and exit of the steam. The equilibrium valves have the upper disc 12 inches diameter, or 113 inches area, the lower disc $10\frac{1}{2}$ inches diameter, or $86\frac{1}{2}$ inches area, lift of steam valves $1\frac{5}{8}$ inches; of exhaust valves $1\frac{7}{8}$ inches. Steam pipes, $12\frac{1}{2}$ inches diameter; exhaust, 13 inches; condenser, 40 inches; air pump, $33\frac{1}{2}$ inches, and 3 feet 6 inches stroke. Beam 21 feet 6 inches long between the end centres, or over three times the stroke; and $3\frac{1}{2}$ feet deep in the middle, or $\frac{1}{6}$ the length. Main centre of wrought iron 12 inches diameter in the beam and 9 inches in the bearings. Spur fly wheel 24 feet 5 inches in diameter, with 230 cogs on the rim, 14 inches broad, 4 inches pitch; the rim is in 10 segments and has a sectional area of 200 square inches.

Fig. 158.

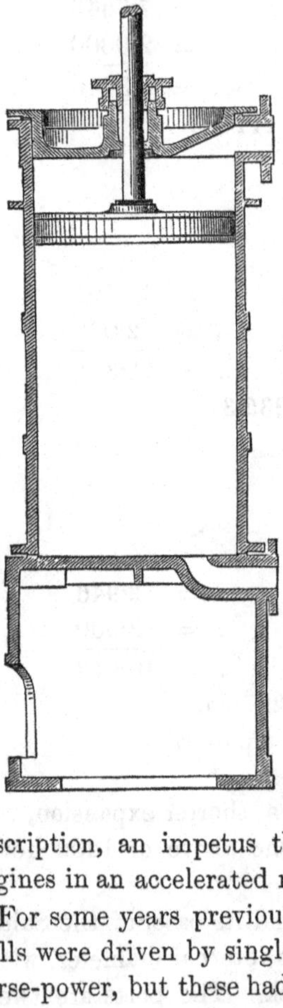

It is now more than thirty years since it was found desirable to increase the power of the steam engines employed in manufacture, and instead of engines of from 20 to 50 nominal horse-power, as much as 100, and in some cases 200 horse-power were required to meet the demand. To keep pace with the rapid extension of our manufactures, not only was the power itself doubled, and in some cases quadrupled, but a new class of men was brought into existence as mechanical engineers, and these, with the facilities afforded by new constructions and improvements of tools, gave to the manufacture of steam engines, and machinery of every description, an impetus that in a few years produced steam engines in an accelerated ratio of ten to one.

For some years previous to the great demand for power, the mills were driven by single engines, some as much as 50 or 60 horse-power, but these had soon to give place to others of much greater force, or, what was found to answer much better, two were employed coupled together as described above. Working in pairs, they were found to afford greater uniformity of action from the cranks being placed at right angles. Again, it was found that the speed of 240 feet per minute, considered as the maximum by Watt, was insufficient with the increasing demand for power, and speeds of 320 to 350 feet per minute are now become general. In some of the old engines, however, with such an increase of speed, the breakages became so numerous as to cause a retrograde movement, and a return to the old speed.

The increase of speed was, however, inadequate to meet the requirements for power in many cases, and the next resource was to increase the pressure of the steam. Unfortunately many of the boilers and engines were not calculated to withstand the forces to which they were thus subjected, and the result was an increase of the number of breakages and explosions to an extent that was ruinous to life and property. The ultimatum of all this was to increase the number of steam engines with an entirely new description of boiler, calculated to withstand higher pressures, and maintain the speed required to work the engine up to the required standard of power.

In the above statement I do not mean to attach blame to any person in his attempts to increase the power of the steam engine to meet the demands of the mill. On the contrary, the majority of manufacturers were against an increase of speed or pressure on account of the dangers they entailed, and the heavy responsibilities attached to them when the lives of workpeople were at stake, and it required a long series of years, in which I advocated the use of high-pressure steam, before the reluctance of the manufacturers was overcome. That is, however, now accomplished, and along with an improved principle of construction in boilers, the steam engine is no longer, when worked with steam of only one-fourth the pressure, what it used to be. To what extent the pressures may yet be carried, and how far the steam may be expanded, is a question still open for solution. But judging from what has already been done, the inference is that we have not as yet attained the maximum pressure, nor the rate of expansion calculated to afford the greatest economy in the use of steam as a source of power.

To accomplish the increase of pressure no change has taken place in the engine itself, beyond the strengthening of the parts, and the substitution of wrought-iron and steel for parts which were before considered sufficiently strong of cast-iron.

Where additional power was required in mills which could not be obtained by increasing the speed and pressure of the old engines, horizontal, high-pressure non-condensing engines were sometimes introduced, and these, in the manufacturing districts, are commonly called *Thrutchers*. These Thrutchers have been largely employed in Staleybridge and the surrounding district, and although not of great value on the score of economy, they are, nevertheless, important as auxiliary to the larger condensing engines. They are generally attached to the main gearing or first motion wheels, and the steam, which enters their cylinder at 50lbs. pressure, exhausts into the cylinder of the condensing engine, and is there expanded and worked over again. This system of double action would appear favourable to the expansive process, but unfortunately the distance

Fig. 159.

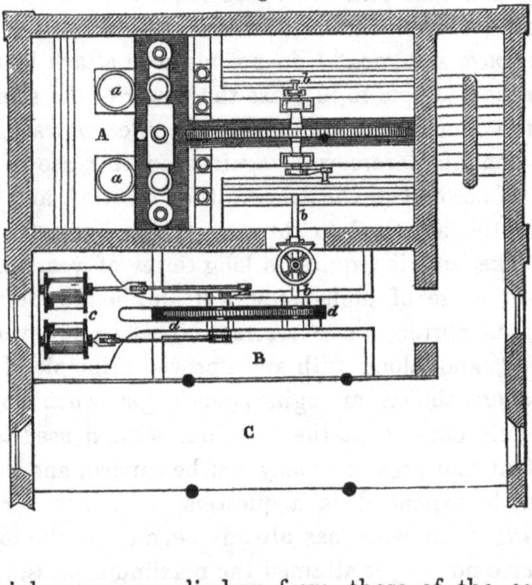

of the high-pressure cylinders from those of the condensing engine, and the consequent loss of heat by radiation and the retardation from friction, including the complicated nature of the connections, &c. is so great as to neutralise or destroy the economy of fuel which would otherwise have been secured.

There is, however, a considerable saving in original cost, which to a certain extent balances this drawback and renders the Thrutchers valuable as an auxiliary power. In cases where engines are overloaded and where it is impossible for want of space to erect new engines on the first principle of construction, the horizontal non-condensing engines are admissible, and may be used with advantage.

The annexed sketch (fig. 159) will explain the mode of connecting the horizontal high-pressure with the vertical condensing engines. At A are shown a pair of 60 horse-power engines with the cylinder *a a*, and fly-wheel *e e*. At B are the double high-pressure engines with their cylinders *c*, and their connection with the main shaft *b b*, by a spur fly-wheel and pinion *d d*. In this way all the four engines are united; a certain portion of the lower part of the mill being occupied with the auxiliary engines B,

Fig. 160.

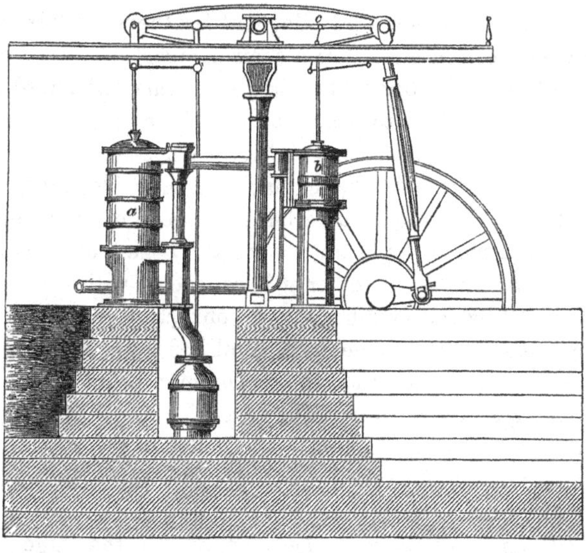

and with a new set of boilers to supply the steam at the requisite pressure.

The great deficiency of power severely felt in many establishments, has been supplied in some cases by another method, without resorting to the erection of new engines, and that is by *McNaughting* the old ones. This process, the invention of Mr. McNaught, consists in increasing, or even nearly doubling, the power of a condensing engine by the introduction of a high-pressure cylinder attached to the same working beam, that is, provided the beam is strong enough to bear the increase of strain. This system of Mr. McNaught is simple and as effective as the Thrutchers, but it has unfortunately the same drawbacks as regards loss of heat and expenditure of fuel, although it must be admitted that the power of the engine is increased to a large degree. The mode of *McNaughting* is shown in fig. 160, where *a* is the cylinder of the condensing engine, *b* the high-pressure cylinder mounted on a pedestal to a convenient height and worked half stroke from the main beam at *c*. This it will be observed is a simple process, as it does not interfere with the main gearing of the mill, and in other respects also it has advantages over the Thrutchers already described.

Fig. 161 shows in perspective view the engine-house and engines erected for Messrs. W. Bailey and Brothers of Staleybridge, for driving a cotton mill. From their compactness, short stroke, and regularity of motion, they are sometimes preferred in the manufacture of cotton to beam engines of the common construction. The working beam or great lever L L L, is, as it were, split into two, one of the halves being placed on each side of the engine, but united at the middle by a large gudgeon or main centre L. At the cylinder end the beam is worked by the crosshead *p*, and side rods L R; and at the other it is connected with the crank of the fly-wheel by the crosshead and connecting rod L K L. The moving mass of the engine is thus placed lower and the whole rendered more compact than in the common beam engine. The stroke being short, the variations of power occur at shorter intervals, and hence the motion transmitted to the machinery of the mill is rendered as uniform as possible through the agency of the immense fly-wheel and the coupling of the two engines. The striking peculiarity of these engines is the large geared fly-wheel W W W W, formed of toothed segments, receiving the power of both engines, correcting its irregularities, and giving it out directly at its periphery, and at a high velocity, to the first motion shaft of the mill Y Y. Not only is the requisite speed of revolution very quickly attained in this way, but all intermediate trains of wheels are entirely dispensed with, and durability with simplicity secured in the highest degree. The pinion *g* which receives the power from the fly-wheel is geared with hornbeam teeth as an additional precaution for rendering the motion perfectly smooth and noiseless.

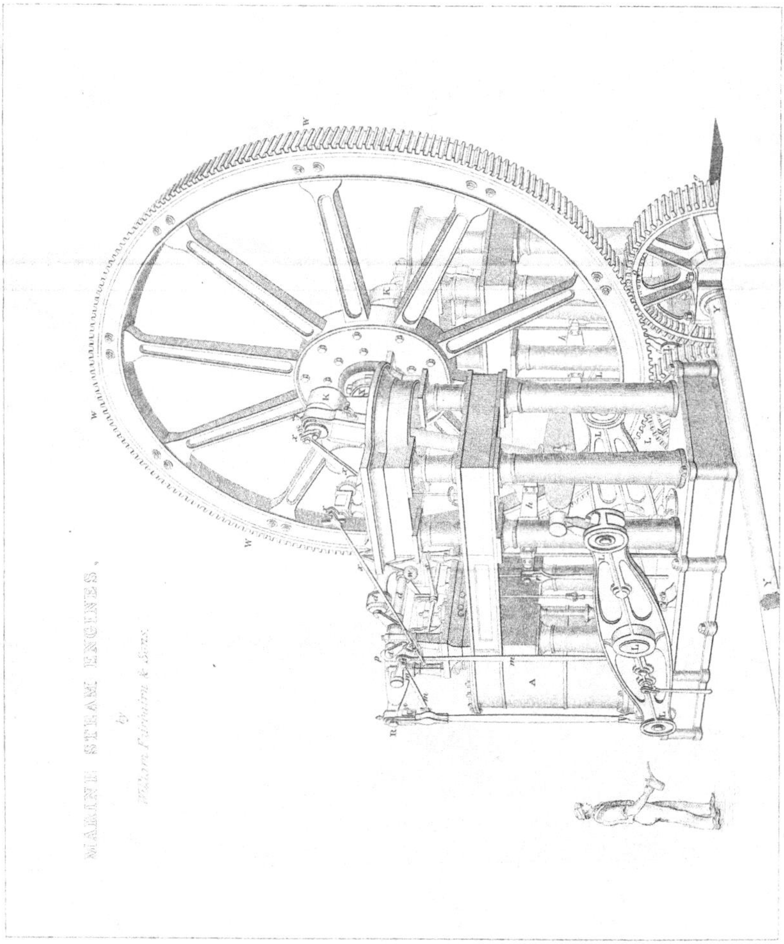

The material originally positioned here is too large for reproduction in this reissue. A PDF can be downloaded from the web address given on page iv of this book, by clicking on Resources Available'.

The steam pipe conducts the steam into an outer steam jacket round the cylinder A A; from this it enters the valve boxes similarly constructed to those of the Saltaire engines, but with short D valves instead of the equilibrium valves with which those engines are fitted. The piston is packed with metallic rings, and the steam ports are formed in the cover and bottom plate of the cylinder. The condenser is placed below the bottom valve chest, and near it the air-pump worked by a cross-head seen at *h*.

The valves receive motion by the following arrangement; a stud in the crank pin K carries round a small radius rod X X on an axis concentric with the crank; a smaller crank on this axis has a length equal to half the throw of the valve, or equal to that which would be given to the ordinary eccentric; and by a bar *x x* similar to the eccentric rod, the valve is moved by this lesser crank in the same way as by an eccentric; *m m m m* are the links of the parallel motion, *w* the governor, *s* the throttle valve.

From the above description it will be seen that the marine engine has some advantages over the long stroke beam engine as applied to mills. At an early period of my own practice I introduced it on an extensive scale, and there are numbers now at work, exclusive of those erected for Messrs. Bailey and Brothers, that are performing an efficient duty, and giving entire satisfaction.

With regard to economy of fuel, the marine engine is equal to the beam engine when worked on the same principle of expansion with equilibrium valves. It has besides the advantage of taking up less room in the mill, and having the whole of the action upon a lower basement, to which the frame of the engine is securely bolted. Every other engine, whether vertical, having a downward motion, or horizontal, has an advantage over the beam engine as regards space, and the distribution of its force through its organic parts direct upon the solid foundation.

But notwithstanding these advantages of marine and direct acting engines, they have not gained upon the old plan of a stationary beam engine for employment as the chief prime mover in mills. The reason that the beam engine has not been supplanted, appears to be, that its simplicity of construction and the facility of getting to every part in case of repairs being necessary, give it a superiority over every other form, however perfect and compact. Besides, the engineers (or *engine tenters*, as they are called in the manufacturing districts) find there is less trouble in cleaning, and there is therefore a desire on their part to have the old construction in preference to every other. From these considerations the old Boulton and Watt form of engine, strengthened and improved by being adapted to work expansively, is now the favourite, and is likely to maintain its ground as long as steam is depended on as the source of power in mills.

In the consideration of steam as a prime mover, it would be unjust to omit to notice the modification of Woolf, so extensively used on the Continent, where fuel is expensive, and where the greatest economy in its use has been an object of serious consideration.

For the last half century Woolf's engine has been preferred in France and other countries on the continent of Europe, and this has arisen from the fact that until the last fifteen years the single cylinder engine has been worked with low pressure steam only, without expansion. Now it is evident that the single cylinder engine worked with full steam throughout the stroke, will require a larger expenditure of fuel than another engine worked expansively. Thus the double cylinder or compound engine, in which high pressure steam was employed, expanded through three-fourths of the stroke, appeared to effect a considerable saving of fuel; but taking both engines worked alike, with steam of the same pressure similarly expanded, as is now the case in the best single cylinder engines, there appears to be no advantage in the compound over the simple single cylinder engine. On the contrary, there is a loss in the original cost of the engine, and the complexity of the one as compared with the other. I have therefore no hesitation in recommending the single cylinder engine worked expansively, as an efficient competitor of the compound engine.

Fig. 161 is a view of the two cylinders and valve chests of a Woolf's engine. The small or high pressure cylinder is shown at A, 23 inches in diameter and 6 feet stroke, and the large or low pressure cylinder at B, 40 inches in diameter and 8 feet stroke; their contents being as 1 to 4. The steam is brought from the boiler by a pipe not shown in the drawing, which admits the steam into the annular passage *c c*, whence it passes into a valve chest D of the ordinary construction. This valve chest communicates by the passages *f* and H with the two valve chests C C of the large or low pressure cylinder B, and the exhaust steam from A passes into these and is expanded to four times its bulk in the larger cylinder, which has no direct communication with the boiler. The steam from the upper side of the piston of the high pressure cylinder passes to the lower part of the larger cylinder, and in this way a simultaneous upward or downward motion of the pistons in the two cylinders is secured. After the steam has been thus expanded, and the work so economised, it passes by the pipe E E to the condenser.

The valves of the large cylinder are equilibrium valves, like those of the Saltaire engines, but differently moved. A reciprocating motion is given to a crank *k* by means of an eccentric on the fly-wheel axis. This motion is communicated by a connecting rod and bell crank to the two rods *l* and *m*, which slide with freedom in a vertical direction. Upon these rods at suitable heights are fixed the arms $l' \, l''$ and $m' \, m''$, by which the valves in the upper and lower chests are actuated; l' and m'' lifting the valves for the admission of the steam into the cylinder, and l'' and m' those for the exhaust. The valve of the high pressure cylinder is similarly worked by an arm connected with the rod *d*, not shown in the drawing.

Before closing our notice of the varieties of steam engines, we have to describe the horizontal condensing engines in which the cylinder is placed horizontally upon a cast iron frame or bed plate. This arrangement presents all the features of cheapness and concentration which we noticed in the high pressure *Thrutchers* already described. The condensing engine, however, works to 12 lbs. or 13 lbs. under the atmospheric pressure, and thus economises part of the work of the steam which is lost in the *Thrutchers*, with the additional disadvantage that they work against a considerable back pressure. It is for these reasons that the high pressure non-condensing engines are not in demand where a large amount of power is required. They are, however, simple and effective, excepting as regards economy of fuel. In some cases they are preferable to the condensing engine, and that is in small establishments and as auxiliaries to water wheels when the supply of water fluctuates, and a small engine is needed when the supply is deficient. In large esta-

Fig. 161.

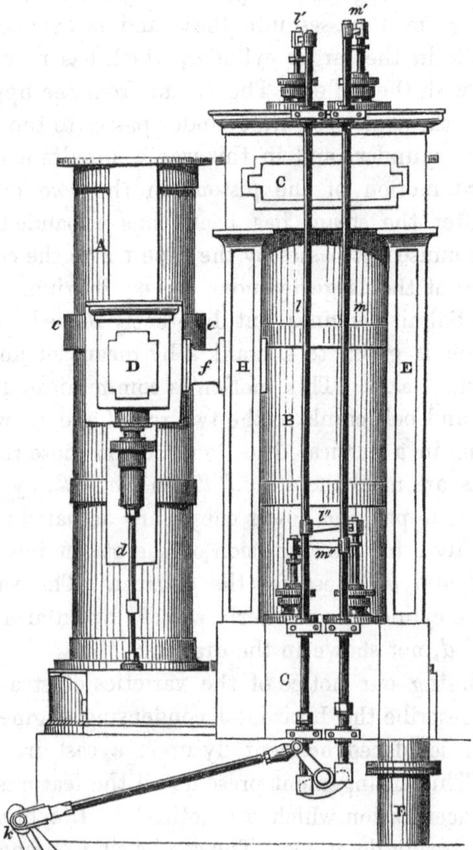

blishments, and more especially where coal is dear, the high pressure engine must give place to the condensing. The horizontal condensing engine presents some of the advantages of both classes of engines, as it is economical as regards fuel, and at the same time is lighter, more compact, requires less space, and costs less money than a beam engine. These are its merits. Its drawbacks are the horizontal position of the cylinder, involving unequal wear of the parts, and the tendency of the cylinder to become oval.

Fig. 162 represents a horizontal condensing engine, in which C is the cylinder, *s s* slide bars carrying the cross-head on the piston rod, from which is worked the fly wheel F F by the connecting rod *r r*, and the bell crank *b b*, by a short link. This bell crank transmits motion to the piston of the air-pump *a*,

Fig. 162.

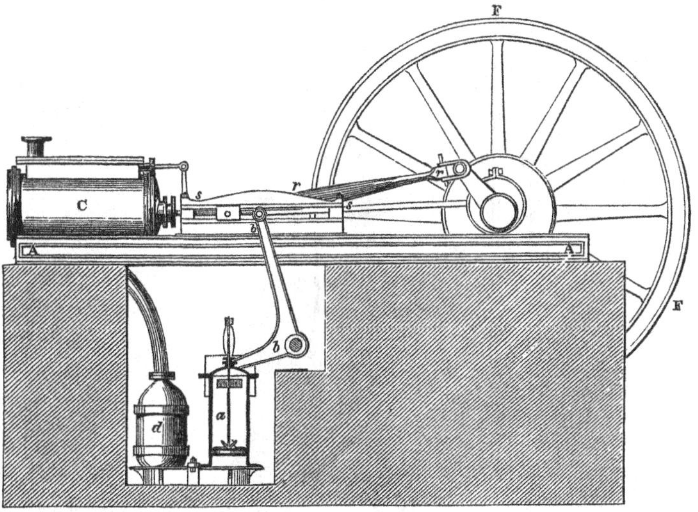

communicating in the ordinary manner with the condenser *d*. The working parts of the engine are firmly fixed on the iron bed plate A A.

High Pressure Engines.

Engines working without condensation, are now frequently employed as auxiliaries, and where the amount of power required is not large. It will not be necessary to describe here all their varieties, but they may be briefly enumerated as, 1st. Horizontal engines, like the last described, but without the condenser; 2nd. Columnar engines, in which the action is vertical, the framing being in the form of a hollow cylindrical column; 3rd. Vertical engines, with the cylinder placed above the crank

Fig. 163.

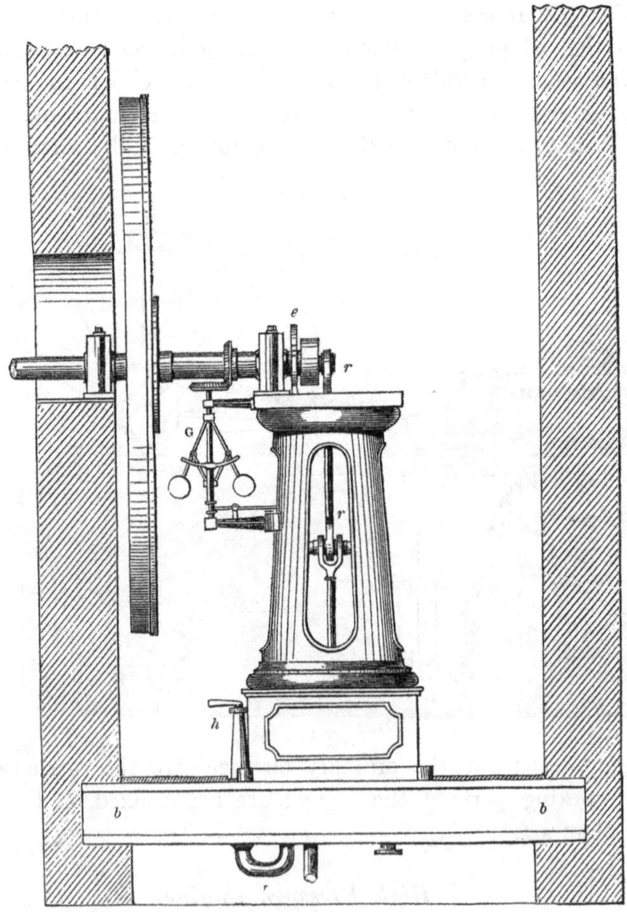

and working downwards; 4th. Oscillating engines, in which the crank is worked direct from the piston rod, and the cylinder oscillates upon trunnions near its centre, to allow of the requisite vibration of the crank; 5th. Steeple engines, in which the piston rod carries a cross-head, from which the connecting rod works downwards to the crank.

One example of the best of these varieties will be sufficient for our purposes in this place. Figs. 163 and 164 represent the columnar engine, which from its simple, compact, and neat form, is probably superior to most other constructions; it combines

Fig. 164.

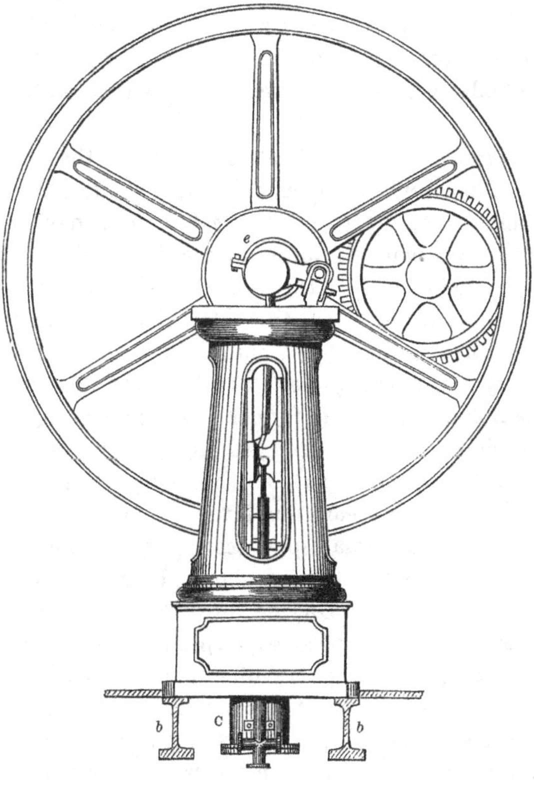

several advantages from its vertical position, and the ease with which it is supported on two iron beams, *b b*, built into the walls of the engine-house for that purpose. In this way the necessity for heavy foundations is entirely done away with, and the boiler may be placed immediately below the engine, in order to be close to its work, and to save space. The annexed drawings represent a front and side elevation of an engine of this description of six horses' power, but the same principle has been successfully extended to engines of thirty horses' power. The piston rod is cottered in the usual way into a cross-head, carrying blocks which slide in fixed guides on each side of the columnar framing of the engine; the connecting rod *r r* is attached to this cross-head, and also to the crank overhead. C is the cylinder with its valve chest, the valves being of the short D construction, worked by the eccentric *e* on the crank shaft. On the other side of the column, the pump for supplying the boiler with water, is worked by a stud on the piston cross-head.

The governor G, attached at the side of the column, is worked by a bevil wheel, upon the crank shaft, and the pipe for supplying steam *f f* enters the valve chest, having a stop valve worked by a handle *h* placed at the side of the column.

In most of the engines hitherto described, the motion of the engine has been transmitted direct from the periphery of the fly-wheel to the gearing of the mill. This plan I introduced nearly twenty-five years ago, and applied both to water wheels and to steam engines; in the former case, the gearing is usually internal, as shown in the plates I., II., III., and IV., and in steam engines it is usually external, as shown in plates V. and VII. But this rule is by no means absolute, as in part of the Deanston water wheels external spur gearing was employed; and in the columnar engine figured above, I have shown an internal geared fly-wheel. On the introduction of this method of obtaining direct the necessary speed in the first motion shaft, by gearing the periphery of the fly-wheel, I met with opposition on all sides, and it was not till a wheel of thirty-six tons weight had been constructed for a pair of engines of 240 horses' power, that the more sceptical were convinced that the regularity of motion was not impaired, and that the train of geared wheels had been abandoned, with a material saving of power, and economy of prime cost. This system of connecting the prime mover with the machinery of the mill, greatly simplifies the motion, by attaining the required velocity at once, without the aid of speed gearing necessary where the motion is taken from the fly-wheel shaft. This system has now become universal in stationary steam engines.

The duty of engines, or amount of work done for a given quantity of coal consumed, has gradually improved as the engine itself has been modified. Smeaton's table of the effect of fifteen atmospheric engines at work at Newcastle in 1769 gives a mean of 5,590,000 lbs. raised one foot per bushel of coals per hour. This is equivalent to an expenditure of 29·76 lbs. of coal per horse power per hour. In Smeaton's own engine, erected at Long Benton, an improved duty of 9,450,000 lbs. raised one foot per bushel of coal was obtained, equivalent to an expenditure of 17·6 lbs. per horse power per hour. In Watt's engines the expenditure of fuel was further reduced to so large an extent that the payment for them was made proportional to their economy, one-third of the annual saving of fuel obtained by their use being paid during the term of the patent. In the earliest of Watt's engines without expansion the expenditure appears to have been about $8\frac{1}{2}$ lbs. to 9 lbs. per horse power per hour. At the present time, in condensing engines working expansively, a duty as high as 2·6 lbs. of coal per horse power per hour has been obtained, or $11\frac{1}{2}$ times the amount of work for the same consumption of fuel as in the early atmospheric engines.

The following tables have been carefully compiled by Mr. H. Harman, late chief inspector of the Association for the Prevention of Boiler Explosions in the Districts of Lancashire and Yorkshire, from the extensive returns furnished to that association. They show most significantly the progressive economy arising from the use of high-pressure steam, and from a long expansion.

TABLE FOR THE YEAR 1858-9.

Description of Engine.	Observed Pressure of Steam per Square Inch in the Boilers.															Maximum consumption of coal per horse power per hour.	Minimum consumption of coal per horse power per hour.
	15 lbs. and under.			16 lbs. to 30 lbs.			31 lbs. to 45 lbs.			46 lbs. to 60 lbs.			Above 60 lbs.				
	No. of Engines.	Indicated horse power.	Average consumption per horse power per hour.	No. of Engines.	Indicated horse power.	Average consumption per horse power per hour.	No. of Engines.	Indicated horse power.	Average consumption per horse power per hour.	No. of Engines.	Indicated horse power.	Average consumption per horse power per hour.	No. of Engines.	Indicated horse power.	Average consumption per horse power per hour.		
Condensing - -	20	969	10·6	193	24674	6·4	88	12338	5·7	6	679	3·8	—	—	—	17·0	2·6
Non-condensing -	—	—	—	3	36	14·0	10	509	9·9	6	399	8·7	—	—	—	23·0	5·5
Working compound	—	—	—	12	822	6·4	105	10340	5·8	201	23688	4·9	49	6796	3·9	11·0	3·0
Condensing. Steam cut off before $\frac{1}{4}$ stroke	—	—	—	37	4297	6·1	15	2320	5·2	2	175	5·2	—	—	—	9·6	2·6
Condensing. Do. $\frac{1}{4}$ to $\frac{1}{2}$ stroke	13	671	10·3	135	18837	6·1	75	10183	5·6	4	504	4·2	—	—	—	17·0	2·9
Condensing. Do. later than $\frac{1}{2}$ stroke - -	6	245	10·6	16	1577	6·7	—	—	—	—	—	—	—	—	—	16·7	3·0

TABLE FOR THE YEAR 1859-60.

Description of Engines.	Observed Pressure of Steam per Square Inch in the Boilers.															Maximum consumption of coal per horse power per hour.	Minimum consumption of coal per horse power per hour.
	15 lbs. and under.			16 lbs. to 30 lbs.			31 lbs. to 45 lbs.			46 lbs. to 60 lbs.			Above 60 lbs.				
	No. of Engines.	Indicated horse power.	Average consumption per horse power per hour.	No. of Engines.	Indicated horse power.	Average consumption per horse power per hour.	No. of Engines.	Indicated horse power.	Average consumption per horse power per hour.	No. of Engines.	Indicated horse power.	Average consumption per horse power per hour.	No. of Engines.	Indicated horse power.	Average consumption per horse power per hour.		
Condensing - - -	4	203	8·3	35	4228	5·8	19	2480	5·8	2	79	4·5	—	—	—	9·8	3·4
Non-condensing -	—	—	—	1	15	—	5	332	9·4	3	194	6·8	—	—	—	11·9	5·1
Working compound	—	—	—	1	146	5·0	15	1938	4·5	18	1935	5·1	5	1012	4·1	8·0	3·0
Condensing. Steam cut off before $\frac{1}{4}$ stroke	1	28	—	15	1572	5·7	6	878	5·0	2	80	4·5	—	—	—	8·0	3·4
Condensing. Do. $\frac{1}{4}$ to $\frac{1}{2}$ stroke	1	51	6·8	16	2429	6·0	13	1605	5·8	—	—	—	—	—	—	8·6	3·6
Condensing. Do. later than $\frac{1}{2}$ stroke - - -	2	124	9·8	4	236	6·0	—	—	—	—	—	—	—	—	—	—	—

In reference to the apparent superiority of the compound engines in the last table, Mr. Harman observes "that owing to increased friction, &c., engines which have been compounded invariably indicate more horses' power than before, the machinery remaining the same; hence arises an advantage, apparent and not real, in calculating the consumption of fuel. . . . . . Consideration has led me to conclude that the gross amount of power exhibited by compound diagrams as at present calculated is fallacious."

CHAP. VIII.

ON BOILERS.

VESSELS for the generation of steam for supplying steam engines are of a great variety of forms, and are usually denominated boilers. These vessels require great care and judgment in their construction, in order that the fuel may be most economically applied, the waste and nuisance of smoke avoided, and the enormous force which steam is capable of exerting at high temperatures, safely restrained.

The boiler is, in fact, to the steam engine what the living principle is to animated existence. Like the stomach, it requires food to maintain the temperature, circulation, and constant action, which constitute the energy of the steam engine as a motive power. To keep up the temperature we have to feed, stoke, and replenish the furnace with fuel, and we may safely consider it a large digester, endowed with the functions of producing that supply of force required in the maintenance of the action of the steam engine.

Fig. 165.

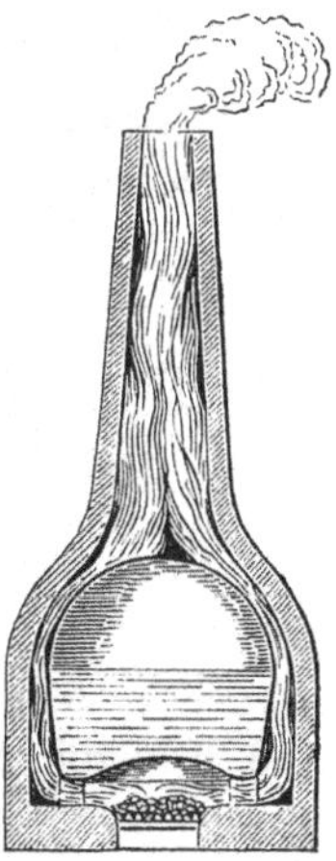

The boiler has undergone great changes of form and construction to adapt it to use. At first it was hemispherical, fig. 165, as when employed by Newcomen, which shape was retained for many years with certain modifications. Subsequently it was altered by Watt to the form of a parallelopipedon with a semi-cylindrical top, as shown in fig. 166. This form of boiler was extensively used by Watt in the early stages of his steam engines, and continued to take precedence of every other description of vessel employed for the production of steam. It was, however, modified by the introduction of a central flue, and a slight modification of its exterior shape to enable it to withstand greater pressures. Fig.

167 represents the improved form of boiler, in which, in addition to the curvature introduced along the bottom, the sides

Fig. 166.

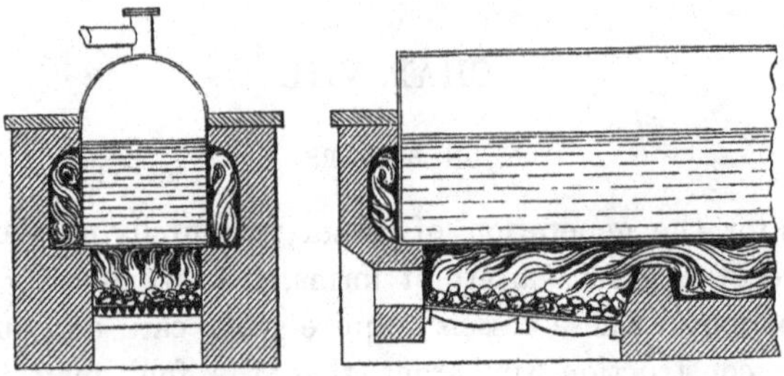

were also constructed in that form, the better to resist the pressure of the steam, which at that time was increased from 7 lbs. to 10 lbs. on the square inch.

Fig. 167.

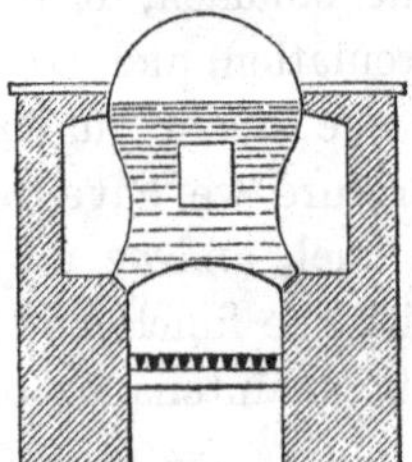

Simultaneously with these improvements, Woolf and Hornblower introduced high pressure boilers of the cylindrical form, some with hemispherical ends, fig. 168, and others with flat ends having a cylindrical flue through the centre, fig. 169. Boilers of this sort were extensively used in Cornwall, where the pumping of the mines by steam engines on Woolf's plan required a pressure varying from 30 lbs. to 40 lbs. per square inch. The same description of boiler was adopted by Watt in his pumping engines, first erected in Cornwall, and worked expansively on

Fig. 168.

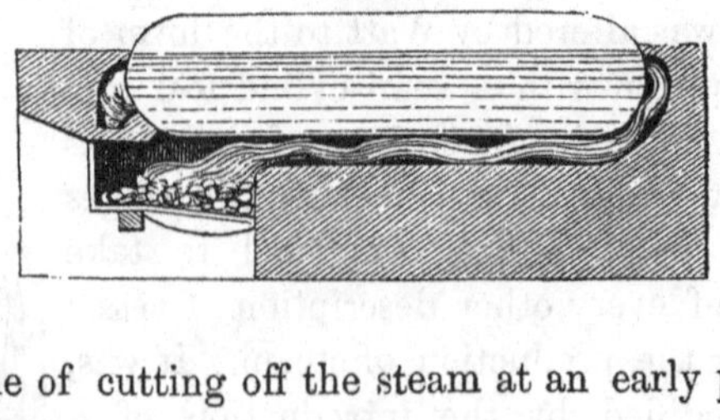

the principle of cutting off the steam at an early point in the stroke.

It was in the Cornish districts that the first great improvements in boilers took place. The high price of coal became an important item in the use of pumps for draining the mines, and every measure of economy was resorted to for a reduction of the cost. The result was the introduction of the cylindrical boiler, fig. 169, with a central flue, and a system of premiums to the superintendents for every bushel of coal saved in raising a given weight of water out of the mines. This system of premiums worked well in Cornwall, and I apprehend the steam engines of those districts are still worked with greater economy than in any other part of the kingdom. The Cornish miners pay more attention to their engines, are more careful of their boilers, and are stimulated to a more rigid economy than in any other part of the kingdom. They are never short of boiler space, and never force their fires or increase the power of their engines without increasing the capacity of their boilers. These conditions give to the Cornish engines the advantages which are lost sight of in other districts, to such an extent, in some instances, as to increase the consumption beyond all reasonable bounds. Of late years great improvements have been effected in this respect, and further progress in the same direction will doubtless lead to similar results in the economical use of steam.

Fig. 169.

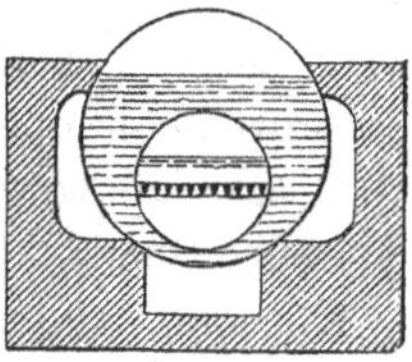

Exclusive of the Cornish principle of construction, boilers have been introduced of a cylindrical form, with a central ellip-

Fig. 170.

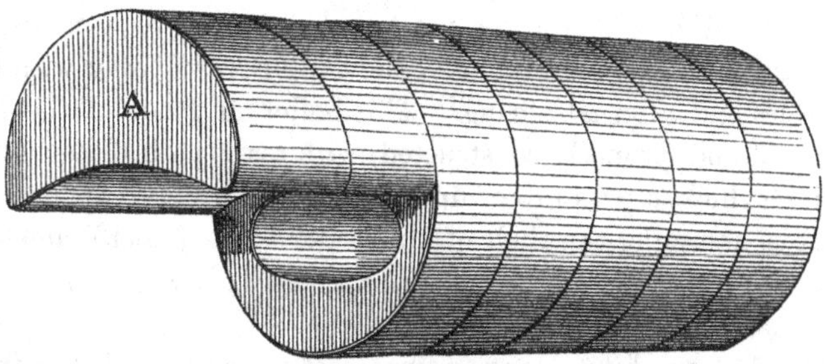

tical flue, and with the bottom cut away at one end to a distance of 8 feet, as shown in fig. 170, to admit a large furnace under that part. It will be seen that this boiler from the large concave arch at A, and the elliptical flue, inherits all the defects of the waggon form, fig. 167, and could not therefore be employed without danger of explosion, at pressures above 12 lbs. per square inch. From its peculiar shape it took the name of the *Whistlemouth*, or *Butterley* boiler, and with its internal flue, through which the products of combustion pass, it presents a large heating surface, and was for several years considered an improvement upon Boulton and Watt's boiler. Like its predecessors it gave way to others better calculated to generate high pressure steam.

The next improvement was the Cornish boiler, with a furnace

Fig. 171.

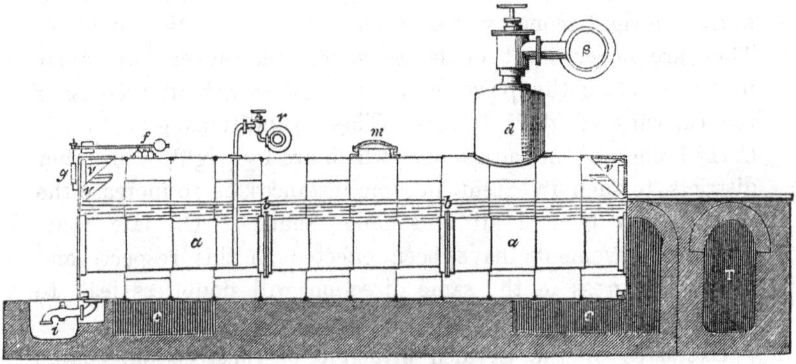

in a large cylindrical flue at one end, fig. 169, but this, from the large diameter of the flue, was liable to explode from collapse, and led to the strongest and most perfect boiler yet constructed for stationary purposes, namely, the double flued boiler with two furnaces and alternate firing. This boiler, if made of the best material and properly constructed, will resist with plates $\frac{3}{8}$th of an inch thick, a pressure of upwards of 300 lbs. per square inch (that is, assuming the shell of the boiler to be 7 feet diameter), and as it is now almost universally adopted, we shall give a fuller account of its proportions.

Fig. 171 exhibits a longitudinal section, and fig. 172 a front view, and a cross section of a double-flued stationary boiler, as I have been accustomed to construct it. It was originally devised with a view to alternate firing in the two furnaces, in order to prevent the formation of smoke; it consists of an external cylindrical shell with flat ends, and has two similar cylindrical flues, *a a*, passing through the water space of the boiler. The shell is six feet to seven feet in diameter, and twenty to thirty feet in length, and composed of $\frac{5}{16}$ to $\frac{3}{8}$ inch plates, riveted with lap joints, according to the pressure it is required to stand. The flues are usually two feet six inches, or two feet nine inches in diameter, leaving a space of about six inches all round, for the circulation of the water. These flues are strengthened by ribs *b b*, to prevent collapse, according to

Fig. 172.

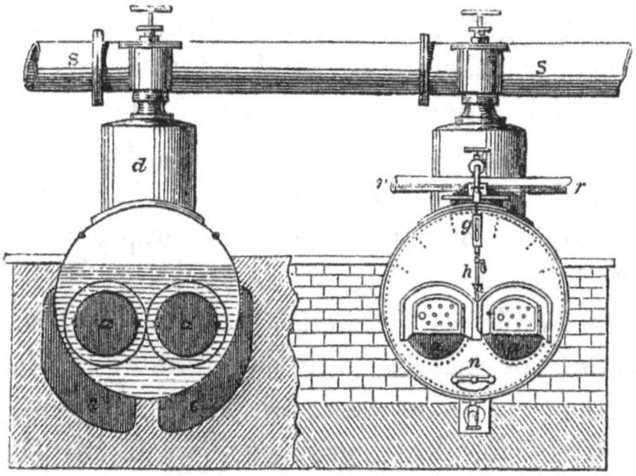

the principles developed in my researches on that subject.* The boiler has flat ends, stayed by triangular gussets *v*, *v*. The furnaces are within the flues, and the products of combustion after passing through these, unite and pass beneath the boiler, towards the front, in the flues *c c*, where they turn and pass back again on the other side to the chimney flue T.

The ordinary fittings of these boilers consist of a stem dome *d*, on which is placed a nozzle or stop valve, communicating with the large steam feed pipe S, with which all the boilers communicate. Man-holes *m n*, are fitted to the boiler for cleaning and examination. The pipe *r* brings the feed water from the hot well of the engine in most cases, whence it passes

* Vide "Useful Information for Engineers," Second Series, pp. 1—45.

through a small stop valve down to the bottom of the boiler near the furnace end. Safety valves are shown at *f*, two having fixed weights, and a third being pressed down by a spring balance *g*. At the bottom of the boiler is a mud cock *i*, and there are usually a steam gauge for registering the pressure, and a glass water gauge *h*, for indicating the amount of water in the boiler.

Another form of boiler frequently employed for mills, is in part multitubular. There are two furnaces in two cylindrical flues, precisely similar to those in the preceding boiler, but immediately beyond the furnace; these flues unite into one chamber, in which the gases mix, and thence the gaseous products pass through about a hundred small tubes, three inches in diameter, and about eight feet long. They then circulate in brick work flues beneath the boiler, and pass to the chimney as in the double flued boiler. The mixing chamber, from its elliptical form is weak, but to remedy this defect it is stayed by three vertical water tubes, riveted to the flat sides of the ellipse. Ten boilers of this description supply the steam for the four 100 horse power engines at Saltaire; their principal dimensions are as follows:— Shell 7 feet diameter, 24 feet long, and $\frac{5}{16}$ thickness of plates. Flues containing furnaces 9 feet long, and 2 feet 9 inches in diameter. Mixing chamber 8 feet long, small tubes 7 feet long, grates $2\frac{1}{2}$ feet by $6\frac{1}{2}$ feet.

The heating surface in one of these boilers is as follows:—

Area of furnace flues	135	square feet.
„ of mixing chamber	102	„
„ of three vertical tubes	28	„
„ of small tubes	550	„
„ of exterior flues	285*	„
Total	1100	„

Area of firegrate, $33\frac{1}{2}$ square feet, being in the ratio of 1 to 32.

Messrs. J. and W. Galloway have patented a boiler in which a number of vertical, conical, water tubes, five or six inches in diameter at bottom, and twice as much at top, are introduced

* The area of the flues, 285 feet, under the exterior of the boiler is of little value, as the greater portion of the heat is absorbed by the time it has passed through the three-inch tubes.

into an elliptical flue passing through the boiler. The flame and heated gases circulate amongst these tubes and impinge against their sides. The dimensions of one of these boilers is as follows:—Length 24 feet; diameter 7 feet; greatest diameter of main flue, 5 feet 7 inches; the flue contains 21 vertical water tubes, acting as stays to prevent collapse, $11\frac{1}{2}$ inches in diameter at top, and 6 inches at bottom. These tubes are welded and placed zig-zag fashion, so that a man may creep along each side of the flue to clean or examine it. The two furnaces are each $7\frac{1}{2}$ feet long, $2\frac{3}{4}$ feet diameter.

Another boiler, known as the French or elephant boiler, is sometimes used. It consists of three cylindrical tubes with hemispherical ends, one larger than the other two, and placed above them. The smaller boilers communicate with the upper one by conical water tubes. The furnace is under the lower tubes, and the gases after passing the length of these, return underneath the larger boiler above, round which they circulate three times.

Messrs. Dunn and Co. manufacture what they term a retort boiler, in which the steam is generated in a number of retorts, or cylindrical tubes, about 9 feet long by 18 inches in diameter, placed *transversely* to the furnace. These all communicate with a large steam chamber above. This boiler is chiefly intended for exportation, being light and convenient for carriage in new countries without roads, or the usual means of conveyance.

Computation of the Power of Boilers.—In our attempts to give any definite rules on this question, we must state that there is hardly any branch of practical science so exceedingly anomalous and unsatisfactory as that of boiler power. In fact, there is no definite rule for our guidance, on the contrary, the whole is a jumble of guesses, and for years I have laboured in vain to reduce our past experience to something like a system; or to some reliable and definite rules calculated to guide us to correct results. This however appears to be impossible, as we are in a constant state of transition with a long vista of speculations before us, which seem likely only to lead to the point from which we started. It is like the smoke question, where every man is his own doctor, and promises much, while nothing is done. The only sound and definite principles of construction we have arrived at, pertain to the locomotive boiler, where the area of heating surface, capacity of fire box and grate area necessary to generate, with the aid of the blast pipe, the required quantity of steam, are well ascertained. But in land and marine boilers we have not as yet come to an agreement, and probably for this reason, that we have not the energy of an artificial draught, which in the locomotive increases or diminishes in proportion to the speed of the engine, and the strength of the blast respectively. Now this is not the case with the condensing engine, either on shore or afloat, and notwithstanding that there are many efficient and well-proportioned boilers doing their work well, we are nevertheless deficient in the knowledge of which under any given circumstances are the best construction and the most economical proportions.

We are still in want of an experimental investigation calculated to supply data on this subject. The trials hitherto made have been too partial and under too variable circumstances to be relied upon. As it is we must be content to take them as they now exist, under the hope that time may elicit greater certainty in the improved conditions now in progress. On some future occasion it is possible we may return to the subject, as it is one of deep importance in forwarding the manufacturing interests of the country.

Horses Power of Boilers.—The horses power of boilers is dependent in part on the capacity of the boiler itself, in part on the heating surface, and in part on the area of grate and the consumption of coal per hour. The common rule for estimating the horse power, is as follows:—Calculate the "*effective section*" of the boiler by adding to the diameter of the boiler the diameters of any internal flues and multiplying by the length of the boiler, and divide the product by the constant 5·5, 5·75, or 6, according to the practice of different engineers.

For condensing engines I have usually allowed about twelve square feet of "effective section" for each nominal horse power of the engine, although in practice many conditions necessitate the alteration of this proportion to suit circumstances. Now, as engines are at present constructed, working at from two to three times their nominal horses power, this is equivalent to an allowance of 5 square feet of "effective section" per indicated horse power, and hence agrees approximately with the rule given above. But this empirical rule is not at all to be relied upon, as it gives erroneous results with boilers of different forms and proportions.

The true method of calculating the proper proportion of boiler for any given engine is, however, to estimate the actual amount of steam required, which can easily be done with the aid of the tables, already given, of the weight and density of steam. Then provide a boiler capable of evaporating that weight of water, according to the data obtained in experiments with boilers of the particular construction employed. Some data of this kind will be given below. It being borne in mind that more heat is required, and less water evaporated with a given weight of coal, the higher the pressure at which the steam is employed.

Area of Heating Surface.—The total area of metal exposed to the flame and hot gases is called the *total heating* surface of the boiler, and is usually expressed in terms of the grate-bar surface. This unit of comparison has, however, been rendered ambiguous by the employment of another unit called the *efficient heating surface.* The efficient heating surface is obtained by deducting from the total heating surface one-half the area of vertical portions, and one-half the area of horizontal cylindrical flues, on the supposition that the vertical heating surfaces and the under side of flues and tubes act less efficiently in absorbing heat than horizontal surfaces above the flame.

A common allowance of effective heating surface for stationary boilers has been 10 to 15 square feet per square foot of grate area, and one square foot of grate is required per nominal horse power of the engine. I have usually allowed 16 or 17 square feet of effective heating surface; and in Cornish boilers 25 square feet is allowed. In general practice it will, however, be found that such a proportion as 17 will better serve the interests of the employers of steam engines than the extreme limits of 1 in 10 or 1 in 25; at least this is the best proportion for cylindrical flued boilers. The limits which define the amount of efficient heating surface are on the one hand the temperature of the gases escaping into the chimney, which should be as low as possible, and on the other the temperature of the boiler bottom, at which soot is deposited. If the gases escape at a higher temperature than is necessary to create a sufficient draught, heat is wasted by dissipation in the atmosphere, in consequence of insufficient heating surface. On the other hand, if the boiler is unduly increased, so that part of the heating surface is coated with soot, and the absorption of heat prevented, not only is boiler space wasted, but heat is lost by radiation.

In the Saltaire boilers the proportions of the heating surface may be estimated as follows:—

	Total heating surface in sq. ft.	Efficient heating surface in sq. ft.
Furnaces	135	68
Mixing chambers	102	51
Vertical tubes	28	14
Three-inch tubes	550	275
Exterior flues	285	192
	1100	600

Area of firegrate 33·5 square feet.

That is, 17 square feet of effective and 32 square feet of total heating surface per square foot of grate.

Again, in a double-flued tubular boiler, 30 feet long, 7 feet diameter, with two flues each 2 feet 8 inches in diameter, we have the following proportions:—

	Total heating surface.	Efficient heating surface.
Internal flues	504	252
Exterior flues	390	318
	894	570

Area of grates = 33 square feet.

Hence, there would be 27 square feet of total heating surface, and 17 feet of effective heating surface, per square foot of grate area.

Boiler Capacity.—In my practice I have always advocated large boilers. I have said before that boilers of limited capacity, when overworked, must be forced, and this forcing is the gangrene which corrupts and festers the whole system of operations. Under such circumstances perfect combustion is out of the question, and every attempt at economy fails. Usually with flued boilers I have allowed 15 to 20 cubic feet of boiler space per indicated horse power after deducting the flues. Mr. Armstrong contends for 27 cubic feet, of which one half is steam, and the other half water room. I have allowed one-third for steam and two-thirds for water where the boiler is fitted with a dome. When the steam-room is too small, the boiler primes, or water is carried over from the boiler with the steam.

Area of Grate-bar Surface.— The area of the grate depends upon the quantity and quality of the fuel to be burnt. In Cornish boilers, in which the combustion is slowest, only 6 to 10 lbs. of fuel are burnt per square foot of grate bar per hour; and in ordinary factory boilers about 14 to 16 lbs. is the average quantity. In marine boilers the combustion is still more rapid, and in locomotives it rises as high as 40 to 120 lbs. per square foot per hour.

The grate bars are ordinarily made to slope, to facilitate the pushing back of the fuel which has been partially coked on the dead plate. This slope varies from 1 in 5 to 1 in 25, being in cylindrical flued boilers somewhat restricted by the form of the flue. The firegrate terminates in a brick bridge, over which the flame and products of combustion pass into the flues. These bridges distribute the flame over the boiler bottom, and cause an eddy which facilitates the mixture and combination of the gaseous products.

Mr. D. K. Clarke has very carefully investigated the relations of grate-bar surface, heating surface, and consumption of coal, and has arrived at the following relations:—1st. For a given area of grate the total hourly consumption of fuel should vary as the square of the total heating surface; that is to say, if the heating surface be doubled, the total consumption of fuel might be increased four times, whilst the same evaporative efficiency would be maintained. 2nd. For a given extent of heating surface, the total hourly consumption should vary inversely as the area of the grate. For instance, if the grate surface were increased to twice the area, the total hourly consumption of fuel should be reduced to one-half, in order to maintain the same efficiency. 3rd. For a given hourly consumption of fuel, the area of the firegrate will vary as the square of the heating surface in maintaining the same efficiency. For example, if twice the heating surface be employed, the grate may be extended to four times. Conversely if half the heating surface be removed, the grate must be reduced to one-fourth of its area. It is apparent from these relations, as Mr. Clarke has observed, that a superfluous size of grate is detrimental to the power of the boiler, unless at a sacrifice of fuel. On the contrary, an extension of heating surface adds a still greater proportion to the power of the boiler, whilst the efficiency of the fuel is maintained. The general formula embodying these relations is $F = C\frac{H^2}{G}$, in which F is the quantity of fuel consumed per hour, H the area of heating surface, G the grate area, and C a constant varying for each kind of boiler.

Grates for burning wood require to be constructed on different principles from those for the consumption of coal. In this case, from the rapid ignition of the material, the furnace must be constructed capaciously, whilst at the same time the area for the admission of air must be reduced. In Russia, where nearly the whole of the coal used in manufacture is imported from this country, it is usual to have the boilers constructed on the same principle as has already been described. It, however, sometimes happens, as in the case of the late war, that the supply of coal ceases, and the owners of mills are in this emergency under the necessity of burning wood, which even in Russia at the present time is more expensive than imported coal. When driven to its use, all that is done is to remove the coal grate and furnace bars, and substitute an iron gridiron, laid on the bottom of the internal flues, which increases the capacity of the furnace and decreases the grate area. The boiler is then as efficient with wood as it was before with coal. In other cases the wood is supplied by a hopper, in which it descends as it burns away at the bottom.

Evaporative Power of Boilers.—Good coal liberates in combustion sufficient heat to evaporate from 14 to 15½ lbs. of water, and good coke to evaporate about 13 lbs. of water per pound of the fuel. Wood evaporates only 6 to 7½ lbs. per pound of fuel.

The actual evaporation in engine boilers falls far short of this theoretical result, owing to the heat carried off by the chimney, imperfect combustion, radiation, &c.

In 1858 a report was published by Mr. Armstrong, Mr. Longridge, and Mr. Richardson, detailing the results of extensive experiments on the evaporative power of steam coals. These experiments were made with a multitubular boiler, with two furnaces and 135 tubes, 5½ feet long and 3 inches in diameter. With this boiler they first determined a standard of evaporative power when the boiler was worked on the ordinary system, every care being taken to obtain the maximum of work out of the boiler, by keeping the fires clear and by frequent stoking. No air was admitted except through the firegrates. As the economic effect of the fuel increases when the ratio of the firegrate surface to the absorbing surface is diminished, they adopted two sizes of firegrates, and obtained in consequence two standards of reference. With the larger firegrate the amount of work done by the boiler per hour was greatest, but this was accomplished at a relative loss of economic value of the fuel, as compared with the smaller grate. The one gave the standard of maximum evaporative power of the boiler,—the other the standard of economic effect of the fuel. The grate areas were 28½ and 19¼ square feet respectively. The heating surface of the boiler was 749 square feet. The results obtained are given in the following table:—

	Firegrate 28½ sq. ft.		Firegrate 19¼ sq. ft.	
	A.	B.	A.	B.
Economic value, or pounds of water evaporated from 212° by 1 lb. of coal	9·41	11·15	10·06	12·58
Rate of combustion, or pounds of coal burned per hour per square foot of grate . . .	21·15	19·00	21·00	17·25
Rate of evaporation per square foot of firegrate per hour, in cubic feet of water, from 60° .	2·62	2·93	2·909	2·995
Total evaporation per hour in cubic feet of water from 60°	74·80	79·12	56·01	57·78

The columns marked A give the general results, much smoke being often evolved; those marked B, the mean of the best results obtained in the experiments when making no smoke. The coal employed in these experiments, viz., the Hartley's, is very superior to that ordinarily employed in factory boilers.

By an apparatus constructed by Mr. Wright, of Westminster, the same experimenters determined the absolute heating effect of this coal, and of some similar coal from Wales, to be as follows:—

Welsh coal	14·30 lbs. from 212°
Hartley coal	14·63 „ „

To give some idea of the practical economic effect of coal in stationary engine boilers, we may transcribe here some results obtained with care by Mr. John Graham, of Manchester, not as completely applying to ordinary practice, but as affording useful guidance when taken in conjunction with the preceding results on a better description of coal. The water was measured by a meter.

	Pounds of water evaporated from 212° by 1 lb. of coal.
Boiler with two internal furnaces, known as "breeches boiler" .	6·88
Waggon boiler	10·26
Cylindrical boiler with external furnace	7·54
Butterley boiler	9·72

Mr. Longridge has found 6 to 7½ lbs. of water evaporated from 62° Fahrenheit, and converted into steam at 20 to 55 lbs. pressure, per pound of coal, by two flued boilers.

Mr. Rankine's formula for the efficiency of ordinary stationary boilers without a feed-water heater and with chimney draught is,

$$\frac{E'}{E} = \frac{\frac{11}{12} S}{S + \frac{1}{2} F} \ldots (1);$$

where E' is the available evaporative power of one pound of fuel in a boiler furnace, E the theoretical evaporative power, S area of heating surface, and F the number of pounds of fuel burnt per hour per square foot of grate. But the reader must be referred to his own work on the steam engine for a discussion of the constants to be employed under different circumstances.

Strength of Boilers.—To be of maximum strength, both the external shell and the internal flues should be as far as possible cylindrical. Where this is impossible and flat surfaces are necessary, careful staying by gussets or longitudinal stays is essential to safety. The cylindrical portions of boilers can be very easily proportioned to the steam they have to bear by the formulæ which will be given below. Certain restrictions are placed upon the proportions of boilers by the nature of the riveted joints. That these may be steam and water tight under pressure, and at the same time not unnecessarily weakened by rivets, it has been found best to use plates of about $\frac{5}{16}$ or $\frac{3}{8}$ inch thick, and plates of other dimensions are very seldom employed in the construction of boilers. Thick plates are inefficiently riveted, thin ones inefficiently caulked, and this restricts the available thickness for the plates within nearly the limits which have been stated. It is necessary, therefore, in proportioning boilers, having given the working pressure, to choose the diameter which is suitable for such a thickness of plates, lessening the diameter for high-pressure boilers and increasing it for low-pressure ones. Length does not affect the strength in vessels subject to internal pressure, and hence the diameter is the variable quantity over which we have most control in proportioning the external shell. But the flues, as will be shown, decrease in strength with their length, and this dimension is in that case more easily modified than the diameter.

The general equation expressing the resistance of thin hollow vessels to internal strain is, for spherical vessels,

$$P = \frac{4\,c\,t}{d} \text{(nearly)} \ldots (2);$$

and for cylindrical vessels, bursting longitudinally,

$$P = \frac{2\,c\,t}{d} \ldots (3).$$

This equation gives the bursting pressure in lbs. per square inch, when the thickness of the plates c in inches, the tenacity of the joints t in lbs. per square inch, and the diameter d in inches, are given.

Thus, for a boiler 7 feet or 84 inches diameter, $\frac{3}{8}$ inch or ·375 inch thick, and with joints having a tenacity of 34,000 lbs. per square inch, the bursting pressure

$$= P = \frac{2 \times 34000 \times \cdot 375}{84} = 303\tfrac{1}{2} \text{ lbs.}$$

The value of t for various materials is given in the following table:—

Without joints:	
Wrought-iron plates	50,000
Steel plates	100,000 to 130,000
Copper, sheet	30,000
Glass	4200 to 6000
With joints:	
Wrought-iron plates, double-riveted	35,700
Wrought-iron plates, single-riveted	28,600
Wrought-iron boiler plates, with single joints, crossed	34,000

In the case of well-constructed wrought-iron stationary boilers, I have been accustomed to take $t = 34000$, and in this case the bursting pressure of cylindrical vessels is, as was taken above,

$$P = \frac{68000\,c}{d} \ldots (4).$$

But with boilers the factor of safety is ordinarily taken at six, or the working pressure is not allowed to exceed $\frac{1}{6}$th of the bursting pressure, and in this case the maximum strain on the iron per square inch of section is 5666 lbs. Putting p for the safe working pressure,

$$p = \frac{11333\,c}{d} \ldots (5);$$

or in the case of the seven-foot boiler taken above,

$$p = \frac{11333 \times \cdot 375}{84} = 50{\cdot}6 \text{ lbs.},$$

equivalent to one-sixth of $303\frac{1}{2}$, the bursting pressure.

For half-inch plates we get from formula (5),

$$p = \frac{5666\,c}{d}, \text{ and } d = \frac{5666}{p};$$

for three-eighths inch plate we have similarly,

$$p = \frac{4250}{d}, \text{ and } d = \frac{4250}{p};$$

and lastly for five-sixteenths inch plate,

$$p = \frac{3541}{d}, \text{ and } d = \frac{3541}{p}.$$

That is, in words, to find the safe working pressure of a boiler, divide 5666 for $\frac{1}{2}$-inch plates, 4250 for $\frac{3}{8}$-inch plates, and 3541 for $\frac{5}{16}$-inch plates, by the diameter in inches. Similarly, to find the safe diameter for a given working pressure, divide the same numbers by that working pressure in lbs. per square inch.

For flues subjected to an external pressure, I have deduced experimentally a formula the data of which are given in "Useful Information for Engineers," Second Series. Putting P for the collapsing pressure and p for the safe working pressure as before, C thickness of plates in inches, D diameter in inches, L length in feet,

$$P = 806300 \frac{C^{2\cdot19}}{L\,D} \dots (6);$$

$$p = \frac{P}{6} = 134400 \frac{C^{2\cdot19}}{L\,D} \dots (7).$$

For practical purposes we may substitute for the power 2·19 the square of the thickness. But it is better to employ a table of logarithms, when we get

$$P = 1\cdot5265 + 2\cdot19 \log. (100\,C) - \log. L\,D.$$

Thus, for example, to find the collapsing pressure of a flue 10 feet long, 36 inches in diameter, and composed of ½-inch plates, we have approximately,

$$P = 806300 \times \frac{(\frac{1}{2})^2}{36 \times 10} = 560 \text{ lbs.};$$

or, more accurately, by logarithms,

$$\log. P = 1\cdot5265 + 2\cdot19 \log. 50 - \log. 360 = \log. 502 \text{ lbs.}$$

The safe working pressure of this flue would be $\frac{502}{6} = 74$ lbs.

This formula shows that with vessels subjected to external pressure the strength varies inversely as the length. That is, a flue 20 feet long will be only one-half the strength of one 10 feet long. This remarkable law enables us to proportion the strength of boiler flues with great ease. By introducing rigid angle or T iron ribs riveted round the exterior of the flues, we virtually decrease the length and increase the strength in the same proportion. Two or three such rings on the flues of boilers, constructed of plates equal in thickness to those of the shell, will usually render the resistance to collapse equal to that of bursting.

Accessories of Boilers. The Feed Pump.—Boilers require replenishing with water in proportion to the waste by evaporation. For this purpose, in the early boilers, working at very low pressures, an open stand pipe was employed with a valve at top, for the admission of the water from a reservoir regulated by a float in the boiler. With the increase of pressure at which steam engines are worked, this stand pipe has been abandoned and replaced by the feed pump, either attached to the engine or worked by a donkey engine attached to the boiler. The capacity of this pump must be such as to discharge into the boiler two to three times the quantity of water required by the engine in the shape of steam. The ample tables we have already given of the density of steam will enable this to be calculated with perfect ease. We have only to find the volume of steam required by the engine at each stroke (depending on the rate of expansion at which it works), and the pressure of the steam being known we have to seek its weight in the table of density, page 215, and provision must be made for the discharge of two or three times this quantity into the boiler at each stroke of the engine.

Back Pressure Valves.—To prevent accident in case of stoppage or fracture of the feed apparatus, there should always be placed on the feed pipe between the boiler and the regulating valve a self-acting valve to prevent the return of the water. Supposing the feed pipe accidentally broken, the water in the boiler would be forced back by the pressure of the steam, and expose the boiler to injury by overheating. In such a case the back pressure valve is of great service in preventing the escape of the water when acted on by the pressure of the steam.

Feed-water Heating Apparatus.—When the products of combustion escape into the chimney at an elevated temperature, the heat may be utilised by the employment of water-tubes through which the feed water is introduced on its way to the boiler. Of the arrangements adopted for this purpose the best is that of passing the feed water through a wrought-iron pipe or supplementary boiler, placed in the main flues immediately behind the boiler, where the water is heated to the boiling point after leaving the pump. A more complete apparatus is that of Mr. Green, of Wakefield, known as the "Fuel Economiser." It consists of a series of upright tubes through which the feed water passes on its way to the boiler, and is heated above the boiling point, and steam in part generated. The formation of soot on the pipes was the source of the ill success of previous attempts in this direction. This difficulty Mr. Green has overcome by an apparatus of scrapers or cleaners, consisting of rings encircling the pipes, and maintained in constant but slow motion by chains and pulleys driven by a belt from the engine. With this apparatus it is found that when the waste gases escape at a temperature of 400° to 500°, the feed water can be heated to an average of 225°, the temperature of the gases after leaving the pipes being reduced to 250°. To produce this effect 10 square feet of heating surface are provided for each horse power.

Water Gauges.—Every boiler should be supplied with a glass tube, fixed in suitable stuffing boxes, and open at the top and bottom to the boiler to show the level of the water. Gauge cocks at various levels are sometimes employed as supplementary to the glass gauge: both are necessary.

Steam Gauges for indicating the pressure of the steam are also indispensable to the safe working of the boiler. For low pressures an open mercury column is employed on the principle of that used by Regnault in his experiments, and in some cases, to bring the indications within a small compass, the fall of the mercury in the cistern rather than its rise in the smaller tube is observed. To avoid the inconvenient length of the open mercury column, the air gauge has been used, in which the mercury in its rise condenses the air in a closed glass tube. This gauge, accurate and sensitive, has yet the fault that the indications decrease in length as the pressure increases, and there is also some difficulty in preserving the quantity of air in the gauge constant. Recently Mr. Allan has overcome these difficulties by the use of a conical air chamber so arranged that the indications of the gauge shall be uniform at all pressures, and the air can be renewed at any instant. In M. Bourdon's gauge a curved metallic tube, communicating at one end with the steam boiler, and at the other closed, is used. The curvature of this tube decreases with the increase of pressure in its interior, and the closed end being free to move is connected with an arrow moving over a graduated arc and marking the pressure. In Schaeffer and Budenberg's gauge the pressure acts on a flat corrugated plate of steel which expands and raises a rack acting on a toothed wheel, and carrying a similar arrow to indicate the amount of pressure. In Smith's gauge a flat spiral spring is used, against which the pressure acts through the medium of a plate of india-rubber. All these gauges should be fixed on the boiler with a siphon in which water from the boiler may condense. In this way the pressure in the boiler is transmitted through the column of water, and at the same time the gauge is unaffected by the temperature of the steam.

Safety Valves.—I usually place three safety valves on boilers, as shown in fig. 173; two of these have fixed weights on their levers, and the other is pressed down by a spring balance, and serves to regulate the working pressure of the boiler. The two larger valves for a fifty horse boiler have each an area of 12 square inches. The third is of only 5 square inches area. These valves are fixed to a common valve seating. The bearing surfaces of the valves are made either flat, conical, or spherical. Flat valves have a tendency to blow off at too low a pressure, from the steam getting between the bearing surfaces. These valves should always be open to the atmosphere that they may be seen.

Fig. 173.

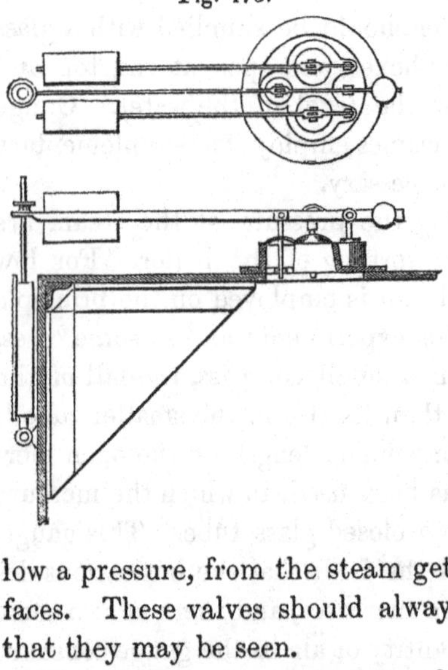

Man Holes are required to obtain access to the boiler for purposes of examination and cleaning. In double-flued boilers one must be placed beneath the flues as well as above them. *Mud Cocks* are placed near the bottom of boilers for the discharge of water and sediment.

Fusible Plugs are portions of metal fusing at a temperature not greatly exceeding the maximum working temperature of the steam, and fixed in that portion of the boiler most liable to be overheated from deficiency of water. These plugs are of pure lead or of an alloy of bismuth, lead and tin, according to the temperature they are required to melt at, and they are thought to prevent danger by relieving the pressure of the boiler, and putting out the fire before the plates are injured by overheating. These plugs, however, tend to lose their fusibility, and to become coated with a protecting coat of oxide or sediment, which prevents the communication of heat. They are not a very reliable provision.

Plans for the prevention of Smoke.—Amongst the earliest of these we may class those which depend on mechanical means for the supply of the fuel.

Of this class is the earliest patent for smoke prevention taken out by James Watt in 1785. By this plan the fire is supplied from above downwards by a reservoir of fuel in contact with the burning mass, the combustion of which is supported by a strong lateral current of air passing direct through the fire to a flue on the other side aided by a slight downward current beside or through the fuel, which last descends by its own gravity as it is consumed. For the purpose of intercepting and completing the combustion a clear fire is maintained at the entrance into the flues, so that the products of the first fire, being subjected to the intense heat of the second and mingled with atmospheric air, may be effectually consumed.

Apart from the external reservoir, we owe to Watt the dead plate very generally adopted in stationary boilers. The fresh fuel is thrown upon the dead plate, where it gradually cokes the more volatile constituents distilling over and being consumed by the bright fire beyond. Then the coked fuel is pushed back on to the bars and a new supply introduced in front. This plan, where proper provision is made for the supply of the necessary quantity of air, obviates the production of smoke as effectually as many more complicated contrivances.

The succeeding patentees of the principle of mechanical feeding as a substitute for hand labour, have followed two different plans. Some have made the grate itself to carry forward the fuel, either by revolving horizontally or by rolling forward longitudinally, the grate-bars being connected together to form an endless chain, or by the oscillation of the alternate bars causing the thrusting forward of the fuel by what has been called a peristaltic movement. Others have made the grate stationary and have used fans revolving horizontally to distribute the fuel over the grate-bars. In all these cases the coal is supplied slowly and uniformly from a hopper. There is no doubt that the uniform distribution of the fuel over the whole surface of the grate-bars, so far as it is secured by these systems, must be to a large extent advantageous in the diminution of smoke and economy of fuel. At one time they were extensively used, but the complication and expense of the apparatus has led to their general abandonment and the return to hand-feeding.

Other plans for the prevention of smoke depend on a double furnace with alternate firing.

Double furnaces patented by Mr. Losh were in use as early as 1815, and in various modifications have been employed ever since. The principle of double furnaces within the same boiler was first introduced by myself; and the plan adopted has already been described as the double-flued boiler. The two flues enable the stoker to fire alternately, and so maintain a more uniform generation of steam than with a single flue, and the flame passing from one flue mingling with the gases from the other, assists in their combustion. I believe that this simple system of alternate firing, when conjoined with the requisites of the economical generation of steam, viz. plenty of capacity in the boiler, sufficient admission of air, and, what is quite as necessary, careful and attentive stoking, will effect the prevention of smoke without any costly apparatus, so far as that is possible with any given description of fuel. There is this further advantage in double furnaces, that the air required for combustion is necessarily variable. Now a double furnace tends to equalise the supply. The two furnaces fed alternately will not require a maximum or a minimum quantity at the same time, and as the two currents of gaseous products mingle, the surplus air of the one furnace will supply the deficiencies of the other. In this way the tendency is to compensate the supply and demand, and prevent waste from too large or too small a quantity in either furnace.

Others in seeking the prevention of smoke have introduced an additional supply of air over the fire.

Mr. C. Wye Williams was one of the earliest, as he has been the most pertinacious and consistent, advocate of the introduction of a large additional volume of air into the furnace, and we have to thank him for the labour he has expended in proving the necessity for air as one of the prime conditions of economy of fuel and success in the prevention of smoke. Mr. Wye Williams contends for a uniform admission of cold air to the furnace, relying upon frequent thin feeding to equalise the needs of the furnace. The peculiar principle of his plan is the mechanical division of the air by causing it to enter the furnace through what he terms a diffusion plate, or partition perforated with numerous small apertures. This is usually placed behind the bridge where the gases needing combustion pass into the flues. There is no doubt this is a convenient method for the introduction of air, and has in many instances effectually prevented the formation of smoke.

Mr. Syme Prideaux contends for a variable admission of air, greatest when the fuel is first thrown on, and decreasing to the ordinary supply through the grate-bars as the fire burns clear. For this purpose he constructs his furnace doors with metal Venetians, which open by a self-acting apparatus when the fuel is supplied. They then gradually close at a regulated speed, altogether independent of the care of the fireman. The air entering through the door is, by an arrangement of plates, warmed as it enters the furnace, and carries back the heat radiating from the door.

All these systems are more or less effective, but I am inclined to think that a judicious engineer with a careful stoker or fireman, will effect all the objects to be attained with the means placed at his disposal, in a well constructed boiler of sufficient capacity, and with a simple furnace such as has been described in the foregoing chapter, as completely as can be done by any one of the numerous nostrums held forth as the only antidotes for smoke, and promising great economy of fuel.

CHAP. IX.

ON WINDMILLS.

Atmospheric disturbances causing wind have from a high antiquity been employed as a motive power, and probably the earliest application of this force was the propulsion of ships by sails. Amongst the most primitive races, long before their intercourse with civilisation, this power was applied in the navigation of small vessels; and the ancient Phœnicians, Greeks, and Romans were all of them well acquainted with this mode of employing the force of the wind for purposes of human industry. It is to be regretted that we have no records of the time when it was applied as a motive power in mills; this event is lost in the oblivion of the past, and it was not till early in the thirteenth century that we find the Dutch and French employed in the construction of windmills adapted to the wants of an energetic and industrious population. These times were marked by a growing intelligence that encouraged and fostered inventive talent, and the Dutch millwrights and engineers were long celebrated for their skill and knowledge in every art that had for its object the improvement of the industrial resources of the people.

It was from Holland that our knowledge of windmills and wind as a motive power was first imported, and it is within my own recollection that the whole of the eastern coasts of England and Scotland were studded with windmills; and that for a considerable distance into the interior of the country. Half a century ago, nearly the whole of the grinding, stamping, sawing, and draining was done by wind in the flat countries, and no one could enter any of the towns in Northumberland, Lincolnshire, Yorkshire, or Norfolk, but must have remarked the numerous windmills spreading their sails to catch the breeze. Such was the state of our mechanism sixty years since, and nearly the whole of our machinery depended on wind, or on water where the necessary fall could be secured. Now both sources of power are also abandoned in this country, having been replaced by the all-pervading power of steam. This being the case, we can only give a short notice of wind as a motive power, considered as a thing of the past.

Two sorts of windmills have been employed in this country, namely the horizontal and the vertical, the sails in the one revolving on a vertical axis, in the other upon a horizontal axis, depressed or raised at a certain angle to the horizon. The first of these has been very little used; the latter kept its ground against all competitors until it was supplanted by its more energetic opponent in the shape of steam.

The vertical windmill consists of a tower, near the top of which is an axle carrying four vanes or sails set in a plane inclined about ten degrees to the vertical. The vanes are also inclined to the plane in which they revolve, their inclination varying from the axis to the extremity of the arms. They are made light and filled with thin plates of wood, or are covered with canvas. Thus the wind, blowing perpendicularly to the plane of revolution of the arms, impinges obliquely upon the broad sails, and a rotatory movement is generated, which, transmitted by bevel gearing, works the millstones, scampers, and machinery of every sort contained in the mill.

The mill-sails require to be placed perpendicularly to the direction of the wind, and for this purpose, in the older mills, the whole upper part of the tower containing the machinery is turned round by manual labour. In more modern constructions, however, a dome or cap carrying the sails is fixed on the summit of the tower, and is turned by a self-acting fly with four or more oblique vanes, similar to a smoke jack, which, acted upon by the changing currents of wind, gives motion to the cap of the tower, carrying round with it the wind axis and sails, keeping them perpendicular to the direction of the wind. Such a mill is shown in fig. 174, where D is the cap moving on rollers, *s* the shaft carrying the sails S S, and the bevel wheel *a a*, gearing into another bevel wheel *b*, on the millstone shaft. The wind acting on the fan F, communicates motion to the bevel wheel and spur pinion *e*, which, acting on a spur wheel or rack fixed on the summit of the tower, causes the revolution of the cap. The sails of the fan are constructed so that when they lie in

Fig. 174.

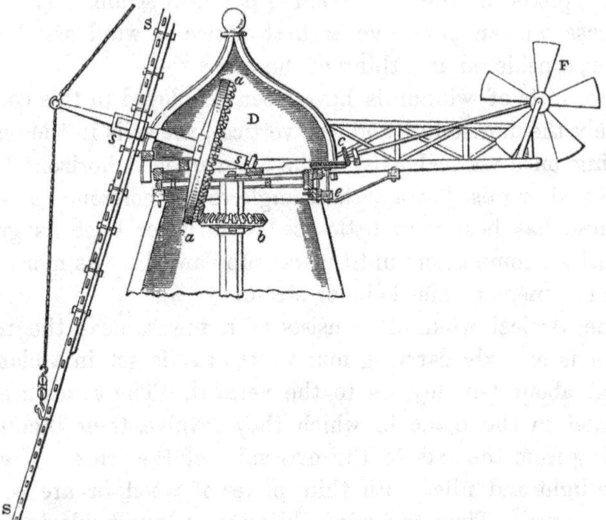

the plane of the wind, they are not affected; but as the wind shifts, it strikes them obliquely, and causes the revolution of the cap till they are again in the plane of the wind.

Of experiments upon windmills by far the most important are those of Smeaton, communicated to the Royal Society in 1759. The inclination of the sail to the plane of revolution he found should vary in the following ratio, where the radius is supposed to be divided into six equal parts, and the angle of the sail given at each point:—

	Angle with the plane of motion.
0	— centre.
1	18°
2	19
3	18 middle.
4	16
5	12½
6	7 extremity.

This inclination of the sail to the plane of revolution is known as its *weather*. Sails before Smeaton's time were simple parallelograms; he found, however, that advantage was gained by adding a triangular sail to the leading edge of the radius or whip *a a*, so placed that the sail was broadest at its periphery. The extreme breadth of the sail, *c b*, was then made equal to one third of the radius or whip, and of this $\frac{5}{8}$, or $\frac{5}{24}$ of the radius, was the breadth of the ordinary sail, *a b*, and the remaining $\frac{3}{8}$, or $\frac{3}{24}$ of the radius, was the breadth of the triangular leading sail, *a c*, as shown in fig. 175. The ordinary length of the whip of the sail is 30 feet.

Fig. 175.

Regulation of the Speed of Windmills.—This is best effected in the case of windmills with cloth sails, by a plan of Mr. Bywater, in which a series of racks and pinions cause the cloth to roll or unroll according to the strength of the wind.

Another plan, suggested by Mr. (now Sir William) Cubitt at the beginning of this century, is shown at fig. 176, applied to sails which have movable boards or thin plates instead of sail-cloth. *a a* is the whip; *b*, the axis on which the sails are carried, and which is hollow to receive the rod *c c*. At the extremity of this is a rack, *c d*, gearing in a pinion, *e*, which is connected with a pulley over which is hung a weight so as to press the rod *c c* outwards with a constant force. *g*, *g*, *g*, are the boards which form the surface of the sail, and which are connected together so as to open or shut like the bars of a Venetian blind. On the last board of each sail is a toothed segment, in

Fig. 176.

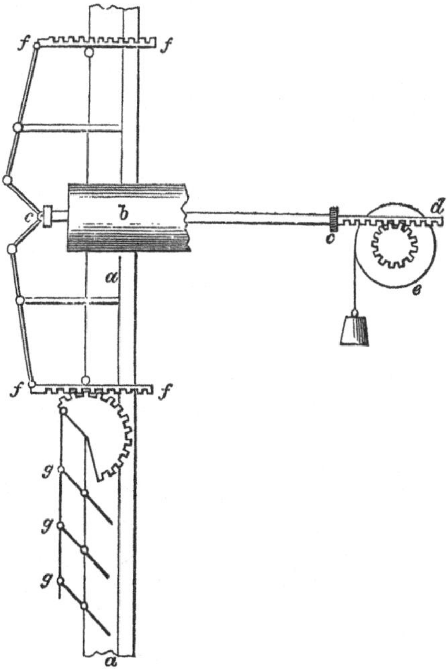

which works a rack, *f f*, connected by levers with the rod *c c*, as shown. By this arrangement the force of the wind, as it varies, opens or shuts the boards of the sail, so as to keep the total pressure on the sails equivalent to the force exerted by the balance-weight hung over the pulley *e*.

APPENDIX.

EXPERIMENTS ON MR. THOMSON'S VORTEX WHEEL AT BALLYSILLAN, TO DETERMINE ITS EFFICIENCY.

Abstract of Data and Calculations according to which the Vortex was designed:—Total Height of Fall=24 feet. Standard quantity of Water=420 Cubic feet per Minute. Calculated Speed, 292·5 Revolutions per Minute.

Radius of Friction Brake, 4 ft. 2 ins. Circumference, 26·18 feet. In the Table the quantity of Water passing over the Weir is calculated by means of very refined data arrived at by MM. Poncelot and Lesbros, and submitted to Academy of Sciences in 1829.

No. of Experiment.	Total weight on Cord of Break in lbs.	Duration of each Experiment.	No. of revolutions of Shaft during Experiment.	No. of revolutions of Shaft per Minute.	Total height of Fall in feet.	Height in Feet of Water over Weir. Length of Weir =3 ft.	No. of Cubic Feet of Water per minute over Weir.	No. of cubic feet of water per min. supplied to the wheel, 1·39 being deducted to weight the break.	Work given out at Friction Break in Foot lbs. per Minute.	Horse Power given out by Wheel.	Work due to the fall of Water in Foot lbs. per minute.	Efficiency of the Wheel or Work given out by Wheel. Work due to Water being 100.	Remarks entered at the time of the Experiments, relating to the supposed accuracy of the experiments.
		min. sec.											
1	46·31	5 34	1800	323·3	23·73	·718	355·8	354·4	392000	11·88	523900	74·81	Satisfactorily accurate.
2	46·31	5 38	1800	319·5	23·71	·718	355·8	354·4	387400	11·73	523500	73·99	Satisfactorily accurate.
3	53·31	6 11	1800	291·0	23·72	·718	355·8	354·4	406100	12·31	523700	77·55	Good.
4	60·31	7 9	1800	251·7	23·71	·718	355·8	354·4	397400	12·04	523500	75·91	Water supposed to be running over.
5	53·31	6 12	1800	290·3	23·71	·718	355·8	354·4	405100	12·27	523500	77·39	One of the best: no water lost.
6	39·31	5 26	1800	331·2	23·21	·718	355·8	354·4	340800	10·33	512400	66·51	Some water lost at water case, and water too low part of time.
7	46·31	6 9	1800	292·7	23·77	·665	318·1	316·7	354900	10·75	469000	75·66	Good.
8	46·31	6 13	1800	289·5	23·75	·665	318·1	316·7	351000	10·63	468600	74·90	Good: one of the best.

Remarks on the above Experiments.—It is to be observed that as the experiments happened to be made in dry weather, the *stream in none* of them supplied the *standard* supply of water for which the wheel was particularly adapted. Even with the diminished quantity of water, the efficiencies experimentally found were very high. Also, from information received after the time of the Experiments, it happened that during the Experiments the joint rings of the Vortex were not screwed properly close to the rings of the wheel, and that on their being afterwards screwed close, the power of the wheel was sensibly increased. It is therefore probable that still higher efficiencies are attainable than those shown in the above Experiments.

END OF THE FIRST VOLUME.

LONDON
PRINTED BY SPOTTISWOODE AND CO.
NEW-STREET SQUARE

MILLS AND MILLWORK

PART II.

LONDON

PRINTED BY SPOTTISWOODE AND CO.

NEW-STREET SQAURE

TREATISE

ON

MILLS AND MILLWORK

PART II.

ON MACHINERY OF TRANSMISSION

AND THE

CONSTRUCTION AND ARRANGEMENT OF MILLS

COMPRISING TREATISES ON WHEELS, SHAFTS, AND COUPLINGS; ENGAGING AND DISENGAGING GEAR; AND MILL ARCHITECTURE; AND ON CORN, COTTON, FLAX, SILK, AND WOOLLEN MILLS: TO WHICH IS ADDED A DESCRIPTION OF OIL, PAPER, AND POWDER MILLS, INCLUDING A SHORT ACCOUNT OF THE MANUFACTURE OF IRON

BY

WILLIAM FAIRBAIRN, ESQ., C.E.

LL.D. F.R.S. F.G.S.

CORRESPONDING MEMBER OF THE NATIONAL INSTITUTE OF FRANCE, AND OF THE ROYAL ACADEMY OF TURIN: CHEVALIER OF THE LEGION OF HONOUR: ETC. ETC.

LONDON

LONGMAN, GREEN, LONGMAN, ROBERTS, & GREEN

1863

PREFACE.

In the first part of this work I endeavoured to give a succinct account of nearly fifty years experienced in the profession of a mill architect, millwright, and mechanical engineer. My professional career commenced just at a time when the manufacturing industry of the country was recovering from the effects of a long and disastrous war, and I was enabled, from this circumstance, to grow up with and follow out consecutively nearly the whole of the discoveries, improvements, and changes that have since taken place in mechanical science. These discoveries have been numerous and invaluable in contributing to the developement of our industrial resources, the diffusion of knowledge, and the extension of trade and commerce throughout the globe.

It will not be necessary to repeat what steam, gas, and electric telegraphs have effected both on sea and land in the same time, and how much we are indebted to these agencies for the abundant comforts, luxuries, and enjoyments which we now possess, as compared with the age in which our fathers lived. It will be found on enquiry that in mills, where these agencies are employed, and where the manufacture of cotton, silk, flax, and wool are carried on, are some of the elements to which we are indebted for the numerous advantages which enter into the improved state of our social existence. To mills, therefore, I have directed my attention, and in this volume I have endeavoured to follow up more in detail the principles of construction and other serviceable data to which, I trust, the intelligent student may refer with some prospect of advantage.

On prime movers as comprised in water-wheels, turbines,

steam-engines, &c., I must refer the reader to the first part of this work. The present volume is chiefly directed to what is known by the name of mill-gearing; and in Section IV. Chapter I., will be found an elaborate treatise on wheels, exhibiting the relations of diameter, pitch, width, and formation of teeth, including formulæ for calculating the strength, proportions, &c., to be observed in the construction of spur and bevel gear. Also tables of the proportions of wheels, pullies, &c., computed from data founded upon experiments and tested in actual practice, which in some respects I believe to be more convenient and comprehensive than any hitherto published. In the same Section I have devoted a chapter to the strengths and proportions of shafts, including rules and tables for calculating their resistance to strains produced by pressure, torsion, &c., and these, with the proportions of journals, friction, lubrication, and other conditions, constitute the contents of Chapter II.

Chapter III. treats of the couplings of shafts, engaging and disengaging gear, and those connections by which motive power may be conveyed to a considerable distance from the prime mover, and by which all the necessary changes of stopping and starting machines may be effected at one part of the mill without detriment or interference with the machinery of any other part.

The first chapter of Section V. embraces a short treatise on mills and mill architecture, with illustrations, suggestions, and improvements to be employed in the construction of those edifices. I have been induced to refer to this subject from the fact, that in former times anything like architecture as applied to mills was unknown and greatly neglected; and there was a total disregard of taste or design until late years, when a few examples of architectural construction were afforded by the introduction of slight cornices and pilasters, showing that it was possible at a small cost to relieve the monotony of a large brick surface, and bring the structure within the category of light and shade. This to some extent introduced a better style of building; and on this subject I have given a few examples for the guidance of the millwright and engineer.

Chapter II. Section V. treats exclusively of corn mills; and as these constructions are chiefly in the hands of the millwright, I have been more particular in directing attention to the buildings as well as the machinery. In this department will be found several examples and illustrations of the best constructions, from those with two to others with thirty-six pairs of stones, including all the necessary machinery for cleansing, grinding, dressing, &c.

I have also given a description of the floating-mill erected for the Government during the late Crimean war; and this, with numerous details of elevators, Archimedean screw, creepers, &c., calculated to make the mills self-acting, comprise the treatise on grinding corn.

Chapters III. IV. V. and VI. are descriptive of mills for the manufacture of the textile fabrics, as comprised in cotton, woollen, flax, and silk mills. These chapters are directed more to the process of manufacture and less to details than those on corn. They, however, contain illustrations and examples of each kind of manufacture taken from mills of my own construction. They show the arrangement, but are not descriptive of the machinery, as machine making for the separate purposes of manufacture is now a distinct trade, and does not therefore enter into that of the millwright. Having omitted the machinery—which may be found in other works on that subject—I have introduced a description of the different processes as they exist in each kind of manufacture; and considering that the mechanical arrangements of which this volume treats apply generally to every description of spinning mill, its moving power, wheels, shafts, &c., being nearly the same in each, it is not necessary to multiply examples in cases where the details closely approximate and become almost identical in form and construction. I have therefore left to others the task of describing the machinery; but I have given to oil, paper, and powder mills separate chapters. The importance of these different branches of industry establishes the necessity of their introduction, as also that of iron, which of all others is most intimately connected with the prosperity of our national industry. From the improvements in the manufacture of iron we derive advantages and facilities for construction which did not exist in former days; and it is not unreasonable that we should from this cause look forward to increased improvements in our mills, and a corresponding augmentation of the industrial resources of the nation.

In the production of this work, I have had the assistance of my former Secretary, Mr. W. C. Unwin, now resident at Kendal; and subsequently of his successor, Mr. E. W. Jacob, who has prepared the drawings for this volume.

MANCHESTER:

August 18, 1863.

CONTENTS.

SECTION IV.

ON MACHINERY OF TRANSMISSION.

CHAPTER I.

CHAP. III.

ON COUPLINGS FOR SHAFTS AND ENGAGING AND DISENGAGING GEAR.

SECTION V.

THE ARRANGEMENT OF MILLS.

CHAPTER I.

CHAP. II.

CHAP. III.

CHAP. IV.

CHAP. V.

CHAP. VI.

CHAP. VII.

CHAP. VIII.

CHAP. IX.

CHAP. X.

APPENDIX I.

APPENDIX II.

LIST OF PLATES.

ERRATA.

Page 102, line 1, *for* 'shaft' *read* 'shafts'
„ 113, „ 25, *for* 'features' *read* 'feature'
„ 171, „ 22, *for* 'has' *read* 'have'
„ 187, „ 10, *for* 'was' *read* 'were'
„ 187, „ 11, *for* 'and' *read* 'which'
„ 188, „ 10, *for* 'heated by steam of the circular form' *read* 'of the circular form heated by steam'
„ 219, „ 31, *for* 'Bundley' *read* 'Brindley'

Plates IX. and X. should have been numbered I. and II., but they were subsequently inserted.

A TREATISE

ON

MILLS AND MILL-WORK.

SECTION IV.

ON MACHINERY OF TRANSMISSION.

CHAPTER I.

ON WHEELS AND PULLIES.

THE elementary principles of motion by rolling contact and by wrapping connectors have been explained in Section II., so that in the present Chapter we have only to examine in detail the methods of applying these principles and their respective advantages, and especially the mode of constructing wheels in gear, so that the resulting motion shall most nearly approach the condition of perfect rolling contact.

We saw in the preliminary Chapter that there were two methods of transmitting power through trains of wheel-work, the first being through the agency of wrapping connectors, and the second by rolling contact.

Wrapping Connectors.—Considerable difference of opinion exists as to the best and most effective principle of conveying motion from the source of power to the machinery of a mill. The Americans prefer leather straps,* and large pullies or

* I have selected the word *strap*, instead of *belts* or *bands*, as a term more generally applied to wrapping connectors in the northern districts.

riggers. In this country, and especially in the manufacturing districts, toothed wheels are almost universally employed. In some parts of the South, and in London, straps are extensively used; but in Lancashire and Yorkshire, where mill-work is carried out on a far larger scale, gearing and light shafts at high velocities have the preference. Naturally, I am of opinion that the North is right in this matter, and that consistently, as I was to a great extent the first to introduce that new system of gearing which is now general throughout the country, and to which I have never heard any serious objection. I have been convinced by a long experience that there is less loss of power through the friction of the journals, in the case of geared wheel-work, than when straps are employed for the transmission of motive power. Carefully-conducted experiments confirm this view, and it is therefore evident which mode of transmission is, as a general rule, to be preferred.

There are certain cases in which it is more convenient to use straps instead of gearing. With small engines driving saw-mills, and some other machinery where the action is irregular, the strap is superior to wheel-work, because it lessens the shocks incidental to these descriptions of work. So, also, when the motive power has been conveyed by wheel-work and shafting to the various floors of a mill, it is best distributed to the machines by means of straps.

In some of the American cotton factories, however, there is an immense drum on the first motion, with belts or straps from two to three feet wide, transmitting the power to various lines of shafting, and these in turn through other pullies and straps, giving motion to the machinery. From this description it will be seen that the whole of the mill is driven by straps alone, without the intervention of gearing.

The advantages of straps are, the smoothness and noiselessness of the motion. Their disadvantages are, cumbrousness, the expense of their renewal, and the necessity for frequent repairs. They are inapplicable in cases where the motion must be transmitted in a constant ratio, because, as the straps wear slack, they tend to slip over the pullies and thus lose time. In other cases, as has been observed, this slipping becomes an advantage, as it reduces the shock of sudden strains and lessens the danger of breaking the machinery.

Very various materials are employed for straps, the most serviceable of all being leather spliced with thongs of hide or cement. Gutta percha has been employed with the advantage of dispensing with joints, but it is affected by changes of temperature, and it stretches under great strains. Flat straps are almost universally employed, in consequence of the property they possess of maintaining their position on pullies, the edge of which is slightly convex (fig. 177). Round belts of catgut or hemp are sometimes used, running in grooves, which are better made of a triangular than a circular section—so that the belt touches the pulley in two lines only, tangential to the sides of the groove; in this case the friction of the belt is increased in proportion to the decrease of the angle of the groove.

Fig. 177.

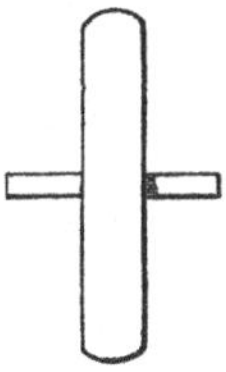

The strength of straps must be determined by the work they have to transmit. Let a strap transmit a force of n horses' power at a velocity of v feet per minute, then the tension on the driving side of the belt is $\frac{33000\,n}{v}$ lbs. independent of the initial tension producing adhesion between the belt and pulley. For example, let v be 314·16 feet per minute, or the velocity of a 24-inch pulley at 50 revolutions per minute, and let 3 horses' power be transmitted; then $\frac{33000 \times 3}{314{\cdot}16} = 312$ lbs., the strain on the pulley due to the force transmitted.

The following table has been given for determining the least width of straps for transmitting various amounts of work over different pullies. The velocity of the belt is assumed to be between 25 and 30 feet per second, and the widths of the belts are given in inches. With greater velocities the breadth may be proportionably decreased.

TABLE I.—APPROXIMATE WIDTHS OF LEATHER STRAPS, IN INCHES, NECESSARY TO TRANSMIT ANY NUMBER OF HORSES' POWER.

Horses' Power.	Smallest Diameter of Pulley in Feet.								
	1	2	3	4	5	6	7	8	10
1	3·6	1·8	1·2	—	—	—	—	—	—
2	7·2	3·6	2·4	1·8	1·4	—	—	—	—
3	10·8	5·4	3·6	2·7	2·1	1·8	1·5	—	—
4	14·4	7·2	4·8	3·6	4·8	2·4	2·0	1·8	1·4
5	18·0	9·0	6·0	4·5	3·6	3·0	2·5	2·2	1·8
7	25·2	12·6	8·4	6·3	5·4	4·2	3·5	3·7	2·5
10	36·0	18·0	12·0	9·0	7·2	6·0	5·1	4·5	3·6
12	43·2	21·6	14·4	10·8	8·6	7·2	6·1	5·4	4·3
14	—	25·2	16·8	12·6	10·0	8·4	7·1	6·3	5·0
16	—	28·8	19·2	14·4	11·5	9·6	8·2	7·2	5·7
18	—	32·4	21·6	16·2	12·9	10·8	9·2	8·1	6·4
20	—	36·0	24·0	18·0	14·4	12·0	10·2	9·0	7·2
25	—	45·0	30·0	22·5	18·0	15·0	12·8	11·2	9·0
30	—	—	36·0	27·0	21·0	18·0	15·0	13·0	10·0
40	—	—	48·0	36·0	28·0	24·0	20·0	18·0	14·0
50	—	—	—	45·0	36·0	30·0	25·0	22·0	18·0
60	—	—	—	—	43·0	36·0	30·0	27·0	21·0
70	—	—	—	—	—	42·0	35·0	31·0	25·0
80	—	—	—	—	—	—	41·0	36·0	28·0
100	—	—	—	—	—	—	51·0	45·0	36·0

Toothed Wheels.—The second method of communicating motion is by rolling contact, as explained in the preliminary Chapter.* But, in practice, the adhesion between the surfaces is seldom sufficient to communicate the necessary power, and hence various contrivances—such as the wheel and trundle, and toothed wheels—have been substituted. The general equations for velocity, ratio, &c. are the same as if the wheels rolled on each other at the pitch circles, but in fact each tooth slides upon its fellow. The determination of the best forms of these teeth so that the friction shall be a minimum and the motion uniform, is one of the most important contributions of applied mathematics to practical engineering.

Of the introduction of toothed wheels and toothed gearing we know very little. Hero of Alexandria, who wrote two centuries before our era, speaks of toothed wheels and toothed bars in a

* Vol. i. page 46.

way which seems to indicate that he was not altogether ignorant of this method of transmitting motion. Later forms are figured in great variety in the different collections of mechanical appliances of the sixteenth and seventeenth centuries.

Spur gearing is employed where the axes on which the wheels are placed are parallel to one another. The smaller wheel in a combination of this sort is termed the pinion. Annexed (fig. 178) is a pinion from Ramelli (A.D. 1588), which, from its form, may be surmised to be of metal. The principle on which spur gearing is constructed is primarily the communication of motion through the rolling of two cylinders on one another. The teeth are introduced to prevent slipping, and thus to insure the regular communication of the motive power.

Fig. 178.

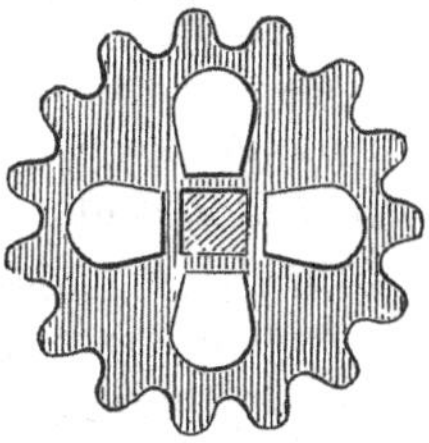

In the older wooden wheels, the teeth were usually formed of hard wood, and driven into mortises on the periphery of a wooden wheel. The pinions were generally replaced by trundles, in which cylindrical staves, fixed at equal distances round the periphery of two discs, were driven by the teeth of the wheel.

The mortise wheels are still retained in countries where iron is expensive, and even in this country they are employed in a modified form. Iron pinions, with wooden cogs fixed in the periphery, are used to receive the motion from the flywheels of engines, with a view to reduce the noise and to increase the smoothness of the motion; and many millwrights prefer, in all cases where large wheels are required to run at high velocities, to make one of them a mortise-wheel, with wooden cogs.

There does not appear to have been much improvement in the construction of wood and iron gear since it was first introduced by Mr. Rennie; the only exception being the introduction of a machine for cutting out the form of the teeth,* which in

* Mr. Smiles states, in his 'Lives of the Engineers,' that Brindley, more than a century ago, invented machinery for the manufacture of tooth and pinion wheels, 'a thing,' as stated by the author, 'that had not before been attempted, all such wheels having, until then, been cut by hand, at great labour and cost.'

those days was done by hand, with keys or wooden wedges fitting into dovetails in the 'shanks' of the cogs, as shown at *a*, fig. 179, on the concave side of the rim; now they are made with an iron pin driven through the cog, close to the rim, as at *b*. The iron pinion or wheel intended to work in contact with the wood teeth was, up to a recent date, turned and carefully divided to the epicycloidal form, and then chipped and filed with great exactitude, in order to fit accurately into the wooden teeth of the driving wheel. In all the corn mills of the present day, and where great speed is required, the same attention to accuracy is observed in wood and iron gear as at former times.

Fig. 179.

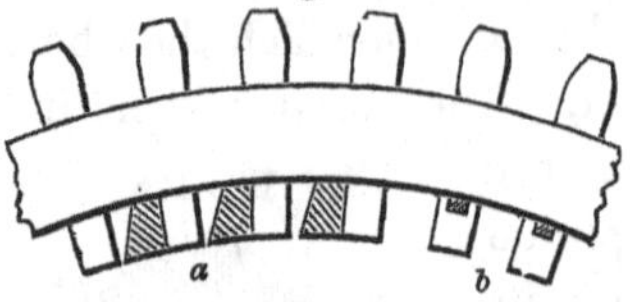

The greatest advance in the application of gearing resulted from the introduction, at the end of the eighteenth century, of cast-iron in place of wood. The credit of the introduction of this material is usually given to Smeaton, who began to use cast-iron in the construction of the Carron Rolling Mill, in 1769.

Fig. 180. Section.

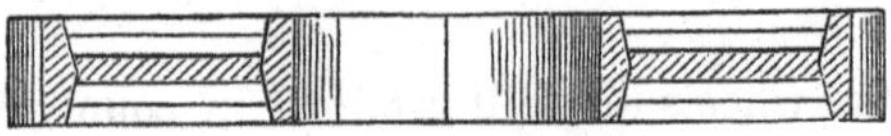

Fig. 181. Plan.

But the late Mr. John Rennie, when at Boulton and Watt's, in 1784, was probably the first to carry the use of cast-iron into all the details of mill-work. Figs. 180, 181 are copied from the original designs for the Soho Rolling Mill, dated 1785. But the Albion Corn Mills, built about the same time (1784-5), may be considered as the earliest instance of the entire replacing of wood by cast iron for the bevel and spur wheels and shafts. This was effected by the same distinguished engineer.

Where the shafts of the wheel and pinion are not parallel to each other, various forms of conical trundles and bevel wheels are employed. The simplest plan is probably the face wheel and trundle, shown in fig. 182, which have been employed from a very early period, and which, if made of metal, take the form of the crown wheel and pinion, fig. 183. Where the axes are not at right angles, conical trundles have been used, one of which is figured in Bessoni (A.D. 1578).

Fig. 182.

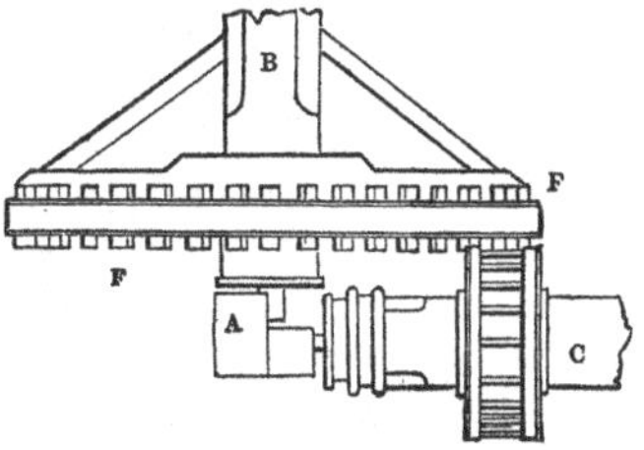

The most perfect arrangement, however, is that in which two wheels called bevel wheels are employed, constructed in the form of frustra of cones. These were not introduced till the middle of the last century, the principles of the construction of the teeth being due to Camus (A.D. 1752). Fig. 184 shows a bevel wheel designed for the Rolling Mill at Soho, by the late Mr. Rennie, in 1785.*

Fig. 183.

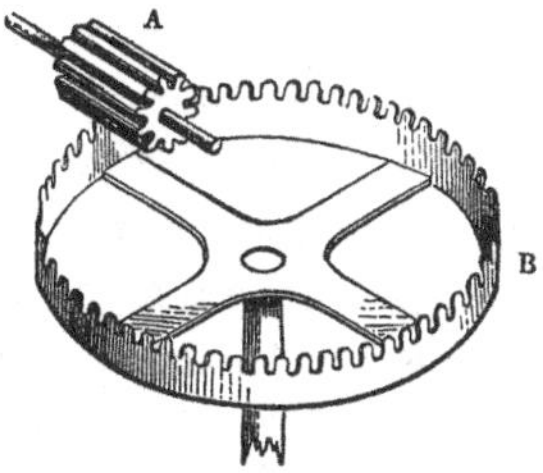

* It is evident from the shape of the eye of these wheels, figs. 180, 181, and 184, that they were intended for wooden shafts, and that cast-iron had not been in use much before that time. At an earlier period, Mr. W. Murdock, of Soho, had a cast-iron bevel wheel, which was considered the first introduced into Scotland, many years previous to the above date. Mr. Smeaton also had introduced iron wheels at Carron in 1754, and afterwards at a mill at Belper, in Derbyshire. (See Smiles's 'Life of Rennie,' page 138.)

Fig. 184.

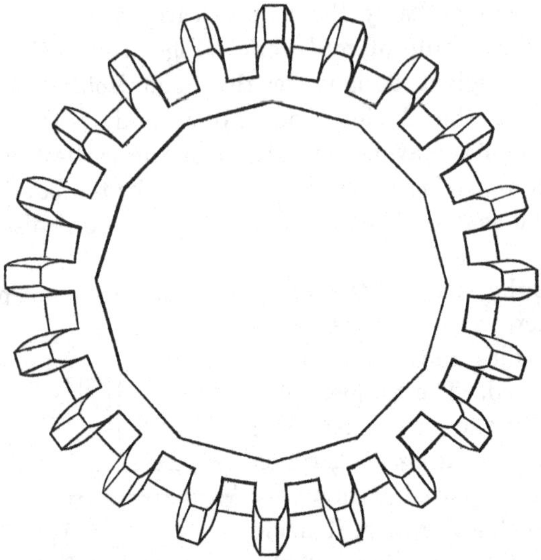

The smoothness and economy of wheel-work depend entirely upon the accuracy of the curvature of the individual teeth which gear with one another. Two chief defects result from imperfections in their construction: first, the motion communicated to the driven wheel is irregular, increasing and diminishing alternately as each tooth passes the line of centres; and, second, there is an unnecessary friction between the teeth in gear, resulting not only in loss of power, but also causing a great and destructive wear in the teeth and journals. These defects can only be avoided by reducing, as far as practicable, the size of the teeth, and by the adoption of true principles in setting out their curvature in the original model.

To the first cause alone a large part of the perfect action of modern machinery of transmission is to be attributed; but there is moreover no doubt that, in practice, even where true principles have not been adopted, a considerable approach has been made to such forms as theory requires. Now, with certain limitations, it is known that if any form of tooth be taken for one wheel, there can be found another tooth which will work correctly with it. But there are certain forms which, being susceptible of accurate mathematical determination, are most convenient for the purpose. Camus, in 1752, was the first to work out the properties of epicycloidal and hypocycloidal curves when employed in the construction of the teeth of spur and bevel gearing. De la Hire adopted the same form. Euler, in 1760, and Kaestner, in 1771, investigated in a similar manner the properties of the involute. Since their time, Ferguson, Buchanan, Hawkins, Rennie, and Airy, have all contributed to perfecting the mathematical theory. And Professor Willis, amongst other important additions, has shown how a close approximation to a true form may be made by the adoption of a system of circular arcs.

From 1788, when Rennie completed the Albion Mills, to the present time, wood and iron gear have been in general use for high velocities, and for every description of machinery where smoothness and accuracy of motion were required. Mr. Rennie was the first to introduce this system; and in most cases he made the driver, or large wheel, with wood cogs, and the driven, or pinion, of iron "chipped and turned"—that is, every tooth of the iron wheel was carefully divided in the pitch, having first been turned on the fane and the ends of the teeth, and drawn to the epicycloidal form. They were then chipped with the hammer and chisel, and accurately filed to the required dimensions and forms. The same process was applied to the wooden teeth; and these wheels, when duly prepared, were keyed on their respective shafts, and securely fixed in contact in the mill.

The chipping and filing process has of late years been superseded by a cutting machine, which effects the same purpose, with less risk of error; and *the good old system of a penny an inch*, as practised in Rennie's time, has been exploded, much to the discomfiture of the old millwrights, who adhere with great tenacity to the hammer and chisel. Fig. 185 shows the cutting machine as constructed by Messrs. Peter Fairbairn and Co., of Leeds.

The object of this machine is not only to pitch and trim the teeth of a large spur or other wheel, but to turn the face and sides of the segments previously, when bolted to the arms.

When used as a lathe for turning, the parts in use are as follows: B is a large headstock, carrying a hollow spindle (C),

Fig. 185.

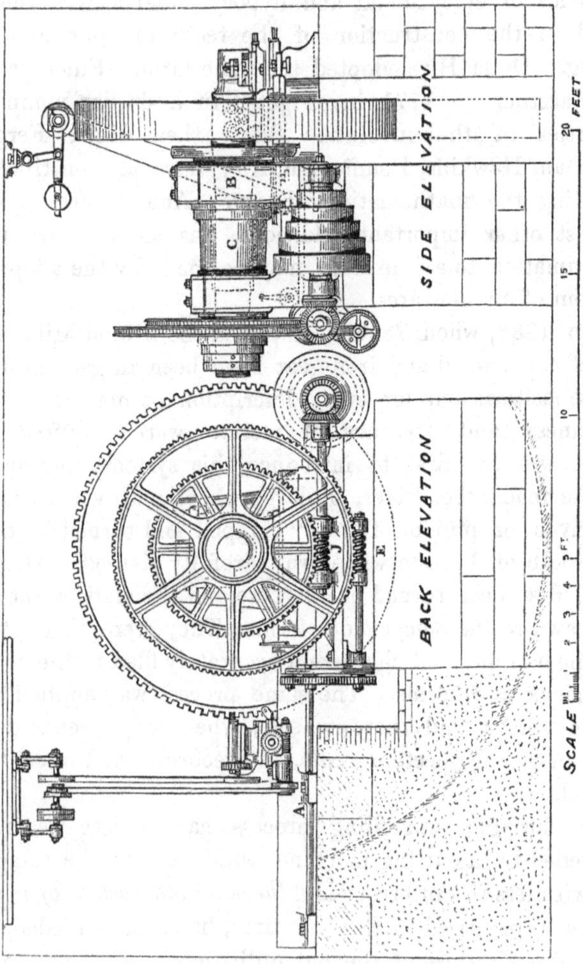

through which is inserted a mandrill upon which the wheel to be cut and turned is keyed. Provision is made for carrying the other end of this mandrill by a loose fixing. The hollow spindle is driven (with the wheel upon it) by a worm wheel (J) which is made to run loose on the spindle, but which is now by a lock bolt connected to the larger worm wheel or dividing wheel (E), the worm of which is now thrown out, and which is keyed firmly on the spindle. The necessary speeds are given by the five-speed cone and mitre gear. The tool for turning is held in an ordinary slide rest, which moves transversally on a saddle, which slides and is fastened in the T grooves of two strong beds (A), firmly secured to masonry, and between which the wheel revolves.

When used for pitching and trimming, the lock bolt connecting the two worm wheels is removed, and the pitch is given by the train of change wheels and division plate (A). The place of the slide rest is now taken by a headstock carrying two cutters, one for roughing, and the other for finishing.

The finishing rose-cutter is the counterpart of the space between the teeth, and is transversed across, making both sides of the tooth alike.

The remainder of the arrangement will be obvious from the sketch. The same machine can be also readily arranged for cutting worm-wheel teeth, or for bevel gear.

The best form which can be given to the teeth of wheels is that which will cause them to be always, in regard to the power they mutually exert, in equally favourable situations, and, consequently, will give the machine the property of being moved uniformly by a power constantly equal. This would be accomplished by simple rolling contact, which corresponds with the case in which the teeth are infinitely small.

Definitions.

1. Spur gearing is that in which the pitch lines of the driving and driven wheel are in the same plane (fig. 186).

Fig. 186.

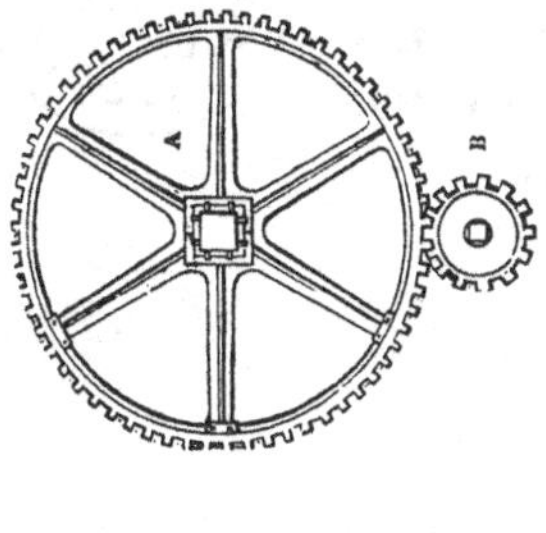

Fig. 187.

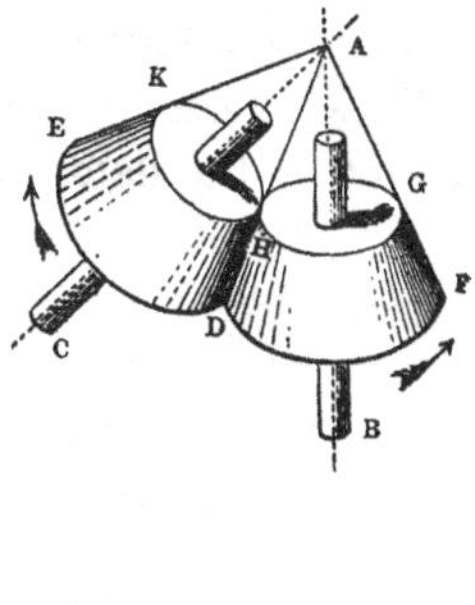

2. Bevel gearing is that in which the planes of the pitch lines of the driving and driven wheel are inclined to each other. In practice, they are in most cases at right angles (fig. 187).

3. Of two wheels in gear, the lesser is called the pinion.

4. When two wheels are in gear, a straight line joining their centres is called the line of centres.

5. If the line of centres be divided into two parts, proportionally to the number of teeth in the wheel and pinion, these parts are called the proportional or primitive radii of the wheel and pinion.

6. The radii of the circles which limit the extremities of the teeth are called the true radii.

7. If, from the centres of the wheel and pinion, circles be drawn with radii equal to the primitive radii, so that they touch one another in the line of centres, these circles are called the pitch lines of the wheel and pinion respectively.

8. The acting surface of a tooth, projecting beyond the pitch circle, is called its face; that enclosed within the pitch circle, its flank.

9. The pitch of a wheel is the distance measured along the pitch circle from the face of one tooth to the corresponding face of the next; it includes, therefore, the breadth of a tooth and space. For two wheels to work in gear, the pitch must be the same in each.

10. Racks are toothed bars in which the pitch line is a straight line.

11. In annular wheels the teeth are cut on the internal edge of an annulus, or ring (fig. 188).

Fig. 188.

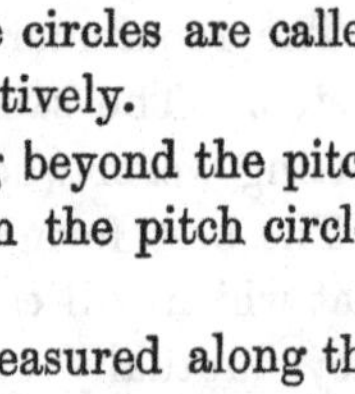

In fig. 189, B F is the line of centres; F A, A B, the primitive radii of the wheel and pinion respectively; A K L and A M N the pitch lines; K L and M N, the pitch; P L, the face; and Q L the flank, of the tooth.

Fig. 189.

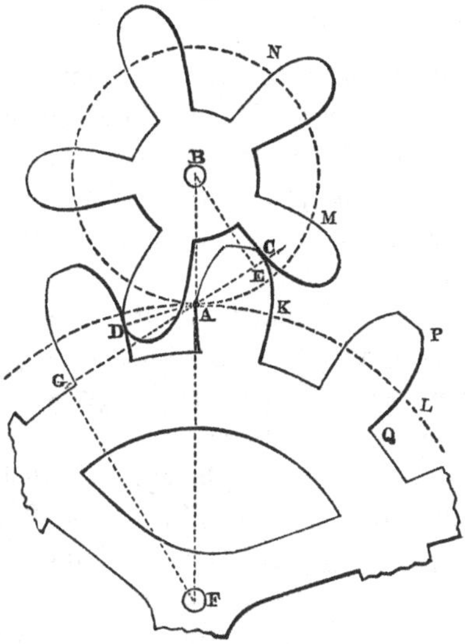

The Pitch of Wheels.

We have seen that the pitch of a wheel is the length of an arc of the pitch line comprising a tooth and space. Millwrights ordinarily measure the pitch as a chord of this arc, and, except in pinions with very few teeth, the two measurements sensibly coincide.

Having the diameter of a wheel, and the number of teeth, the pitch may be found, as follows:

Let D be the diameter of a wheel, N the number of teeth, and p the pitch; then, as 3·1416 D = the circumference of the circle,

$$p = \frac{3{\cdot}1416\ \text{D}}{\text{N}}$$

or approximately,

$$= \frac{22\ \text{D}}{7\ \text{N}}$$

Conversely, if the pitch of a wheel be given, and the number of teeth, then the diameter may be found,

$$\text{D} = \frac{p\ \text{N}}{3{\cdot}1416} = \frac{7\ \text{N}\ p}{22} \text{ nearly.}$$

And if the pitch and diameter of a wheel be given, then the number of teeth may be found,

$$N = \frac{3 \cdot 1416\,D}{p} = \frac{22\,D}{7\,p} \text{ nearly.}$$

But since a wheel must contain a whole number of teeth, N may never be a mixed number. If, therefore, this equation gives N with a fraction, a wheel cannot be constructed of that diameter and pitch. In this case, however, by slightly increasing or decreasing either the diameter or the pitch, the necessary conditions may be complied with.

In practice it is convenient to limit the number of pitches, with a view to the reduction of the number of patterns required for casting. Thus the following series gives all the most ordinary pitches of my own practice:—

Spur flywheels, 5, $4\frac{1}{2}$, 4, $3\frac{1}{2}$, $3\frac{1}{4}$, 3, $2\frac{1}{2}$, 2, $1\frac{1}{2}$ inches.

Spur and bevel wheels, 5, $4\frac{1}{2}$, 4, $3\frac{1}{2}$, $3\frac{1}{4}$, 3, $2\frac{3}{4}$, $2\frac{1}{2}$, $2\frac{1}{4}$, $2\frac{1}{8}$, 2, $1\frac{3}{4}$, $1\frac{5}{8}$, $1\frac{1}{2}$, $1\frac{3}{8}$, $1\frac{1}{4}$, $1\frac{1}{8}$, 1, $\frac{7}{8}$ inches.

Wheels of smaller pitch than this are not used in mill-work; but in machines, &c., the following pitches would probably be sufficient, viz.:—

1, $\frac{3}{4}$, $\frac{5}{8}$, $\frac{1}{2}$, $\frac{3}{8}$, $\frac{1}{4}$ inch.

The value of $\pi = \frac{22}{7}$ ordinarily employed is not very accurate; hence it is convenient to calculate beforehand the values of $\frac{p}{3 \cdot 1416}$ and $\frac{3 \cdot 1416}{p}$ for the most useful pitches.

The following table gives these values:—

Pitch in inches.	3·1416 / Pitch.	Pitch / 3·1416.	Pitch in inches.	3·1416 / Pitch.	Pitch / 3·1416.
5	0·6283	1·5915	$1\frac{3}{4}$	1·7952	0·5570
$4\frac{1}{2}$	0·6981	1·4270	$1\frac{5}{8}$	1·9264	0·5141
4	0·7854	1·2732	$1\frac{1}{2}$	2·0944	0·4774
$3\frac{1}{2}$	0·8976	1·1141	$1\frac{3}{8}$	2·2848	0.4377
$3\frac{1}{4}$	0·9666	1·0345	$1\frac{1}{4}$	2·5132	0·3978
3	1·0472	0·9548	$1\frac{1}{8}$	2·7924	0·3580
$2\frac{3}{4}$	1·1333	0·8754	1	3·1416	0·3182
$2\frac{1}{2}$	1·2566	0·7958	$\frac{7}{8}$	3·5904	0·2785
$2\frac{1}{4}$	1·3963	0·7135	$\frac{3}{4}$	4·1888	0·2386
2	1·5708	0·6366	$\frac{5}{8}$	5·0265	0·1988
$1\frac{7}{8}$	1·6755	0·5937	$\frac{1}{2}$	6·2832	0·1591

RULE 1.—Given the pitch and number of teeth in a wheel to find its diameter.

Multiply the number of teeth by the constant in the third or sixth column of the preceding table corresponding to the pitch.

RULE 2.—Given the pitch and diameter of a wheel to find the number of teeth.

Multiply the diameter by the constant in the second or fifth column of the table corresponding to the pitch.

If this rule gives a mixed number, or whole number and fraction, a wheel cannot be constructed, as before said. The most convenient way of proceeding in that case will be to take the nearest whole number to the number given by the rule, and, using Rule 1, find a new diameter which will differ but slightly from the one previously assumed. This new diameter must be taken for the pitch circle in constructing the wheel.

Thus, suppose it required to find the diameter of a wheel of 2 inches pitch and 150 teeth. By Rule 1, we have $D = 150 \times 0{\cdot}6366 = 95\frac{1}{2}$ inches $=$ 7 ft. $11\frac{1}{2}$ inches.

Or, required the number of teeth in a wheel of 3 inches pitch and 9 feet diameter. By Rule 2: $N = 108 \times 1{\cdot}0472 = 113{\cdot}097$. Here the wheel will contain very nearly 113 teeth; but if we wish to know more accurately the diameter of a wheel of 3 inches pitch and 113 teeth, we find by the 1st Rule, $D = 113 \times 0{\cdot}9548 = 107{\cdot}89$ inches $=$ 8 feet $11\frac{9}{10}$ inches. That is, a wheel of exactly 9 feet could not be constructed with a 3-inch pitch, but one of 8 feet $11\frac{9}{10}$ inches might and would contain 113 teeth.

Professor Willis has employed another method of graduating the sizes of wheels. Suppose the diameter, instead of the circumference, to be divided into as many equal parts as the wheel has teeth, and let one of these parts be called the diametral pitch of the wheel, to distinguish it from the common or circular pitch. Let M be the diametral pitch, so that

$$\frac{D}{N} = M$$

and let a series of values be taken for M in simple fractions of an inch, so that

$$M = \frac{1}{m}$$

where N and m are always whole numbers.

The ordinary values of m are 20, 16, 14, 12, 10, 9, 8, 7, 6, 5, 4, 3, 2, 1, which include wheels in which the circular or common pitch varies from $\frac{1}{8}$ inch to 3 inches, as shown in the following table, given by Professor Willis: —

Value of m.	Circular Pitch in inches and decimals.	Circular Pitch to nearest $\frac{1}{16}$th.	Value of m.	Circular Pitch in inches and decimals.	Circular Pitch to nearest $\frac{1}{16}$th.
3	1·047	1	9	·349	—
4	·785	$\frac{3}{4}$	10	·314	$\frac{5}{16}$
5	·628	$\frac{5}{8}$	12	·262	$\frac{1}{4}$
6	·524	$\frac{1}{2}$	14	·224	—
7	·449	$\frac{7}{16}$	16	·196	$\frac{3}{16}$
8	·393	$\frac{3}{8}$	20	·157	$\frac{1}{8}$

This system is convenient where wheels of small pitch are employed, and involves less calculation than the common system.

Since $\frac{D}{N} = M$, we have $M = \frac{p}{3 \cdot 1416}$; therefore, in the previous table (p. 14) the quantities in the third and sixth columns are the diametral pitches corresponding to the circular pitches in the first column, and the numbers in the second column are the corresponding values of m. In fact, this scheme differs from the first simply by expressing in small whole numbers the quantity $\frac{3 \cdot 1416}{p}$ instead of p.

The following table (pages 18 and 19) gives the relation of diameter, pitch, and number of teeth, for wheels of from $\frac{1}{2}$ inch to 5 inches pitch, and of from 12 to 200 teeth. Intermediate numbers may be found by direct proportion, by multiplying the number given for a wheel of half or a third of the number of teeth by two or three, or by adding together the diameters given for two wheels the sum of whose teeth is the number required. For an odd number of teeth, add the number given at the head of the table as many times as may be necessary to the diameter for a wheel of the nearest number of teeth given.

The Principles which Determine the Proper Form of the Teeth of Wheels.

The problem which presents itself in the construction of the teeth of wheels, is to discover the curvature which they should have in order that they shall revolve through the action of the teeth in precisely the same manner as they would by the rolling of the circumferences of their pitch lines.

The general principle by which this uniformity of motion is secured is as follows :—When wheels in gear act on each other so that a line perpendicular to the common tangent of the surfaces of the teeth at the point of contact passes always through the point where the pitch circles cut the line of centres, they will exert mutually the same force, move with uniform velocity, and be of true figure.

Or, in other words, the teeth will be rightly constructed when a line drawn from the point of contact of the pitch circles to the point of contact of two teeth is a normal to the surfaces in contact in all positions of the wheel and pinion.

Thus, let fig. 189 represent a wheel and pinion in gear, and let B A, A F be the primitive radii, and therefore A K L and A M N the pitch lines. Then if the teeth touch in C and D, and the lines A C, A D be always perpendicular to the common tangent to the touching parts, the teeth will be of true figure.

Epicycloidal Teeth.

The epicycloid is the curve traced by a fixed point in the circumference of a circle, which rolls over or within the circumference of another circle, or on a straight line. Thus, let the circle A B C be fixed, and let the circle C D E roll over its circumference, then a point C in the circumference of this the generating circle will describe an epicycloid C, C′, C″, C‴, C⁗, without the circle A B C. Similarly, a point F on the circumference of a generating circle F G, rolling within the circumference of A B C, will describe an interior epicycloid or hypocycloid F, F′, F″, F‴.

The remarkable properties of the epicycloid which determine its fitness for describing the teeth of wheels are : 1st, when the generating circle is half the diameter of the base circle, and

TABLE SHOWING THE RELATION OF PITCH, DIAMETER, AND NUMBER OF TEETH.

Number of teeth in wheel.	Pitch in Inches.															
	½	¾	1	1¼	1½	1¾	2	2¼	2½	2¾	3	3¼	3½	4	4½	5
For each tooth add	·1591	·2386	·3182	·3978	·4774	·5570	·6366	·7135	·7958	·8754	·9548	1·035	1·114	1·273	1·427	1·592
12	1·91	2·86	3·82	4·77	5·73	6·68	7·64	8·56	9·55	10·50	11·46	12·41	13·37	15·28	17·12	19·10
13	2·07	3·10	4·14	5·17	6·21	7·24	8·28	9·28	10·35	11·38	12·41	13·45	14·48	16·55	18·55	20·69
14	2·23	3·34	4·46	5·57	6·68	7·80	8·91	9·99	11·14	12·26	13·37	14·48	15·60	17·83	19·98	22·28
15	2·39	3·58	4·77	5·97	7·16	8·36	9·55	10·70	11·94	13·13	14·32	15·52	16·71	19·10	21·41	23·87
16	2·55	3·82	5·09	6·37	7·64	8·91	10·19	11·41	12·73	14·01	15·28	16·55	17·83	20·37	22·83	25·46
17	2·70	4·06	5·41	6·77	8·12	9·47	10·82	12·13	13·53	14·88	16·23	17·59	18·94	21·64	24·26	27·06
18	2·86	4·30	5·73	7·17	8·60	10·03	11·46	12·84	14·32	15·76	17·19	18·62	20·05	22·92	25·69	28·65
19	3·02	4·54	6·05	7·56	9·08	10·58	12·10	13·56	15·12	16·63	18·14	19·66	21·17	24·19	27·11	30·24
20	3·18	4·77	6·36	7·96	9·55	11·14	12·73	14·27	15·92	17·51	19·10	20·69	22·28	25·46	28·54	31·83
21	3·34	5·01	6·68	8·36	10·02	11·70	13·37	14·98	16·71	18·38	20·05	21·72	23·40	26·74	29·97	33·42
22	3·50	5·25	7·00	8·76	10·50	12·25	14·01	15·70	17·51	19·26	21·01	22·76	24·51	28·01	31·39	35·01
23	3·66	5·49	7·32	9·16	10·98	12·81	14·64	16·41	18·30	20·13	21·96	23·79	25·62	29·28	32·82	36·60
24	3·82	5·73	7·64	9·55	11·46	13·37	15·28	17·12	19·10	21·01	22·92	24·83	26·74	30·56	34·25	38·20
25	3·97	5·97	7·96	9·96	11·94	13·93	15·92	17·84	19·90	21·89	23·87	25·86	27·85	31·84	35·68	39·79
30	4·77	7·16	9·55	11·93	14·32	16·71	19·10	21·41	23·87	26·26	28·64	31·04	33·42	38·21	42·81	47·74
35	5·57	8·35	11·14	13·92	16·71	19·50	22·28	24·97	27·85	30·64	33·42	36·21	38·99	44·57	49 95	55·70

40	6·36	9·54	12·73	15·91	19·10	22·28	25·46	28·54	31·83	35·02	38·19	41·38	44·56	50·94	57·08	63·66
45	7·16	10·74	14·32	17·90	21·49	25·07	28·65	32·11	35·81	39·39	42·97	46·55	50·13	57·30	64·22	71·62
50	7·96	11·93	15·91	19·89	23·87	27·85	31·83	35·67	39·79	43·77	47·74	51·73	55·71	63·67	71·35	79·58
55	8·75	13·12	17·50	21·88	26·26	30·64	35·01	39·24	43·77	48·15	52·51	56·90	61·28	70·04	78·49	87·53
60	9·55	14·32	19·09	23·87	28·64	33·42	38·20	42·81	47·75	52·52	57·29	62·07	66·85	76·40	85·62	95·49
65	10·34	15·51	20·68	25·86	31·03	36·21	41·38	46·38	51·72	56·90	62·06	67·24	72·42	82·77	92·76	103·45
70	11·14	16·70	22·27	27·85	33·42	38·99	44·56	49·94	55·71	61·28	66·84	72·42	77·99	89·13	99·89	111·41
75	11·93	17·89	23·86	29·84	35·81	41·78	47·75	53·51	59·69	65·66	71·61	77·59	83·56	95·50	107·03	119·36
80	12·73	19·09	25·46	31·82	38·19	44·56	50·93	57·08	63·66	70·03	76·38	82·76	89·13	101·87	114·16	127·32
85	13·52	20·28	27·05	33·81	40·58	47·35	54·11	60·65	67·64	74·41	81·16	87·93	94·70	108·23	121·30	135·28
90	14·32	21·47	28·64	35·80	42·97	50·13	57·29	64·21	71·62	78·78	85·93	93·11	100·27	114·60	128·43	143·23
95	15·11	22·66	30·23	37·79	45·36	52·91	60·47	67·78	75·60	83·16	90·70	98·28	105·84	120·96	135·57	151·19
100	15·91	23·86	31·82	39·78	47·74	55·70	63·66	71·35	79·58	87·54	95·48	103·45	111·41	127·32	142·70	159·15
110	17·50	26·24	35·00	43·76	52·51	61·27	70·03	78·48	87·54	96·29	105·03	113·80	122·55	140·05	156·97	175·07
120	19·09	28·63	38·18	47·74	57·28	66·84	76·39	85·62	95·50	105·05	114·58	124·14	133·69	152·78	171·24	190·98
130	20·68	31·02	41·36	51·72	62·06	72·41	82·76	92·75	103·45	113·80	124·12	134·50	144·83	165·52	185·51	206·90
140	22·27	33·40	44·54	55·70	66·84	77·98	89·12	99·89	111·41	122·56	133·67	144·83	155·97	178·25	199·78	222·81
150	23·86	35·79	47·73	59·67	71·61	83·55	95·49	107·03	119·37	131·31	143·22	155·18	167·12	190·98	214·05	238·73
160	25·45	38·18	50·91	63·65	76·38	89·12	101·86	114·16	127·33	140·06	152·77	165·52	178·26	203·71	228·32	254·64
170	27·04	40·56	54·10	67·63	81·16	94·69	108·22	121·29	135·29	148·82	162·32	175·87	189·40	216·44	242·59	270·56
180	28·64	42·95	57·28	71·60	85·93	100·26	114·59	128·43	143·24	157·57	171·86	186·21	200·54	229·18	256·86	286·47
190	30·23	45·33	60·46	75·58	90·71	105·83	120·95	135·57	151·20	166·33	181·41	196·56	211·68	241·91	271·13	302·39
200	31·82	47·72	63·64	79·56	95·48	111·40	127·32	142·70	159·16	175·08	190·96	206·90	222·82	254·64	285·40	318·30

Fig. 190.

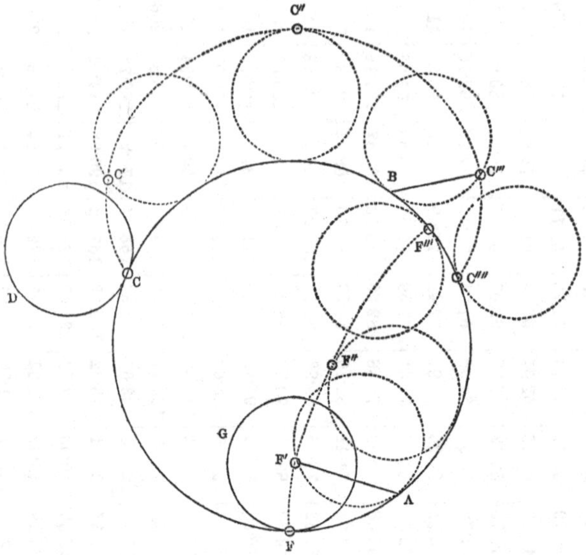

rolls within it, the hypocycloid is a straight line forming a diameter of the base; 2nd, if through the points of contact of the generating circle and the base, and the point describing the epicycloid, straight lines be drawn, these straight lines will be perpendicular to the curvature of the epicycloid at these points. Thus, for example, B C‴ drawn from the point of contact B to the describing point C‴, is a normal to the curve at that point; and similarly A F′ is a normal to the curve at F′.

Fig. 191.

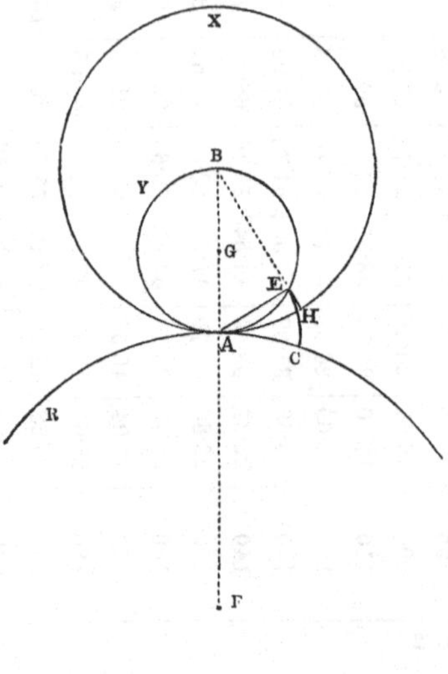

Suppose in the same plane three circles R X Y (fig. 191), which touch each other in the point A, and whose centres F B G are consequently in a straight line. Let one of these circles be made to revolve round its centre, and force the other two to turn round their centres, which we suppose to be fixed, moving these circles by the point of continual contact A, common to the three circumferences; it is evident that all the parts of the circumference of the circle made to revolve will be applied in succession to every part of the circumferences of the other two circles, in the same manner as if the two circles R and X remained immovable, while the third, Y, revolved on the circumferences of the other two. Hence, if we suppose a style fixed to the circumference of the circle Y, movable round its centre, the three circles having been obliged to turn by the motion of the one which has carried along the other two; when the style is at E, if each of the two arcs A C and A H be made equal to the arc A E, the style will have described on the movable plane of the circle R, on the exterior part of which it revolves a portion C E of an epicycloid, and on the movable plane of the circle X, within which we may consider it to revolve, a portion E H of a hypocycloid. (*Camus.*)

These two epicycloids traced out at the same time by the style E affixed to the circle Y, will touch each other in the point E; for the straight line A E drawn through A, where the generating circle Y touches its bases R and X, will be a normal to the two epicycloids. The same will be true in every position of the circles, viz. that the epicycloid and hypocycloid will have a common normal passing through A. Hence, if E C and E H be the faces of two teeth on the wheel and pinion R and X respectively, the condition of uniform motion already given will be complied with, the teeth will be of true form, and if the hypocycloid E H be moved by the epicycloid E C, or *vice versâ*, the wheel and pinion R and X will move precisely as if they rolled together at their pitch circles.

Wheels usually have their teeth constructed of such a form, that the flanks or parts within the pitch circle are bounded by straight lines radii of the pitch circles. Bearing in mind the property already stated, that the hypocycloid described by a generating circle of half the diameter of the base is a straight line forming a diameter of the base, we may so arrange our generating circle in describing the teeth of wheels as to comply with the above rule. By taking a generating circle Y of diameter equal to the radius of the base X, the hypocycloid E H will be part of a radius of X; or, in other words, a radius B H of X will always touch the epicycloid C E described without the circle R, by a generating circle Y, of a diameter equal to the radius of X. And the angle B E A being the angle of a semicircle, will always be a right angle. That is, the perpendicular to the straight line B H, at the point of contact with the epicycloid E C, will always pass through A.

We have hitherto supposed the circles moved by contact at the point A, in order to explain the generation of the epicycloid C E and straight line E H; but if we suppose these already described, the former being fixed to the circle R, and the latter to the circle X; then if E H roll by contact on the epicycloid C E, it will move the circle R precisely in the same manner as if the circle were moved by contact at A.

Construction of Epicycloidal Teeth.

Since every tooth in a wheel is of precisely the same form, it is sufficient to construct a single pattern tooth of true epicy-

Fig. 192.

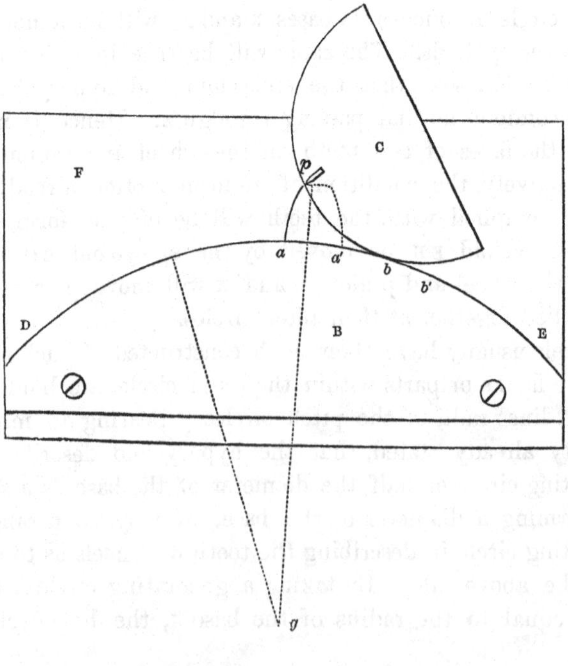

cloidal curvature, which may be used in setting out all the other teeth.

First method, when the generating circle is the same for wheel and pinion, the face of the tooth an epicycloid, and the flank a hypocycloid.

Construct two templets A and B (figs. 192, 193) having their faces arcs of the pitch circle of the wheel for which the tooth is required, and a third templet C cut to an arc of the intended generating circle of the epicycloid. Fix a steel tracing point *p* in the edge of the templet C, and for convenience a board F, on which to draw the tooth, may be fixed beneath the templet B. Mark off on the board F (fig. 192) the pitch circle of the wheel D E, and take distances *a b*, *b c* equal to the pitch of the teeth, and distances *a a′*, *b b′* equal to the thickness of the teeth. If then

Fig. 193.

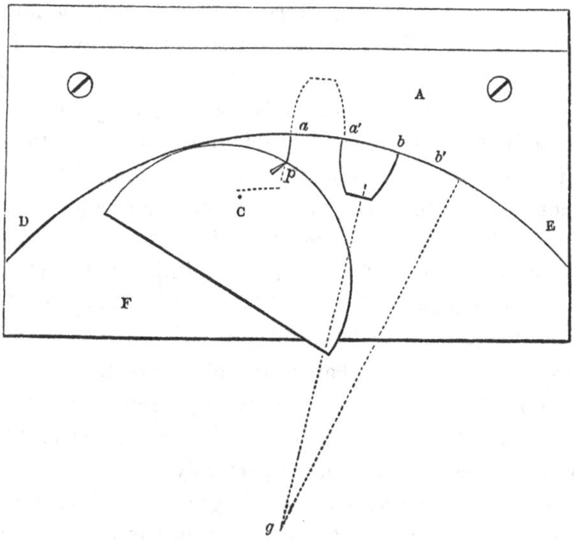

the templet C be placed touching B, and with the tracing point *p* coinciding with one of the marks as *a*, and be rolled towards E, the point will trace out an epicycloid *a p* on the board F, which will form one face of the tooth. Next let the point *p* be made to coincide with *a′*, and the templet C be rolled towards D, the other face of the tooth will be described.

To draw the flanks, the templet A must now be fixed on the board F, with its face in contact with B; remove B and describe hypocycloids (fig. 193) from a and a', by rolling C on the inside of the pitch circle.

The length of the teeth is usually fixed as a proportional part of the pitch, but the least necessary length may be found experimentally by replacing the templet B on the board F, and making p coincide with a, roll C towards E till it touches B in b, the corresponding face of the next tooth; mark then the position of the tracing point, and through this point draw an arc from the centre g of the wheel: this arc will mark the extremity of the tooth, and the arc $g p$ will be the true radius of the wheel.

This process, which, though complicated in description, is very easy in practice, must be repeated with two templets cut to the pitch circle of the pinion, the same generating circle C being employed; a similar pattern tooth will thus be found for the pinion, which will work with that already found for the wheel. The usual custom in practice is for the millwright first to describe the epicycloidal and hypocycloidal forms of the teeth required in the wheel and pinion; he then constructs two model teeth, one for the wheel and the other for the pinion, and from these he determines the true curves, and by means of his compasses transfers the same to the wheels or patterns on which these forms are to be impressed. The generating circle, it may be observed, must not exceed in size the radius of the pinion, or it would give rise to a weak form of tooth, thinner at the root than at the pitch circle.

Second method, where two generating circles are employed, in order that the flanks of the teeth may be straight lines radii of the wheel and pinion respectively.

It is the usual practice of millwrights to make the parts of the teeth of wheels within the pitch circles radii of the wheel. Now, we have seen that a hypocycloid described by a generating circle equal in diameter to the radius of the wheel would be a diameter of the wheel. If, therefore, the flank of the tooth of the wheel and the face of the tooth of the pinion be described by a templet cut to a radius equal to half that of the wheel and the flank of the tooth of the pinion and face of that of the wheel be described by a templet cut to a radius equal to half that of the pinion, then these teeth will work together truly, and will have radial flanks.

Since it is unnecessary to describe the flanks of such teeth by templets, there will be needed only one templet cut to the pitch circle of each wheel, but templets of two generating circles are required. In other respects the method is identical with that already described. The great defect of this method is, that neither the wheel nor pinion will work accurately with a wheel or pinion of any other diameter than that for which they were originally made, and thus a vast number of wheel patterns must be made to fulfil the requirements of practice; whereas wheels described by the previous method will work equally well with all other wheels the teeth of which have been described by the same generating circle—it being understood that only the parts of teeth *without* the pitch circle of the wheel roll on the parts *within* the pitch circle of the pinion, and those without the pitch circle of the pinion on those within the pitch circle of the wheel.

Hence Professor Willis has been led to suggest that for a given set of wheels a constant generating circle should be taken to describe both the parts without and within the pitch circles of the whole series, instead of making that circle depend on the diameters of the wheels. In this case the first solution must be employed, and the flanks of the teeth will not be straight; but the great advantage is gained, that any pair of wheels in the series will work together equally well.

To determine the proper size of the generating circle, we must remember that a tooth of weak form is produced when the generating circle is greater than half the diameter of the wheel. Hence the generating circle may be best made of a diameter equal to the radius of the smallest pinion of the series which are to work together.

The Rack is the extreme case of a wheel, or may be considered as a wheel of infinite radius. It may be described by either of the methods above, only noting that, if the second method be employed, the generating circle which traces the face of the teeth of the wheel becomes a straight line, and the epicycloid becomes an involute.

If the teeth of a series of wheels and of a rack be described by the same generating circle, any of the wheels will work with equal accuracy into the rack.

Involute Teeth.

The Involute, the curve traced by a flexible line unwinding from the circumference of a circle, is called an involute.

Fig. 194.

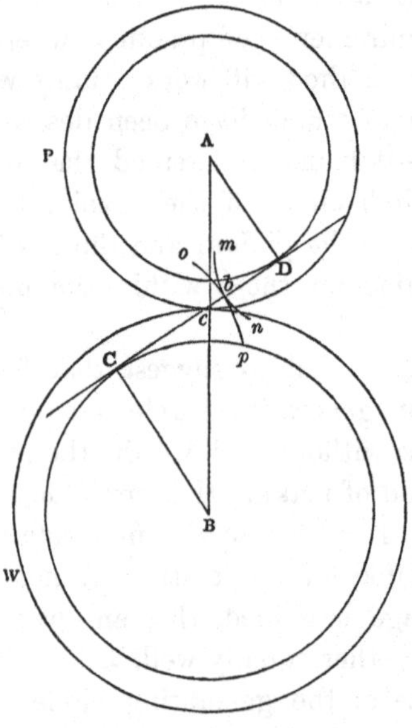

Let P and W (fig. 194) be the pitch lines of a wheel and pinion, and let A and B be their centres. From A and B describe two circles D C, with radii A b and B b of the wheel and pinion respectively; so that

A c : B c :: A D : B C

Let $m\,n$ and $o\,p$ be two involute curves described by flexible lines unrolling from the circles D and C respectively, and touching at b. Then if bC, bD be drawn tangents to the circles at the points D and C, they are also in one straight line, because they are both normals to the curves at b. It may also be shown that the line C D intersects A B in c, where the pitch lines touch. Hence we have found two curves such, that the line perpendicular to their common tangent passes in all positions of the wheel and pinion through c, which is the sufficient condition of their uniform motion, if moved by the sliding of the curves instead of by contact at c. Hence, if the wheels be constructed with teeth formed to these involute curves, they will work with perfect regularity of motion.

In practice, the chief condition to be observed is to diminish the pressure on the axes, which is the chief defect of this form of teeth. The common tangent should be drawn through c, making an angle with A B, not deviating more than 20° from a right angle. Involute wheels have the double advantage that they work equally well if, through the wear of the brasses, the wheels have receded from one another; and any involute wheels of the same pitch and similarly described — that is, having the common tangent to the base circles passing through the point of contact of the pitch lines; or, in other words, base circles proportional to the primitive radii — will work together.

Mr. Hawkins, the translator of *Camus*, first proposed a simple instrument for describing the teeth of wheels to an involute curve. It consists of a straight piece of watch-spring *a b* (fig. 195), with a screw at one end, and filed away at the

Fig. 195.

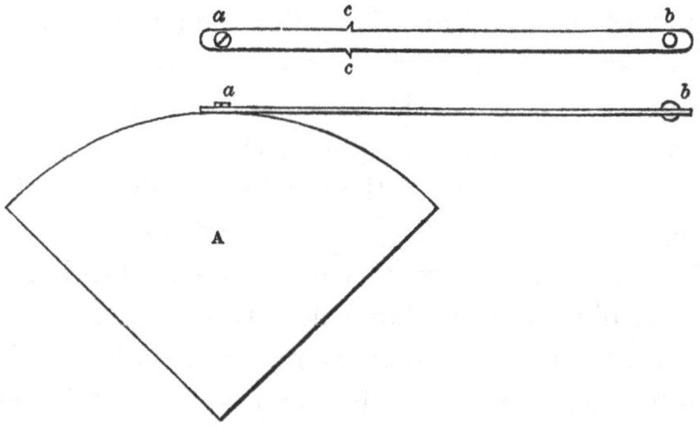

edges so as to leave two teeth or tracers, *c c*, projecting from the edges of the watch-spring. At *b* a bit of wire is put through, and riveted, so as to form a knot by which the spring can be firmly held and stretched, as it is unwound from the base on which the involute is generated. This watch-spring is screwed to the edge of a templet A, curved to the radius of the base circle of the involute; and this being placed so that its centre coincides with the centre of the wheel, and revolved to bring one of the tracing points *c* in succession to each of the points at which corresponding faces of the teeth cut the pitch line, a series of involute curves may be described by unfolding the watch-spring, whilst keeping it firmly stretched tangentially to the sector to which it is fixed. The sector A must then be turned over, and the involutes of the opposite faces of the teeth struck in a similar manner.

Another plan is to employ a straight ruler instead of the watch-spring, a tracer being fixed in its edge. This shows that the involute is an epicycloid generated by a straight line. The ruler must be kept in contact with the base circle, and the tracer brought in succession to all the points in which the faces of the teeth cut the pitch line.

Hence, to describe a wheel with involute teeth, the line of centres must be drawn and divided proportionally to the number of teeth in the wheel and pinion. Draw the pitch line; divide the pitch line into the same number of equal parts as there are teeth in the wheel, and at these points mark out the thicknesses of the teeth all round. Draw the tangent to the base circles, making an angle of about 80° with the line of centres, which will give the radius of the base circle drawn touching it. A templet must be made to this radius, and then the involutes may be drawn by either of the preceding methods.

Allowance must be made to permit free play of the teeth in the spaces, the teeth being somewhat shorter than the distance between the bases of the involutes. But wheels of this figure require but little play in the engagement.

In the case of racks, the rack-teeth are bounded by straight lines perpendicular to the tangent drawn from the point where the pitch lines touch, to the base circle from which the involutes of the wheel are struck. If the teeth of the rack be made rectangular—that is, bounded by lines perpendicular to the pitch line—the involute must be struck from a base circle equal to the pitch circle of the wheel. In the former case there is a downward pressure on the rack; in the latter, the teeth of the wheel touch those of the rack in a single point—namely, the pitch-line of the latter.

Professor Willis's Method of Striking the Teeth of Wheels.

In practice, the custom of describing the teeth of wheels as arcs of circles, has, from its simplicity, been generally adopted. The methods already given, however simple, when adopted in the formation of a single tooth, become tedious in their application to wheels of large size; and to this must be added the imperfect comprehension of their advantages by the millwrights charged with the task of designing wheel patterns.

Circular arcs struck at random, according to the judgement of the millwright, are often employed; and even where better principles have been introduced, it is common, after describing a single tooth accurately, to find by trial a circular arc nearly corresponding with its curve, and to employ this in marking out the cogs of the required wheel.

Seeing the advantages of the circular arc, and believing that it is not objectionable if only the employment of it is guided by true principles, Professor Willis has rendered this great service to practical mechanics—he has shown how, by a simple construction, the arcs of circles may be found, which, used in the construction of the teeth of wheels, will work truly on each other.

Fig. 196.

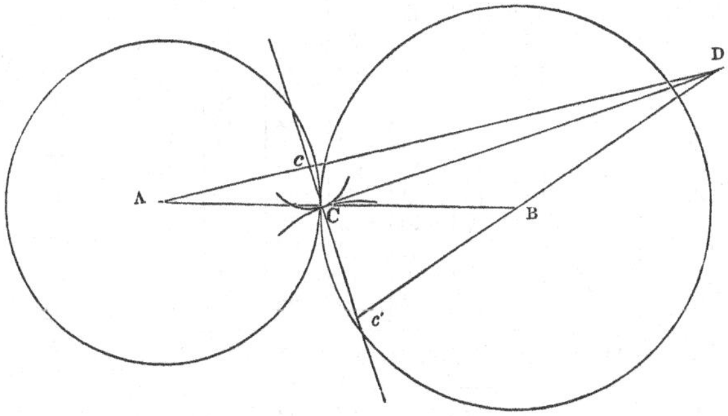

Let A B (fig. 196) be the centres of a wheel and pinion, and C the point of contact of the pitch circles on the line of centres. Through C draw c C c' at any angle with A B. Assume c as the centre from which to describe an arc for a tooth of the wheel a. Draw C D perpendicular to c C c', and from A through c draw A c D, meeting C D in D. Lastly, from D through B draw D B c', meeting c C c' in c'. Then a small arc drawn from c with radius c C as a tooth for the wheel a, will work correctly with a small arc drawn from c', with a radius c' C as a tooth for the wheel B.*

Professor Willis recommends 75° 30′ as the best magnitude of the angle A C c, so that Cos. $75° 30' = \frac{1}{4}$. If this angle be constant in a set of wheels, any two will work truly together.

For the easier description of these teeth, Professor Willis has

* Willis's 'Principles of Mechanism,' p. 123.

Fig. 197.

Tables showing the place of the Centres upon the Scales

Centres for the Flanks of Teeth

Number of Teeth	Pitch in inches: 1	1¼	1½	1¾	2	2¼	2½	3
13	129	160	193	225	257	289	321	386
14	69	87	104	121	139	156	173	208
15	49	62	74	86	99	111	123	148
16	40	50	59	69	79	89	99	191
17	34	42	50	59	67	75	84	101
18	30	37	45	52	59	67	74	89
20	25	31	37	43	49	56	62	74
22	22	27	33	39	43	49	54	65
24	20	25	30	35	40	45	49	59
26	18	23	27	32	37	41	46	55
30	17	21	25	29	33	37	41	49
40	15	18	21	25	28	32	35	42
60	13	15	19	22	25	28	31	37
80	12	...	17	20	23	26	29	35
100	11	14	...	...	22	25	28	34
150	...	13	16	19	21	24	27	32
Rack	10	12	15	17	20	22	25	30

Centres for the Faces of Teeth

Number of Teeth	1	1¼	1½	1¾	2	2¼	2½	3
12	5	6	7	9	10	11	12	15
15	...	7	8	10	11	12	14	17
20	6	8	9	11	12	14	15	18
30	7	9	10	12	14	16	18	21
40	8	...	11	13	15	17	19	23
60	...	10	12	14	16	18	20	25
80	9	11	13	15	17	19	21	26
100	...	...	...	...	18	20	22	...
150	...	...	14	16	19	21	23	27
Rack	10	12	15	17	20	22	25	30

The figure is of half the linear dimensions of the original

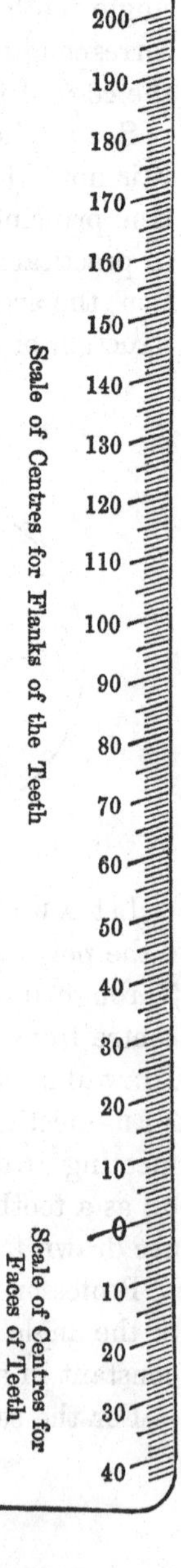

invented the Odontograph, a simple instrument of graduated card or wood, by which the position of the centres and radii of the arcs of the teeth can very easily be found. This instrument* is of the form shown in fig. 197, of half its proper lineal dimensions. It has the bottom edge bevilled off at an angle of 75°. The point where this would cut the right-hand edge is the zero of the scales. These scales are graduated to twentieths of an inch, to avoid fractional parts in the tables, and depart in each direction from the zero, the upper being that employed in finding the centres of the flanks of the teeth or parts within the pitch circle, and the lower for finding the centres of the faces of the teeth or parts without the pitch circle. Tables are given on the odontograph for finding the graduation on the scale corresponding to any given pitch and number of teeth. For intermediate pitches, not given in the table, or for wheels of greater size, the corresponding numbers can be found by simple proportion. For wheels of only twelve teeth the flanks are straight, and form parts of radii of the pitch circle.

Fig. 198.

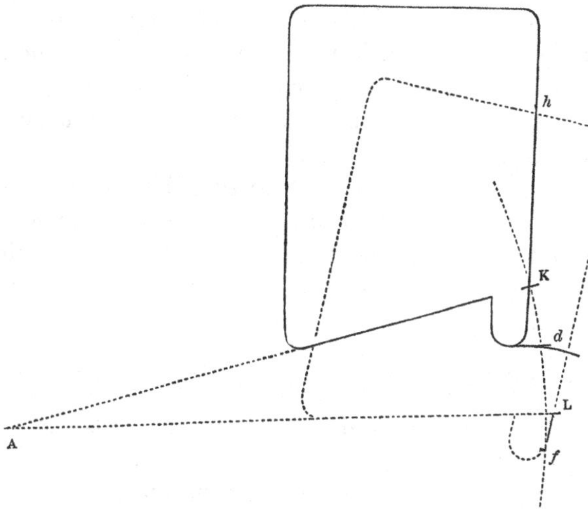

In fig. 198, let A be the centre of a wheel, K *d* L the pitch line. Set off K L equal to the pitch, and bisect it in *d*. Draw

* Professor Willis's Odontograph may be obtained of Messrs. Holtzapfel of London.

radii A K, A L. Place the odontograph with its bevilled edge on the radius A K, and zero of the scale on the pitch line. Then look out in the table of centres for the flanks of teeth, the number corresponding to the pitch, and required number of teeth, and mark off this point *h* from the scale of centres for the flanks of teeth. Then remove the odontograph, and similarly place it on the radius A L. Find in the table of centres for the faces of the teeth the number corresponding to the pitch and number of teeth in the wheel, and mark it off at *f*, on the scale for centres of the faces of teeth. Then describe two arcs from *h* and *f*, with *h d* and *f d* as radii; these will form the side of a tooth. Then, from *d* let the pitch line be marked off into as many equal spaces as there are teeth in the wheel, and these be divided proportionally to the widths of the teeth and spaces. Through *h* and *f*, with radii A *h* and A *f*, draw circles. Take *h d* as a radius, and, placing one foot of the compass on the divisions of the pitch line, and the other in the circle drawn through *h*, describe a series of arcs forming the flanks of the teeth. Similarly with radius *f d*, and one leg of the compass on the circle drawn through *f*, describe the faces of the teeth.

For an annular wheel the same rules apply, only that the part of the curve which is face in a spur wheel becomes the flank in an annular wheel, and *vice versâ*. For a rack, the pitch line is straight, and A K, A L are parallel and perpendicular to it, at a distance equal to the pitch.

As these odontographs may be purchased in a very convenient form, with tables for their use, and also with tables of the widths of teeths, and spaces and length of teeth within and without the pitch circle, it is not necessary to describe them in further detail here.

General Form and Proportions of Teeth of Wheels.

On Plate IX. have been drawn a series of wheels and racks to illustrate the general form of the teeth of wheels. The pitch in figs. 1, 2, 3, and 4 is one inch, and that in fig. 5 is $2\frac{1}{2}$ inches.

In figs. 1, 2, 3, and 4 the wheel is 19·1 inches diameter; in fig. 5 it is 13 feet diameter.

Fig. 1 represents the form of the teeth on Professor Willis's system, the curves being arcs of circles. Fig. 2 gives the form

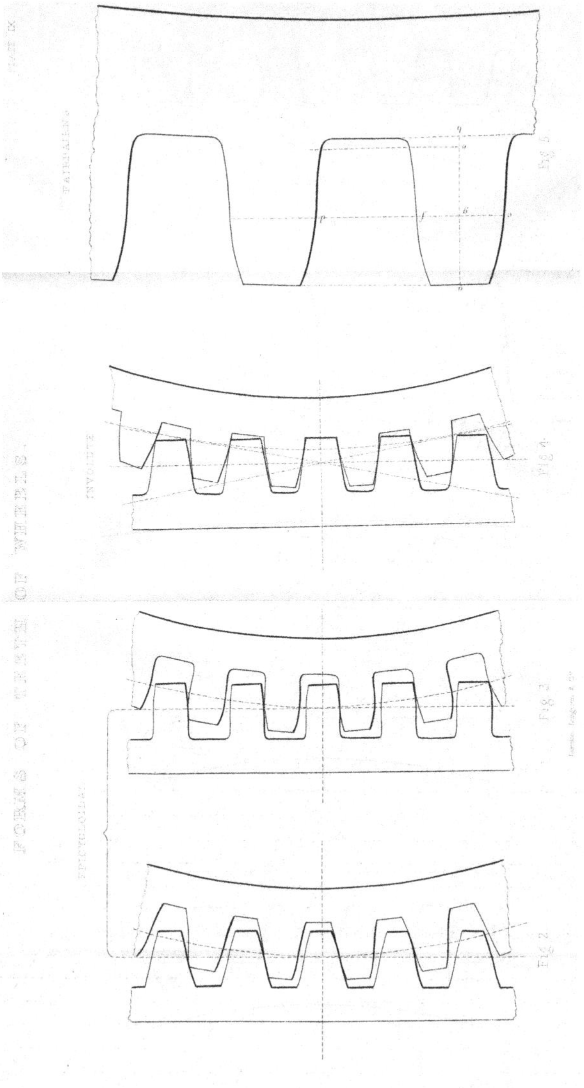

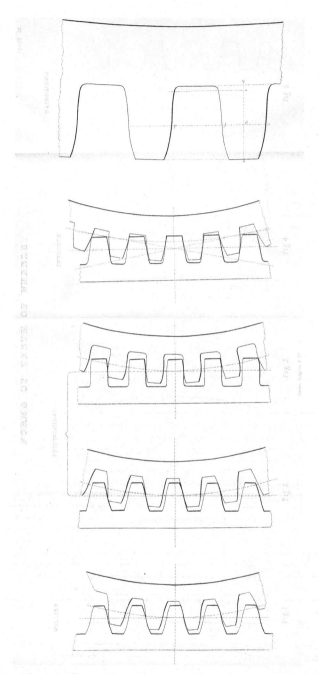

The material originally positioned here is too large for reproduction in this reissue. A PDF can be downloaded from the web address given on page iv of this book, by clicking on 'Resources Available'.

PLATE X

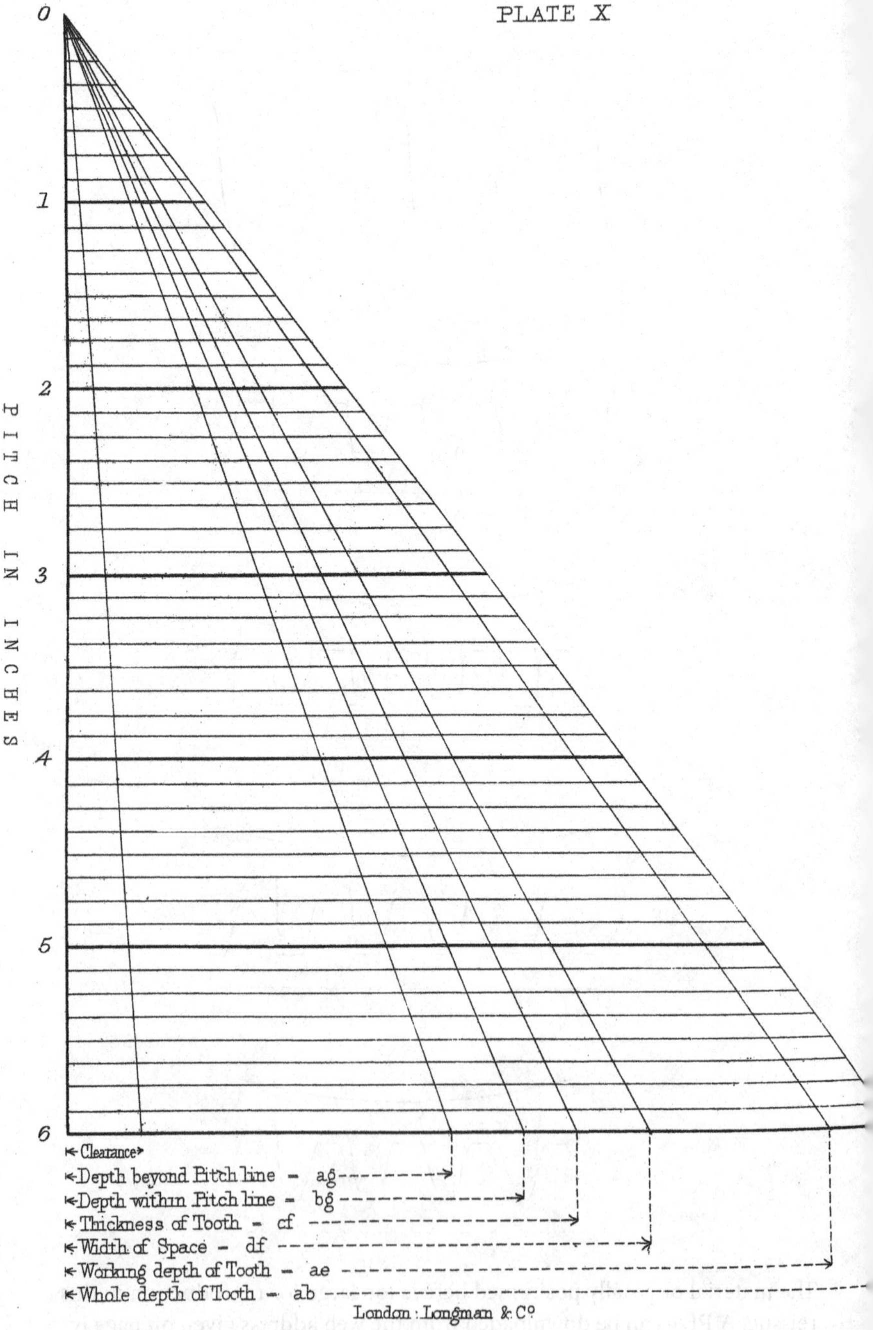

London: Longman & Co.

of epicycloidal teeth, struck by a single generating circle rolled without the pitch circle for the faces, and within it for the flanks. This is the best system, as any pair of wheels so struck, with the same generating circle and of equal pitch, will work together. Fig. 3 shows the common form of epicycloidal teeth, the flanks being straight. In this case the faces of the rack are struck by a generating circle half the diameter of the wheel, and the faces of the wheel, being obtained by a generating circle of infinite diameter or straight line, become involutes. Fig. 4 gives the form of teeth described as involutes, the curve being continuous, and, in the case of the rack, a straight line perpendicular to the tangent to the base circle. In these teeth it is possible to work with very little play. They are a good form for wheel and rack working together, the pressure on the journals being in this case less objectionable. Fig. 5 shows the teeth of a large wheel, traced from one of my own patterns, to exhibit the form and proportion which practice has shown to be desirable.

In these teeth the pitch $c\,d$ being $2\frac{1}{2}$ inches, the depth of the tooth or distance $a\,b$ is $\frac{1.5}{8}$ths or $\frac{3}{4}$ths of the pitch. The proportions of the parts may be given as follows:—

				Proportional Part.		Inches.
Pitch	=	$c\,d$	=	1·00	=	$2\frac{1}{2}$
Depth	=	$a\,b$	=	0·75	=	$1\frac{7}{8}$
Working depth	=	$a\,e$	=	0·70	=	$1\frac{3}{4}$
Clearance	=	$e\,b$	=	0·05	=	$\frac{1}{8}$
Thickness	=	$c\,f$	=	0·45	=	$1\frac{1}{8}$
Width of space	=	$f\,d$	=	0·55	=	$1\frac{3}{8}$
Play or $f\,d$, $c\,f$	=		=	0·10	=	$\frac{1}{4}$
Length beyond pitch line	=	$a\,g$	=	0·35	=	$\frac{7}{8}$

Taking these proportions, we may construct a scale which shall give directly the corresponding numbers for any pitch. Taking a vertical line, and dividing it into eighths of an inch, we get the scale of pitches (Plate X.). Draw lines perpendicular to this, and on any one of them mark off a series of distances equal to the clearance, depth, thickness &c. of the teeth corresponding to that pitch. Through o and these points draw the lines shown in the figure; they will divide the lines corresponding to all other pitches in the same proportion.

It is usual to allow a greater amount of clearance in small

wheels than is necessary in large ones. Very varying proportions have been given by different millwrights, $\frac{1}{10}$th, $\frac{1}{12}$th, $\frac{1}{15}$th and $\frac{1}{20}$th of the pitch having been used in different circumstances, even with the best mill-work. In the scale (Plate X.) this has to a certain extent been taken into account; $\frac{1}{10}$th of the pitch is allowed in smaller wheels, decreasing to $\frac{1}{15}$th in the largest; hence the lines are not absolutely straight, but are slightly curved, except that for the whole depth of the tooth, which quantity has been assumed to vary directly as the pitch.

Assuming that this scale represents with sufficient accuracy the proportions which practice shows to be best in average cases, we may construct a table for the guidance of the millwright. From this he must vary in cases where it appears necessary to allow more for defects of workmanship, or to permit less "backlash;"* it being understood that the table will only apply in cases where the teeth are formed with an approximation to the true mathematical figure.

In wood and iron gear where the teeth are carefully cut, very little if any clearance is necessary, as they work much better when the tooth of each wheel fills their allotted spaces. It is,

TABLES OF PROPORTIONS OF TEETH OF WHEELS FOR AVERAGE PRACTICE.

Pitch.	Clearance and play.	Depth beyond pitch line.	Depth within pitch line.	Working depth.	Whole depth.	Thickness of tooth.	Width of space.
½	·06	·16	·22	·32	·38	·22	·28
¾	·08	·25	·33	·50	·58	·33	·42
1	·10	·335	·435	·67	·77	·45	·55
1¼	·12	·42	·54	·84	·96	·56	·69
1½	·13	·51	·64	1·02	1·15	·68	·82
1¾	·14	·60	·74	1·20	1·34	·80	·95
2	·16	·685	·845	1·37	1·53	·92	1·08
2¼	·17	·775	·945	1·55	1·72	1·04	1·21
2½	·19	·86	1·05	1·72	1·91	1·15	1·35
2¾	·20	·95	1·15	1·90	2·10	1·27	1·47
3	·22	1·04	1·26	2·08	2·30	1·39	1·61
3¼	·23	1·13	1·36	2·26	2·49	1·51	1·74
3½	·25	1·215	1·465	2·43	2·68	1·62	1·88
3¾	·26	1·305	1·565	2·61	2·87	1·74	2·01
4	·28	1·39	1·67	2·78	3·06	1·86	2·14
4½	·31	1·565	1·875	3·13	3·44	2·09	2·40
5	·34	1·745	2·085	3·49	3·83	2·33	2·67
5½	·37	1·925	2·295	3·85	4·21	2·56	2·93
6	·40	2·10	2·50	4·20	4·60	2·80	3·20

* A technical expression for *reaction* on the back of the teeth.

however, different where wheels have to gear together direct from the foundry, where the teeth are not unfrequently deranged in the act of moulding in the sand.

This table gives the number to the nearest hundredth of an inch It may be converted into the ordinary scale of eights by the following table:—

	Thirty Seconds of an Inch.									
	1	2	3	4	5	6	7	8	9	10
Corresponding Decimal.	·031	·062	·094	·125	·156	·188	·219	·250	·281	·3125

As, unfortunately, decimal scales are not yet much used by millwrights, the following table has been prepared, giving the numbers in the preceding table in thirty seconds of an inch, such changes being made as will reduce as much as possible the errors of employing this rough standard. The former table is to be preferred where it can be used, but in other cases the following one may be relied on. The left-hand figures in each

Table Giving the Proportions of the Teeth of Wheels in Inches and Thirty Seconds of an Inch.

Pitch, inches.	Clearance.	Depth beyond the pitch line.	Depth within the pitch line.	Working depth.	Whole depth.	Thickness of tooth.
½	0″ 2	0″ 5	0″ 7	0″ 10	0″ 12	0″ 7
¾	0 3	0 8	0 11	0 16	0 19	0 10
1	0 3	0 11	0 14	0 22	0 25	0 14
1¼	0 4	0 13	0 17	0 26	0 30	0 18
1½	0 4	0 16	0 20	1 0	1 4	0 21
1¾	0 4	0 19	0 23	1 6	1 10	0 25
2	0 5	0 22	0 27	1 12	1 17	0 29
2¼	0 5	0 25	0 30	1 18	1 23	1 1
2½	0 5	0 28	0 33	1 24	1 29	1 5
2¾	0 6	0 31	0 37	1 30	2 4	1 8
3	0 7	1 1	1 8	2 2	2 9	1 12
3¼	0 7	1 4	1 11	2 8	2 15	1 16
3½	0 8	1 7	1 15	2 14	2 22	1 20
3¾	0 8	1 10	1 18	2 20	2 28	1 23
4	0 9	1 12	1 21	2 24	3 1	1 27
4½	0 10	1 18	1 28	3 4	3 14	2 3
5	0 11	1 24	1 35	3 16	3 27	2 10
5½	0 11	1 30	1 41	3 28	4 7	2 18
6	0 12	2 4	2 16	4 8	4 20	2 25

column are inches, the right-hand ones thirty seconds of an inch, the denominators of the fraction being omitted.

Bevel Wheels.

Hitherto we have considered only that case of toothed wheels in which the pitch lines are in one plane. We have now to examine the modifications which are necessary when the axes of

Fig. 199.

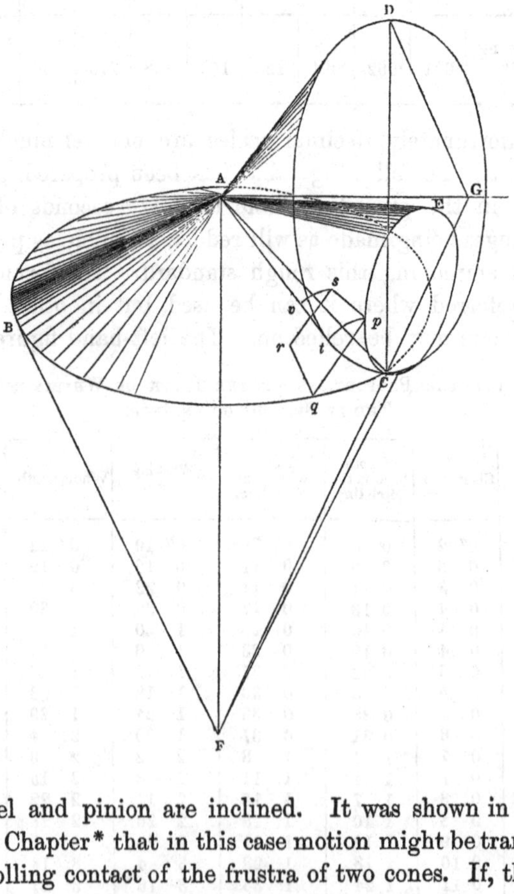

the wheel and pinion are inclined. It was shown in the preliminary Chapter* that in this case motion might be transmitted by the rolling contact of the frustra of two cones. If, therefore,

* Mills and Mill-work, Vol. I., p. 46, § 68, 69.

teeth be applied to these frustra, in the same manner as in spur gearing they are attached to cylindrical surfaces, bevel gearing will be formed acting on the same principles of sliding contact which we have already discussed.

Let A B C, A C D (fig. 199) be two cones rolling in contact; take any other cone A E C also rolling in contact with A B C, in the line A C. As these cones roll together, the generating cone A E C will describe an epicycloidal surface *p q r s* on the outside of the cone A C D, and a hypocycloidal surface *p t v s* on the inside of the cone A C D. These surfaces will touch in the line *p s*, and will have a plane normal to their common tangent passing through A C. If, therefore, these surfaces be attached respectively to the cones A B C, A C D, and the motion of one cone be communicated to the other through the sliding contact of these surfaces, the motion will be uniform, as if the cones were driven by rolling contact at A C.

The curves *p t*, *p q*, lie in reality on the surface of a sphere of a radius equal to A C; but in practice, in bevel wheels, a small frustrum of a cone, tangential to the sphere at the circumference of the pitch line, is substituted for the spherical segment. Thus draw F C G (fig. 199) perpendicular to A C, cutting the axes of the cones in F and G. Let these lines revolve over the pitch lines of the cones and describe the narrow frustra. Then the epicycloidal surfaces may, without sensible error, be supposed to lie in these frustra, and to be generated there by the revolution of a generating circle C E.

Imagine the surface of these frustra to be unwrapped so as to lie in one plane, they will form parts of circular annuli. Thus let A B C, A C D (fig. 200), be two conical frustra; draw F C G as before, perpendicular to the line of contact A C. From G, with radii G H, G C and G K, describe the circles K L, C M, H N; and from F, with radii F K, F C, F H, describe similar circles K P, C Q, H R; then the surfaces K P R H and K L N H will be developements of the frustra C D, C B. Let these be treated as spur wheels, and C Q, C M being treated as the pitch lines, let teeth be described by a describing circle in the method already explained for epicycloidal or other teeth. If, then, the plane on which these have been described, and which we suppose of drawing paper or other flexible material, be cut along the arcs K P, H R, K L, H N,

Fig. 200.

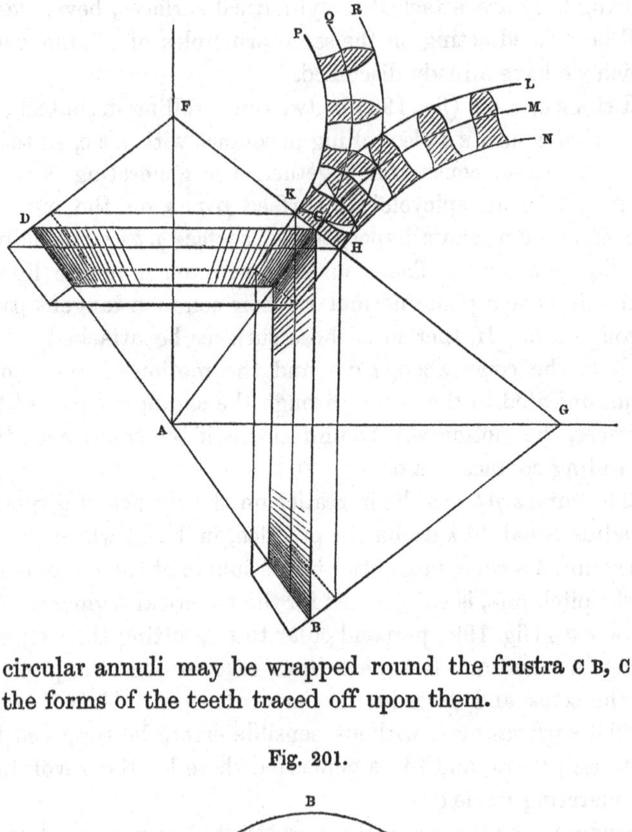

the circular annuli may be wrapped round the frustra C B, C D, and the forms of the teeth traced off upon them.

Fig. 201.

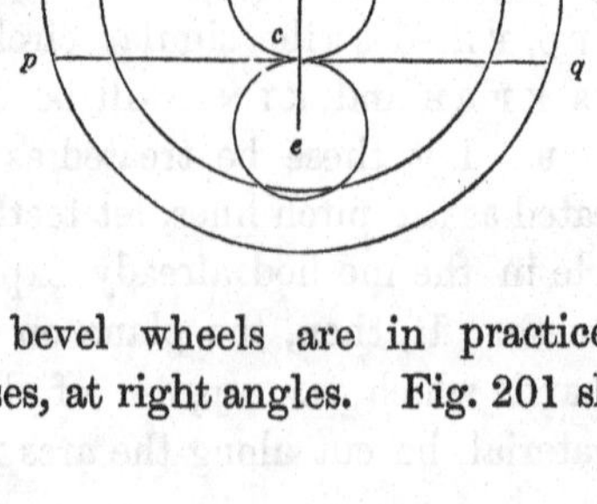

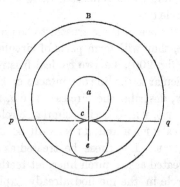

The axes of bevel wheels are in practice, in the great generality of cases, at right angles. Fig. 201 shows such a pair of bevels, with the frustra of the extremity of the teeth developed in the manner described.

Skew Bevels.

When two axes or shafts, which have to be connected by bevel wheels, do not meet in direction, it is usual, as stated in the preliminary Chapter,* to introduce an intermediate bevel wheel with two frustra. But the same object can more easily be accomplished by adopting skew bevels.

Let B *p* *q* (fig. 201) be the place of one of the two frustra, *a* its centre, and *a* *e* the shortest distance between the axis of

Fig. 202.

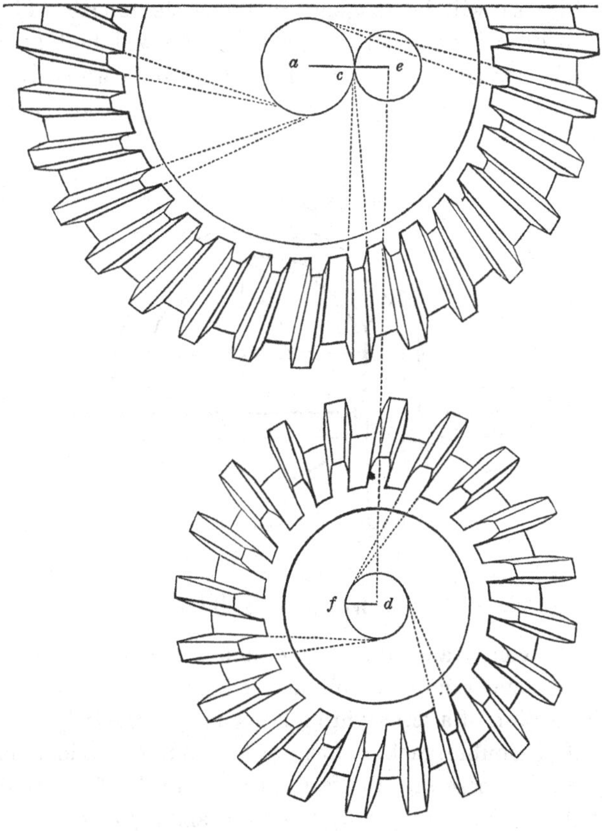

* Mills and Mill-work, Vol. I., p. 47, § 70, 71.

B *p q*, and the axis of the wheel to be connected with it. Divide *a e* in *c*, so that *a c* : *e c* :: mean radius of A B C : mean radius of frustrum working with A B C. Draw *c p q* perpendicular to *a e*, then *c p* or *c q* is the line of action of the teeth, according to the direction in which the teeth are laid out in the pinion.

Figure 202 shows two wheels laid out in this manner; *a e*, as before, is the eccentricity or shortest distance between the two shafts, and is divided in *c* proportionally to the mean radii of the wheels; with centre *a* and radius *a c* describe a circle, and draw *e d* perpendicular to *a e*. Take *d f* = *c e*, then *d* will be the centre of the other wheel. From centre *d*, with radius *f d*, describe a circle. Then the directions of all the teeth in A B C will be tangents to the circle described about *a*, and the directions of all the teeth in D E F will be tangents to the circle described about *f*. Fig. 203 shows two such wheels in gear, the eccentricity permitting the shafts to pass each other.

Fig. 203.

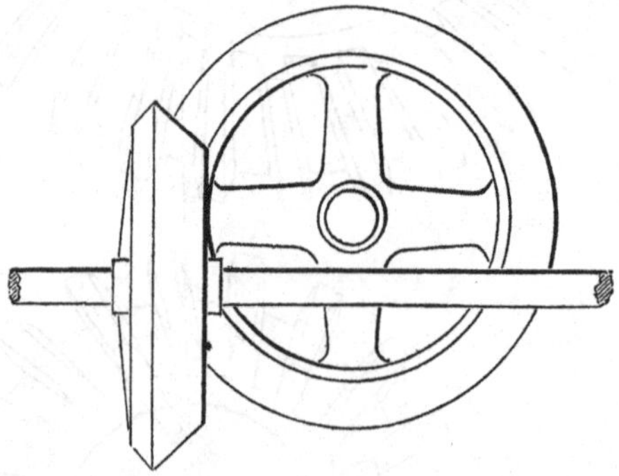

The Worm and Wheel.

By this contrivance the motion of a screw is communicated with great smoothness to oblique teeth on a spur wheel.

The section of a screw through its axis is precisely similar to that of a double rack. Let A B be such a section, and for simplicity suppose that the form of the threads of the screw has been determined by one of the rules already given for racks.

Fig. 204.

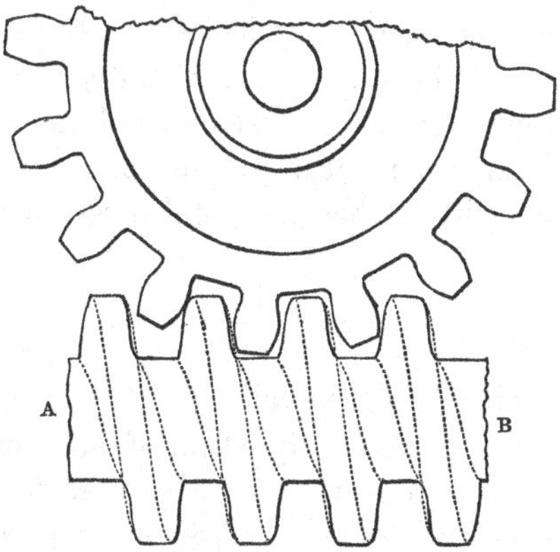

Then the teeth of the wheel C D E may evidently be formed so as to work with the centre section of the screw. Now the effect of the revolution of the screw is precisely similar to that of the racks, and the sections of the threads of the screw will appear to travel from end to end, in the same way as a rack pushed forwards in the same direction. If, therefore, it is sufficient that the wheel teeth be in contact with the screw at one point only, the teeth of the wheel may be made oblique, but straight, the obliquity being equal to the pitch of the screw. This is the usual practice of millwrights. If, however, the teeth are required to be in contact with the entire breadth of the tooth, the outline of the tooth must vary in every section of the wheel, and the process of describing these teeth becomes very complex. Practically, the difficulty has been overcome by first making a pattern screw of steel, notched in the threads to convert it into a cutting instrument. The wheel is then roughly cut out, and being fixed in a frame, the screw is used to cut out the spaces between the teeth to their true form.

Strength of the Teeth of Wheels.

The pressure on the teeth varies directly as the horse-power transmitted and inversely as the velocity of revolution. Thus if one wheel transmit 5 horse-power and another 10 horse-power at the same velocity, the strain on the latter will be twice that on the former. Or, again, if two wheels transmit the same power, but one at a velocity of 100 feet per minute, and the other at only 25 feet per minute, the strain on the former will be only one-fourth that on the latter.

Let v be the velocity in feet per second, H the number of horse-power transmitted, then the total pressure on the wheels will be —

$$P = \frac{550\,H}{v}$$

where P is the statical pressure in lbs.

For example, suppose the fly-wheel of an engine to be 24 feet in diameter, and to work into a pinion 5 feet diameter. And let the work transmitted be 150 horse-power. Then, if the wheel makes 25 revolutions per minute, the periphery will move at a velocity of $\frac{75 \cdot 4 \times 25}{60} = 31 \cdot 4$ feet per second; and the statical pressure on the teeth will be $\frac{550 \times 150}{31 \cdot 4} =$ 2627 lbs.

In addition to statical pressure, however, a different element has to be taken into account, namely, the impacts due to sudden accelerations or retardations of speed. The allowance which must be made to prevent accident from this cause varies exceedingly in different kinds of machinery. It is great in the gearing of rolling mills for instance, and in all machinery in which the strains are irregular.

Fig. 205.

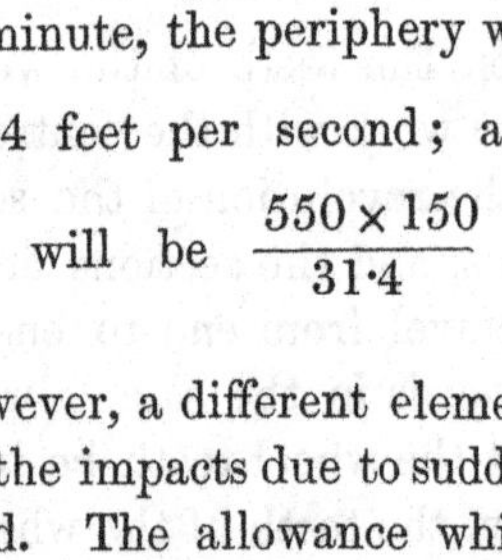

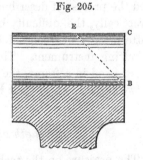

In calculating the strength of the tooth, it has been usual to consider it as a short beam fixed at one end, and having the whole of the pressure applied along the extremity of the tooth. But there is a position in which the teeth may be subjected to a severer stress still; owing to the wear of brasses and teeth, we cannot calculate upon the strain bearing always on the whole breadth of the tooth. The pressure may not only come on to the extremity of a tooth, but, if any obstruction come in between the teeth, it may be thrown entirely upon one corner of the tooth. In such a case it may be shown, by the rules of maxima and minima, that if E C = C B, the greatest stress will be near the line E B.

Tredgold has expressed the strength of a tooth on this supposition by the formula

$$W = \frac{f\,d^2}{5}$$

where d is the thickness of the tooth. To allow for wear, however, he adds one-third, so that

$$W = \frac{f\,d^2\,(1 - \frac{1}{3})^2}{5} = \frac{f\,d^2}{10{\cdot}25}$$

In cast-iron $f = 15{,}300$, and hence

$$d = \sqrt[2]{\frac{W}{1500}}$$

Or in words, the thickness necessary for the tooth or inches is equal to the square root of the stress on the tooth in pounds divided by 1500. Hence Tredgold has computed the following table, the breadths of the teeth being deduced, on the principle that the stress should not exceed 400 lbs. per inch breadth :—

Table of Thickness, Breadth and Pitch of Teeth of Wheels.

Stress in lbs. at the pitch line.	Thickness of teeth in inches.	Breadth of teeth in inches.	Pitch in inches.
400	0·52	1	1·1
800	0·73	2	1·5
1,200	0·90	3	1·9
1,600	1·03	4	2·2
2,000	1·15	5	2·4
2,400	1·26	6	2·7
2,800	1·36	7	2·9
3,200	1·46	8	3·0
3,600	1·56	9	3·3
4.000	1·64	10	3·4
4,400	1·70	11	3·6
4,800	1·78	12	3·7
5,200	1·86	13	3·9
5,600	1·93	14	4·0
6,000	2·00	15	4·2

To use this table when the horses' power transmitted by the wheel are known, the reader must refer to the table on page 47.

Elsewhere Tredgold has given a rule of the following description:—

$$d = \tfrac{3}{4}\sqrt{\frac{\text{H}}{v}} \text{ for cast-iron,}$$

where d is the requisite thickness of a tooth to transmit a force of H horses at a velocity v feet per second.

Hence Tredgold's last rule for the thickness of cast-iron teeth is as follows: — "Find the number of horses' power transmitted by the wheel, and divide that number by the velocity in feet per second of the pitch line of the pinion or wheel; extract the square root of the quotient, and three-fourths of this root will be the least thickness of cast-iron teeth for the wheel or pinion." From this he derives a second rule for the pitch, which manifestly depends on the thickness of the tooth, namely, multiply the thickness of the tooth by 2·1, and the product will be the pitch. The same result may be obtained by inspection from the tables I have given at pages 34, 35. Wooden teeth he recommends to be made of twice the thickness of cast-iron ones. But one-and-a-half times the thickness is a sufficient allowance.

A writer in the "Engineer and Machinists' Assistant" deduces another but equally simple rule for the thickness of teeth; he assumes the relation

$$t = c\sqrt{\text{W}};$$

where t is the thickness of the tooth W, the pressure on the tooth and c a constant, depending on the nature of the material. Let then a be the strength of a bar 1 inch long, 1 broad, and 1 thick. Then, to support a weight W by a bar of a length l, and breadth b,

$$t = \sqrt{\frac{\text{W} \times l}{a \times b}};$$

suppose the breadth of the tooth to be fixed at twice its length;

$$t = \sqrt{\frac{\text{W}l}{a \times 2l}} = \sqrt{\frac{\text{W}}{2a}}.$$

Taking $a = 8000$ lbs. for cast-iron, $2a = 16,000$ lbs., but as this is the breaking weight, the safe working-pressure will be only 1600 lbs., and the thickness of the tooth for safe working will be for cast iron:

$$t = \sqrt{\frac{W}{1600}} = 0{\cdot}025\sqrt{W}.$$

Where W being given in lbs. t is found in inches. Similarly for other materials he obtains;

$$c = {\cdot}035 \text{ for brass,}$$
$$= {\cdot}038 \text{ for hard wood.}$$

For example, in the wheel assumed at p. 41, W was found to be 2627 lbs. Hence the necessary thickness of the tooth, if of cast iron, would be ${\cdot}025\sqrt{2627} = 1{\cdot}28$ inches. Referring to the tables of the relation of pitch &c. we find that the wheel must be of $2\frac{3}{4}$ inches pitch, the teeth of 2·1 inches length, and the breadth of the wheel $2{\cdot}1 \times 2 = 4{\cdot}2$ inches at the least. By Tredgold's latter rule, the thickness of the teeth for the same wheel would be $t = \frac{3}{4}\sqrt{\frac{150}{31{\cdot}4}} = 1{\cdot}41$ inches; the pitch $= 2{\cdot}1 \times 1{\cdot}41 = 3{\cdot}0$ inches, and the breadth $= \frac{2627}{400} = 6\frac{1}{2}$ inches.

Bearing in mind that $W = \frac{550\,H}{v}$, where H is the maximum horse-power transmitted, and v the velocity of the pitch line of the wheel in feet per second, we may give these formulæ in a more convenient form:

$$t = x\sqrt{\frac{H}{v}}.$$

Where $x = 0{\cdot}587$ for cast-iron,
" $= 0{\cdot}821$ for brass,
" $= 0{\cdot}891$ for wood.

Conversely, if a wheel having teeth t inches thick be given, the horse-power it is capable of transmitting is given by the formula:

$$H = \frac{t^2 v}{x^2}.$$

Where $x^2 = 0{\cdot}344$ for cast-iron,
" $= 0{\cdot}674$ for brass,
" $= 0{\cdot}795$ for wood.

From the following table the pressure at other velocities, and with another amount of horse-power, may be obtained by interpolation, remembering that the pressure varies inversely as the former, and directly as the latter. To this we have appended another table, giving the horses' power, which can be safely transmitted by wheels of different pitches when proportioned according to the table at page 34. The last of these tables has been calculated on the assumption that 400 lbs. per inch is the greatest working stress which is consistent with durability in ordinary cases.

RELATION OF HORSES' POWER TRANSMITTED AND VELOCITY AT THE PITCH CIRCLE TO PRESSURE ON TEETH.

Number of horses' power transmitted.	Velocity in feet per second.									
	1 ft.	3 ft.	5 ft.	7 ft.	9 ft.	11 ft.	13 ft.	15 ft.	20 ft.	25 ft.
lbs.	lbs.	lbs.	lbs.	lbs.	lbs.	lbs.	lbs.	lbs.	lbs.	lbs.
1	550	183	110	79	61	50	42	37	28	22
2	1,100	367	220	157	122	100	85	73	55	44
3	1,650	550	330	236	183	150	127	110	83	66
4	2,200	733	440	314	244	200	169	146	110	88
5	2,750	917	550	393	306	250	212	183	138	110
10	5,500	1,833	1,100	786	611	500	423	367	275	220
15	8,250	2,750	1,650	1,179	917	750	635	550	413	330
20	11,000	3,667	2,200	1,571	1,222	1,000	846	733	550	440
25	13,750	4,583	2,750	1,964	1,527	1,250	1,058	917	688	550
30	16,500	5,500	3,300	2,357	1,833	1,500	1,269	1,100	825	660
40	22,000	7,333	4,400	3,143	2,444	2,000	1,692	1,467	1,100	880
50	27,500	9,167	5,500	3,928	3,055	2,500	2,115	1,833	1,375	1,100
60	33,000	11,000	6,600	4,714	3,667	3,000	2,538	2,200	1,650	1,320
70	38,500	12,833	7,700	5,500	4,278	3,500	2,962	2,567	1,925	1,540
80	44,000	14,667	8,800	6,285	4,889	4,000	3,385	2,933	2,200	1,760
90	49,500	16,500	9,900	7,071	5,500	4,500	3,808	3,308	2,475	1,980
100	55,000	18,333	11,000	7,857	6,111	5,000	4,231	3,667	2,750	2,200
110	60,500	20,167	12,100	8,643	6,722	5,500	4,654	4,033	3,025	2,420
120	66,000	22,000	13,200	9,423	7,333	6,000	5,077	4,400	3,300	2,640
130	71,500	23,833	14,300	10,214	7,944	6,500	5,500	4,767	3,575	2,860
140	—	25,667	15,400	11,000	8,556	7,000	5,923	5,133	3,850	3,080
150	—	27,500	16,500	11,786	9,167	7,500	6,346	5,500	4,125	3,300
160	—	29,333	17,600	12,571	9,778	8,000	6,769	5,867	4,400	3,520
170	—	31,167	18,700	13,357	10,389	8,500	7,192	6,233	4,675	3,740
180	—	—	19,800	14,143	11,000	9,000	7,615	6,600	4,950	3,960
190	—	—	20,900	14,929	11,611	9,500	8,038	6,967	5,225	4,180
200	—	—	22,000	15,714	12,222	10,000	8,462	7,333	5,500	4,400
300	—	—	33,000	23,571	18,333	15,000	12,692	7,700	8,250	6,600
400	—	—	44,000	31,428	24,444	20,000	16,923	8,067	11,000	8,800
500	—	—	55,000	39,285	30,555	25,000	21,154	8,433	13,750	11,000

TABLE SHOWING THE PITCH AND THICKNESS OF TEETH TO TRANSMIT A GIVEN NUMBER OF HORSES' POWER AT DIFFERENT VELOCITIES.

Pitch in inches.	Thickness of teeth in inches.	Velocity at Pitch line in feet per second.												
		1	3	5	7	9	11	13	15	20	25	30	40	50
		H. P.	H. P.	H. P.	H. P.	H. P.	H. P.	H. P.	H. P.	H. P.	H. P.	H. P.	H. P.	H. P.
½	·22	·14	·42	·71	·99	1·26	1·5	1·8	2·1	2·8	3·5	4·2	5·6	7·1
¾	·33	·32	·95	1·59	2·22	2·75	3·5	4·1	4·7	6·3	7·9	9·5	12·7	15·9
1	·45	·59	1·77	2·95	4·12	5·31	6·5	7·7	8·8	11·8	14·7	17·7	23·6	29·5
1¼	·56	·91	2·74	4·56	6·38	8·22	10·0	11·9	13·7	18·2	22·8	27·4	36·5	45·6
1½	·68	1·34	4·03	6·72	9·41	12·09	14·8	17·5	20·2	26·9	33·6	40·3	53·7	67·2
1¾	·80	1·86	5·58	9·31	13·03	16·74	20·5	24·2	27·9	37·2	46·5	55·8	74·4	93·1
2	·92	2·46	7·38	12·31	17·23	22·14	27·1	32·0	36·9	49·2	61·5	73·8	98·4	123·1
2¼	1·04	3·14	9·43	15·72	22·01	27·29	34·6	40·9	47·2	62·9	78·6	94·3	125·7	157·2
2½	1·15	3·85	11·54	19·23	26·92	34·62	42·3	50·0	57·7	76·9	96·1	115·4	153·8	192·3
2¾	1·27	4·69	14·07	23·45	32·82	42·21	51·6	61·0	70·3	93·8	117·2	140·6	187·5	234·5
3	1·39	5·62	16·84	28·08	39·31	50·42	61·8	73·0	84·2	112·3	140·4	168·5	224·6	280·8
3¼	1·51	6·63	19·88	33·14	46·40	59·64	72·9	86·2	99·4	132·6	165·7	198·8	265·1	331·4
3½	1·62	7·63	22·89	38·15	53·40	68·67	83·9	99·2	114·4	152·6	190·7	228·9	305·1	381·5
3¾	1·74	8·80	26·40	44·00	61·61	79·20	96·8	114·4	132·0	176·0	220·0	264·0	352·0	440·0
4	1·86	10·06	30·17	50·29	70·40	90·50	110·6	130·7	150·8	201·1	251·4	301·7	402·3	502·9
4½	2·09	12·70	38·09	63·49	88·89	114·27	139·7	165·1	190·5	254·0	317·5	380·9	507·9	634·9
5	2·33	15·78	47·35	78·91	110·47	142·05	173·6	205·2	·236·7	315·6	394·5	473·5	631·3	789·1
5½	2·56	19·05	57·15	95·26	133·36	171·45	209·6	267·7	285·8	381·0	476·3	571·5	762·0	952·6
6	2·80	22·79	68·37	113·96	159·54	205·11	250·7	316·3	341·9	455·8	569·7	683·7	911·6	1,139·6

TABLE SHOWING THE BREADTH OF TEETH REQUIRED TO TRANSMIT DIFFERENT AMOUNTS OF FORCE AT A UNIFORM PRESSURE OF 400 LBS. PER INCH.

Number of horses' power transmitted.	Velocity of pitch line in Feet per second.										
	1	3	5	7	9	11	13	15	20	25	30
	ins.	ins.	ins.	ins.	ins.	ins.	ins.	ins.	ins.	ins.	ins.
1	1·4	0·5	0·3	0·2	0·15	0·12	0·11	0·09	0·07	0·05	0·05
2	2·7	0·9	0·6	0·4	0·3	0·25	0·21	0·18	0·14	0·11	0·09
3	4·1	1·4	0·8	0·6	0·5	0·38	0·32	0·27	0·21	0·17	0·14
4	5·5	1·8	1·1	0·8	0·6	0·50	0·42	0·37	0·28	0·22	0·18
5	6·9	2·3	1·4	1·0	0·8	0·62	0·53	0·46	0·35	0·28	0·23
10	13·7	4·6	2·7	2·0	1·5	1·3	1·06	0·92	0·69	0·55	0·46
15	20·6	6·9	4·1	2·9	2·3	1·9	1·6	1·37	1·03	0·83	0·69
20	34·2	9·2	5·5	3·9	3·1	2·5	2·1	1·83	1·38	1·10	0·92
25	41·2	11·4	6·9	4·9	3·8	3·1	2·6	2·3	1·72	1·38	1·14
30	55·0	13·8	8·2	5·9	4·6	3·8	3·2	2·8	2·06	1·65	1·38
40	68·8	18·3	11·0	7·9	6·1	5·0	4·2	3·7	2·75	2·20	1·83
50	82·5	22·9	13·7	9·8	7·6	6·3	5·3	4·6	3·44	2·75	2·29
60	96·2	27·5	16·5	11·8	9·1	7·5	6·4	5·5	4·1	3·30	2·75
70	110·0	32·1	19·2	13·8	10·7	8·8	7·4	6·4	4·8	3·85	3·21
80	123·8	36·7	22·0	15·7	12·2	10·0	8·5	7·3	5·5	4·40	3·67
90	137·5	41·2	24·8	17·7	13·8	11·2	9·5	8·2	6·2	4·95	4·12
100	151·2	45·8	27·5	19·6	15·3	12·5	10·6	9·1	6·9	5·50	4·58
110	165·0	50·4	30·3	21·6	16·8	13·7	11·6	10·1	7·5	6·05	5·04
120	178·8	55·0	33·0	23·5	18·3	15·0	12·7	11·0	8·3	6·6	5·50
130	—	59·8	35·7	25·5	19·8	16·3	13·8	11·9	8·9	7·2	5·98
140	—	64·2	38·5	27·5	21·4	17·5	14·8	12·8	9·6	7·7	6·42
150	—	68·7	41·3	29·4	22·9	18·7	15·8	13·7	10·3	8·2	6·87
160	—	73·3	44·0	31·4	24·5	20·0	16·9	14·6	11·0	8·8	7·33

CHAPTER II.

ON THE STRENGTH AND PROPORTIONS OF SHAFTS.

THE system of transmitting power from a common centre to a large number of machines, at some distance, is comparatively modern. In the operations of spinning and weaving by a consecutive series of machines, placed in rows, shafting became essential for distributing the power of the common prime mover. At first, the machines were brought as close to the prime mover as possible; and the early construction of mills—when the water-power was divided into separate falls—must be fresh in the recollection of many persons now living. In some cases, before the introduction of the steam engine, it was the custom to have a separate water-wheel to every machine, thus splitting up the power into as many parts as there were machines, or pairs of machines, to drive. In process of time, it was found more convenient, on the score of economy, to husband the water and concentrate the prime movers; hence one large water-wheel was constructed, around which the machinery was arranged, either in rows or otherwise, as best suited the work to be performed.

This principle, of the concentration of the motive power, destroyed the old system of separate buildings, and led to the employment of a large number of machines for the various processes of manufacture in one building. From this we derive the Factory system, in which any number of processes are carried on, the machinery being distributed over the different floors of a large building, and receiving motion from a single prime mover at a convenient distance. In this way, the power is conveyed by lines of shafting coupled together in lengths, adapted to the bays or divisions of the building. At first, the buildings were short, and shafting of great length was not required; gradually, more and more machines were concentrated in the same building, and shafting of 200 or 300 feet in length became necessary. To show to what an extent this system has been carried, it may be mentioned, that in the large mills at Saltaire, the shafting, if placed in a single line, would extend for a distance of more than two miles. This progress has been chiefly due to the introduction of the steam engine, in place of water-wheels, because the available power is no longer limited by the circumstances of the locality in which the mill is placed.

This concentration of a great number of machines in one building is peculiar to the Factory system; and in the present highly-improved state of mechanical science and its application to the production of textile fabrics, it has become essential to economy in the manufacturing processes, that they should be carried on in the same building. Spinners and manufacturers are fully aware of the advantages peculiar to this system of concentration, so much so, that out of what would formerly have been considered a mere fractional saving, large profits and large fortunes are now made. In fact, the amalgamation of the different processes under one management and under one roof, gave rise to the shed system, where the operations of the manufacture of cotton are carried on under what is called the "*saw-tooth*" roof, in order to bring the whole on the ground-floor under one inspection.

1. *The Material of which Shafting is constructed.*

The selection of the material for shafting is of great importance, and the uses to which it is to be applied require careful consideration. Formerly wood, with iron hoops and gudgeons, was universally employed; then cast-iron was introduced and subsequently wrought-iron has in most cases superseded both. Wood, indeed, has become obsolete; but cast-iron is as good as, if not superior to, wrought-iron, in certain cases. The main and vertical shafts of a mill are generally of cast-iron, both on account of its cheapness, and its high resistance to torsion. The vertical shafts, which convey the power from the first motion wheels to the different rooms of the mill, are more rigid and less subject to vibration when of cast-iron; even the main horizontal shafting, when of large dimensions, is, if substantially fixed, quite as good, when of the same material, and much cheaper than wrought-iron. Where the shaft is exposed to impact, or any irregularity of force, wrought-iron has the superiority; but in other cases, when the castings are sound and good, cast-iron may be employed with perfect safety.

The dimensions required for a shaft, transmitting any given force, will depend on the resistance of the material of which it is composed. Consequently, the selection of material must be determined by the necessity for strength. Shafts may be considered as subject to two forces: a force producing simple flexure, arising from their own weight, the weight of the wheels and pullies, and the strain of the belts; and a twisting force or torsion, arising from the power transmitted. If the flexure be great, the brasses will be much worn, vibration becomes considerable, and the disintegration of the machinery goes on in an accelerating ratio; it is therefore necessary to proportion shafting to the simple weight and direct transverse strain it has to sustain, so as to reduce the flexure within exceedingly narrow limits. In addition to this, the shafting, having to transmit a torsive force, must at least be capable of transmitting it without danger of rupture. In long and light shafting the tendency to flexure is usually greater than that to rupture by torsion; the former consideration will therefore determine the size of the shaft. In short axles, etc., the danger from flexure almost disappears, and the strength of the shaft is determined by its resistance to torsion only. In all cases both conditions must be complied with, if security and permanence are to be obtained.

2. *Transverse Strain.*

Resistance to Rupture. The general formula for resistance to rupture, in the case of a bar or beam supported at each end and loaded in the centre, is

$$\mathrm{W} = \frac{adc}{l} \quad . \quad . \quad . \quad . \quad (1.)$$

where W is the load in the centre, a the area of a section of the bar, perpendicular to the length; d the depth of the bar, and l its length. In this case c is derived from experiment, and is constant for similar bars or beams.

For rectangular bars this formula becomes,

$$W = \frac{c\,b\,d^2}{l} \quad \ldots\ldots\ldots \quad (2.)$$

where b is the breadth and d the depth.

The value of c, for rectangular bars found by Mr. Barlow, for various materials, is given in the following table. In applying these numbers to calculations, it must be remembered that a and d are to be taken in inches, and l in feet; then c, the centre breaking-weight, is found in lbs.

When the beam is supported at one end and loaded at the other, the formula is

$$W = \frac{c\,b\,d^2}{2l} \quad \ldots\ldots\ldots \quad (3.)$$

Value of c for different Materials.

	lbs.
English Malleable Iron	2050
Cast Iron	2548
Oak	400
Canadian Oak	588
Ash	675
Pitch Pine	544
Red Pine	447
Riga Fir	376
Mar Forest Fir	415
Larch	280

In my own experiments * I found the value of c for cast-iron to range from 1606 to 2615, the mean value being about 2050, as given above for malleable iron. Wrought-iron ranges from the value given above to 3000 lbs.

For cylindrical shafts supported horizontally the ultimate resistance to rupture is about

$$W = \frac{15000\,d^3}{l} \text{ for wrought-iron,}$$

$$= \frac{19000\,d^3}{l} \text{ for cast-iron,}$$

* *On the Application of Cast and Wrought-Iron to Building Purposes*, p. 74 et seq.

where W is the centre-breaking weight in lbs., d the diameter, and l the length between supports in inches, the shaft being supported at the ends and loaded in the middle.

If the cylindrical shaft be loaded at one end and supported at the other, these formulæ become

$$W = \frac{7500 d^3}{l} \text{ for wrought-iron,}$$

$$= \frac{9500 d^3}{l} \text{ for cast-iron.}$$

If a beam be uniformly loaded over its entire length, it will sustain twice the load that would break it if placed at the centre.

If the load be placed at any point intermediate between the centre and the ends, the breaking weight may be found by the following rule:— Divide four times the product of the distance in feet, of the weight from each bearing, by the whole distance in feet, and the quotient may be substituted for l in the formulæ above. That is, if x and y be its distances in feet from the two bearings respectively;

$$l = \frac{4xy}{(x+y)}$$

From these rules the strength of shafts may be calculated, in all the cases of ordinary practice, where the tendency to transverse fracture has to be guarded against, making the actual strength at least five to ten times the strain to be carried. In shafting, however, it is not usually the transverse rupture, but the flexure produced by lateral stress, which limits the size of the shaft;—stiffness in fact becomes, in these cases, a more important element than strength.

The following formula has been given for the deflection of bars or beams loaded at the centre and supported at the ends:—

Let, d be the depth in inches;
b the breadth in inches;
L the length between supports in feet;
W the load in lbs.;
δ the deflection at the centre in inches:
M the modulus of elasticity;

then:— $$W = \frac{M b d^3 \delta}{432 L^3}; \text{ and if } b = d, \text{ —}$$

$$d^4 = \frac{432\,\text{L}^3\,\text{W}}{\text{M}\,\delta} \text{ or } d = \sqrt[4]{\frac{432\,\text{L}^3\,\text{W}}{\text{M}\,\delta}} \quad . \quad . \quad (4.)$$

Or, in words, multiply the product of the load in lbs., and the cube of the length in feet, by 432, and divide by the product of the modulus of elasticity and the deflection assumed in inches; the fourth root of the quotient will be the side of a shaft of square section which would deflect δ inches with a weight of W lbs. placed at its centre.*

The following table gives the values of the modulus of elasticity for various materials:—

	Modulus of elasticity in lbs.
Cast-iron	13,000,000 to 22,907,000
,, ,, mean	17,000,000
Malleable iron	24,000,000 to 29,000,000
Steel	29,000,000 to 42,000,000
Brass	8,930,000
Tin	4,608,000
Ash	1,600,000
Beech	1,353,600
Red pine, mean	1,700,000
Spruce, mean	1,600,000
Larch	900,000 to 1,360,000
English oak	1,200,000 to 1,750,000
American oak	2,150,000

For a cylindrical shaft, the same formula will apply with another constant. I am not aware that this has been experimentally ascertained, but it has been given approximately as 734. Hence, for cylindrical shafts,

$$d^4 = \frac{734\,\text{L}^3\,\text{W}}{\text{M}\,\delta} \text{ or } d = \sqrt[4]{\frac{734\,\text{L}^3\,\text{W}}{\text{M}\,\delta}} \ldots (5).$$

In the work just quoted, these formulæ have been simplified, by fixing a maximum value for δ, the deflection. The writer assumes that, with shafting, the deflection ought never to exceed $\frac{1}{100}$ of an inch for every foot length of the shaft. Sub-

* *Engineer and Machinist's Assistant*, p. 135, from which formulæ (4), (5), (6), to (11), and (23), in their present convenient form for practical use, have been quoted. The fundamental formula, however, is due to Young (*Nat. Philos.* vol. ii. art. 326), and to Tredgold (Strength of Cast Iron, p. 208).

stituting this value, and also the numerical value of the modulus of elasticity, he obtains the following formulæ: —

1. *For wood,* — taking M generally = 1,500,000, and $\delta = \frac{L}{100}$ inches.

Then, for square shafts, d being the depth of the side of the square —

$$d^4 = \frac{L^2 W}{35} \dots (6).$$

And for round shafts, d being the diameter in inches —

$$d^4 = \frac{L^2 W}{20} \dots (7).$$

2. *For cast-iron* — taking M = 18,000,000 lbs. and L as before —

For square section, $d^4 = \frac{L^2 W}{412} \dots (8).$

For round section, $d^4 = \frac{L^2 W}{240} \dots (9.)$

3. *For wrought-iron* — taking M = 24,500,000 lbs. and δ as before —

For square section, $d^4 = \frac{L^2 W}{567} \dots (10).$

For round section, $d^4 = \frac{L^2 W}{334} \dots (11).$

By transposition, the formulæ given above become, —

For wood —

Square section, $L = \sqrt{\frac{35 d^4}{W}} \dots (12).$

$$W = \frac{35 d^4}{L^2}$$

Round section, $L = \sqrt{\frac{20 d^4}{W}} \dots (13).$

$$W = \frac{20 d^4}{L^2} \dots (14).$$

For cast-iron —

Square section, $L = \sqrt{\frac{412 d^4}{W}} \dots (15).$

$$W = \frac{412 d^4}{L^2} \dots (16).$$

$$\text{Round section, } L = \sqrt{\frac{240\, d^4}{W}} \;\ldots\; (17).$$

$$W = \frac{240\, d^4}{L^2} \;\ldots\; (18).$$

For wrought-iron —

$$\text{Square section, } L = \sqrt{\frac{567\, d^4}{W}} \;\ldots\; (19).$$

$$W = \frac{567\, d^4}{L^2} \;\ldots\; (20).$$

$$\text{Round section, } L = \sqrt{\frac{334\, d^4}{W}} \;\ldots\; (21).$$

$$W = \sqrt{\frac{334\, d^4}{L^2}} \;\ldots\; (22).$$

When the weight is uniformly distributed over the length of the shaft, the general formula is

$$d^4 = \frac{270\, L^3\, W}{M\, \delta} \quad \text{or} \quad d = \sqrt[4]{\frac{270\, L^3\, W}{M\, \delta}} \;\ldots\; (23).$$

Substituting in this equation the same values of M and δ as before, we obtain the following formulæ: —

For wood— $d^4 = \frac{L^2\, W}{56}$ for square shafts.

$d^4 = \frac{L^2\, W}{32}$ for round shafts.

For cast-iron— $d^4 = \frac{L^2\, W}{666}$ for square shafts.

$d^4 = \frac{L^2\, W}{383}$ for round shafts.

For wrought-iron— $d^4 = \frac{L^2\, W}{907}$ for square shafts.

$d^4 = \frac{L^2\, W}{521}$ for round shafts.

The following tables for cast and wrought-iron round shafting, are calculated from the formulæ (9) and (11) for weights placed at the centre of a shaft supported at each end. In using them for cases in which the weight is distributed along its length, as in the case of the weight of the shaft itself, it must be remembered that a distributed weight produces $\frac{5}{8}$ths of the deflection of the same weight placed at the centre.

TABLE 1. — RESISTANCE TO FLEXURE. WEIGHTS PRODUCING A DEFLECTION OF $\frac{1}{1200}$TH OF THE LENGTH IN CAST-IRON CYLINDRICAL SHAFTS.

Length between supports in feet.	Diameter of Shaft in Inches.																		
	1	1½	2	2½	3	3½	4	4½	5	5½	6	7	8	9	10	12	14	16	18	20
	lbs.	lbs.	lbs.	lbs.	lbs.	lbs.	lbs.	lbs.	lbs.	lbs.	lbs.	lbs.	lbs.	lbs.	lbs.	tons.	tons.	tons.	tons.	tons.
5	9·6	49	154	375	778	1,441	2,458	3,937	6,000	8,785	12,442	23,050	39,322	62,986	96,000	88·9	164·6	280·9	449·9	685·7
6	6·7	34	107	261	540	1,000	1,707	2,734	4,166	6,100	8,640	16,007	27,307	43,741	66,660	61·7	114·3	195·1	312·4	476·2
7	4·9	25	78	191	397	735	1,254	2,008	3,061	4,482	6,348	11,760	20,062	31,673	48,980	45·3	84·0	143·3	229·5	349·8
8	3·7	19	60	146	304	563	960	1,538	2,344	3,431	4,860	9,004	15,360	24,604	37,500	34·7	64·3	109·7	175·7	267·9
9	—	15	47	116	240	445	758	1,215	1,852	2,711	3,840	7,115	12,136	19,440	29,630	27·5	50·8	86·7	138·9	211·6
10	—	12	38	94	194	360	615	984	1,500	2,196	3,110	5,763	9,830	15,747	24,000	22·2	41·2	70·2	112·5	171·4
11	—	10	32	77	160	298	508	814	1,240	1,815	2,571	4,763	8,124	13,014	19,834	18·4	34·0	58·0	93·0	141·7
12	—	8	27	65	135	251	427	683	1,042	1,525	2,160	4,002	6,827	10,935	16,667	15·4	28·6	48·8	78·1	119·1
13	—	—	23	56	115	213	364	583	888	1,299	1,840	3,410	5,817	9,317	14,201	13·1	24·4	41·5	66·6	101·4
14	—	—	19	48	99	184	314	502	765	1,121	1,587	2,940	5,016	8,034	12,245	11·3	21·0	35·8	57·4	87·4
15	—	—	17	42	86	160	273	437	666	976	1,383	2,561	4,369	7,000	10,667	9·9	18·3	31·2	50·0	76·2
16	—	—	15	37	76	141	240	384	586	858	1,215	2,251	3,840	6,151	9,374	8·7	16·1	27·4	43·9	66·9
17	—	—	13	33	67	125	213	341	519	760	1,076	1,994	3,401	5,448	8,305	7·7	14·2	24·3	39·0	59·3
18	—	—	12	29	63	112	192	309	471	686	960	1,779	3,034	4,860	7,407	6·9	12·7	21·7	34·7	52·9
19	—	—	11	26	54	100	173	277	423	615	868	1,606	2,738	4,397	6,700	6·2	11·4	19·6	31·3	47·9
20	—	—	10	23	48	90	154	246	375	550	778	1,441	2,458	3,937	6,000	5·5	10·3	17·5	28·1	42·9

TABLE 2. — RESISTANCE TO FLEXURE. WEIGHTS PRODUCING A DEFLECTION OF $\frac{1}{1200}$TH OF THE LENGTH IN WROUGHT-IRON CYLINDRICAL SHAFTS.

Length between bearings in feet.	Diameter of Shaft in Inches.																		
	1	1½	2	2½	3	3½	4	4½	5	6	7	8	9	10	12	14	16	18	20
	lbs.	lbs.	lbs.	lbs.	lbs.	lbs.	lbs.	lbs.	lbs.	lbs.	lbs.	lbs.	lbs.	lbs.	tons.	tons.	tons.	tons.	tons.
5	13·4	68	214	522	1,082	2,005	3,420	5,478	8,350	17,314	32,078	54,724	87,654	133,600	123·6	229·1	391·0	626·0	954
6	9·3	47	148	363	751	1,392	2,375	3,804	5,799	12,024	22,277	38,003	60,871	92,780	85·9	159·1	272·0	435·0	663
7	6·8	35	109	267	552	1,023	1,745	2,795	4,260	8,834	16,367	27,920	44,721	68,163	63·1	116·9	200·0	319·4	487
8	5·2	26	84	204	423	783	1,336	2,140	3,262	6,764	12,531	21,376	34,240	52,188	48·3	89·5	153·0	244·6	373
9	—	21	66	161	334	619	1,055	1,691	2,577	5,344	9,900	16,890	27,053	41,235	38·1	70·7	121	193·2	295
10	—	17	53	130	223	501	855	1,370	2,087	4,329	8,020	13,681	21,913	33,400	30·9	57·3	97·7	156·5	231
11	—	14	44	108	188	414	707	1,132	1,725	3,577	6,628	11,306	18,110	27,603	25·5	47·3	80·7	129·4	197
12	—	12	37	91	160	348	594	952	1,450	3,006	5,569	9,500	15,218	23,195	21·5	39·8	67·9	108·7	166
13	—	10	31	77	149	297	507	810	1,235	2,561	4,745	8,095	12,966	19,763	18·3	33·9	57·8	92·6	141
14	—	—	27	68	138	256	436	699	1,065	2,208	4,091	6,980	11,180	17,041	15·8	29·2	49·9	79·9	122
15	—	—	24	60	122	223	380	609	928	1,924	3,564	6,080	9,740	14,844	13·7	25·5	43·4	69·6	106
16	—	—	21	51	106	196	334	535	816	1,691	3,133	5,344	8,560	13,047	12·0	22·4	38·1	61·1	93·2
17	—	—	19	46	95	173	296	474	722	1,498	2,775	4,734	7,582	11,557	10·7	19·8	33·8	54·2	82·5
18	—	—	17	42	84	157	268	430	655	1,359	2,518	4,296	6,880	10,488	9·7	18·0	30·6	48·3	74·8
19	—	—	15	37	76	141	240	386	589	1,221	2,261	3,858	6,179	9,419	8·7	16·1	27·5	43·5	67·2
20	—	—	13	33	68	125	213	342	522	1,082	2,005	3,420	5,478	8,350	7·7	14·3	24·4	39·1	59·6

From the foregoing it will be seen, that the weights given in the tables are correct indications of the load required in the centre to produce a deflection of the $\frac{1}{1200}$ of the length of the shaft.* This fraction is not however the universal standard among millwrights; on the contrary, there appears to be no recognised standard in practice, by which the deflection from a given weight can be ascertained, and although $\frac{1}{1200}$ may, in many cases, give a larger area with increased weight, in shafts that are not heavily loaded in the middle, nevertheless it is important that the shafts, when loaded as above, should not bend more than $\frac{1}{1200}$ of their length. In cases where the load is light and equally distributed, lighter and smaller shafts would suffice.

The following tables give the deflection of cylindrical shafts with their own weight:—

Table 3.—Deflection arising from the Weight of the Shaft. Cast-Iron Cylindrical Shafts.

Length between bearings in feet.	Diameter of Shaft in Inches.								
	1	2	4	6	8	10	12	14	16
	ins.	ins.	ins.	ins.	ins.	ins.	ins.	ins.	ins.
5	·004	·001	·000	·000	·000	·000	·000	·000	·000
10	·067	·017	·004	·002	·001	·001	·001	·000	·000
15	·338	·085	·021	·009	·005	·003	·002	·002	·001
20	1·067	·267	·067	·029	·017	·011	·007	·005	·004
25	2·603	·651	·163	·073	·041	·026	·018	·013	·010

Table 4.—Deflection arising from the Weight of the Shaft. Wrought-Iron Cylindrical Shafts.

Length between bearings in feet.	Diameter of Shaft in Inches.								
	1	2	4	6	8	10	12	14	16
	ins.	ins.	ins.	ins.	ins.	ins.	ins.	ins.	ins.
5	·003	·001	·000	·000	·000	·000	·000	·000	·000
10	·050	·013	·003	·001	·001	·001	·000	·000	·000
15	·256	·064	·016	·007	·004	·003	·002	·001	·001
20	·808	·202	·051	·022	·013	·008	·005	·004	·003
25	1·972	·493	·123	·055	·031	·020	·013	·010	·008

The above tables clearly indicate the deflection of shafts of different lengths by their own weight, and will be a guide to the

* This standard is the one assumed by Tredgold (Strength of Cast Iron, p. 210).

millwright in calculating the distance of the bearings between which they revolve. It is important in shafting, when extended in long ranges, that there should not be any serious deflection, either from the weight of the shaft, or lateral stress; I have always found that a stiff shaft, although heavier in itself, is lighter to retain in motion than a smaller one which bends to the strain.

3. *Torsion.*

In addition to the lateral flexure from transverse forces, shafting is subjected to a wrenching or twisting, from the power transmitted acting tangentially to its circumference. This causes one end of the shaft to revolve in relation to the other end, through a smaller or greater angle, known as the angle of torsion, and, if sufficient force be applied, this angle increases till the resistance of the material is overcome, and the shaft gives way.

Coulomb laid the basis of our knowledge of the resistance to torsion of cylindrical bodies, and he verified his theoretical deductions by admirably-contrived experiments, on a small scale. He showed that in wires where the diameter is small in relation to the length, the angles of torsion are in proportion to the length, and reciprocally proportional to the moment of inertia of the base of the cylinder in relation to its centre. He also discovered that each wire acquired a permanently acceleration-varying torsion, according to the degree in which it departed from its primitive position, and that these permanent torsions have no fixed relation to the temporary torsions, coexisting with the application of the moving force. With the same wire he found the torsion to be in proportion to the force applied; with the same length and force inversely as the fourth power of the diameter.

These deductions are expressed by the following formula:—

$$\theta = \frac{2\,\mathrm{R}}{\pi\,\mathrm{G}} \times \frac{\mathrm{W}l}{r^4}$$

where θ is the angle of torsion, r the radius, and l the length of the wire, R the leverage at which the weight W acts, and G the modulus of torsion for the material; being about $\frac{2}{5}$ths of the modulus of elasticity.

In 1829 a paper was communicated to the Royal Society by Mr. Bevan, containing experimental determinations of the modulus of torsion for a large number of substances, of which the most important are given below.

Let δ be the deflection of a prismatic shaft of a given length l when strained by a given force w in lbs., acting at right angles to the axes of the prism and at a leverage r; let d be the side of the square section of the shaft, l, r, δ, d, being in inches.

$$\delta = \frac{r^2 lw}{d^4 \text{T}}$$

where T is the modulus of elasticity in the following table.

If the transverse section of the prism be a parallelogram, let b be the breadth and d the depth, then Mr. Bevan gives the formula—

$$\delta = \frac{(d+b)\, lr^2 w}{2 b d^3 \text{T}}$$

If the torsion be required in degrees (Δ), then let $\rho = 57{\cdot}29578$,

$$\Delta = \frac{r \rho l w}{d^4 \text{T}}, \text{ for square shafts.}$$

For example,

$$\Delta = \frac{rlw}{31000 d^4} \text{ for wrought-iron and steel,}$$

$$= \frac{rlw}{16600 d^4} \text{ for cast-iron.}$$

A very careful experimental study of the effect of torsion on various materials has been made by Mr. M. G. Wertheim, and was presented to the Académie des Sciences in 1855. The general results at which he has arrived may be stated as follows:—

1. The total angle of torsion consists of two parts, of which one is purely temporary, whilst the other persists after the force has ceased to act. It is not possible to assign the limit at which the permanent torsion begins to be sensible, nor has it any fixed relation to the temporary torsion; it augments at first very slowly, afterwards more rapidly, till the bar breaks.*

* We have many practical instances of this tendency to rupture which at first appear only temporary, but a continuation of the same action, particularly in long

TABLE 5.—VALUES OF MODULUS OF TORSION ACCORDING TO MR. BEVAN.

Material.	Specific gravity	Modulus of torsion. (T).	
		lbs.	
Ash	—	20,300	
Beech	—	21,243	
Elm	—	13,500	
Scotch fir	—	13,700	
Hornbeam	·86	26,400	
Larch	·58	18,967	
English oak	—	20,000	
Memel pine	—	15,000	
American pine	—	14,750	
Teak	—	16,800	Old and partially decayed.
Teak, African	—	27,300	
Iron, English wrought	—	1,775,000	(Mean.)
Steel	—	1,753,000	(Mean.)
Iron (cylindrical)	—	1,910,000	
,, ,,	—	1,700,000	
,, (square)	—	1,617,000	
,, ,,	—	1,667,000	
,, ,,	—	1,951,000	
Cast-iron	—	940,000	
,, ,,	—	963,000	
,, ,,	—	952,000	
,, ,,	—	951,600	(Mean.)
Bell metal	—	818,000	

2. The temporary angles are not rigorously proportional to the moments of the forces applied.

3. The mean angles of torsion are not rigorously proportional to the length of the bar, increasing, although very slightly, in proportion to the length, as the bars are made shorter.

4. The interior cavity of all hollow homogeneous bodies diminish by torsion, and this diminution is proportional to the

ranges of shafts, in process of time, developes itself in the form of a permanent deterioration which ultimately leads to fracture. This was strikingly exemplified in a range of shafts, 220 feet long, tapering from three inches diameter at the driving end, to two inches diameter at the other.

The work done by these shafts was uniform throughout, but it was soon found that the shaft had made nearly 1·16 revolutions at the driven end of the room, before it began to move at the other. The result was a continued series of jerks or accelerated and retarded motion, injurious to the machinery, and destructive to the work it had to perform, It was, moreover, injurious to the shafts, particularly in the middle, where the twist was severely felt, and would have led to rupture, but from the circumstance that they had to be renewed with a stiffer and stronger range.

length and to the square of the angle of torsion for unity of length.

5. For cylindrical bodies Mr. Wertheim gives the following formulæ:—

Let ψ be the mean temporary angle of torsion, for $p = 1$ kilogramme, and $l = 1$ mètre;

$p =$ the sum* of the two weights producing torsion and constituting a couple in kilogrammes;

$R =$ the leverage at which the weight p acts;

$l =$ length of the bar subject to torsion, in millimètres;

$r =$ the exterior radius of the section of the bar, in millimètres;

$r_1 =$ the interior radius of hollow bars, in millimètres;

$E =$ the modulus of elasticity of the material obtained from experiments on tension.

Then, for solid bars:—†

$$\psi = \frac{16}{3} \cdot \frac{180}{\pi^2} \cdot \frac{pR}{E} \cdot \frac{l}{r^4}$$

and for hollow cylinders

$$\psi = \frac{16}{3} \cdot \frac{180}{\pi^2} \cdot \frac{pR}{E} \cdot \frac{l}{r^4 - r_1^4}$$

In the following experiments, $p = 1$ kilogramme, $R = 247{\cdot}5$ millimètres, $l = 1000$ millimètres.

RÉSUMÉ OF EXPERIMENTS ON CYLINDERS OF CIRCULAR SECTION.

—	Material.	Radius r.	Coefficient of elasticity, E.	Mean angle of torsion. By formula.	Mean angle of torsion. By experiment.
		mm.		° ′ ″	° ′ ″
1	Iron	8·220	17,805	0 17 46·1	0 17 52·1
2	Iron	5·501	„	1 28 0·8	1 26 31·3
3	Cast-steel . . .	5·055	19,542	1 53 12·0	1 51 13·4
4	Copper . . .	5·031	9,395	3 59 59·1	3 54 6·0
5	Glass	3·535	6,200	24 51 56·0	24 15 34·7
6	Glass	3·4225	„	28 18 2·0	28 30 14·0

* In Mr. Wertheim's experiments equal weights, acting in opposite directions at the same leverage were hung one on each side of the bar, subjected to torsion.

† The above formulæ may be used with English measures, E being taken from English tables, if p be given in lbs. and r, l, and R in inches.

RÉSUMÉ OF EXPERIMENTS ON THE TORSION OF HOLLOW CYLINDERS OF COPPER.

	External radius (r).	Internal radius (r_1).	Coefficient of elasticity from tension (E).	Angle of torsion (ψ).	
				By formula.	By experiment.
				° ′ ″	° ′ ″
53	11,525	10,021	10,917	0 17 30·2	0 20 0·6
54	7,082	4,955	10,444	1 12 18·3	1 16 52·9
55	6,047	30,315	10,276	4 9 4·0	4 6 54·7
7	5,602	24,665	9,665	2 37 40·4	2 33 38·2
8	45,605	2,478	9,855	6 11 10·3	6 0 53·8
9	36,955	2,471	10,645	15 9 14·4	15 42 37·3

The accordance, in these tables, between the formulæ and the experiments is very satisfactory, especially considering that the value of E cannot be determined with perfect accuracy. The errors do not generally exceed $\frac{1}{50}$th, and the observed angles are smaller than those found by calculation, except in the case of the cylinders 9, 53, and 54.

For bars of elliptical section M. Wertheim has deduced the formula

$$\psi = \frac{8}{3} \cdot \frac{180}{\pi^2} \cdot \frac{p\text{R}}{\text{E}} \cdot \frac{l(c_1^2 + c_2^2)}{c_1^3 c_2^3}$$

where c_1 and c_2 are the two semiaxes of the ellipse, the other letters remaining as before.

RÉSUMÉ OF EXPERIMENTS ON THE TORSION OF ELLIPTICAL BARS.

	Material.	Semiaxes.		Coefficient of elasticity by tension (E).	Mean angle of torsion (ψ).	
		c_1.	c_2.		By formula.	By experiment.
		mm.	mm.		° ′ ″	° ′ ″
11	Cast steel . .	7,105	3,697	19,085	2 13 56·7	2 10 55·4
12	,, . .	9,900	25,075	,,	4 18 0·1	4 13 18·2
13	Copper . . .	7,062	3,669	9,634	4 32 56·7	4 30 41·2
14	,, . . .	9,875	2,498	,,	8 38 11·2	8 54 33·9

For bars of rectangular section the formula becomes

$$\psi = \frac{180}{\pi^2} \cdot \frac{1}{2} \cdot \frac{p\text{R}}{\text{E}} \cdot \frac{l(a^2 + b^2)}{a^3 b^3}$$

But it is necessary to apply a coefficient of correction c to the calculated angle such that if ψ_1 be the calculated angle of torsion, and ψ_2 the angle found by experiment, then $c=\frac{\psi_1}{\psi_2}$. This coefficient varies with the ratio $\frac{a}{b}$ of the sides of the bar; thus, when $l=500$ millimètres, and the section was 36 millimètres square.

$\frac{a}{b}$	1	2	4	8
Value of coefficient	0·8971	0·9617	0·9520	0·9878

It varies also with the ratio $\frac{l}{b}$ and with the moment of the couple $p\,\mathrm{R}$.

For the ultimate resistance of cylindrical shafts to rupture by torsion, Professor W. J. M. Rankine gives the following formula:*

Let l denote the length in inches of the lever, such as a crank, at the end of which a wrenching or twisting force is applied to an axle. Let W be the working load in pounds, multiplied by a suitable factor of safety (usually six); then

$$\mathrm{W}l=\mathrm{M}$$

is the wrenching moment in inch pounds.

For a solid axle let h be its diameter; then

$$\mathrm{M}=\frac{fh^3}{5{\cdot}1} \text{ and } h=\sqrt[3]{\frac{5{\cdot}1\,\mathrm{M}}{f}}$$

For a hollow axle let h_1 be the external, and h_0 the internal diameter in inches; then

$$\mathrm{M}=\frac{f(h_1^4-h_0^4)}{5{\cdot}1\,h_1}=\frac{fh_1^3}{5{\cdot}1}\cdot\left(1-\frac{h_0^4}{h_1^4}\right)$$

$$\text{and } h_1=\sqrt[3]{\left\{\frac{5{\cdot}1\,\mathrm{M}}{f\left(1-\frac{h_0^4}{h_1^4}\right)}\right\}}$$

The values of the modulus of wrenching f are—

for cast iron	about	30000,
for wrought iron	„	54000,

* Manual of Applied Mechanics, p. 355. Manual of Steam Engine, p. 78.

and taking six as the factor of safety, if we put the working moment of torsion in the formulæ instead of the wrenching moment, we may put instead of f

for cast iron 5000,
for wrought iron . . . 9000.

Hence we get for W, the working stress, with solid shafts,

$$W_1 = \frac{5000\, h^3}{5{\cdot}1\, l} = \frac{980\, h^3}{l} \text{ for cast iron} \quad . \quad . \quad . \quad (2.)$$

$$= \frac{9000\, h^3}{5{\cdot}1\, l} = \frac{1765\, h^3}{l} \text{ for wrought iron.} \quad . \quad . \quad (3.)$$

On this principle I have calculated the following tables (pages 68, 69), giving the safe moment of torsion for cylindrical cast and wrought iron shafts, and also the working stress to which they may be subjected at the circumference of pullies or wheels of various diameters. In cases where the horses' power transmitted by a shaft is given instead of the stress, the latter may be found by the table on page 47.

The greatest angle of torsion, which it is safe to allow in a line of shafting, is determined by the extension of the material within the elastic limits. If $\frac{1}{1200}$th of the length be assumed as the maximum extension with the safe working load, then the shaft must be so proportioned that the angle of torsion is less than that given by the following formula

$$\psi = \frac{2284\, L}{1000\, d} \quad . \quad . \quad . \quad . \quad . \quad . \quad (4.)$$

where L is the length of the shaft in feet, d its diameter in inches, and ψ the angle of torsion in degrees.

It is convenient to estimate the ultimate resistance of shafts to torsion, not only as a statical pressure acting at a leverage, but also in horses' power. Now the stress resulting from the transmission of power must evidently increase in proportion to the power, and decrease in proportion to the velocity. A shaft will transmit 100 horses' power at 80 revolutions a minute with no more stress than it would transmit 50 horses' power at 40 revolutions, or 25 horses' power at 20 revolutions. Hence the

TABLE 6.—SAFE WORKING TORSION FOR CAST-IRON SHAFTS.

Diameter of Shaft in inches.	($W_1 l$) Working moment of torsion in inch lbs.	Working Stress in lbs. at the following Radii, or W_1.										
		6 ins.	9 ins.	12 ins.	15 ins.	18 ins.	21 ins.	24 ins.	27 ins.	30 ins.	42 ins.	60 ins.
1	980	163	109	81	64	54	46	41	36	32	23	16
1½	3,309	552	368	276	220	184	158	138	122	110	79	55
2	7,843	1,307	871	653	522	435	372	326	290	261	186	131
2½	15,320	2,553	1,702	1,276	1,020	851	730	638	567	510	365	255
3	26,470	4,411	2,941	2,206	1,764	1,470	1,260	1,103	980	882	630	441
3½	42,030	7,005	4,670	3,502	2,802	2,335	2,002	1,751	1,556	1,401	1,001	701
4	62,740	10,457	6,971	5,228	4,182	3,485	2,988	2,614	2,323	2,091	1,494	1,046
4½	89,350	14,891	9,928	7,445	5,956	4,964	4,254	3,722	3,309	2,978	2,127	1,489
5	122,550	20,425	13,617	10,212	8,170	6,809	5,834	5,106	4,539	4,085	2,917	2,043
5½	163,110	27,185	18,123	13,592	10,874	9,061	7,766	6,796	6,041	5,437	3,883	2,718
6	211,760	35,293	23,528	17,646	14,116	11,764	10,084	8,823	7,842	7,058	5,042	3,529
6½	269,240	44,873	29,916	22,436	17,950	14,958	12,820	11,218	9,972	8,975	6,410	4,487
7	336,280	56,047	37,364	28,023	22,418	18,682	16,012	14,012	12,454	11,209	8,006	5,605
7½	413,600	68,933	45,955	34,466	27,572	22,977	19,696	17,233	15,318	13,786	9,848	6,893
8	501,970	83,661	55,774	41,830	33,464	27,887	23,904	20,915	18,591	16,732	11,952	8,366
8½	602,100	100,350	66,900	50,175	40,140	33,450	28,672	25,088	22,300	20,070	14,336	10,035
9	714,720	119,120	79,413	59,560	47,648	39,706	34,034	29,780	26,471	23,824	17,017	11,912
9½	840,570	140,095	93,396	70,047	56,038	46,698	40,026	35,024	31,132	28,019	20,013	14,010
10	980,000	163,333	108,888	81,666	65,332	54,444	46,666	40,833	36,296	32,666	23,333	16,333
11	1,304,900	217,483	144,988	108,741	86,994	72,494	62,138	54,371	48,329	43,497	31,069	21,748
12	1,694,100	282,350	188,233	141,175	112,940	94,116	80,672	70,587	62,744	56,470	40,336	28,235
13	2,154,000	359,000	239,333	179,500	143,600	119,666	102,570	89,750	79,777	71,800	51,285	35,900
14	2,690,200	448,366	298,911	224,183	179,346	149,455	128,104	112,091	99,637	89,673	64,052	44,837
15	3,308,800	551.466	367,544	275,733	220,586	183,772	157,562	137,866	122,514	110,293	78,781	55,147
16	4,015,700	669,283	445,200	334,641	267,712	222,600	191,224	167,321	148,400	133,856	95,612	66,928
17	4,816,800	802,800	535,200	401,400	321,120	267,600	229,370	200,700	178,400	160,560	114,685	80,280
18	5,718,000	953,000	635,333	476,500	381,200	317,666	272,286	238,250	211,777	190,600	136,143	95,300
19	6,724,600	1,120,766	747,177	560,383	448,306	373,588	320,218	280,191	249,059	224,153	160,109	112,077
20	7,843,000	1,307,166	871,444	653,583	522,866	435,722	375,762	326,791	290,481	261,433	186,881	130,717

TABLE 7.—SAFE WORKING TORSION FOR WROUGHT-IRON SHAFTS.

Diameter of Shaft in inches.	(W_1 *l*.) Working moment of torsion in inch lbs.	Working Stress in lbs. at the following Radii.										
		6 ins.	9 ins.	12 ins.	15 ins.	18 ins.	21 ins.	24 ins.	27 ins.	30 ins.	42 ins.	60 ins.
1	1,765	294	196	147	118	98	84	73	65	59	42	29
1½	5,956	923	662	496	396	331	284	248	221	198	142	99
2	14,120	2,353	1,569	1,176	942	784	672	588	523	471	336	235
2½	27,570	4,595	3,063	2,297	1,838	1,531	1,312	1,148	1,021	919	656	460
3	47,650	7,942	5,294	3,971	3,176	2,647	2,270	1,985	1,764	1,588	1,135	794
3½	75,660	12,610	8,406	6,305	5,044	4,203	3,602	3,152	2,802	2,522	1,801	1,261
4	112,940	18,823	12,549	9,411	7,530	6,274	5,378	4,705	4,183	3,765	2,689	1,882
4½	160,805	26,801	17,867	13,400	10,720	8,933	7,658	6,700	5,956	5,360	3,829	2,680
5	220,590	36,765	24,510	18,382	14,706	12,255	10,504	9,191	8,170	7,353	5,252	3,676
5½	293,600	48,933	32,622	24,466	19,572	16,311	13,980	12,233	10,874	9,786	6,990	4,893
6	381,180	63,530	42,353	31,765	25,412	21,176	18,152	15,882	14,117	12,706	9,076	6,353
6½	484,630	80,771	53,848	40,385	32,308	26,924	23,078	20,192	17,949	16,154	11,539	8,077
7	605,290	100,881	67,254	50,440	40,352	33,627	28,822	25,220	22,418	20,176	14,411	10,088
7½	744,480	124,080	82,720	62,040	49,632	41,360	35,452	31,020	27,573	24,816	17,726	12,408
8	903,530	150,588	100,392	75,294	60,234	50,196	43,024	37,647	33,464	30,117	21,512	15,059
8½	1,083,800	180,633	120,422	90,316	72,254	60,211	51,610	45,158	40,140	36,127	25,805	18,063
9	1,286,500	214,416	142,944	107,208	85,766	71,472	61,262	53,604	47,648	42,883	30,631	21,442
9½	1,513,000	252,166	168,111	126,083	100,866	84,055	72,048	63,041	56,037	50,433	36,024	25,217
10	1,765,000	294,166	196,111	147,083	117,666	98,055	84,048	73,541	65,370	58,833	42,024	29,417
11	2,348,800	391,466	260,966	195,733	156,586	130,483	111,848	97,866	86,988	78,293	55,924	39,147
12	3,049,400	508,233	338,822	254,116	203,292	169,411	139,496	127,058	112,941	101,646	69,748	50,823
13	3,877,100	646,123	430,788	323,091	258,474	215,394	184,624	161,545	143,596	129,237	92,312	64,618
14	4,842,300	807,050	538,033	403,525	322,820	269,016	230,586	201,762	179,344	161,410	115,293	80,705
15	5,955,800	992,633	661,755	496,316	397,054	330,877	283,610	248,158	220,585	198,527	141,805	99,263
16	7,228,200	1,204,700	803,133	602,350	481,880	401,566	344,200	301,175	267,711	240,940	172,100	120,470
17	8,670,100	1,445,016	963,366	722,508	578,006	481,683	412,862	361,254	321,122	289,003	206,431	144,501
18	10,292,000	1,715,333	1,143,555	857,666	686,132	571,777	490,096	428,833	381,185	343,066	245,048	171,533
19	12,104,000	2,017,333	1,344,888	1,008,666	806,932	672,444	576,380	504,333	448,296	403,466	288,190	201,733
20	14,120,000	2,353,333	1,568,888	1,176,666	941,332	784,444	672,380	588,333	522,962	470,666	336,190	235,333

torsion varies as $\frac{\text{H}}{\text{R}}$, where H is the number of horses' power per minute, and R the number of revolutions per minute.

Buchanan's rules for the power transmitted by shafts are*:—

For fly-wheel shafts

$$d = \sqrt[3]{\left\{\frac{\text{H}}{\text{R}} \times 400\right\}}$$

For shafts of waterwheel gearing and other heavy work,

$$d = \sqrt[3]{\left\{\frac{\text{H}}{\text{R}} \times 200\right\}}$$

For shafts of ordinary mill gearing

$$d = \sqrt[3]{\left\{\frac{\text{H}}{\text{R}} \times 100\right\}}$$

An ordinary allowance for wrought iron shafting in practice is

$$d = \sqrt[3]{\left\{\frac{\text{H}}{\text{R}} \times 250\right\}} \quad . \quad . \quad . \quad . \quad . \quad (5.)$$

From the foregoing observations in regard to torsion, and the power of transmission of shafts at different velocities, it is a desideratum of much importance to the engineer, so to proportion shafts in relation to their lengths as well as velocities, as to be within the limits of sensible permanent torsion and flexure,† and at the same time to increase the speeds in a given ratio to the velocities of the machine and the nature of the work it has to execute. In the above disquisition we have only given the law and the safe measure of torsion as regards length and area, but much must still depend on the calculation and judgement of the millwright and engineer; in its application to the character of the work they have to perform, and the resistances they have to overcome.

From formula (5.) the following table (page 71) has been calculated, giving the diameter necessary to transmit from 1 to 150 horses' power at from 10 to 1000 revolutions per minute.

* These rules will be found in the second edition of Buchanan, at pages 328, *et seq.*

† Although we speak of the limits of permanent torsion, we are not prepared to fix these limits, as we find that what produces a permanent set in any material, however minute a fraction it may be, will in process of time, if continued, and often repeated, lead to fracture. This law applies to every description of strain or material, and we may therefore consider that there are limits to endurance, however distant that may be.

TABLE 8. — DIAMETER OF WROUGHT IRON SHAFTING NECESSARY TO TRANSMIT WITH SAFETY VARIOUS AMOUNTS OF FORCE.

Horses' power safely transmitted.	Diameter of Shaft in inches, with the following revolutions per minute.											
	10	25	50	75	100	150	200	250	300	400	500	1000
1	2·92	2·15	1·71	1·50	1·36	1·20	1·09	1·00	·94	·85	·79	·63
2	3·68	2·71	2·15	1·91	1·71	1·49	1·36	1·26	1·20	1·09	1·00	·79
3	4·22	3·11	2·47	2·15	1·95	1·71	1·54	1·44	1·36	1·23	1·14	·91
4	4·64	3·42	2·71	2·35	2·15	1·91	1·71	1·36	1·49	1·36	1·26	1·00
5	5·00	3·68	2·92	2·57	2·32	2·00	1·84	1·71	1·61	1·45	1·36	1·09
10	6·30	4·64	3·68	3·21	2·92	2·57	2·11	2·15	2·03	1·84	1·71	1·36
15	7·21	5·31	4·22	3·68	3·35	2·92	2·66	2·46	2·32	2·21	1·95	1·55
20	7·94	5·85	4·64	4·06	3·68	3·21	2·92	2·71	2·56	2·32	2·15	1·71
30	9·09	6·69	5·31	4·64	4·22	3·68	3·34	3·11	2·92	2·66	2·46	1·95
40	10·00	7·37	5·85	5·10	4·64	4·06	3·68	3·42	3·21	2·92	2·71	2·15
50	—	7·94	6·30	5·51	5·00	4·36	3·96	3·68	3·46	3·14	2·92	2·32
60	—	8·43	6·69	5·85	5·31	4·64	4·22	3·91	3·68	3·35	3·11	2·46
70	—	8·88	7·05	6·15	5·59	4·89	4·44	4·12	3·88	3·52	3·27	2·60
80	—	9·28	7·37	6·44	5·84	5·10	4·64	4·31	4·05	3·68	3·42	2·71
90	—	9·65	7·66	6·69	6·08	5·31	4·83	4·48	4·22	3·83	3·56	2·82
100	—	10·00	7·94	6·93	6·30	5·51	5·00	4·64	4·36	3·96	3·68	2·92
110	—	—	8·19	7·16	6·50	5·67	5·16	4·79	4·51	4·10	3·80	3·01
120	—	—	8·43	7·37	6·69	5·85	5·31	4·93	4·64	4·22	3·91	3·11
130	—	—	8·66	7·57	6·88	6·01	5·46	5·06	4·76	4·33	4·02	3·19
150	—	—	9·09	7·94	7·21	6·30	5·72	5·31	5·00	4·55	4·22	3·35

4. *Velocity of Shafts.*

As the quality of the material employed for the construction of shafts enters largely into the calculation of their strength, so also the velocity at which they revolve becomes an important element in the calculation of the work transmitted by them. In all cases where machinery has to be driven at a high speed, it is advantageous and even essential to run the shafting at a proportionate velocity. If, for example, there are a series of machines running at 500 revolutions per minute, it will be advisable to run the shafts at half that speed, by which means the following very important advantages will be gained.

There will be a great saving in the weight of the shafts, for with a slow motion of 50 revolutions per minute, fully three times the weight would be necessary to transmit the same power. There would also be a saving in original cost in the power absorbed, and in maintenance.

Shafts running at low velocities are cumbersome, heavy, and expensive to repair. They are costly in the first instance, and they block up the rooms of the mill with large drums and pullies, obstructing the light, which, in factories, is a consideration of very great importance.

At the commencement of the present century mills were geared with ponderous shafts, such as those just described. They were generally of cast iron, square, and badly coupled, and the power required to keep them in motion was in some cases almost equal to that required by the machinery they had to drive. In the present improved system, with light shafts accurately fitted and running at high velocities, the work which previously was absorbed in transmission is now conveyed to the machinery of the mill.

I may safely ascribe my own success in life and that of my friend and late partner, Mr. James Lillie, to the saving of power effected by increasing threefold the velocity of the shafting in mills more than forty years ago. The introduction of light iron shafting not only enabled the manufacturer to effect a considerable saving in the original cost, but a still greater saving was effected in power, whilst it relieved the mills from the ponderous wooden drums and heavy shafting then in use, and established an entirely new system of operations in the machinery of transmission.

5. *Length of Journals.**

Another consideration of considerable importance to the smooth and safe working of shafting is the length of the journals. From a number of years' experience I have been led to believe, that with cast iron, one and a half times the diameter of the shaft is the best proportion for the length of the bearing, and with wrought iron, one and three quarters the diameter. On the question of shafts revolving in the steps of plummer blocks and the proportions necessary to effect motion without danger of heating, it is essential (without entering largely into the laws of friction on bodies in contact) that we should ascertain from actual practice and long-tried experience the best form of journals of shafts adapted for that purpose. The lengths proportionate to the diameters have already been given, but we have yet to consider the dimensions of the journals of large shafts where they are small in comparison with the pressure or the weight they have to sustain. Let us, for example, take a fly-wheel shaft and the foot or toe of a line of vertical shaft extending to a height of six or seven stories in a mill filled with machinery, and we have the safe working pressure per square inch as indicated in the last column in the following table:—

Description of Shaft.	Length and diameter of Shaft in ins.	Number of square inches in bearing.	Weight on bearing in lbs.	Weight in lbs. per square inch on bearing.
Fly-wheel shaft wrought iron	18 × 14	252	45,024	178·21
Vertical shaft cast iron .	— × 11	95	23,061	242·70
Horizontal shaft cast iron .	15 × 10	150	6,000	40·00
Horizontal shaft wrought iron	6 × 3	18	540	30·00
Ditto ditto ditto .	2 × 4	8	160	20·00

From the above it will be seen that in fly-wheel shafts the

* Rules for the diameters of gudgeons or journals for those cases in which they are calculated independently of the diameter of the shaft, are given in Mills and Millwork, vol. i. p. 116.

pressure should never exceed 180 lbs. per square inch, and in that of the toes of vertical shafts 240 lbs. per square inch. Even with this latter pressure it is difficult to keep the shafts cool, and it requires the greatest possible care to keep them free from dust or any minute particles of sand or other sharp substances getting into the steps. The feet of vertical shafts also require the very best quality of gun metal for the shaft to run in, and fine limpid oil for lubrication to prevent the toe from cutting. It is, moreover, necessary for the shaft to fit well on the bottom of the step, and not too tight on the sides, and to have a fine polish.

Fig. 206.

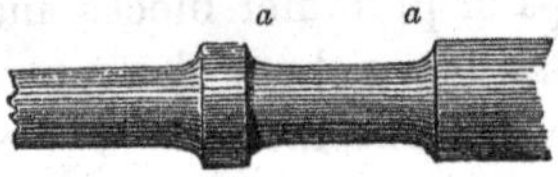

Another point for consideration is the proper form of the journals of shafts, and that is, they should never have the journal turned or cut square down to the diameter, but hollowed in the form shown in the figure at *a a a a*. From a series of interesting experiments it has been shown that the square-cut shaft loses nearly $\frac{1}{5}$ of its strength, and by simply curving out the shaft at the collars in the form described, the resistance to strain is increased $\frac{1}{5}$ or in that proportion.

6. *Friction.*

On the subject of friction much cannot be said. We may, however, adduce a few experiments from Morin and Rivière, which appear to bear out our previous experience of the length of journals.

In the years 1831, 1832, and 1833, a very extensive set of experiments were made at Metz by M. Morin, under the sanction of the French Government, to determine, as nearly as possible, the laws of friction, and by which the following were fully established: —

When no unguent is interposed, the friction of any two surfaces, whether of quiescence or of motion, is directly proportional to the force with which they are pressed perpendicularly together; so that for any two given surfaces of contact there is a constant ratio of the friction to the perpendicular pressure of the one surface upon the other. Whilst this ratio is thus the same for the same surfaces of contact, it is different for different surfaces of contact. The particular value of it in respect to any two given surfaces of contact, is called the coefficient of friction in respect to those surfaces.

When no unguent is interposed, the amount of the friction is, in every case, wholly independent of the extent of the surfaces of contact; so that the force with which two surfaces are pressed together, being the same, their friction is the same, whatever be the extent of their surfaces of contact.

That the friction of motion is wholly independent of the velocity of the motion.

That where unguents are interposed, the coefficient of friction depends upon the nature of the unguent, and upon the greater or less abundance of the supply. In respect to the supply of the unguent, there are two extreme cases, — that in which the surfaces of contact are but slightly rubbed with the unctuous matter, as, for instance, with an oiled or greasy cloth, and that in which a continuous stratum of unguent remains continually interposed between the moving surfaces; and in this state the amount of friction is found to be dependent rather upon the nature of the unguent than upon that of the surfaces of contact. M. Morin found that with unguents (hog's lard and olive oil) interposed in a continuous stratum between surfaces of wood on metal, wood on wood, and metal on metal, when in motion, have all of them very nearly the same coefficient of friction, being in all cases included between ·07 and ·08. The coefficient for the unguent tallow is the same, except in that of metals upon metals. This unguent appears to be less suited for metallic surfaces than the others, and gives for the mean value of its coefficient under the same circumstances ·10. Hence it is evident that where the extent of the surface sustaining a given pressure is so great as to make the pressure less than that which corresponds to a state of perfect separation, this greater extent of surface tends to increase the friction by reason of that adhesiveness of the unguent, dependent upon its greater or less viscosity, whose effect is proportional to the extent of the surfaces between which it is interposed.

Mr. G. Rennie found, from a mean of experiments with different unguents on axles in motion, and under different pressures, that with the unguent tallow, under a pressure of from 1 to

5 cwt., the friction did not exceed $\frac{1}{39}$th of the whole pressure; when soft soap was applied it became $\frac{1}{34}$th; and with the softer unguents applied, such as oil, hog's lard, &c., the ratio of the friction to the pressure increased; but with the harder unguents, as soft soap, tallow, and anti-attrition composition, the friction considerably diminished: consequently, to secure effective lubrication, the nature of the unguent must be accommodated to the pressure or weight tending to force the surfaces together.

Table of Coefficients of Friction under Pressures increased continually up to Limits of Abrasion. By Mr. G. Rennie.

Pressures per Square Inch.	Coefficients of Friction.			
	Wrought Iron upon Wrought Iron.	Wrought Iron upon Cast Iron.	Steel upon Cast Iron.	Brass upon Cast Iron.
32·5 lbs.	·140	·174	·166	·157
1·66 cwts.	·250	·275	·300	·225
2·00 „	·271	·292	·333	·219
2·33 „	·285	·321	·340	·214
2·66 „	·297	·329	·344	·211
3·00 „	·312	·333	·347	·215
3·33 „	·350	·351	·351	·206
3·66 „	·376	·353	·353	·205
4·00 „	·395	·365	·354	·208
4·33 „	·403	·366	·356	·221
4·66 „	·409	·366	·357	·223
5·00 „	—	·367	·358	·233
5·33 „	—	·367	·359	·234
5·66 „	—	·367	·367	·235
6·00 „	—	·376	·403	·233
6·33 „	—	·434	—	·234
6·66 „	—	—	—	·235
7·00 „	—	—	—	·232
7·33 „	—	—	—	·273

From a paper lately read at the Institution of Civil Engineers in London, on the comparative friction of steam engines of different modifications, it appears that, as respects the friction caused by the strain, if the beam engine be taken as the standard of comparison—

The vibrating engine . .	has a gain of	1·1	per cent.
The direct engine with slides	„ loss of	1·8	„
Ditto with rollers . .	„ gain of	0·8	„
Ditto with a parallel motion	„ gain of	1·3	„

It also states, as an opinion, that excessive allowance for friction has hitherto been made in calculating the effective power of engines in general; as it is found practically by experiments with the engines at the Blackwall Railway, and also with other engines, that where the pressure upon the piston is about 12 lbs. per square inch, the friction does not amount to more than $1\frac{1}{2}$lbs.; and also that by experiments with an indicator on an engine of 50 horse-power, at Truman, Hanbury, and Co.'s brewery, the whole amount of friction did not exceed 5 horse-power, or $\frac{1}{10}$th of the whole power of the engine.

7. *Lubrication.*

On this question it is necessary to observe that the durability of shafts, and their easy working, depends on the way in which they are lubricated, and the description of unguent used for that purpose. We have already seen the difference which exists in the coefficient of friction from the use of different kinds of unguents, and we have now to consider what system of lubrication should be adopted to lessen the friction and maintain smooth surfaces on the journals of shafts. In large cotton mills I have known as much as ten to fifteen horses'-power absorbed by a change in the quality of the oil used for lubrication; and in cold weather, or when the temperature of the mill is much reduced (as is generally the case when standing over Sunday), the power required on a Monday morning is invariably greater than at any other time during the week.

It is, therefore, necessary in most mills—particularly those employed in textile manufacture—to retain a uniform temperature, and to employ the best quality of oil for lubricating the machinery, as well as the shafts of the mill.

The best lubricators are pure sperm and olive oils; they should be clean and limpid, and sparingly applied, as it is a profligate waste of valuable material to pour, as is not unfrequently done, large quantities of oil on the bearings, nine-tenths of which run on to the floor, and cover the shafts and hangers with a coat of glutinous matter, that soon hardens, and accumulates nothing but filth.

This process of oiling shafts is generally left to the most negligent and most untidy person in the establishment; and the result is, that every opening for the oil to get to the bearings is plugged up, the brass steps are cut by abrasion, and the necks or journals of the shafts destroyed. In the best regulated establishments this is certainly not the case, as the greatest possible care is observed in selecting the best kinds of oil, and that used with attention to cleanliness and strict economy in its application.

To save power and effect economy in the use of lubricants, several schemes have been adopted for attaining a continuous system of lubrication. None of them appears to answer so well as that which consists of a small cistern, *a*, fig. 207, which contains a quantity of oil, and is fixed on the top of the plummer block. In the centre of the cistern is a tube, which stands a little above the level of the oil; and into this is inserted a woollen thread, with its end descending a short distance below the surface of the oil in the cistern; and when properly saturated, the oil rises by capillary attraction, and flows gently, in very minute quantities, on to the neck of the shaft. From this description it will be seen that the quantity used can be regulated to the greatest nicety, and sufficient to lubricate the bearings without waste. Other plans have been devised for the same object, but none of them seems to answer so well as that just described.

Fig. 207.

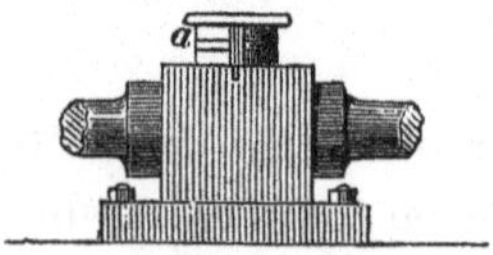

CHAPTER III.

ON COUPLINGS FOR SHAFTS AND ENGAGING AND DISENGAGING GEAR.

In every description of mill where the machinery is spread over a large area, and at a distance from the moving power, it is necessary to have long lines of shafting, revolving at the required velocity. Such lines are seldom made in one piece; short lengths must, therefore, be coupled together, so as to form an unbroken line, extending, in most cases, the whole length of the mill.

When cast-iron shafts were substituted for wood, a square coupling-box, made in one piece, was generally used, so as to slide over the two ends of the shafts, or in two pieces, bolted together, as shown in figs. 208 and 209.

Fig. 208.

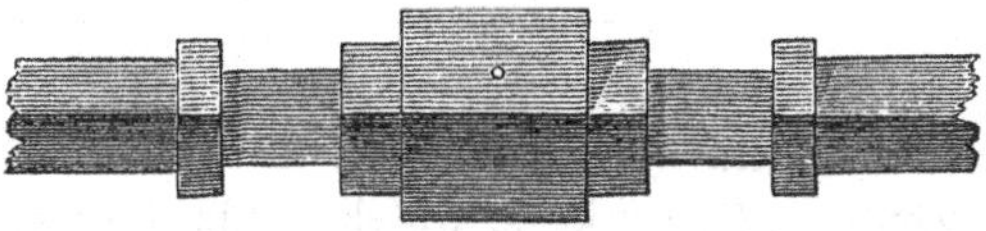

In the former case the box was slipped on loose, and the adjustment was so imperfect that the shafts rose and fell in the box at every revolution, destroying gradually any accuracy of fitting which, in the first instance, had been attained.

Fig. 209.

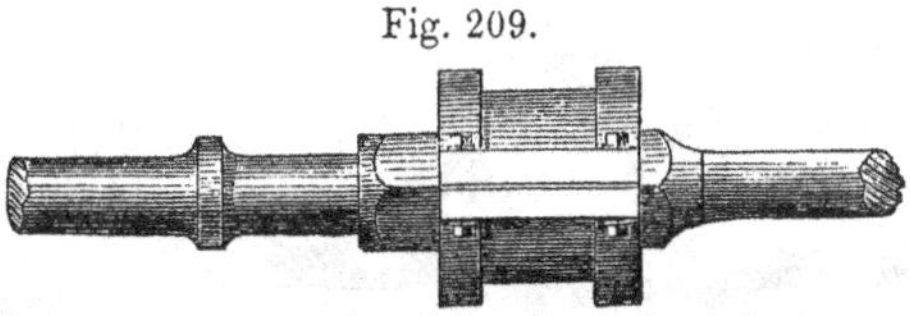

After the square-box coupling came the claw, or two-pronged coupling, made in two parts, wedged, but more frequently keyed on to the ends of the shafts, as shown in fig. 210. This was a great improvement, as the leverage of the bearing parts was

Fig. 210.

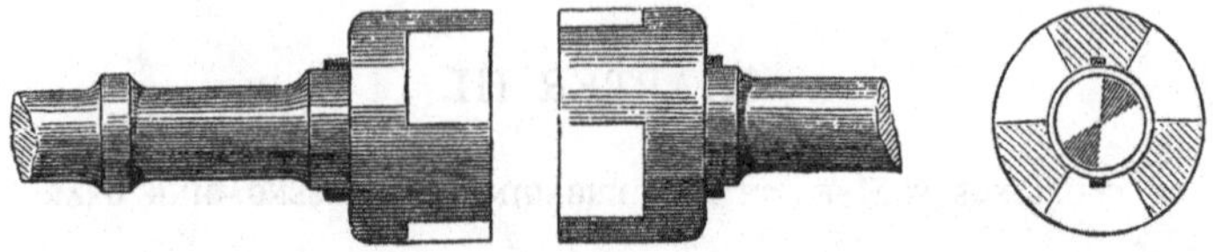

greatly increased, and the coupling, in consequence, became more durable.

A description of half-lap coupling was introduced by the late Mr. Hewes. It was formed by the lapping over a part of the end of each shaft, which was cast square. A square box was also fitted over the two ends, so as to bind them together, and *three* keys were inserted on the top side, as shown in fig. 211.

Fig. 211.

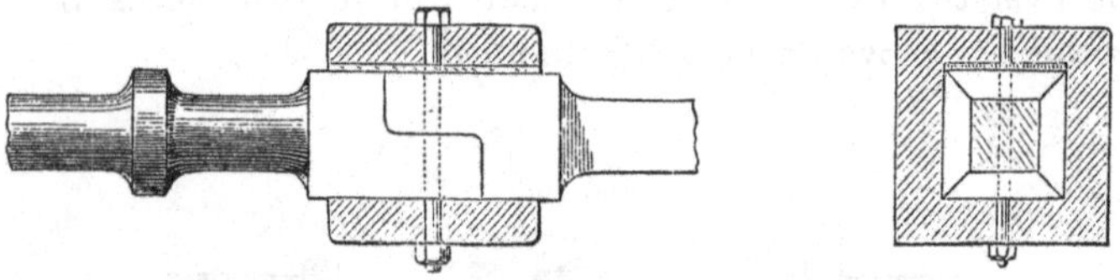

The objections to this coupling were the difficulty of fitting and the loosening of the keys, which made a creaking noise with every revolution of the shaft.

Another coupling, still in use, is the disc. It consists of two discs or flanches, one on the end of each shaft, bolted together by four bolts, as shown in fig. 212. This coupling

Fig. 212.

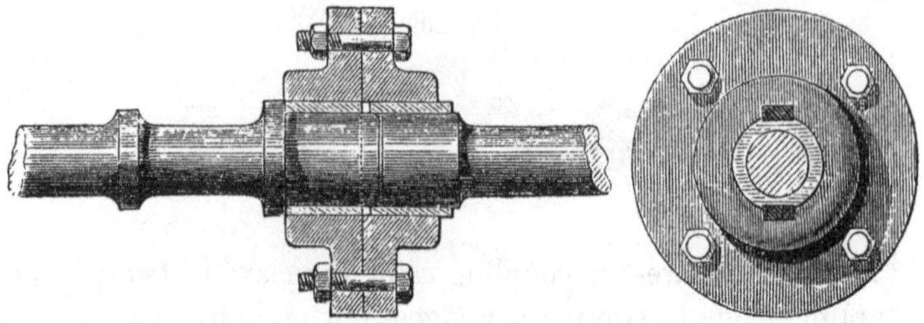

was superior to all the preceding, when properly bored and turned, so as to have its faces accurately perpendicular to the shafting.

The best coupling for general purposes, and the most accurate and durable, is the circular half-lap coupling, introduced into my own works nearly forty years ago. It is perfectly round, and consists of two laps, turned to a gauge, and, when put together by a cutting machine, it forms a complete cylinder, as shown in fig. 213. A cylindrical box is fitted over these, and fixed by a

Fig. 213.

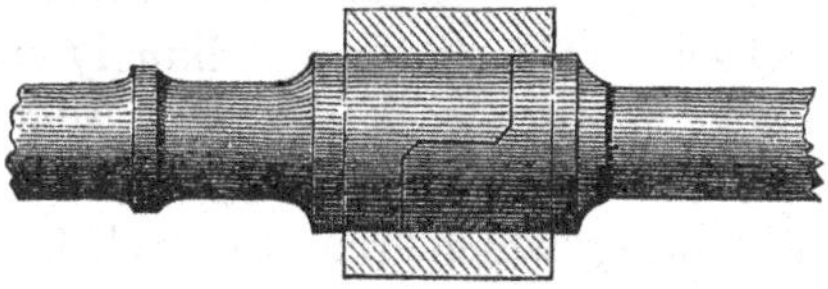

key, grooved half into the box and half into the shaft. The whole is then turned in the lathe to the same centres as the bearings of the shaft, and by this process a degree of accuracy is attained which cannot be surpassed, nor is any other coupling so neat and so well adapted for the transmission of power.

The proportions of this coupling are found by experiment to be—

Twice the area of the shaft is the area of the coupling.

The length of the lap is the diameter of the shaft.

And the length of the box is twice the diameter of the shaft.

These proportions have been found in practice to answer every purpose, both as regards strength and the wear and tear of the joints.

Fig. 214.

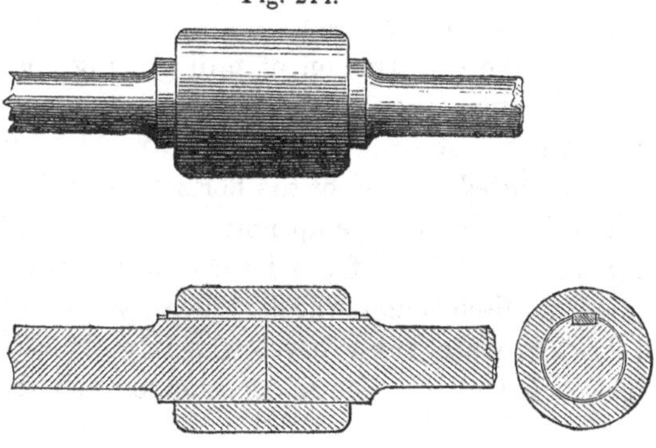

There is another coupling which has come of late years extensively into use, namely, the cylindrical coupling, with butt ends. It has the same proportions as the former, but not so strong nor so durable as the half-lap coupling of the same dimensions, as the entire force of torsion is transmitted through the key; but in cases where strength is not the chief object, it forms a cheap and effective coupling.

8. *Disengaging and re-engaging Gear.*

This is an important branch of mill-work, requiring careful consideration and the utmost exactitude of construction when ponderous machinery has to be started, without endangering the shafts and wheels. This is most strikingly exemplified in the case of powder mills, where trains of edge stones are employed for grinding the gunpowder, and in rolling and callendering machinery, which requires well-fitted friction-clutches to communicate the motion by a slow and progressive acceleration from a state of rest to the required velocity.

It used to be customary in cotton and silk mills to place disengaging clutches at the point of connection of the upright or driving shaft and the main shafting of each room, so that, in case of accident, a room full of machinery could be thrown out of gear at once. But these provisions were found unsteady in practice, and rather tended to increase than to diminish the number of accidents, owing chiefly to the time lost in disengaging, and the breakages which occurred in attempting to place the machinery in gear again, when the engine was running at full speed. It has, consequently, been found safer to have a permanent connection between the main lines of shafting throughout the mill, and signals from each room into the engine-house, in case of accident.

When the construction of mill gearing was less perfect than it is at present, the main shaft driving the machinery in a room was thrown out of gear by a lever, which contained the steps, and supported the end of the horizontal shaft and wheel, which geared into that on the upright shaft, as shown in fig. 215, with a rope at the end of the lever *a* to pull it out of gear. This mode of disengaging wheels was very ineffective, as in many mills there are three bevel wheels gearing into that on the upright *b*, and it becomes complicated and dangerous to have

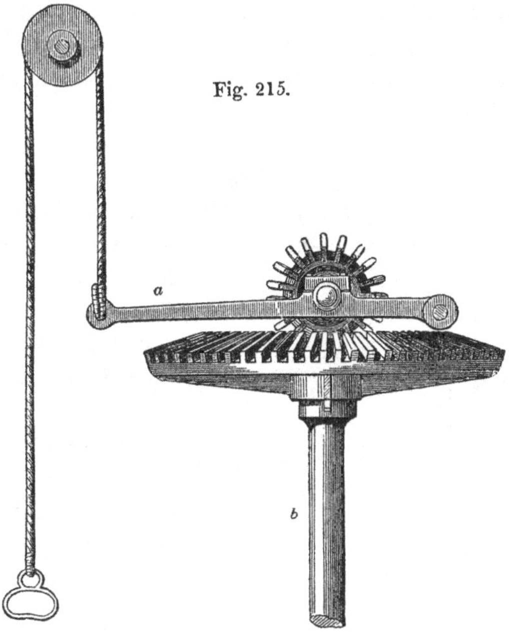

Fig. 215.

movable levers to each. To remedy these defects, standards or plummer-blocks, with a movable slide *e*, fig. 216, in which the end of the shaft revolved, was introduced. To the top

Fig. 216.

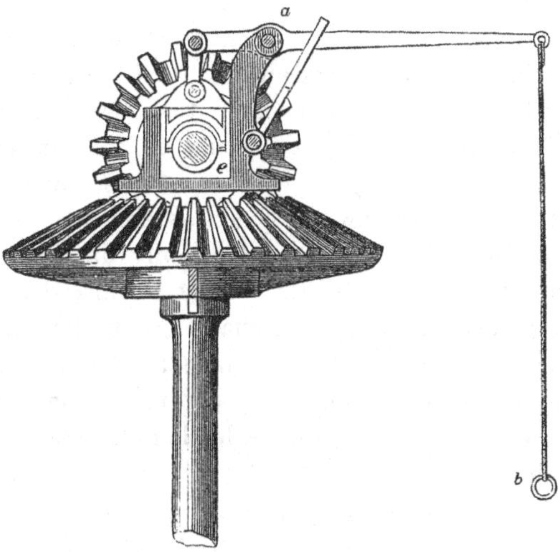

of this slide was attached a lever *a*, with a handle *b*, by which it could be drawn out of gear; and the link *c*, falling along with the lever, retained the shaft out of gear until the mill was stopped.

All these contrivances were, however, found inoperative on a large scale, as the shafts and wheels got out of order; and it was ultimately found essentially necessary to make them stationary, by screwing the plummer-block down to the frame which connects the shafts and wheels.

Fig. 217.

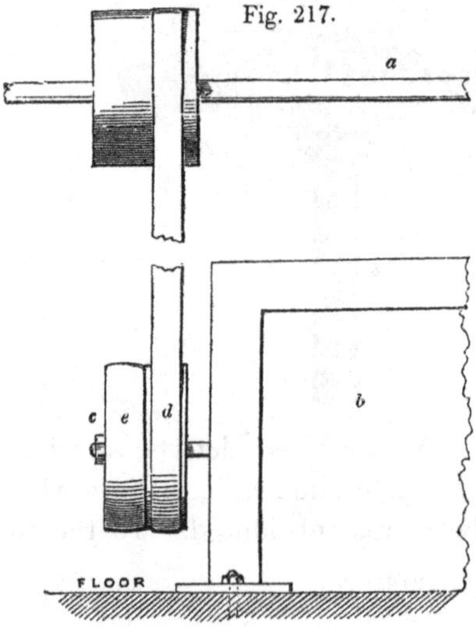

Several devices have been employed for the purpose of rapidly engaging and disengaging machines from the driving shaft. The best of all are the fast and loose pulleys, with a travelling strap. Thus, in fig. 217, *a* is the driving shaft, acting upon two pulleys *e* and *d*, fixed on the driving spindle of the machine *b*; one of them, *d*, is keyed fast, and the other runs loose. When the machine is at work the strap is on the fast pulley *d*, and when it is necessary to stop, it is moved by a forked lever on to the loose pulley *e*, which revolves with the strap without acting on the machine. The machine is thrown into gear with equal ease by moving the strap on to the fast pulley *d*. Once on either of the pulleys, the strap is held in position without any danger of moving by the slight curvature of the pulley, as already explained. The forked lever must act on that side of the strap which runs towards the pulleys, and not on that which leaves them.

A second and equally effective process for starting or stopping machinery is shown in fig. 218. A leather strap is hung loosely over the driving and driven pulleys *a* and *b*, so that, left to itself, the friction is not sufficient to communicate motion to the

Fig. 218.

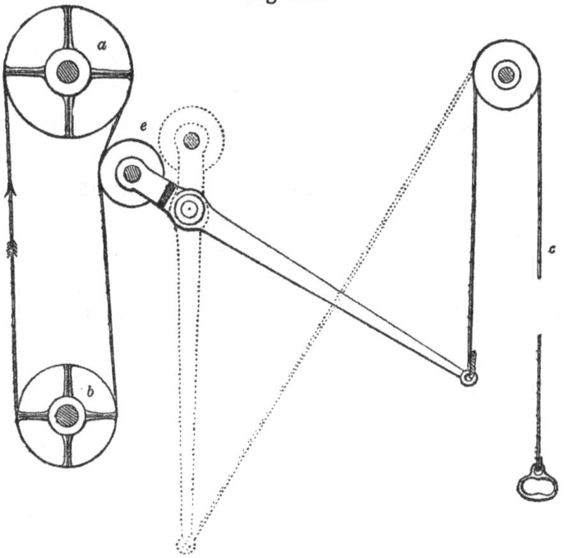

pulley on the shaft *b*; but a tightening pulley fixed on a suitable lever *e* is forced against it by pulling the rope *c*, which bends the strap tightly upon the pulley *b*, and gives motion to the machine. This arrangement is in general use for sack teagles in corn mills, and for some other purposes. The same effect is sometimes produced by the sack teagles being fixed on the lever, and, by raising one end, the strap is tightened, and the barrel which raises the load is caused to revolve.

Fig. 219.

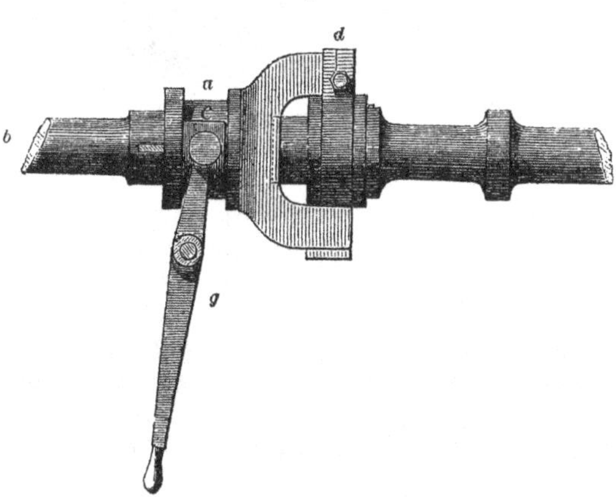

The clutch most in use for throwing into gear heavy calendering machines is a clip friction hoop, which consists of a sliding box *a*, with two projecting horns on the driving shaft *b*. These horns, when slid forward by a lever *g*, working in the groove *c*, come in contact with the friction hoop *d*, which embraces a groove in a second box, keyed upon the shaft of the machine. The instant the machine receives the shock of engagement, the clip *d* slides in its groove, until the friction overcomes the resistance, and the callender attains the speed of the driving shaft. The object of the friction clip is to reduce the shock of throwing the clutch suddenly into gear, as without this precaution any attempt to move instantaneously a powerful machine from a state of rest to a state of motion would break it in pieces.

Friction cones are also much used for this purpose, and when carefully executed with the proper angle are safer than the clutch just described. The objection to the friction clutch is, that the whole driving power is thrown on the clip at once; whereas, with the cones, the parts can be brought into contact with the greatest nicety, and the friction regularly increased to any degree of pressure. Fig. 220 shows this description of disen-

Fig. 220.

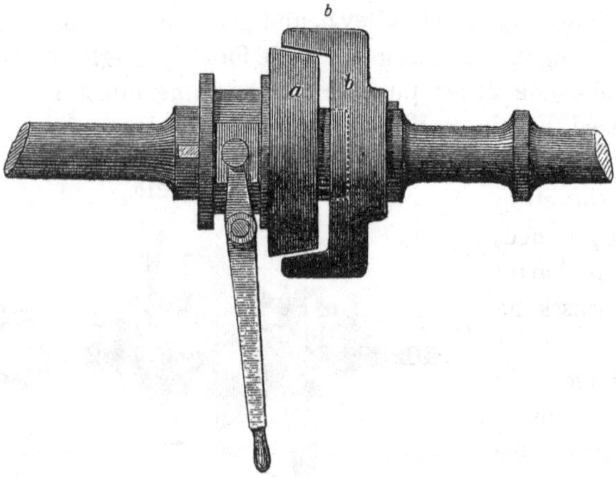

gaging gear: *a* is the male sliding cone, worked by a lever in the usual way, *b* the female cone, keyed on the driven shaft, and the two surfaces, when brought into contact, communicate the required motion with perfect safety.

Machines driven by friction, and requiring to be frequently

stopped, are very numerous. Some of the lighter description are driven by a vertical shaft, *b*, fig. 221, supporting a horizontal disc, which communicates motion to the wheel *a*, rolling on its surface, and gives the necessary motion to the machine. The advantage of this friction-wheel is, that the velocity of the machine may be increased or diminished at pleasure by moving the wheel *a* nearer to or farther from the edge of the disc.

Fig. 221.

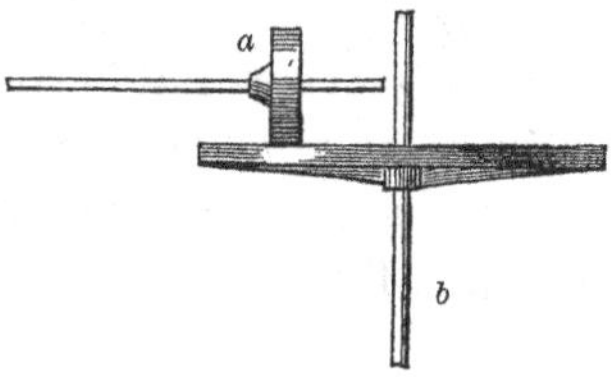

Fig. 222 is another combination of discs suitable for couplings with only one bearing. The disc *b* is keyed on one shaft, and is recessed on the face, to receive the smaller disc, *c*; this disc is sunk flush with the face of the other, and is screwed tightly

Fig. 222.

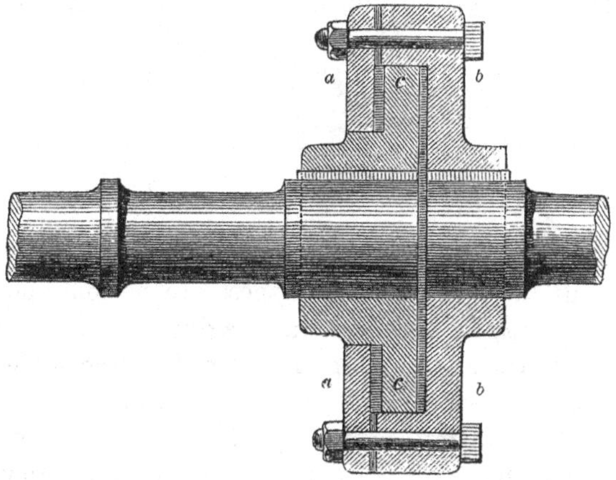

up to it by means of the ring *a*, which is bolted to the disc *b*, and secures that marked *c*. Between the three plates, *a*, *b*, and *c*, annular pieces of leather are interposed, which bring them all to a proper bearing.

This combination, termed a friction coupling, is useful for preventing breakage of the connections in case of a sudden stoppage or reversal of the motion. It is plain that the holding power of the coupling depends simply upon the lightness with which the discs are screwed together, and the consequent frictional force of the surfaces of leather and metal.

Besides these more permanent forms of couplings, there are other contrivances adopted when the object to be attained is the engagement and disengagement of certain parts of the machinery or gearing during the course of operations.

With the same view of admitting of this disengagement of the connection, in cases of sudden stoppage or reversal, the coupling, fig. 223, is sometimes employed.

Fig. 223.

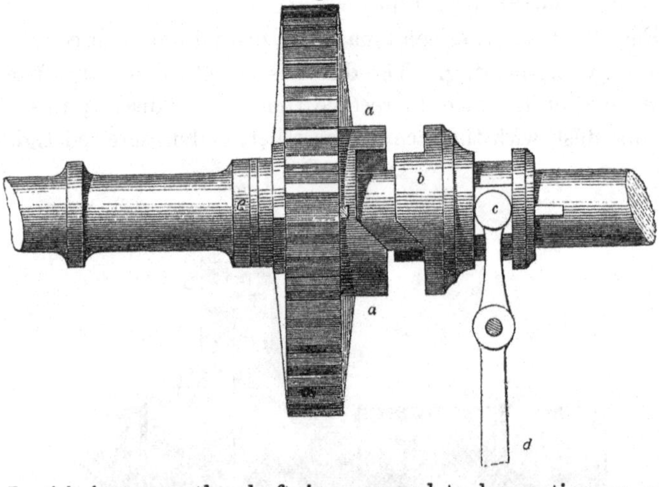

In this instance, the shaft is supposed to be continuous, and the coupling may be termed a disengaging coupling. *a* and *b* are the two parts of the coupling, formed on the acting faces into alternate projections and recesses, such that they correspond with and exactly fit into each other when in gear. The part *a* is, in this example, cast on a spur or bevel wheel, from which the motion of the shaft is supposed to be taken off. Both of the parts *a* and *b* are, to a certain extent, loose on the shaft; the former being capable of moving round on it, though deprived of longitudinal motion by washers and a collar, marked *e*, and the latter being free to slide on the shaft, though prevented from turning on it by a sunk key, which slides in a slot inside the clutch or sliding piece *b*. The mechanism is put into gear by means of the lever *d*, which terminates in a fork with cylindrical extremities *c*; and it is obvious that, by the contact of the flat faces of *a* and *b*, the latter will immediately carry with it the other part at the same speed as the shaft. Supposing, now, that the motion of the wheel *a* is suddenly accelerated, the oblique faces of the couplings immediately fall out of contact, and slide free of each other, leaving the couplings clear, and the shaft free to continue in motion.

In the old form of this contrivance, known as the sliding bayonet clutch, the part *b*, instead of the toothlike projections on the face, had two or more prongs which laid hold of corresponding snugs cast on the face of the part *a*, which, moreover, was usually a broad belt pulley, introduced with a view to modify the shock on the gearing on throwing the clutch into action.

In an older form still, the pulley was made to slide end long on the shaft. A form analogous to this was known as the "lock pulley," a few specimens of which still remain in the older factories. Instead of the end long motion common to the other modes, the parts were "locked" together by a bolt fixed upon the side of the pulley, and which, when shifted towards the axis, engaged with an arm of a cross, of which the part *b*, in the preceding figure, is the modern representative. The bolt was wrought by means of a key and stop, the turning of the key throwing back the bolt, and thereby unlocking and disengaging the pulley. The form of coupling represented by fig. 223 is particularly applicable when the impelling power is derived from two sources—a circumstance which frequently occurs in localities affording water power to some extent, and yet not in sufficient abundance for the demands of the work. The deficiency is usually supplied by a steam-engine; and the two powers are concentrated in the main line of shafting by a coupling of the kind depicted. In cases of this kind, the speed of the shafting being fixed, and the supply of water inconstant, the power of the water-wheel ought to be thrown upon the wheel *a a*, and that of the engine upon the shaft at another point. By this arrangement, the speed of the line can be exactly regulated by working the engine to a greater or less power, according to the supply of water. The proper speed of the water-wheel will likewise be maintained, which is of importance in economising the water power.

"The same form of coupling is also used occasionally for engaging and disengaging portions of the machinery. But for this purpose the object is to obtain a mode of connection by which the motion may be commenced without shock; for, in consequence of the inertia of all material things—that is, the tendency which every portion of matter has, when at rest, to remain at rest, and when in motion, to continue to move—the parts of the mechanism, when acted upon too suddenly by a moving power, are liable to fracture and disarrangement. It is a law in mechanics that when a body is struck by another in motion some time elapses before it is diffused from the point struck through the other parts; consequently, if the parts receiving the blow have not sufficient elasticity and cohesive force to absorb the whole momentum of the striking body till the motion be transmitted to the centre of rotation, fracture of the body struck must necessarily ensue. Hence, in a system of mechanism, any parts intended to be acted upon suddenly by others in full motion ought not only to be strong, but they ought to be capable of yielding on the first impulse of the impelling force with as little resistance as possible, and gradually bring the whole weight into motion. The common mode of driving by belts and pulleys accomplishes this object very satisfactorily. In this the elasticity of the belt comes into action; and being thrown upon the pulley by the strap guide or fork, it continues to slip, till, by the friction between the sliding surfaces, the belt gradually brings the quiescent pulley into full motion. This mode of connection is unexceptional when the power to be transferred is not great; but its application to large machinery is attended with inconvenience." *

In figs. 224, 225, two other forms of clutches are shown, as often used to connect the shafting of different parts of the same mill, where it is not necessary to throw into or out of gear when running at full speed. They consist of a fixed and sliding box, one on each shaft, with teeth or projections which fit in corresponding notches. The sliding box has a groove turned in it, in which a forked lever works, as at *a* (fig. 224) and at *a* (fig. 225), by which it is drawn backwards or forwards as the case may be. The peculiarity of the clutch (fig. 225) is that of

* Extract from Engineer's and Machinist's Assistant, p. 144.

the driving shaft, which, reversed by any accident in its motion, as is not unfrequently the case in starting and stopping the steam engine, the sliding clutch is forced back by the wedge-shaped faces of the projections, and the machinery thrown out of gear.

Fig. 224.

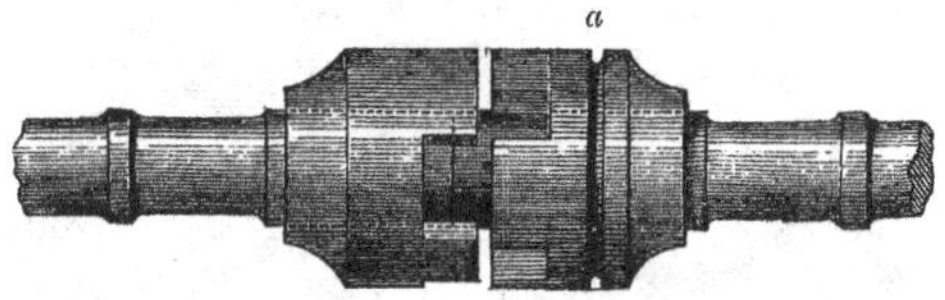

Fig. 225.

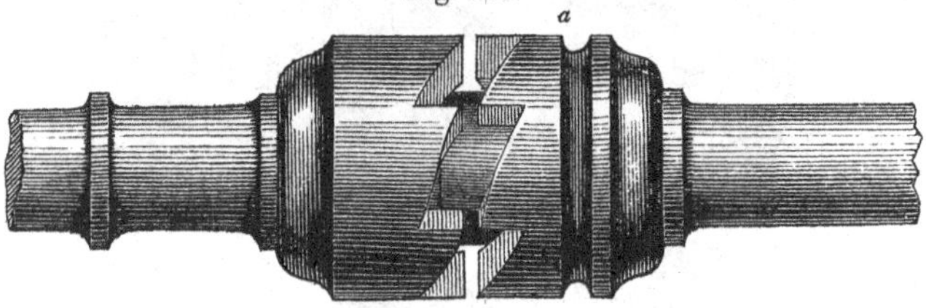

Fig. 226 shows one of these clutches on a small scale, fixed on a line of shafting beneath the floor of a mill. It is placed between two standards *a a*, supporting the ends of the shaft, and the lever *b* working on a pivot at bottom, and having a pin working in the groove of the sliding clutch box, serves for throwing the driven shaft into or out of gear whenever it may be necessary.

Fig. 226.

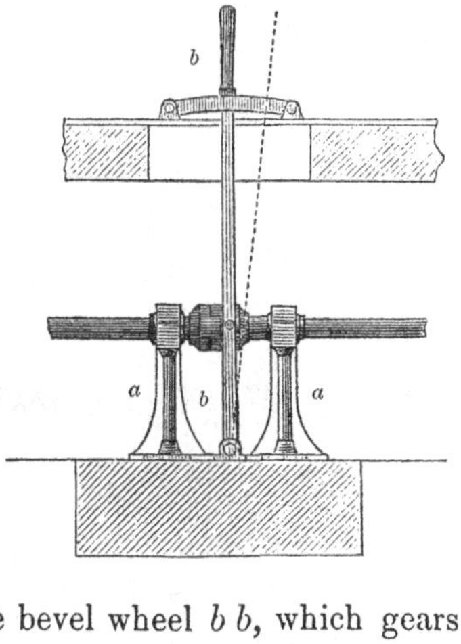

Another ingenious contrivance, I believe invented by Mr. Bodmer, is shown in figs. 227 and 228. It consists of a box *a a* running loosely on the driving shaft *s s*, but carrying the bevel wheel *b b*, which gears into another wheel on the driven shaft, not shown in the figures. Tightly keyed on the driving shaft *s s* is a boss *c c*, with two trunnions, on which slide two friction sectors *k k*; the outer surface is coated with a copper plate, accurately fitting the interior surface of the running box *a a*. The boss *c c* carries

Fig. 227.

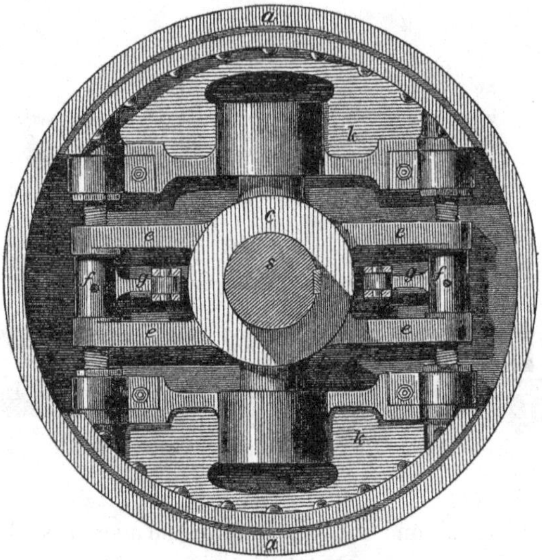

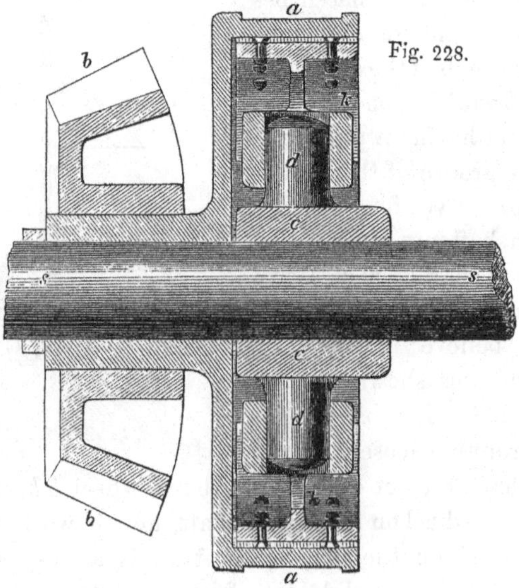

Fig. 228.

also four projections *e e e e*, which serve as guides for four screws, alternately left and right handed, and attached to the nuts *f f* and levers *g g*; these screws act on the extremities of the friction slides *k k*, so that when the levers *g g* are drawn back they are both with equal pressure forced upon the inner surface of the box *a a*. As the pressure can be very regularly and gradually brought on this box through the levers and screws, the motion of the driving shaft *s s* is communicated with perfect regularity, and without shock to the bevel wheel *b b*.

In the above description I have given such examples of engaging and disengaging gear as are most commonly in use. Others of a more complicated character might be cited, but they are not to be recommended as applicable in general practice. The last form, figs. 227 and 228, is, however, specially noticed as suitable for gunpowder mills, where the greatest possible freedom from shocks is essentially necessary.

9. *Hangers, Plummer-blocks, &c., for carrying Shafting.*

Shafting is supported in three ways, viz. on foundation stones in the floor, beneath beams suspended from the ceiling, and to the walls of the mill. This necessitates as many different forms of framework, known as hangers, plummer-blocks, standards, &c.

Fig. 229.

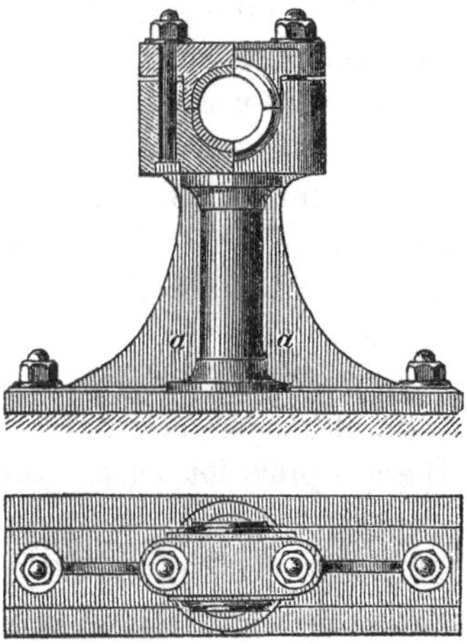

The simplest mode of supporting a range of light shafting is from the floor, and a pedestal suitable for this purpose is shown in fig. 229. It consists of a cast iron base plate and column, with deep wings *a a* cast on to strengthen it free from vibration. The upper portion is hollowed out to receive the lower brass step, and the cap carrying the upper step. When the entire pressure of the shafting is downwards the upper brass bush is omitted, and the cap is cast hollow and kept full of grease, so as to secure the most perfect lubrication of the journal of the shaft.

Fig. 230 shows a pedestal for bolting to a wall, the chief

Fig. 230.

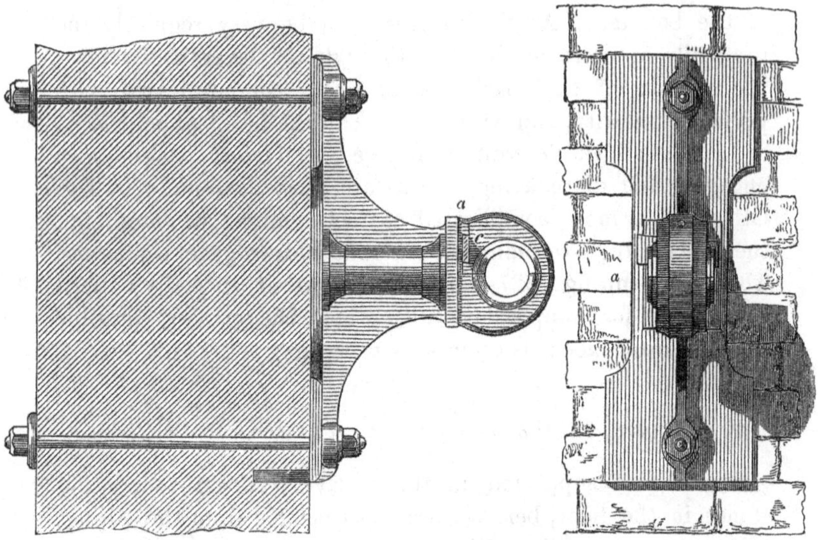

difference being that the cap is now fixed on its inner side by a wedge or cotter (*c*). In this figure a shell cap *a* is shown. If the pull is upwards, and two brasses be required, "lugs" have to be added to the extremity of the pedestal and cap for bolting the two together.

There are various ways of suspending ranges of shafting from the ceiling, according to the means which exist for their attachment. If wooden beams, as s, are present, the hanger has a large plate (*a*), which bolts to the side of the beam, as shown in figs. 231, 232. The caps are fixed by a cotter, as in the previous case.

Figs. 233, 234 show a front and side elevation of another form of hanger for attachment to wooden beams. In this case there is provision for a second line of shafting, at right angles to, and receiving motion from, the primary line. For this purpose a small plummer block is bolted on to a recess at the side of the hanger. The thrust, owing to the pair of bevel wheels which would be placed near this hanger, is no longer simply vertical, and hence two brass steps are placed for the journal of the

Fig. 231.

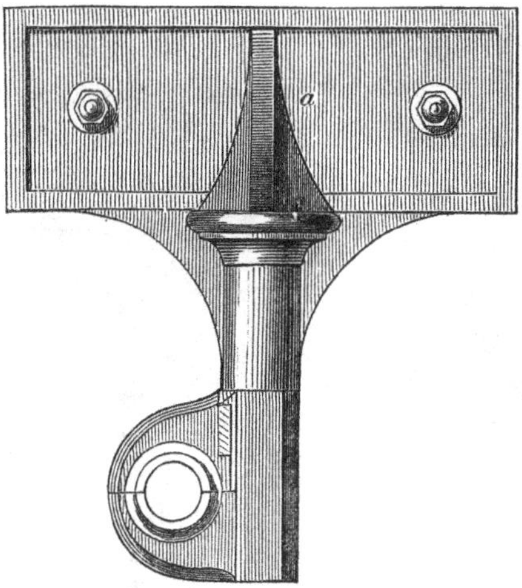

Fig. 232.

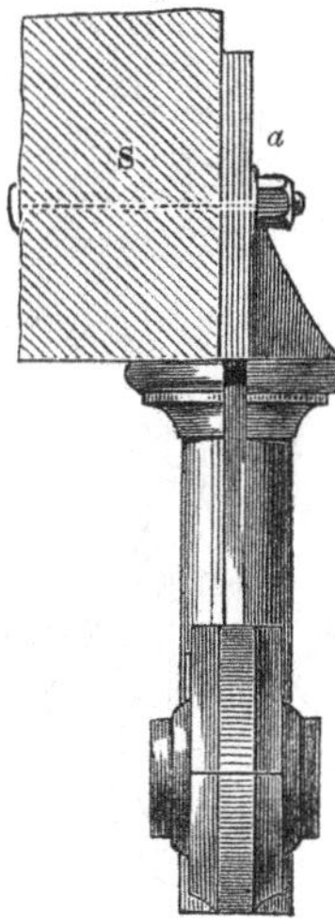

principal shaft, with a bolt at *d*, fig. 233, in addition to the cotter, to keep the cap in its place.

Fig. 233.

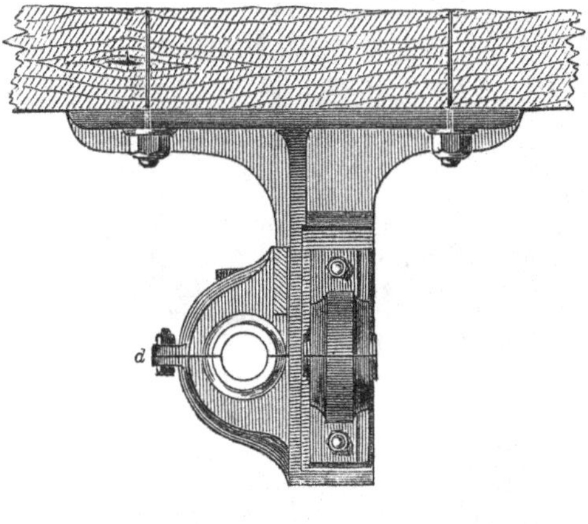

Fig. 234.

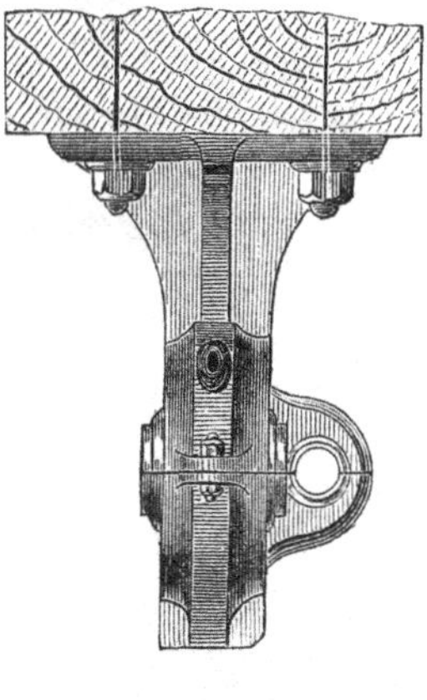

Fig. 235 shows another form of light hanger sometimes employed in weaving sheds, and also in use for supporting shafts in fire-proof mills, being bolted up to the under side of the cast-iron beams, as shown at fig. 237.

Fig. 236.

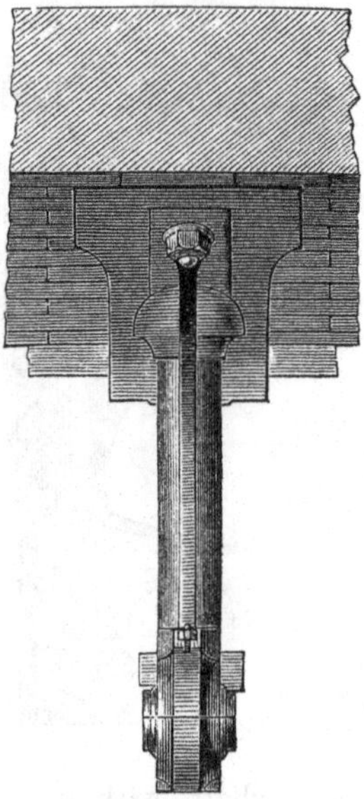

Fig. 235.

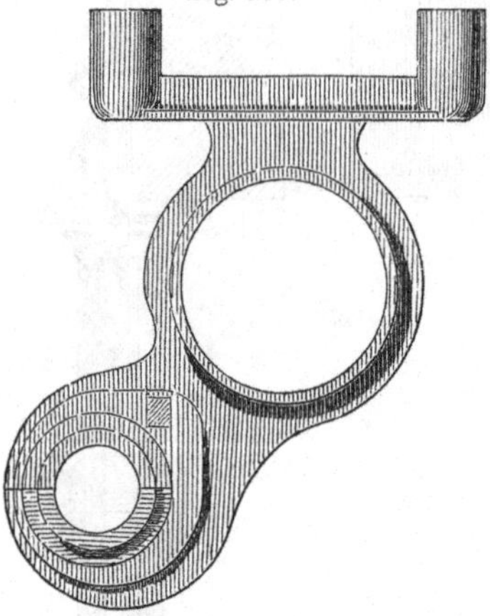

Fig. 237.

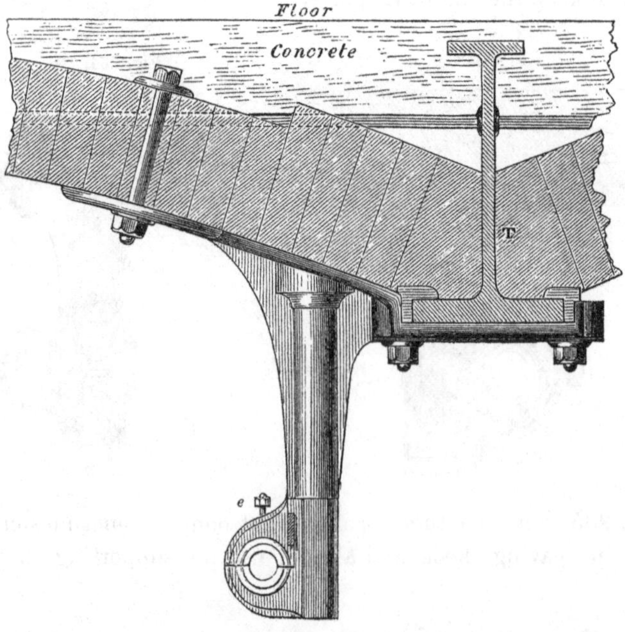

Where greater strength and firmness are required, especially in long hangers in which there is considerable leverage, the arrangement shown in figs. 236, 237 is adopted; the hanger in this case is bolted to a cast-iron beam, and by an extension of the flange plate to the brick arch, which springs from the beam T, it is firmly secured to both beam and floor. At *e* is a screw for tightening the upper brass step on the shaft.

More complicated arrangements are sometimes necessary where two or three ranges of shafting have to be brought in connection with each other by means of bevil or mitre wheels. Figs. 238 and 239 show a front and side elevation of this

Fig. 238.

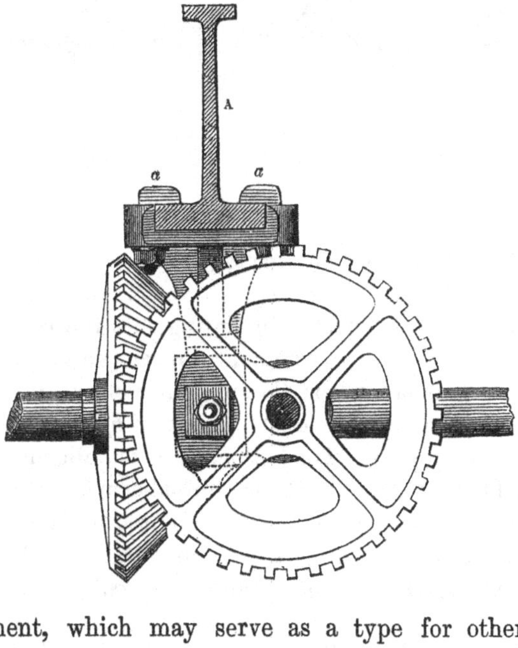

arrangement, which may serve as a type for others. The hanger is attached to a cast-iron beam A, by hooked bolts with nuts beneath the top plate, as shown at *a a*, *care being taken in this attachment not to weaken the flange of the iron beam by boring holes in it.* Double brass steps are necessary in this case for the main line of shafting, and also for two smaller ranges at right angles to it, which revolve in opposite directions, as shown at fig. 239.

Fig. 239.

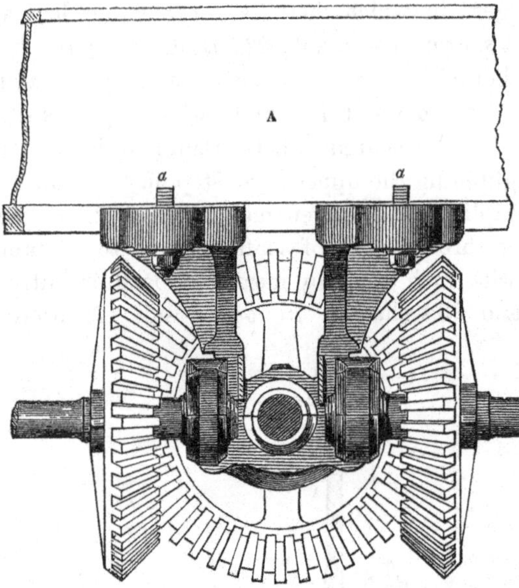

A very frequent case in practice is the connection of two ranges of shafting, at right angles to each other, at the corner of a room. This is effected by letting into the corner of the building a cast-iron frame, commonly known as a wall-box, which serves as a foundation for the plummer-blocks carrying the shafting. Such an arrangement is shown in fig. 240 in elevation, and in fig. 241 in plan. The box *w, w, w,* is built into the wall, and bolted both to it and to the cast-iron beam *b*. It carries two plummer-blocks on a plate firmly supported by brackets. The wall pieces in these two figures are similar, but with a slightly different arrangement of the plummer-blocks.

Irrespective of the various forms of engaging and disengaging apparatus, it will be necessary to consider the position, form, and proportions of the wheels and shafting required in mills where the power is divided and widely distributed. To show the enormous extent to which the concentration of machinery in one building has been carried, I may mention that in mills of my own construction there have been on the average not less

Fig 240.

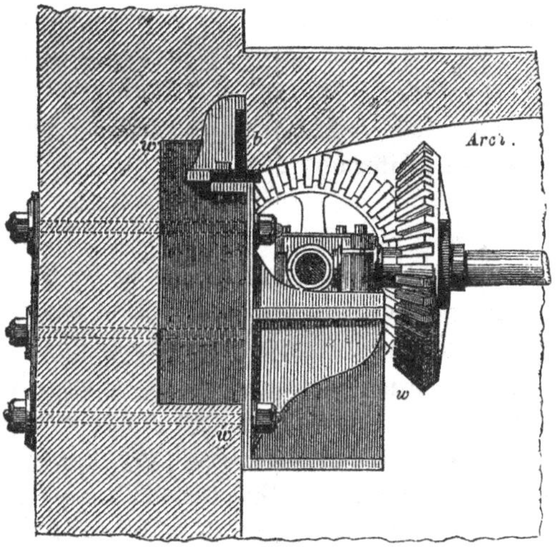

than 450 wheels and 7,000 feet of shafting in motion. In the large mills at Saltaire there are upwards of 600 wheels and 10,000 feet, or two miles, of shafting distributed over an area of

Fig. 241. (Plan.)

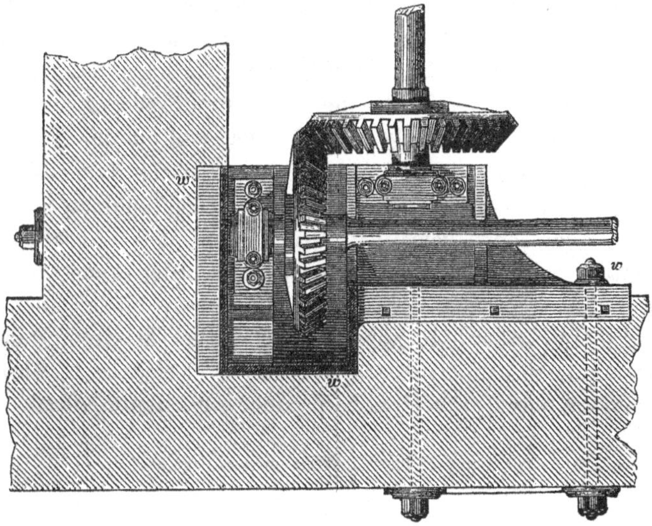

flooring equivalent to 12 acres. In corn mills and iron works, where the machinery is more closely connected with the prime mover, these considerations are of less importance; but in factories for the manufacture of textile fabrics the machinery covers a great extent of surface, and the greatest care is necessary in giving due proportion to the transmissive machinery, in order to secure uniformity of motion at the remotest parts of the mill.

In gearing a mill, the first consideration is the power of the engines, the position of the machinery to be driven, and the strength, diameter, &c., of the first-motion shaft, and other requisites for the transmission of motion in a well-geared mill. It is upwards of twenty years since the fly-wheel was converted into a first motion, and a new system of transmitting the power of the steam engines to the machinery of the mill introduced. Previous to that time it was effected by large spur-wheels inside the mill, now it is taken direct from the circumference of the fly-wheel.* The advantage of this system was the abolishing of the cumbrous first-motion gearing; and the requisite velocity being already present in the fly-wheel, it was only necessary to cast it with teeth, and to take off the power by a suitable pinion at the level most convenient for the purposes of the mill.

In another place I have given general rules for the pitch, breadth and strength of the teeth of wheels. The Table (p. 101), computed from examples which have occurred in my own practice, exhibits the best proportions of spur fly-wheels to secure durability of both wheel and pinion.

It will be observed that the diameters of the fly-wheels are not always proportionate to the power of the engines, nor yet to their respective velocities. In practice, it is impossible to maintain uniformity in this respect, as, in order to meet all the requirements of manufacture, it is necessary to deviate from fixed principles, and to approximate as near as circumstances will admit to the diameters, weights, and velocities of wheels, as may be found convenient to produce a maximum effect.

* Compare Part I., Prime Movers, page 248.

TABLE 9.—DIAMETERS, PITCH, VELOCITY, &c., OF SPUR FLY-WHEELS OF THE NEW CONSTRUCTION.

Nominal Power of Steam Engine. Horse-power.	Diameter of Fly-wheel. Feet.	Inches.	Pitch in Inches.	Breadth of Cog in Inches.	Velocity of Pitch-line per Minute in Feet.
Two 150=300	30	$1\frac{1}{2}$	$4\frac{1}{2}$	16	The velocities vary according to circumstances, from 1,250 to 1,650 feet per minute.
Single = 50	13	$3\frac{3}{4}$	$4\frac{3}{8}$	12	
Two 100=200	24	5	4	14	
Two 80=160	23	4	4	14	
Two 80=160	22	4	4	14	
Single = 60	19	$0\frac{1}{2}$	$3\frac{3}{8}$	12	
Two 70=140	24	5	$3\frac{1}{4}$	12	
Two 70=140	22	$8\frac{1}{4}$	$3\frac{1}{4}$	14	
Two 50=100	21	0	$3\frac{1}{4}$	12	
Two 40= 80	21	0	$3\frac{1}{4}$	10	
Two 45= 90	20	0	$3\frac{1}{4}$	12	
Single = 50	18	$2\frac{1}{2}$	$3\frac{1}{4}$	12	
Two 35= 70	16	$0\frac{3}{4}$	$3\frac{1}{4}$	10	
Single = 40	17	10	3	10	
Two 25= 50	13	10	3	10	
Single = 25	8	$11\frac{1}{2}$	3	12	
Two 20= 40	15	6	$2\frac{1}{2}$	7	
Two 25= 50	15	$4\frac{1}{2}$	$2\frac{1}{2}$	8	
Single 25	15	$4\frac{1}{2}$	$2\frac{1}{2}$	7	
Two 18= 36	13	0	$2\frac{1}{2}$	8	
Single 15	10	0	$2\frac{1}{2}$	7	
Single 18	17	11	2	6	
Single 12	10	0	2	5	

Of late years, the speed of the piston of factory steam engines has been accelerated from 240 to 300, and in some cases to 350 feet per minute. This united to the increased pressure of steam nearly doubles the power of the engines to what they were thirty years ago. The standard speed of a Bolton and Watt 7 feet stroke engine previous to that date, was seventeen and a half strokes per minute.

In closing this section of practical construction, I may state that the couplings, engaging and disengaging gear, including the different forms of hangers, fixings, &c., are taken from my own designs, first introduced as a substitute for the cumbrous attachments that were in general use previous to the years 1820 and 1823.

Having determined the diameter, speed and strength of the fly-wheel, the next consideration is the material, diameter, &c.,

of the main shaft. These are usually of cast-iron, and their diameters depending on the power transmitted through them, and the velocity at which they revolve, will be found by the tables and formula already given. The distribution of the power is usually effected by a vertical shaft, extending from the bottom room, through the various floors of the mill, to the top story; the power being taken off at each stage by a pair of bevil wheels. This arrangement, as shown in fig. 242, represents one engine-house with a section of part of one division of the mills at Saltaire; and this may be considered as a type of other mills adapted for spinning and similar purposes.

It will be observed that there are four divisions in the Saltaire mills—one for the preparatory process, one for the wool combing, another for the spinning, and a fourth for the weaving. All these are driven by four steam engines, each of 100 nominal horses power, but collectively distributing a force through these different departments of upwards of 1,250 horses.

On referring to the drawings, figs. 242 and 243, which represent a cross and longitudinal section of the mill, it will be seen that the vertical shaft A A, is driven direct from the fly-wheel by the horizontal shaft B, giving motion to the machinery in each room as it ascends. It is fixed on a solid pier of ashlar, as shown at fig. 244, page 106, and supported on strong cast-iron plates and bridgetrees, firmly secured by bolts to the foundations below. In each room it is securely fixed, by cast-iron frames and boxes, forming a recess for the bevil wheels, into the wall which divides the engine-house and the rooms above from the mill. This wall is generally made strong and thick, with sufficient weight to resist the action of the wheels prepared to drive the main lines of horizontal shafts with a speed and force equivalent to the work done in each room. In the case of the Saltaire mills this is considerable; nearly 300 horses' power being distributed through the upright shaft alone, the remainder being carried off to the loom shed by a second wheel, working into the bevil wheel *a*, on the horizontal shaft B, but not shown in the drawing. It is important, in mills where powerful steam engines are employed, that the foundations and fixings to which the main shafts are attached are of the most

Fig. 242.

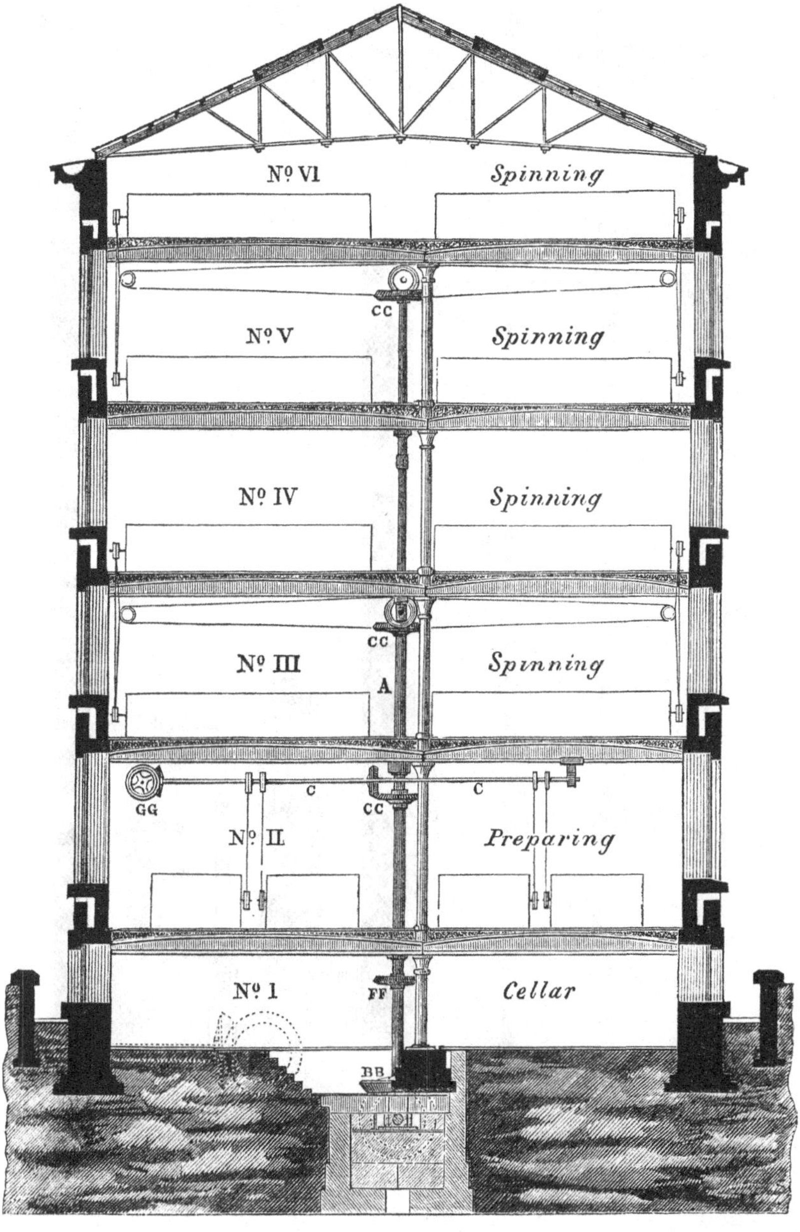
Nº VI
Spinning
CC
Nº V
Spinning
Nº IV
Spinning
CC
Nº III
Spinning
A
GG
C
C
CC
Nº II
Preparing
Nº 1
FF
Cellar
BB

Fig. 243.

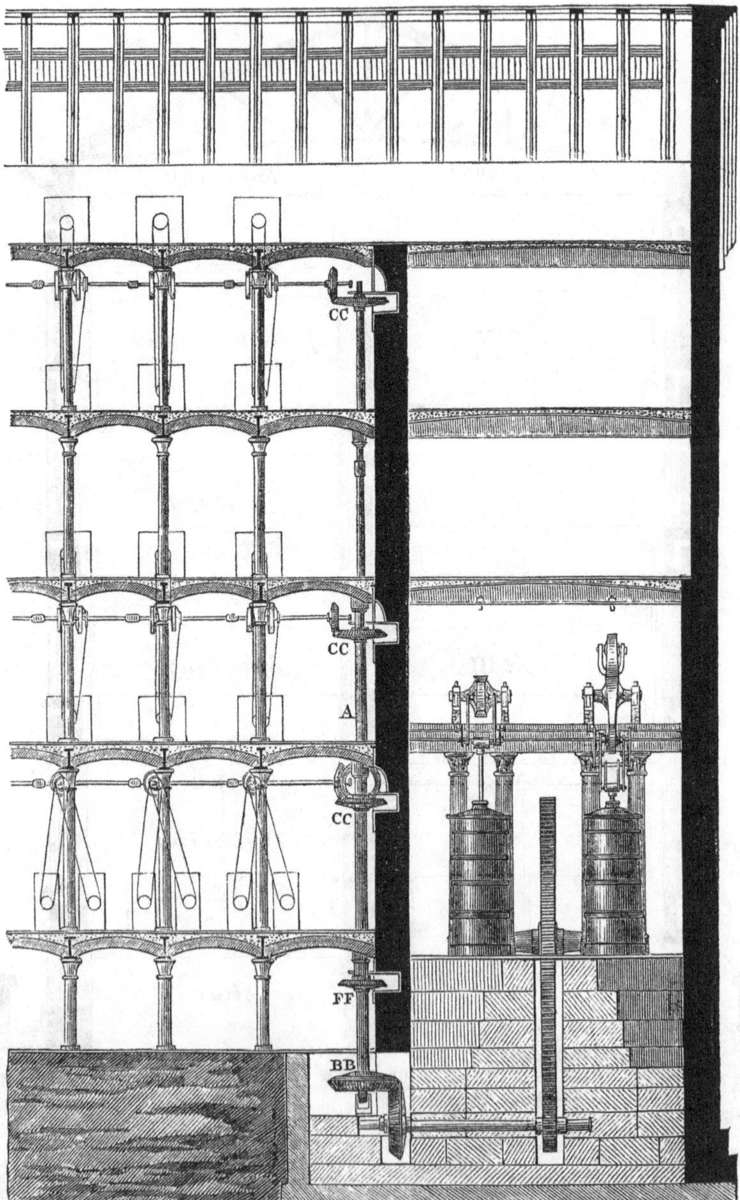

substantial description, and the greatest precaution is necessary in order to secure them from vibration, and to render them perfectly rigid when the whole force of the engines is applied.

In the Saltaire mills, as in many others for the manufacture of cotton, flax, and wool, the preparatory machinery, such as carding, combing, roving, &c., is generally driven by lines of horizontal shafts, or by a series of cross shafts, branching off at right angles from the main line extending down the centre of the room, as shown at *c c* in No. II. room. Nos. III. and IV. rooms are driven by the longitudinal shaft in No. III.; and Nos. V. and VI. by the shaft in No. V. room. On this plan it will be noticed that the spinning machinery is driven by iron pulleys from the horizontal shafts, at a velocity of nearly 200 revolutions per minute, and the straps or belts from those pulleys are directed by means of guide pulleys to the machinery in both rooms. For this purpose, iron boxes are inserted into the arches supporting the floors, for the admission of the straps to the machinery in the upper floor.

It will not be necessary to give the dimensions of the shafts in each room, as these details and calculations must be left to the judgement of the millwright, and the nature of the work they have to perform. Suffice it to observe, that the vertical shaft A is 10 inches diameter through the first two rooms, 8½ inches through the third room, and 6½ inches to the top; the velocity being 94 revolutions per minute.

As respects the couplings for this shaft, we may refer the reader to the Table of Dimensions (page 109) made from couplings actually in use, and which have been found, by experiment, serviceable in every case where strength and durability are required.

Great trouble is sometimes experienced with the foot of the vertical shaft, which, from its weight and the great pressure upon it, has a tendency to heat, unless sufficient bearing-area is allowed and the parts kept thoroughly lubricated. The general arrangements of the footsteps and gearing in large mills are shown in fig. 244: *s s* is the first motion-shaft, and *t t* the vertical shaft; *a a* the bevil wheel on the former, and *b b* the bevil wheel on the latter; *c* a plummer-block for the first motion-

Fig. 244.

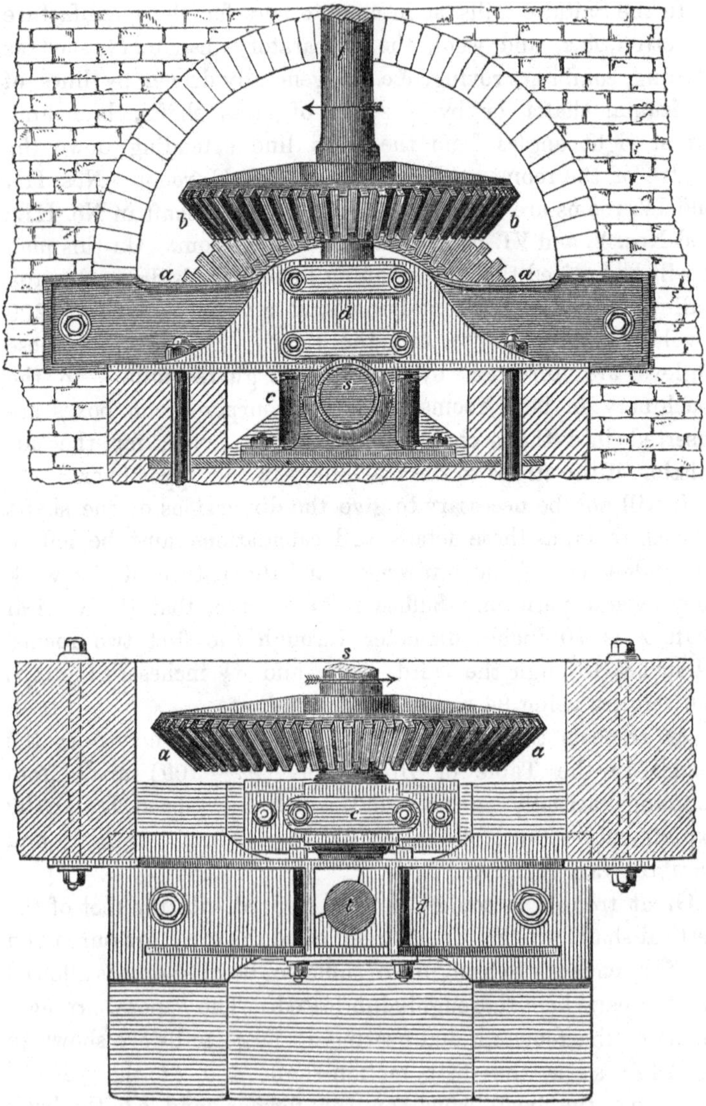

shaft, and *d d* the box containing the brass footstep for the vertical shaft: this box rests on a large base plate, bolted to the foundation stones and to the wall of the engine-house. In order to insure a constant supply of oil to the bearing, it is usual to cut away nearly the whole depth of the footstep, or that portion of the brass in the corner opposite to the thrust of the bevil wheels, as shown in the plan, fig. 244; this cavity is then kept full of oil, and lubricates the shaft throughout at every revolution.* Again, in cotton, woollen, and flax mills, when the first motion and vertical shafts have been duly proportioned to the work they have to perform, it becomes necessary to consider the diameter, speeds, &c., of the light shafting for driving the machinery in the different rooms. The formula given for strength, &c., in a former part of this work, will apply to this description of gearing and mill-work where the length does not exceed 120 feet. In long ranges of shafts, of from 150 to 200 feet in length, where the power applied to the machinery at the end of the room is considerable, it is essentially necessary to increase their strengths in order to prevent torsion or twist. This is a consideration of much importance, and requires careful attention, as long ranges of light shafts are very elastic — they, in some cases, effect nearly a complete revolution at the point of imparted motion before the extreme ends begin to move. The result of the power so irregularly transmitted by the spring of the shafts, resolves itself into a series of accelerated and retarded motions through the whole line of shafts, and imparts to the machinery in one-half of the room a very variable motion. Want of stiffness is a great evil in long lines of shafting, and, as we have already observed, instances are not wanting in which whole lines have been removed entirely from this cause.

The transmission of power to machinery placed at different angles from the line of shafts, which is sometimes the case in old mills, has generally been effected by the universal joint A,

* The reader may compare what is here said of footsteps with that in Part I. pp. 168, 172, on the steps for turbine shafts.

fig. 245, which works moderately well at an obtuse angle, but I have always found in my own practice that bevil wheels, as

Fig. 245.

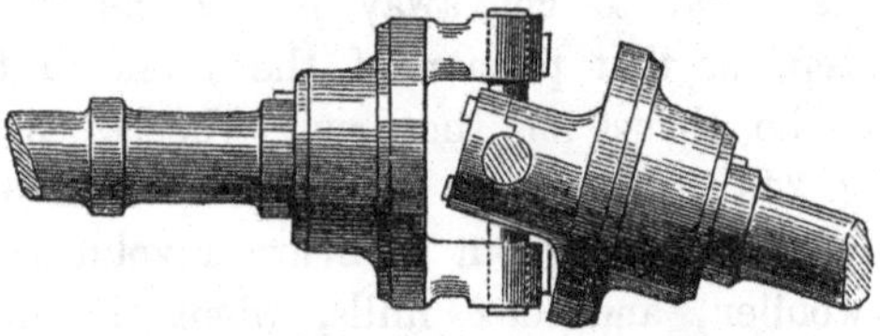

at B, fig. 246, are preferable and more satisfactory. They give much less trouble, and work with greater ease, than the universal joint. Other examples might be given for the guidance of the practical millwright; but, having to discuss these points

Fig. 246.

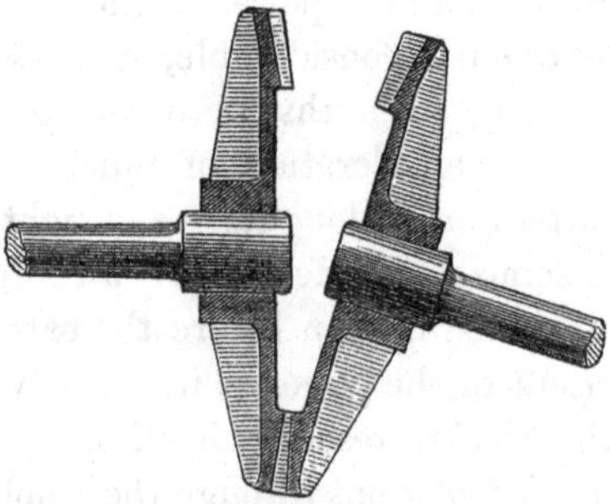

at greater length when we come to treat of the different kinds of mills and different methods of gearing, we must direct the reader to those portions of the work which concern his own immediate practice.

The following Table exhibits the diameter of shafts, length of journals, diameter and proportions of couplings, &c., derived from actual practice, which may be useful to the less experienced millwright and engineer:—

TABLE 10.—LENGTH, DIAMETER, &c., OF COUPLINGS, COUPLING-BOXES, &c.

Diameter of Shaft.	Length of Neck.	Diameter of Coupling.	Length of Lap.	Length of Box.	Diameter of Box.	Thickness of Metal.
$1\frac{1}{2}$ and $1\frac{5}{8}$	3	$2\frac{1}{2}$	2	$4\frac{1}{2}$	$4\frac{1}{4}$	$\frac{7}{8}$
$1\frac{3}{4}$	$3\frac{1}{4}$	3	$2\frac{1}{4}$	5	5	1
2	4	$3\frac{1}{4}$	$2\frac{1}{4}$	$5\frac{1}{2}$	$5\frac{1}{2}$	$1\frac{1}{8}$
$2\frac{1}{4}$	$4\frac{1}{2}$	$3\frac{1}{2}$	$2\frac{3}{4}$	6	6	$1\frac{1}{4}$
$2\frac{1}{2}$	5	4	3	$6\frac{1}{2}$	$6\frac{3}{4}$	$1\frac{3}{8}$
$2\frac{3}{4}$	5	$4\frac{1}{2}$	$3\frac{1}{4}$	7	$7\frac{1}{2}$	$1\frac{1}{2}$
3	$5\frac{1}{2}$	$4\frac{3}{4}$	$3\frac{1}{2}$	$7\frac{1}{2}$	$7\frac{3}{4}$	$1\frac{1}{2}$
$3\frac{1}{4}$	$6\frac{1}{4}$	5	$3\frac{3}{4}$	8	$8\frac{1}{4}$	$1\frac{5}{8}$
$3\frac{1}{2}$	$6\frac{1}{2}$	$5\frac{1}{2}$	$3\frac{3}{4}$	$8\frac{1}{4}$	$8\frac{3}{4}$	$1\frac{5}{8}$
4	7	6	4	$8\frac{1}{2}$	$9\frac{1}{2}$	$1\frac{3}{4}$
$4\frac{1}{2}$	$7\frac{1}{2}$	$6\frac{1}{2}$	$4\frac{1}{2}$	9	$10\frac{1}{2}$	2
5	8	$7\frac{1}{4}$	5	10	$11\frac{1}{4}$	2
$5\frac{1}{2}$	$8\frac{1}{2}$	8	$5\frac{1}{2}$	11	$12\frac{1}{4}$	$2\frac{1}{8}$
6	9	9	6	12	$13\frac{1}{2}$	$2\frac{1}{4}$
$6\frac{1}{2}$	$9\frac{1}{2}$	$9\frac{3}{4}$	$6\frac{1}{2}$	13	$14\frac{3}{4}$	$2\frac{1}{2}$
7	$10\frac{1}{4}$	$10\frac{1}{2}$	7	14	16	$2\frac{3}{4}$
$7\frac{1}{2}$	$11\frac{1}{4}$	$11\frac{1}{4}$	$7\frac{1}{2}$	15	17	$2\frac{7}{8}$
8	12	12	8	$16\frac{1}{2}$	18	3
$8\frac{1}{2}$	$12\frac{1}{2}$	$12\frac{1}{2}$	$8\frac{1}{2}$	17	19	$3\frac{1}{4}$
9	$13\frac{1}{2}$	13	9	18	20	$3\frac{1}{2}$
$9\frac{1}{2}$	14	$13\frac{1}{2}$	$9\frac{1}{2}$	18	21	$3\frac{3}{4}$
10	$14\frac{1}{2}$	14	10	$18\frac{1}{2}$	$22\frac{3}{4}$	$3\frac{3}{4}$
11	15	16	11	20	24	4
12	16	$17\frac{1}{2}$	12	21	26	$4\frac{1}{4}$
13	17	$18\frac{1}{2}$	13	22	$27\frac{1}{2}$	$4\frac{1}{2}$

SECTION V.

ON MILL ARCHITECTURE.

CHAPTER I.

ANCIENT AND MODERN MILLS.

In the early stages of civilisation, when industrial progress was at a low ebb, and in those days when the whole population was trained to war, and a miserable system of tillage existed, mills were little in demand, with the exception of corn and fulling mills; the former to grind oats and barley, and the latter to mill a rough description of serge or blanket. At that period, mill architecture was out of the question, as the dwellings of the retainers, and those employed in the field or manufactory, were mere hovels, and the architecture of the country was confined to churches, public buildings, and the mansions of the barons or lords of the soil. In such a state of society mills were simply sheds, with water-wheels having straight floats, and a long conduit or spout to carry the force of the water descending from the higher fall against the float-boards below. In process of time, as the population increased in numbers and intelligence, new demands for food and clothing were created, and a new description of mills was introduced to meet the requirements of a superior class to those under the feudal system, who were chiefly engaged in war and plunder. At this period mills were improved and enlarged, but there were no attempts at architecture; and, what is still more surprising, the engineers and millwrights of those days appear to have had no idea of the advantages derivable from large water-wheels, but contented themselves with additional wheels to meet the demand of additional work. On these occasions every pair of millstones and every pair of fulling stocks had their separate water-wheels, and these were multiplied according to the demands or necessities of the trade. Fig. 247 represents a plan of the wheels, and the way in which they were arranged, in order to give motion, as was then the fashion, to three pairs of millstones, with a dressing machine and sieves, as the case might require.

Fig. 247.

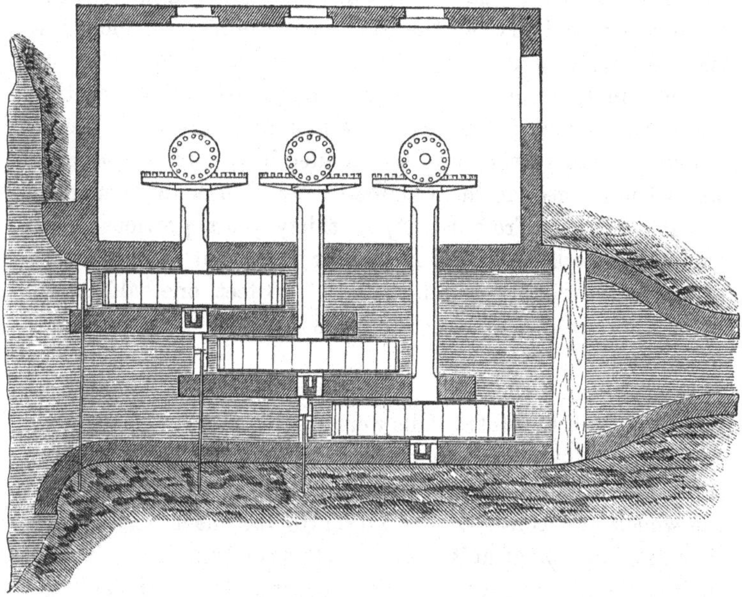

Most of the mills in this country and in other parts of Europe, were of this description up to the middle of the last century; and it was during the days of Smeaton that mills began to assume a better system of classification, and that concentration of machinery and power, which ultimately led to the high state of perfection in which we now find them in this country.

Immediately succeeding Smeaton came the late John Rennie, who built the Albion steam mills, and effected a new system of concentration, totally different to that in use, for corn mills driven by water. The steam engine was, therefore, the first innovator and improver of concentrated power; and Mr. Rennie availed himself of its introduction to effect a new arrangement of the machinery in mills, tending to meet the requirements of a new power concentrated on one point, and diverging in different directions to reach the various machines of the manufactory. This was an important step gained in the classification of the machinery. It caused a change in the construction of water-wheels by increasing their diameter and width, and by making one wheel do the work of two or three on the old plan, the water-wheel being brought to bear upon the machinery in the mill, on the same principle as that of the steam engine. For many years the old plan still lingered in the minds of the millwrights of the last century, and it was not until the year 1824 that an improved system of applying water power to mills was effectively introduced. This was accomplished at the Catrine Cotton Works in Ayrshire, at the above date. From twenty to thirty years previous to that time these works were driven by four wheels at different parts of the building, with a divided fall of 48 feet, one half the fall being appropriated to one of the mills and the other to another upon the lower level.

In the new wheels, a description of which is given at page 126, Vol. I., the two falls were united, and the whole power concentrated in a separate house, equidistant between the two mills, and a line of strong shafts projected at right angles from the wheels and conveyed the power to the machinery in each. From this system of united force great advantages were derived, both in water and steam, which ultimately led to the arrangements now in use, of having both water-wheels and steam-engine erected separate from the mill, thus rendering the establishment free from damp and heat. Water-wheels are now upon a very different system than heretofore, and instead of being sunk in deep trenches in the very centre of the mill, obstructing the different processes of manufacture, and where it is next to impossible to get near them for repairs, they are now erected in a commodious house with broad platforms and galleries, six feet wide, and from which every part of the wheel can be reached.

The concentration of motive power bearing direct from a single point upon mill machinery, led to the introduction of long lines of shafting to reach such machinery at the extreme end of the different flats of the mill, and these, again, as already explained, suggested other improvements in the buildings, particularly in cotton mills, where they are not unfrequently extended to a length of 300 to 350 feet.

It is in the recollection of persons still living, when the operations of carding, spinning, and weaving of cotton, flax, and wool, were chiefly carried on in the farm-house and the cottages of the labouring poor, and it is not more than fifty years since the power-loom was introduced, and became the precursor of future changes which ultimately destroyed almost every vestige of our domestic manufacture. It is not for me to offer an opinion on the effects of these changes on the whole of our industrial population; suffice it to observe, that it has altered the domestic habits of the people, and concentrated within large and substantial buildings great numbers of people employed in the different processes of manufacture, which, formerly distributed over a large surface of country, are now concentrated under one roof. As late as 1784, there were no factories, properly so to speak, but the improvements in cotton machinery introduced by Arkwright and Crompton, suggested enlarged space in the shape of separate buildings; and the large profits arising from the cotton manufacture at that time, enabled the proprietors to build mills, some of them considered of colossal dimensions. At first, these mills were square brick-buildings, without any pretensions to architectural form, as shown at Fig. 248. This

Fig. 248.

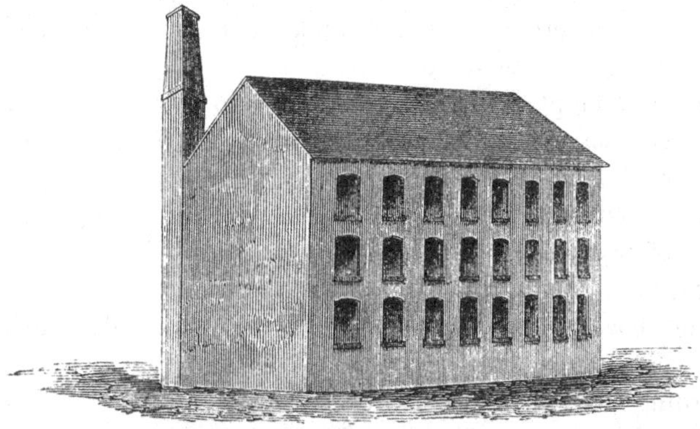

description of building with bare walls was for many years the distinguishing features of a cotton mill, and for many years they continued to be of the same form and character throughout all parts of the country. About the year 1827, I gave designs for a new mill of a different class, and persuaded the proprietor to allow some deviation from the monotonous forms then in general use. This alteration had no pretension to architectural design; it consisted chiefly in forming the corners of the building into pilasters, and a slight cornice round the building, as shown in Fig. 249. This simple change of form gave a new

Fig. 249.

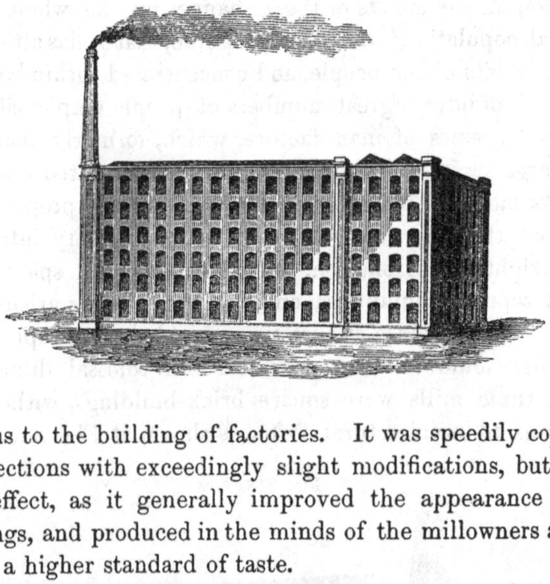

impetus to the building of factories. It was speedily copied in all directions with exceedingly slight modifications, but always with effect, as it generally improved the appearance of the buildings, and produced in the minds of the millowners and the public a higher standard of taste.

These attempts at improvement led to the employment of architects, and we have now, as may be seen from the annexed view of the façade of Saltaire, the factory buildings in this country vieing with our institutions and public buildings as works of art, both in the power and harmony of their parts and the *tout ensemble* of their appearance.

The above style of architecture is not confined to mills of such large dimensions as those of Saltaire, but other forms, according to the taste of the builder, are generally adopted in most mills, greatly to the benefit of the owner, and advantageous in other respects as regards appearance and progressive improvement in the useful arts.

Fig. 250.

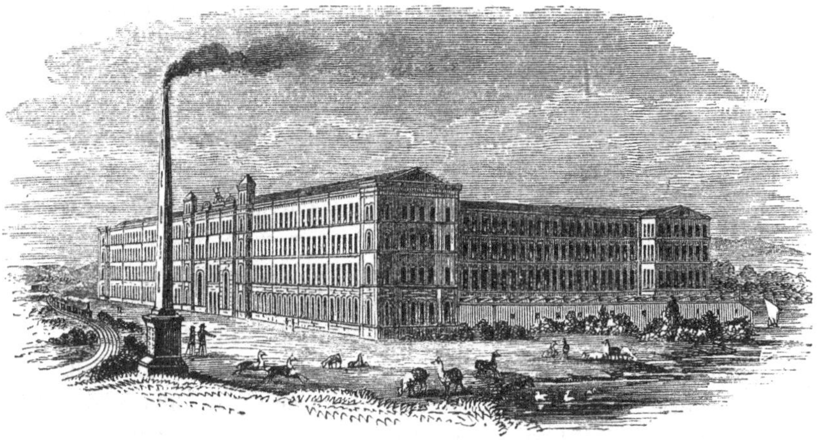

Irrespective of the external appearance, the results of improvements in mill architecture, were manifest in the desire which they induced to elaborate with greater certainty and effect the art of design; to accustom the mind to objects of interest which embodied lines of harmonious proportion, and united in these constructions the taste of the architect and the stability of the engineer. To these may be added the improvements that were introduced in the general arrangements of the buildings, their adaptation for the reception of the different kinds of machinery; security from fire and other requisites for carrying on a large and successful process of manufacture. Contemporaneous with the architectural improvements in mills, the shed principle lighted from the roof, or the "saw-tooth" system, came into operation. It was chiefly adapted for power-weaving, and contained many advantages in having the machines on the ground floor, accompanied with a slight degree of moisture, which was considered beneficial to the processes carried on.

Fig. 251.

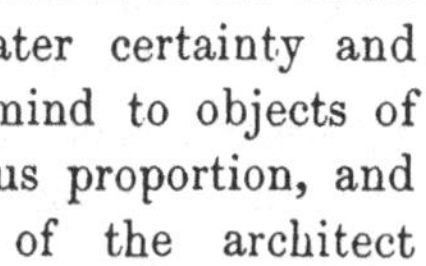

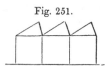

Since the introduction of the shed principle as a convenient building for the reception of cotton machinery, it has been generally adopted for workshops, and various other descriptions of manufactures. The workshops at Woolwich and the Enfield Lock Rifle Factory are upon this principle, which have great advantages where the different processes of manufacture are continuous from one department to another, and where the whole can be carried on by rail or tramway, on the ground floor. It is difficult to estimate the advantages of this description of building for manufacturing purposes; they are, however, considerable, and, where land can be had moderately cheap, it is found superior in many respects, particularly as regards light, to buildings composed of three, four, or more stories.

Another description of building, composed entirely of iron, has been introduced for mill purposes, namely, a combination of cast and wrought iron, as exhibited in a small corn mill, constructed by Messrs. W. Fairbairn, Sons, & Co., Millwall, London, from my own designs in the year 1841. This building was, to the best of my belief, the first iron house constructed in this country, and was from its unique character one of the most successful constructions of its kind. It was formed of hollow cast-iron pilasters composing the standards and framework of the building, which gave it an architectural appearance, and added considerably to its durability and strength. The spaces between the pilasters were filled up with cast-iron plates to the height of the first floor, and the two remaining stories were formed of wrought-iron plates, riveted to the flanges of the pilasters as shown in Figs. 252, 253, 254, and 255, pages 119, 120, 121, and 122. The whole of the vertical sides and ends were covered with a corrugated iron roof, and the floors were supported by iron beams and columns in the usual manner. This building when completed was exhibited at the works, Millwall, and from that time to the present, a large trade in the construction of iron houses, churches, warehouses, &c., has been carried on between this country and the Colonies, India, and America.

CHAPTER II.

ON CORN MILLS.

In the first part of this work was sketched a very brief notice of the antiquity of corn mills, and the state in which grinding went on from generation to generation, until it came down with comparatively little improvement to the present time. During the Jewish period, Moses speaks of the nether and upper mill-stones, and for a succession of ages in Egypt, Greece, and Rome, the quern and some other description of mills, driven by horses or by bullocks, appear to have been in use without change or improvement of any kind. The same may be said of the Middle Ages, which were anything but fruitful of improvement or mechanical invention; and, until the close of the last century, little or no progress was made in the process of grinding, or the development of those principles by which the whole operation is reduced to a system. Corn mills, like every other description of milling, have of late years undergone great changes, and the introduction of steam has given certainty and effect to the mill operations of every description of manufacture, that was inconceivable at a previous epoch. The fact of having motive power at command in every district of the country, mills being no longer dependent upon the state of the wind or the supply of water, has now as nearly as possible supplanted the old wind and water contrivances, and transferred the operation of grinding, with all its necessary improvements, into the very heart of towns. From this certainty of action we derive most of the changes and improvements that are now visible in corn mills, and in every other description of manufacture throughout the country; we have therefore now to show in what these improvements consist, and how they may be maintained upon sounder principles than those known to our forefathers.

The corn mills chosen for illustration are, one of three pairs of stones erected at Constantinople for the Seraskier Halel Pasha in 1842, and the other of thirty pairs of stones for a Russian company at Taganrog, on the borders of the Black Sea. The first was built under conditions that the building should be entirely of iron, that it might not be burnt to the ground by the fires which so frequently occur in the Turkish capital. The other was intended for the purpose of grinding the large supplies of wheat which are grown on the steppes of Southern Russia for the European markets, and also for the supply of bread and biscuits for the Russian navy.

It is now upwards of forty years since the new system of corn mills, having the millstones in a continuous line, was first adopted. For many years it was called in question, and it met with a determined opposition from the old millers and millwrights, who stoutly maintained that the bevil-wheel principle was decidedly inferior to the old plan of stones ranged round a large spur-wheel. Time, however, showed the advantages of the new plan, as it not only proved that the bevil wheels worked as well as the spur, but it gave greatly increased facilities for the operations of the mill in the different processes of cleansing, grinding, dressing, sacking, &c., of the flour, as it came from the machine ready for the baker.

A description of this mill, and of the Old Union Mills, Birmingham, was published some years since by Messrs. Blackie and Son, in their work, entitled 'The Engineers' and Mechanics' Assistant,' and, as that work contains an accurate description of this system of mills, I have extracted from it large quotations, accompanied with improvements, which have been since introduced.

Fig. 252 is a sectional elevation of the mill, the line of section being taken in a longitudinal direction, and exhibiting the position of the stones, the engine, and driving gearing, and of such portions of the subordinate apparatus as are visible on the side of the mill which is exposed to view.

It also exhibits the millstones P P P, sections of the elevators, screw creepers, and the wheat bins or hoppers, Z Z Z.

Fig. 252.

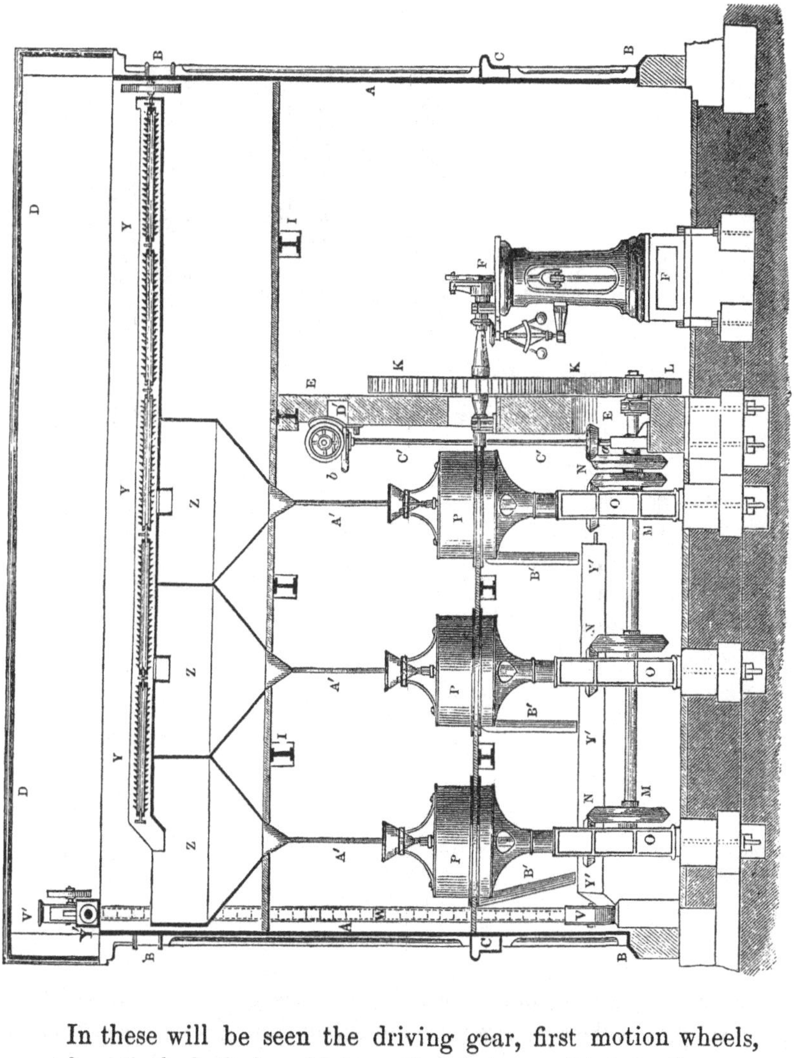

In these will be seen the driving gear, first motion wheels, and vertical shaft, by which motion is transmitted to the machinery in the floors above.

Fig. 253 is a plan of the bottom floor, showing engine and boilers, corresponding to the above, and taken on a horizontal line passing through the lower story of the mill.

Fig. 253.

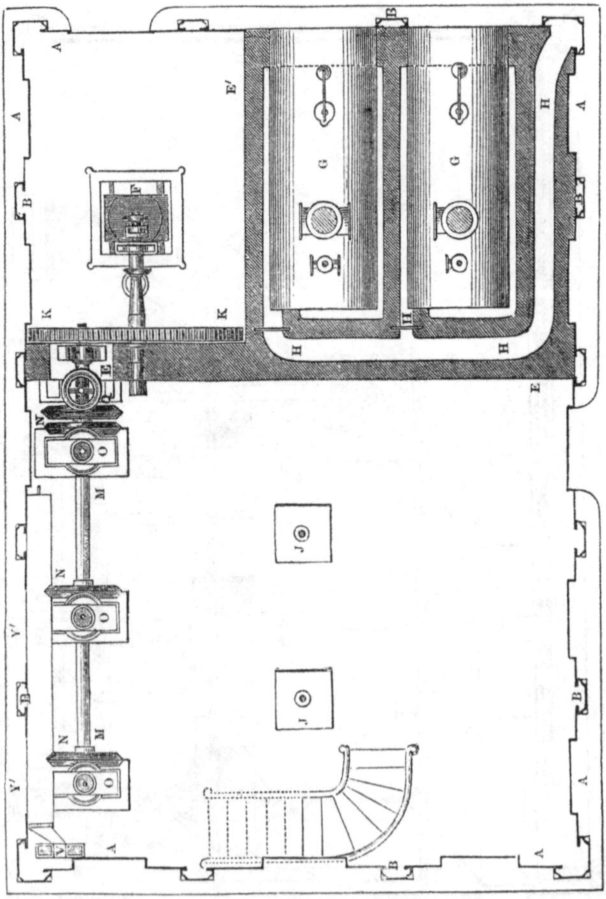

Fig. 254 is a transverse section of the entire mill, in which are shown the garners for undressed and dressed wheat, the mechanism by which it is cleaned and conveyed from the former into the latter, the sack teagle, &c.

Fig. 254.

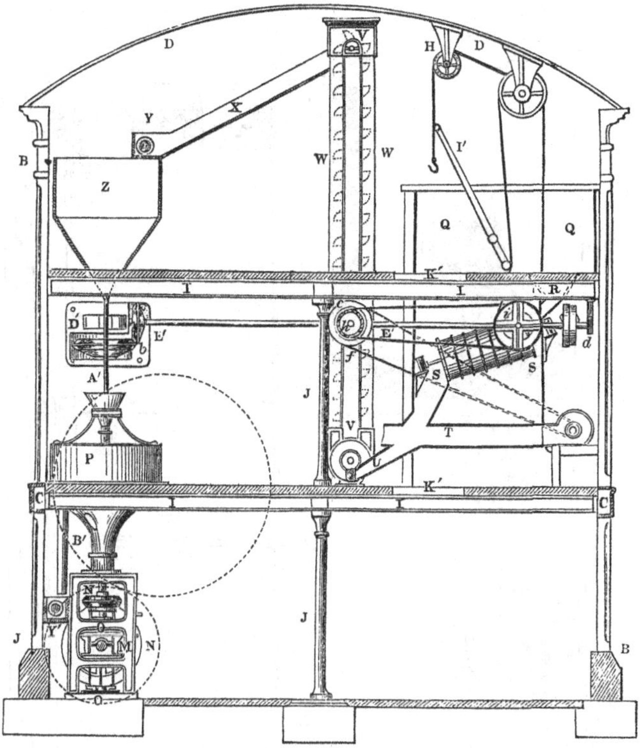

Fig. 255 is a plan of the first floor, corresponding with the plan, fig. 253, the line of section being taken through the first story of the mill.

The house in which this mill is contained consists of an assemblage of plates of sheet iron, A, A, A, of a suitable thickness, consolidated and bound together by the square cast-iron columns, or pilasters, B, B, B, and by the strong cast-iron girders, C, C, C, situated at such a height as to oppose and neutralise the strain of the principal working parts. It is surmounted by an arched roof, D, D, formed of plates of corrugated sheet-iron. A wall of masonry, E, E, is erected in the interior, for the purpose of affording a foundation for the bearings of the heavier gearing of the mill. The motive power is supplied by a high pressure steam-engine, F, of 12-horse

Fig. 255.

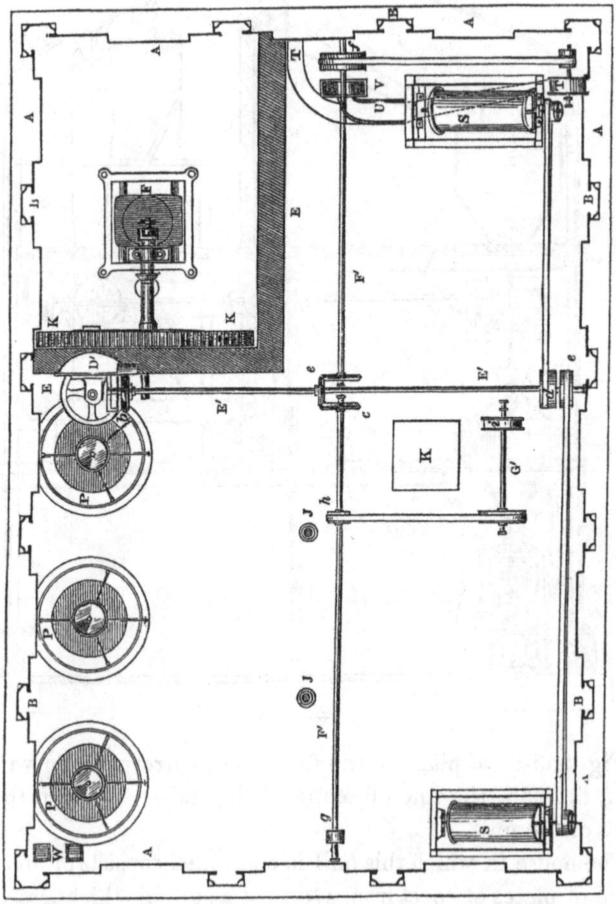

power, the distinguishing feature of which consists in its principal working parts being wholly enclosed within a large cast-iron column. By this arrangement great firmness and stability are imparted to the engine, while the space which it occupies is reduced to the smallest possible dimensions. The boilers, G, G, are situated in an adjoining part of the house, and their flues, H, H, are formed, in the usual manner, of brickwork, abutting on the one hand against the wall E, and on the other against the side of the house itself. Thus the engine and boilers occupy nearly the entire half of the lower story of the mill. The whole erection is strengthened and bound together by the cast-iron beams I, I, I, which pass transversely through the interior of the house, and are supported in the middle of their length by the columns, J, J. On these beams, also, the flooring of the upper and lower flat is disposed.

The fly-wheel K, K, of the steam engine is of that kind denominated spur fly-wheels, from the circumstance of their being formed with teeth, on the exterior of the rim, and thus serving at once to regulate the velocity, and to transmit the power of the steam-engine to which they are attached. The spur fly-wheel, K, K, the diameter of which is 9 feet $3\frac{3}{4}$ inches, gears with the pinion L, of 4 feet $10\frac{3}{4}$ inches diameter; consequently, the velocity of the crank shaft is nearly doubled upon the horizontal shaft M, M, to which the latter is fixed, and which by means of the bevil wheels and pinions N, N, N, gives motion to the stones contained within the mill-stone cases, P, P, P. The shaft M, M, has a bearing upon the wall E, close to the back of the pinion L, and one in each of the standards O, O, O, to which the mechanism necessary for impelling and regulating the action of the stones is attached.

Considerable difference of opinion exists amongst the millwrights of the present day regarding the comparative advantages of spur and bevil gearing as employed for driving grinding machinery. Into this question our limits do not admit of our entering; in our examples we have chosen the latter method, as being that most generally practised. We may, however, be permitted to enumerate a few of the more obvious advantages attending the present system.

1st. It admits of the stones, whatever may be the number employed, being ranged in a straight line instead of in a circle, thereby economising space, and tending to a more convenient and economical disposition of the garners and apparatus by which they are fed.

2nd. It dispenses with the cumbrous and expensive framework necessary for binding together the parts of the system of spur gearing.

3rd. It admits of the employment of wheel-work of a finer pitch, and consequently of a more smooth and equable action, than could be used in the other case. And,

4th. The use of the bevil gearing increases the facility of disengaging at pleasure any pair of stones which may require examination or repair.

We shall now proceed to describe the mechanism of the various processes by which the grain is treated, both previously and subsequently to the process of grinding; and, to avoid repetitions, we shall notice these processes in the order in which they occur.

The corn to be ground is deposited in the upper floor of the mill in the large garner, Q, Q, fig. 254, from which it is conducted through the spout, R, into the screening machine S, S, where it is cleansed of the dust and other extraneous matter which is found more or less combined with it. The corn enters at its upper extremity, and having been thoroughly agitated in its passage through the interior of the machine, and thereby divested of the greater portion of the refuse with which it was mixed, falls into a spout U, at its lower end, which conducts it to the elevator V; being subjected in its passage through this spout to the action of a blast from the fan T, by which the remaining portion of the sand and dust that escapes with the grain, is carried off by a passage leading to the exterior of the house. The grain, after being thus cleansed, is caught by the elevator V, and raised nearly to the summit of the mill, where it is delivered, through an inclined spout X, into the creeper-box Y, Y, by which it is distributed into the feeding garners Z, Z, Z.

The corn which is supplied to the garners Z, Z, Z, falls through the feeding pipes or spouts A′, A′, A′, into the hoppers, by which the grinding apparatus is surmounted. After being reduced into flour, it falls through the pipes B′, B′, B′, into the creeper-box Y′, Y′, by which it is transferred to the elevator V. By this elevator it is again raised to the summit of the house, and carried by means of the creeper Y′ to the dressing machine S′, after passing which the different products are stored up in sacks, or otherwise disposed of, as may be most convenient. The machines and mechanism connected therewith will subsequently be very fully described, and need not here be further alluded to.

The gearing by which the subordinate machinery of the mill is driven, consists, first, of an upright shaft C′, C′, set in motion by a pair of bevil wheels *a*, from the main horizontal shaft M, M. This shaft has its lower bearing in an arched standard embracing the shaft M, and at its upper extremity it is supported by a plummer-block bolted to a double bracket D′, embedded in the wall E. Its motion is here transferred by means of another pair of bevil wheels *b*, to the horizontal shaft E′, E′, passing transversely across the mill. On this shaft are fixed the pulleys *d* and *e*, which drive the screening and dressing machines respectively, and a set of small bevil wheels *c*, *c*, serve to transmit the motion to the longitudinal shafts F′, F′, by which the elevators, creepers, &c., are propelled; as also the short shaft G′, by which the sack teagle is driven.

We subjoin a list of the various wheels, pinions, and pulleys employed in this mill, and the velocities imparted by them to the machines driven by them respectively.

STEAM ENGINE F, 12-HORSE POWER, MAKES FORTY REVOLUTIONS OF CRANK-SHAFT PER MINUTE.

Description of Gearing.	Driver.		Driven.	Result. Revolutions per Minute.
	Diameter.	Revolutions.	Diameter.	
	ft. in.		ft. in.	
Spur pair K, L, .	9 3¾	40	4 10¾	76 on horizontal shaft M.
Bevil pairs N, N, N,	3 6	76	1 10	140 on the stones.
Bevil pair *a* . . .	3 6	76	1 10	140 on upright shaft C′.
Bevil pair *b* . . .	3 0	140	1 9	242 on transverse shaft E′.
Bevil pair *c*, *c* . .	1 1½	242	1 11½	140 on longitudinal shafts F′ F′.
Pulley *d*	1 6	242	1 0	363 on screening machines.
Pulley *e*	1 6	242	1 0	363 on dressing machine S′.
Pulley *f*	2 0	140	0 6	560 on fan T.
Pulley *g*	0 8	140	2 0	46·6 on elevators & creepers.
Pulley *h*	1 0	140	2 0	70 on intermediate shaft G′.
Pulley *i*	1 6	70	2 0	47 on sack teagle H′.

REFERENCES.

A, A, A, the sheet iron sides of the house in which the mill is erected.
B, B, B, columns for supporting and strengthening the structure.
C, C, horizontal beams at the level of the first floor.
D, D, the roof formed of corrugated iron.

E, E, a wall of masonry, affording a foundation for the bearings of the driving gearing, &c.

F, the steam engine by which the mill is driven.

G, G, the boilers.

H, H, the flues and seat of the boilers.

I, I, I, transverse cast-iron beams for supporting the floors, &c.

J, J, columns for supporting these beams.

K, K, the spur fly-wheel of the engine.

L, pinion working into the above.

M, M, the main horizontal shaft.

N, N, N, level mortice-wheels and pinions by which the millstone spindles are driven.

O, O, O, the standards of the grinding machinery.

P, P, P, the millstone cases.

Q, Q, the large garner for uncleaned wheat.

R, spout leading from the garner, Q, to

S, the screening machine.

S′, the dressing machine.

T, a fan attached to the wheat screen.

U, spout leading from the wheat screen to

V, W, the first elevator.

X, passage conducting the grain from the first elevator to

Y, Y, Y, the creeper, by which it is distributed into

Z, Z, Z, the garners for feeding the stones.

A′, A′, A′, the feeding pipes.

B′, B′, B′, pipes by which the flour is delivered into

Y′, Y′, the second creeper-box, conducting it to

V′, W, the second elevator.

a, bevil wheel and pinion giving motion to

C′, C′, the vertical shafts of the mill.

D′, a cast-iron support for the bearings of the vertical and transverse horizontal shafts.

b, bevil wheel and pinion giving motion to

E′, E′, the transverse horizontal shaft.

c, *c*, a set of small bevil gearing, giving motion to

F′, F′, the longitudinal horizontal shaft.

d, *e*, *f*, *g*, *h*, *i*, pulleys for giving motion to the various subordinate machinery of the mill.

G′, intermediate shaft, conveying motion to

H′, H′, the sack teagle.

I′, a lever for stopping and starting the sack teagle.

K′, K′, hatchways by which the sacks are admitted or withdrawn.

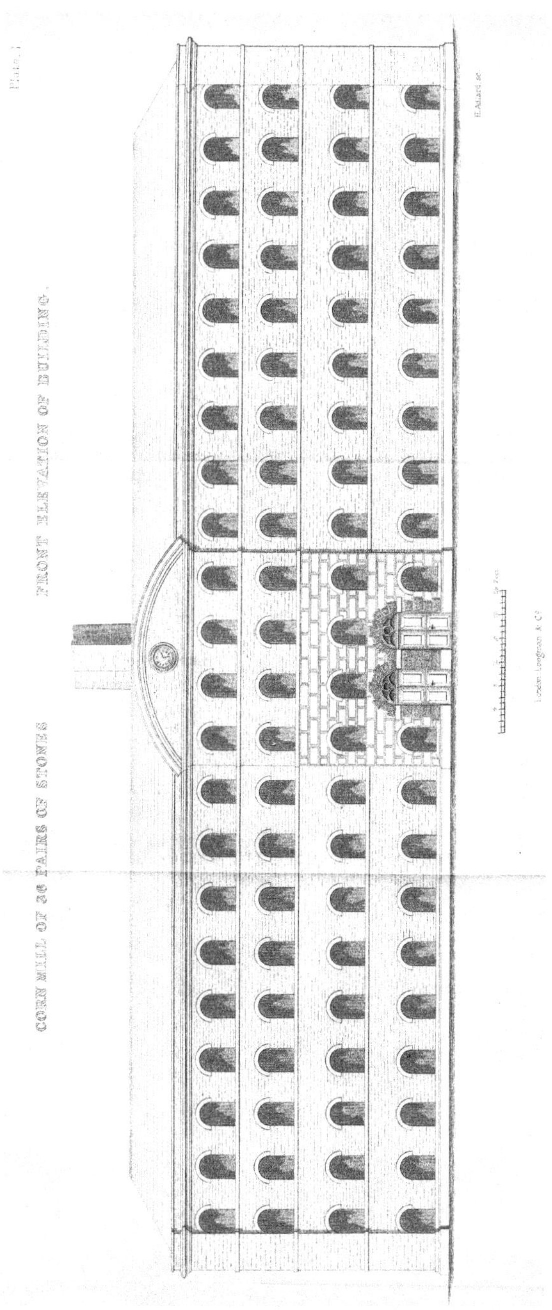

The material originally positioned here is too large for reproduction in this reissue. A PDF can be downloaded from the web address given on page iv of this book, by clicking on 'Resources Available'.

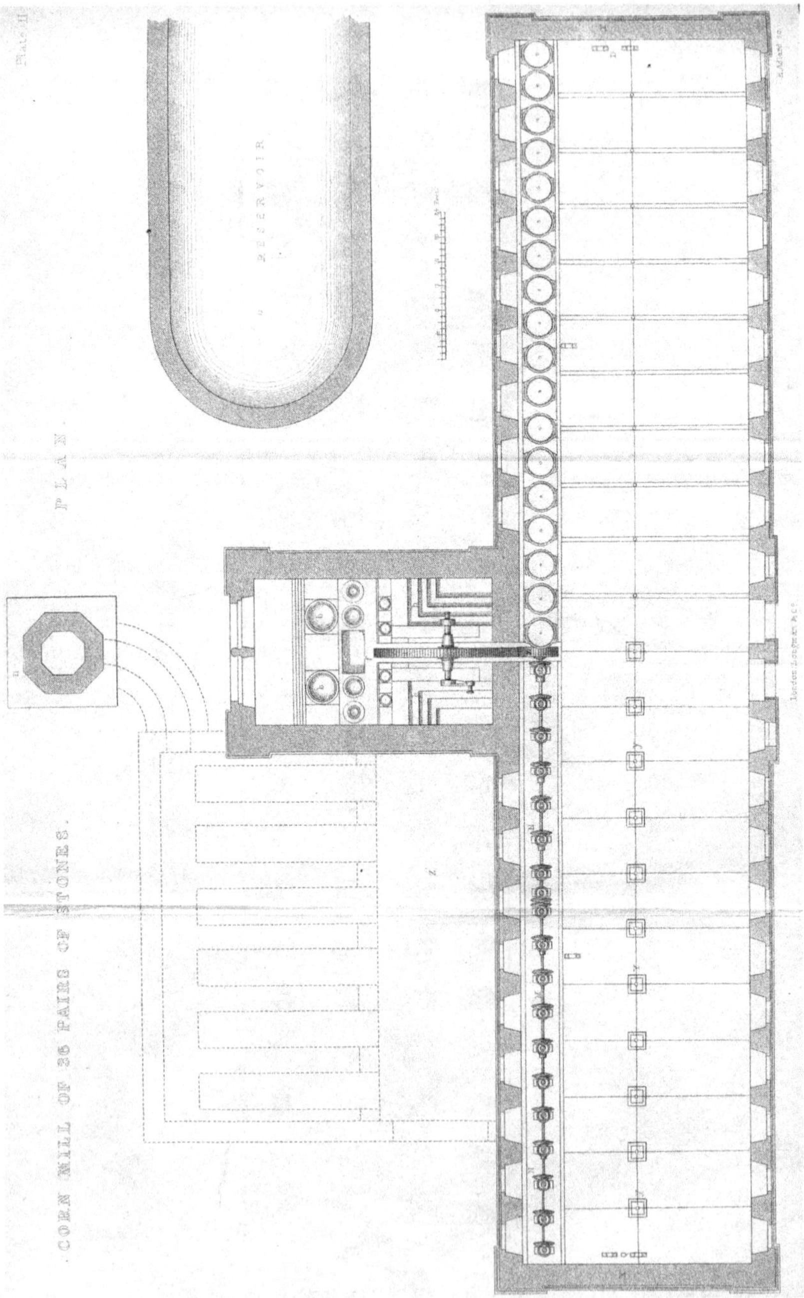

The material originally positioned here is too large for reproduction in this reissue. A PDF can be downloaded from the web address given on page iv of this book, by clicking on 'Resources Available'.

The next example of a corn-mill on a large scale is the recently-constructed mills of Taganrog, on the north shore of the Black Sea. It consists of 36 pair of stones and all the machinery requisite for grinding 180 to 200 bushels of clean dry wheat per hour. It was built for the purpose of a general trade, and a bakery for bread and biscuit for the Russian navy, but the machinery for this latter department has not been erected, and the operations have hitherto been confined to grinding alone.

Plate I. is a front elevation of the mill, and Plate II. a plan showing the position of the engines. The engines work in concert with the cranks at right angles, and the boilers, which are sunk under the surface of the ground, immediately adjoining the engine-house, are fired from the space z'. The chimney w' is placed in a line with the centre of the engine house and the mill. Plate III. is a longitudinal section which exhibits the position of the machinery, millwork, millstones, &c.; and Plate IV. exhibits a transverse section, taken from a line drawn through the centre of the engine house and the flywheel.

The motive power is supplied by two 100 horses power engines united in the flywheel l, 24 feet 5 inches diameter. It is on the new principle of a first motion, with 230 teeth 4 inches pitch, and 14 inches broad on the tooth, and makes 24·7 revolutions per minute. The engines have a stroke of 7 feet, and the steam is supplied by six boilers, each 30 feet 6 inches long and 7 feet diameter. The chimney is octagonal, and placed 40 feet from the engine house, with underground flues and an outlet of 5 feet wide at the top. The walls are 3 feet 6 inches thick at the bottom, and taper to 1 foot 3 inches at a height of 140 feet. The engines are situated behind in the centre of the mill, and the power is given off at both sides of the pinion, which gears into the flywheel at a velocity of 87·7 revolutions per minute. For a distance of 14 feet on each side the main shafts are $8\frac{1}{4}$ inches diameter, for a farther distance they are $7\frac{1}{4}$ inches diameter, and from this point they taper respectively to $6\frac{1}{2}$ and 6 inches diameter on both sides. The distance between the centres of each pair of stones is 5 feet 6 inches, and they are arranged in a straight line running parallel to the walls through the entire length of the building.

The bevil mortice wheels on the main horizontal shaft are 3 feet $4\frac{3}{4}$ inches diameter, and the bevil wheels N N N on the millstone-spindles are 2 feet $1\frac{1}{2}$ inches, and make 140 revolutions per minute. The elevators for meal D D rise perpendicularly at either end of the building to the attic story, and the shafts L L, for driving the subordinate machinery in the different flats, rises to the fifth story, giving off its power to the different machines on its course by bevil wheels. The third story of the mill contains the dressing and bolting machines F F and K K, also the wheat bins V V V for the supply of the corn to the millstones. These wheat-bins are about 11 feet square and 12 feet high, and are supplied by the elevator and creeper Q and U, so that manual labour is entirely dispensed with. The fourth story contains the screening machines T T T, and the garners for the bolting machines H H, these being also supplied by creepers and elevators. The attic story contains the separators S S and the hoists R′ R′, &c., &c.

The wheat enters the mill from the granary at *n′*, and by the elevator Q is carried to the creeper R, by which it is conducted to the separators S S. Having passed through the separators, and the earthy particles, dust and small grain, having been abstracted, it enters the wheat screens T T; here it is subjected to a thorough brushing, and the dust falls through the wire gauze of which the screen is composed. On its way to the creeper U it is subjected to the action of fans J J J, which blow out any remaining dust; it is then conveyed by the creepers to the wheat bins V V V, where it is stored, and supplies itself in sufficient quantity through the feed pipes W W W into the hoppers A A A, and from thence descends to the millstones B B B. After having been ground, it falls through to the meal creepers, by which it is conveyed to the elevators D D, and from thence, by another creeper, to the dressing machines and the garners for the bolting machines H H H. There the meal undergoes the last process, where the flour is separated from the bran and is ready to be stored in sacks for the market.*

* Some few years ago two conical mills were introduced: the one by Mr. Schiele of Oldham, which consisted of a conoid of stone of peculiar form, placed with the base upwards, which merely fitted into a block of stone hollowed out to receive it,

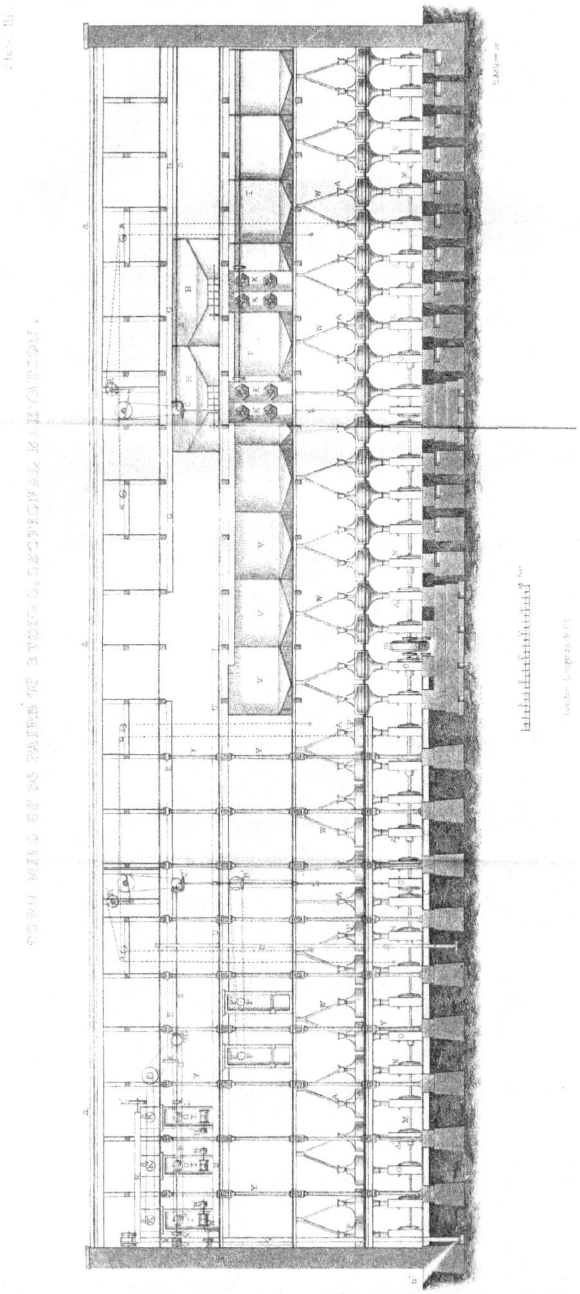

The material originally positioned here is too large for reproduction in this reissue. A PDF can be downloaded from the web address given on page iv of this book, by clicking on 'Resources Available'.

Plate IV

CORN MILL OF 36 PAIRS OF STONES,

ERECTED AT TAGANROG, RUSSIA, IN 1860.

By William Fairbairn & Sons, Manchester.

TRANSVERSE SECTION THROUGH ENGINE HOUSE.

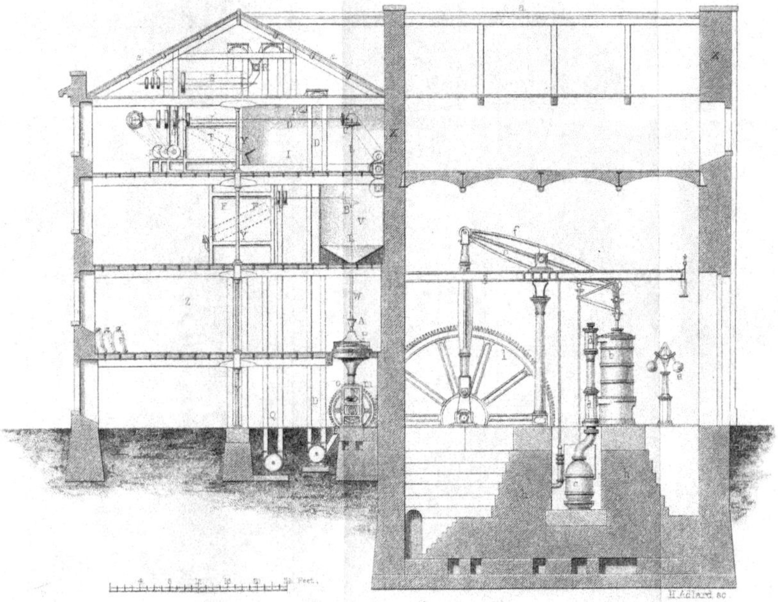

H Adlard sc.

London Longman & Co

The material originally positioned here is too large for reproduction in this reissue. A PDF can be downloaded from the web address given on page iv of this book, by clicking on 'Resources Available'.

This large establishment was originally intended not only for the supply of biscuit for the Russian Navy, but for export in the shape of flour in place of wheat. The Crimean war, the destruction of Sebastopol, and the subsequent treaty of peace, whereby it was agreed that no vessels of war should be retained on the Black Sea excepting those for the protection of commerce, have changed the objects originally contemplated by the erection of these mills, and have caused the company to abandon the Baking department, and confine their operations to the simple process of grinding, dressing, &c.

It is for these reasons that we are necessitated to confine our description to the various processes through which the corn passes; and the construction of the various machines and the mechanical appliances for the saving of manual labour, are described in detail at the end of this section, so that further reference to them here is rendered unnecessary.

In order to make the description as complete as possible, we append a list of wheels and speeds, including the table of references, which will be found serviceable.

and in which it revolved. The form given to these stones was a curve, termed the equitangential tractory, which by revolving on its axis generates the conoid.

The other conical flour mill was the invention of Mr. Westrup. The points of peculiarity in this mill were, that there were two pairs of stones to each mill. Between the two pairs of stones a cylindrical screen of about 2 feet 6 inches high was fixed. The lower instead of the upper stones revolved, and brushes were attached to the spindle, in the space between the two sets of stones, by which the finest flour is brushed through the vertical cylinder. The bottom stones are convex, and make about 250 revolutions per minute; the upper ones are concave, and about 2 feet 6 inches diameter. When the eye-hole is cut out, it leaves some 9 inches of grinding surface, and in that width the bevil of the cone is 4 inches. Cold air is introduced between the stones by means of a fan, which blows out the meal from between them. The stones are fed by means of a hopper placed on one side, with a feed pipe in the top of it, and an upright spindle carrying a dish, which, revolving quickly, evenly distributes the corn.

LIST OF WHEELS AND SPEEDS.

Description of Gearing.	Marks.	Diameter of Wheels and Pulleys. ft.	in.	Number of Revolutions.	Geared into Wheels of Diameter. ft.	in.	Result—Revolutions per Minute.		
Fly wheel	——	24	5	21·7	6	0½	87·7	revolutions of	main driving shaft.
Bevil wheels	A′ A′, &c.	3	4¾	87·7	2	1½	140	,,	millstones and upright shafts.
,,	B′	3	0	140	1	9	242	,,	cross shaft for driving dressing machines.
,,	C′ C′ C′, &c.	2	6	140	2	1½	164·3	,,	shafts.
,,	D′	2	6	164·3	2	1½	192	,,	cross shaft for driving wheat screens, &c.
,,	E′	1	9	192	2	5	140	,,	cross shaft for driving fans and creepers.
Pulleys	F′ F′	2	7½	242	1	3	500	,,	dressing machines.
,,	G′ G′ G′	2	6	192	1	3	380	,,	wheat screens.
,,	H′ H′ H′	2	6	140	0	7	600	,,	fan for screens.
,,	I′ I′ I′	0	9	140	1	6	70	,,	creepers under wheat screens.
,,	J′	0	10	192	2	5½	65	,,	counter shaft for driving separator.
,,	K′ K′ K′	0	8	52	2	3½	20	,,	separators.
,,	L′	1	0	164·3	2	0	82	,,	creepers over the wheat separators.
,,	M′	1	2	164·3	1	8	114	,,	cross shaft for large creeper over wheat garners.
Spur wheels	N′	0	8⅝	114	2	11½	27	,,	large creeper over wheat garners.
Pulleys	O′ O′, &c.	1	0	164·3	2	2¼	75	,,	small creeper over wheat garners.
,,	O O	1	0	164·3	2	2¼	75	,,	small creeper over bolting machines, garners, and cooling room.
,,	P′ P′	1	9	87·7	2	2¼	72	,,	small creeper for meal in ground floor.
,,	Q′ Q′ Q′	0	8	164·3	2	5½	45	,,	elevators for wheat and meal.
,,	R′ R′ &c.	1	2	164·3	3	0	64	,,	chain barrels 10 inches diameter for hoists.
,,	S′ S′	1	2½	164·3	3	0	66	,,	cross shafts for bolting machines.
,,	T′ T′	1	6	66	1	6	66	,,	second cross shaft on bolting machines.
Bevil wheels	U′ U′	0	7	66	1	9	22	,,	bolting machines.
Pulleys	V′ V′	0	6	66	1	6	22	,,	scraper shaft.
,,	W. W′	1	0	22	1	0	22	,,	scrapers.

REFERENCES.

X X, the walls of the mill.

Y Y Y, columns for strengthening the building and supporting the floors.

a a a, the roof of the building.

f, the steam engine by which the mill is driven.

b, the cylinder.

c, the condenser.

d d, the feed pipes.

e, the governors.

g, the main beam.

h h, the foundations.

l, the spur fly wheel of the engine.

m, the pinion gearing into the above.

M M, the main horizontal shaft.

N N, bevil mortice wheels and pinions by which the millstone spindles are driven.

O O, the standard frame and inverted cone supporting the millstones.

P P, the millstone cases.

Q, the elevator for lifting the wheat from the garners to the separator.

R, the creeper to supply the separators.

S S S, the separators.

T T, the wheat screens.

J J, the fan attached to the wheat screens.

U U U, the creeper conveying the grain to

V V V, the wheat bins, and through

W W W, the feed pipes, to

A A A, the hoppers, and to

B B B, the millstones.

C C C, the creeper for conveying the ground corn to

D, the meal elevator.

E E, the creeper for conveying the ground corn to the cooling rooms ready to be supplied as required to

F F, the dressing machines.

G G, the creeper for conveying the ground corn to

H H, the garners of the bolting machines.

K K, the bolting machines.

L L, the shafts for driving the subordinate machinery.

B′, the bevil wheels giving motion to the cross shaft of the dressing machines.

C′, bevil wheels giving motion to
D′, the cross shaft for screening machines.
A′A′, bevil wheels on upright shaft.
I, the dust chamber for the wheat screens.
n, the chimney.
o, the reservoir for supplying the boilers of the engine.
Z, the store room for flour.
Z′, the space from which the boilers are fired.
E′, cross shaft for driving fans and creepers.
F′ F′, pulleys for driving dressing machines.
G′ G′ G′, pulleys for driving dressing machines.
H′ H′ H′, pulleys for driving fans for screens.
I′, pulley for creeper under wheat screen.
J′, counter shaft under separator.
K′ K′ K′, pulleys for separators.
L′, pulley for creeper over wheat screen.
M′, cross shaft for creeper over wheat garners.
N′, spur wheel for large creeper over wheat garners.
O′, pulley for small creeper over wheat garners.
P′ P′, pulleys for meal creeper on ground floor.
Q′ Q′, pulleys for meal elevators.
R′ R′, the chain barrels of the hoists.
S′ S′, cross shafts for bolting machines.
T′ T′, second cross shaft for bolting machines.

During the siege of Sebastopol it was determined, on the urgent recommendation of Assistant-Commissary General Julyan, to effect an arrangement for supplying the troops daily with new bread and fresh flour from the grain of the surrounding country, by providing the means of converting the wheat into flour and baking it upon the spot by a floating mill and bakery. Having been consulted as to the best means of carrying out this proposal, drawings and plans were prepared for the mills and ovens, and two iron screw steamers, subsequently named the Bruiser and the Abundance, were purchased by the Government for adaptation to this purpose, and were fitted with machinery by Messrs. William Fairbairn & Sons, the whole being completed in less than three months.

It is curious to trace the history of the means by which large bodies of men have been supplied with food, and the obligations assumed by states for provisioning armies in times of war. We learn that, in the early period of Roman history, grain was the only article of food issued to the soldiers, and was ground by means of a hand mill, which formed a part of every man's equipment; the flour was simply worked into a paste called *puls*, which constituted the principal food. The constitution of modern armies and the peculiar character of modern warfare render the soldier, however, more dependent upon the cares of the administration than was the case with the ancients; and we have seen how prostrate and helpless they are, when deprived of the resources of a well-conducted and far-seeing commissariat. The French, Spanish, and other continental troops can live upon a moderate allowance of vegetable and farinaceous food, and a lump of oil cake will maintain a Russian for a week; but it is widely different with the English, who become disorganised when their rations fail. Under these circumstances it is a matter of essential importance to maintain a system of daily supply; and hence followed the introduction of the floating mill and bakery.

The arrangement of the floating mill is shown in figs. 256, 257, 258, and 259. Fig. 256 is a longitudinal section of the vessel; fig. 257 is a plan of the machinery, with the decks removed and partly in section; and figs. 258 and 259 are transverse sections of the vessel.

The mill machinery is all driven from the propeller shaft A, fig. 257, which is driven by the engines B; and the whole of the processes are performed without the aid of manual labour. The wheat is stored in the forehold of the vessel, and is raised by an elevator into the screw-creeper C, which conveys it into the corn-dressing machine D, where it is cleaned and winnowed. Thence it is again conveyed by the elevator E and the screw-creeper F into the hoppers G G for feeding the mill-stones H H, by which it is ground. The grain is fed to the stones by the silent feeders I, now in general use in this and foreign countries. After being ground by the mill-stones H, the flour or meal is delivered into the screw-creeper K, which conveys it to the elevator L, by which it is delivered into the flour-dressing machine M; it is here freed from the bran and filled into sacks, having been separated into a fine and coarse quality. This

Fig. 256.

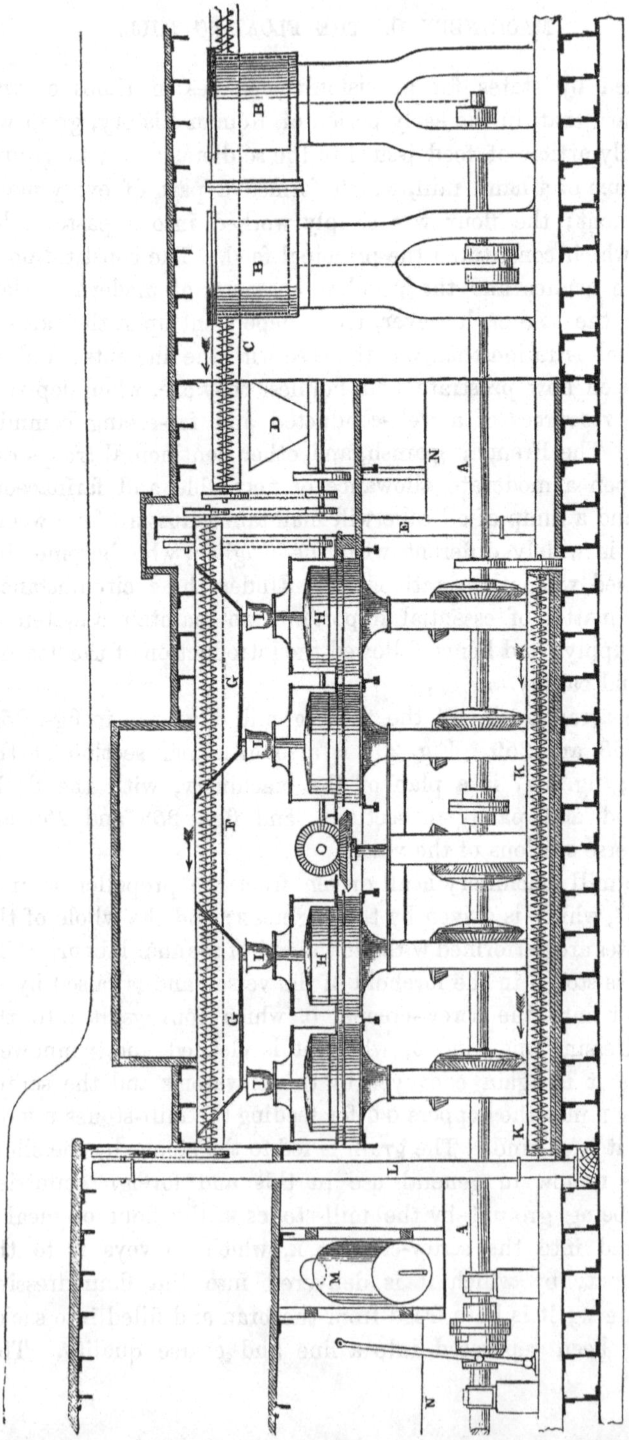

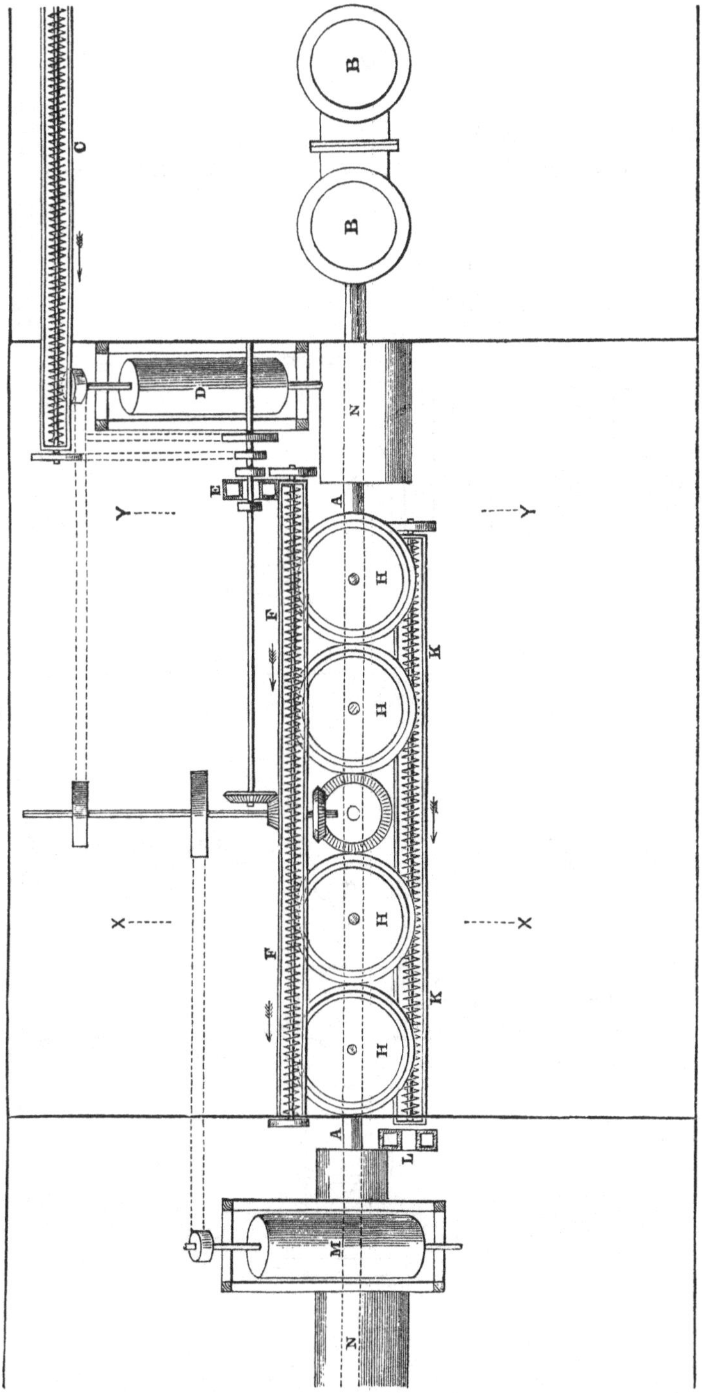

Fig. 257.

Fig. 258.

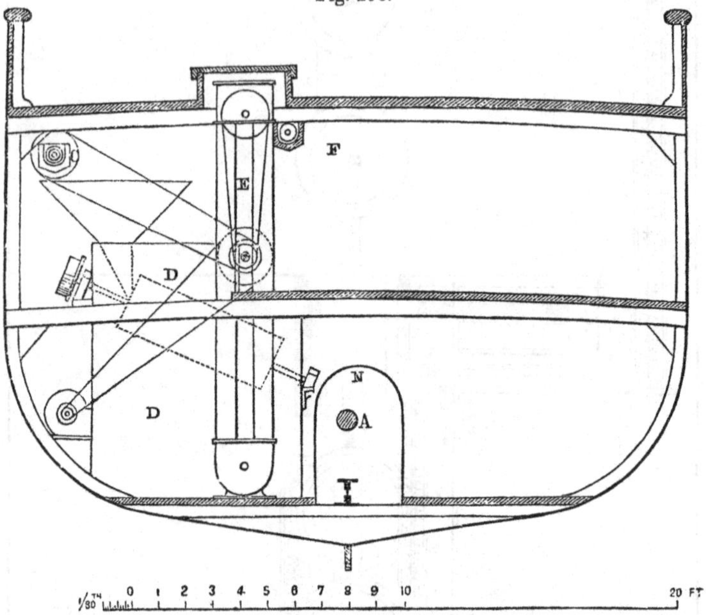

Fig. 259.

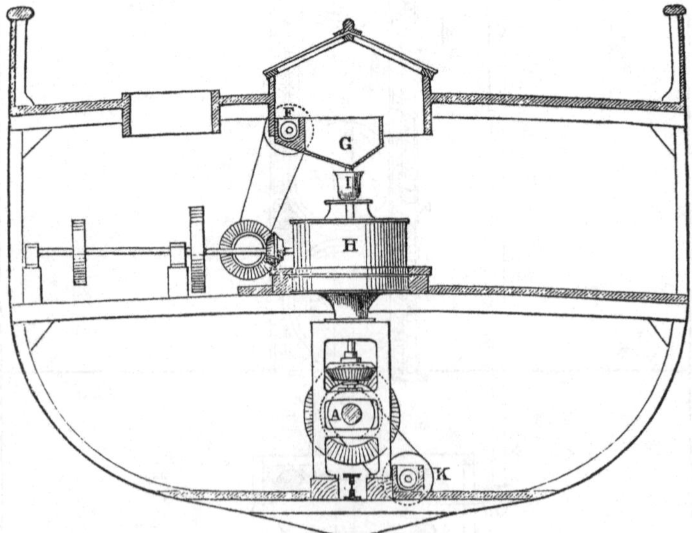

completes the whole process. The propeller shaft A is exposed under the mill-stones, but covered by an iron trough N in the other parts of the vessel.

During the time the vessel was in harbour at Balaklava, the daily produce of flour from this mill was about 24,000 lbs., and that from very hard wheat, full of small gravel, and consequently the more difficult to grind. It was originally intended that the mill should be capable of producing 20,000 lbs. of bread per day, but it proved equal to a considerably larger production; and not the least important of its good qualities was that it never got out of order during the whole period of service in the Black Sea. From the reports made to the Government at home respecting its working, it appears that important advantages were gained by the introduction of this machinery for the use of the troops. There is probably no description of food so essential to the maintenance of health and the recovery of the sick as fresh flour and fresh bread; and the salutary effects produced upon the health of the troops and the number of lives saved in the late war, by the abundant supply of wholesome bread and flour that was poured into the camp during the latter part of the siege, forcibly suggest the necessity of a light portable steam-engine and mill for grinding being constantly attached to the camp, whenever or wherever an army takes the field. This can be done at a very moderate cost, and, in my opinion, no army should attempt to take the field without it. The whole affair would not exceed the weight of a heavy siege gun, such as now accompanies our armies; and there appears no practical difficulty in the way of introducing an engine capable of supplying newly-baked bread, from an oven constructed in the smoke-box of a portable locomotive engine mounted on wheels and prepared to grind at the same time.

The results of the working of the floating corn mill are given in the official reports at 20 tons of flour ground per day of 24 hours when constantly in full work; and 18,000 lbs. of bread in 3 lb. loaves produced daily from the bakery. This rate of work was continued uninterruptedly for many months, and the machinery answered completely the object intended. The total quantity of bread produced in three months from 1st January to 31st March, 1856, was 1,284,747 lbs.; and the expenses of working were 2,017*l.*, or 3*s.* 2*d.* per 100 lbs. of bread made, including the expense of a sea establishment for the vessel, which would not be required where the vessel was stationary. The quantity of flour ground in the same period was 1,331,792 lbs. with 358,172 lbs. of bran, the wheat supplied being 1,776,780 lbs.; the expenses of working were 2050*l.*, or 2*s.* 4*d.* per 100 lbs. of wheat ground, or 3*s.* 1*d.* per 100 lbs. of flour produced. The total cost of the flour produced was therefore about 25*s.* 3*d.* per 100 lbs., the wheat costing about 18*s.* per 100 lbs., and the value of the bran being deducted at 7*s.* per 100 lbs. or less than 1*d.* per lb.

The grinding of wheat was found to be performed quite satisfactorily whilst the vessel was at sea, even in a heavy swell causing an excessive motion, which tried the fitness of the machinery for the work to an unusual degree; the grinding whilst the vessel is performing her voyage being obtained from the same power that propels her. On one occasion, when the vessel was steaming 6½ knots or 7½ miles per hour, 10 sacks of 168 lbs. each or 1680 lbs. of wheat were ground per hour, and the mill was kept in constant work for 35 hours, the men being divided into watches of four hours each; the mill continued working well throughout, and was found to run more regularly than when the screw was disconnected.

The mill machinery of the Bruiser is similar to that ordinarily employed on shore in this country, with such modifications only as were necessary to adapt it to its novel position, and fit it to sustain the constant and varying motion of the vessel at sea. These difficulties were overcome, and the mill was found to answer admirably, grinding, in almost all weathers, at the rate of 20 bushels or 1120 lbs. of flour per hour, and that at a time when the vessel was steaming at 7½ knots or 8½ miles per hour, both the mill machinery and the ship being propelled by the same engines, which were constructed by Messrs. Robert Stephenson & Co., of 80 horses power.

Details of Machinery.

The elevator consists of a long endless chain of small buckets formed of tin plate, and mounted at regular distances upon a leather or canvas band passing over two pulleys inclosed within the cast-iron frames V, V and the wooden boxes W, W.

Fig. 260.

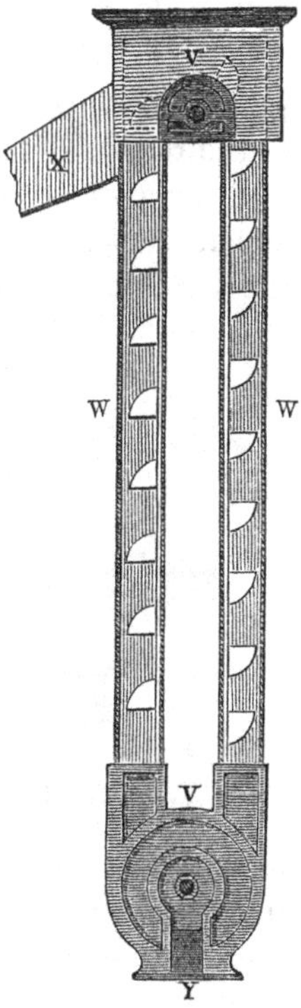

The uppermost of these pulleys is driven at a moderate velocity by a belt, and the buckets passing in succession the opening Y (which is kept constantly supplied with the material to be raised by means of an inclined spout) become each charged with a certain portion, which is carried up one side, and from which it falls through the spout X into the garner or the machine prepared for its reception. From this it will be seen that the buckets, having discharged their contents into the spout X, descend empty on the opposite side, ready to receive or take up their respective load as they pass the feed-spout Y, under the lower pulley.

The lower part of the elevator is shown in fig. 261. It consists of two cast-iron

Fig. 261.

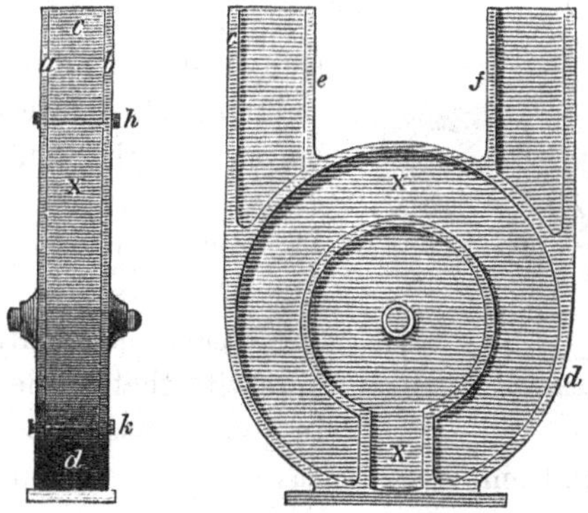

frame plates *a*, *b*, bolted together, about a quarter of an inch in thickness, upon which grooves are cast for the reception of the sheet-iron plates *c*, *d*, and *e*, *f*, which are bent to the shape of the frame plates, and inserted in the groove, forming, when the bolts *h*, *k*, are screwed tight, a neat and compact box.

The creeper, fig. 262, is an archimedian screw worked in conjunction with the contrivance just described, which is applicable to the raising of the grain or flour from a lower to a higher level. For horizontal transport modern millwrights make use of this apparatus, which is an application of a well-known principle for the abridgement of manual labour. This creeper consists of a

Fig. 262.

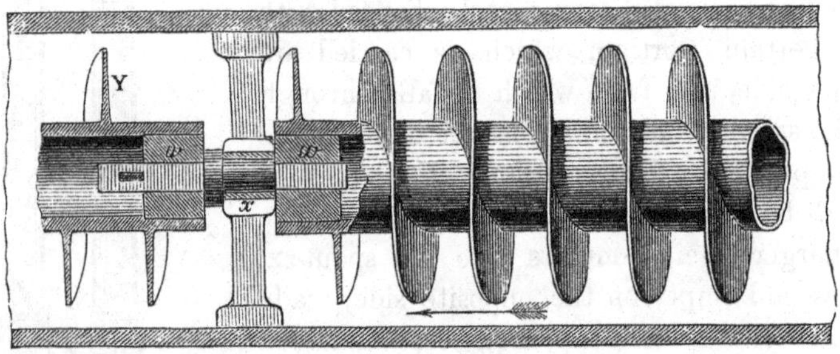

long endless screw with a wide pitch and projecting thin threads, inclosed in a wooden box or trough, of dimensions slightly greater than its own diameter. It is made to revolve on its axis by means of a belt and pulleys, at a velocity corresponding with that of the elevators, and, being restricted from moving longitudinally, the threads, or rather leaves, of the screw force the grain introduced at one end of the trough to the other. The action of the screw in the case of the creeper is identical in its nature with that of the endless screw in giving motion to a worm wheel. The material employed is cast-iron; the creeper is made in

Fig. 263.

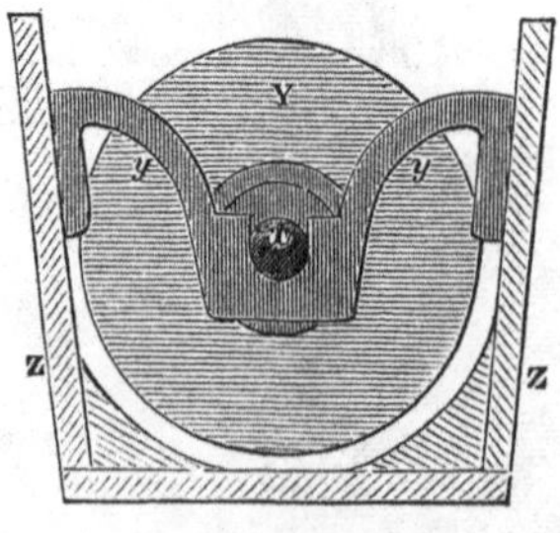

6 feet lengths, each length being in the form of a tube $3\frac{1}{2}$ inches in diameter, and about $\frac{3}{8}$ of an inch thick, with broad leaves or threads cast round it, after the manner of an archimedian screw. The thickness of the threads does not exceed $\frac{3}{16}$ of an inch at the outer extremity.

The different lengths of which the entire creeper is composed are joined together by short wrought-iron studs *x*, *x*, fig. 264, forming also the journals on which it revolves. These are made with square tails fitted into similar holes formed in the

Fig. 264.

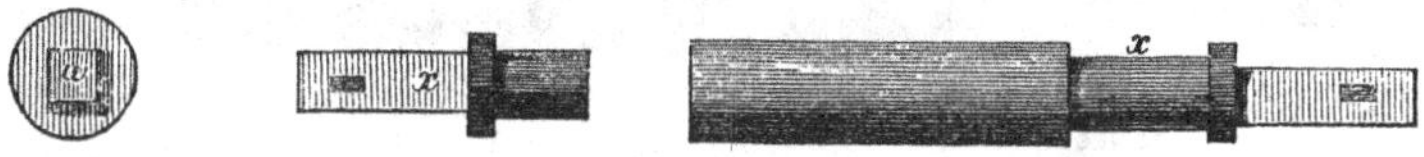

centre of the small cylindrical blocks *w w*, which are carefully turned on their exterior surfaces, and driven into the open ends of the pipes Y Y, figs. 262 and 263, previously bored to the same diameter.

This construction at once insures a strictly rectilinear axis for the entire range, whatever may be its length. The arrow, fig. 262, indicates the direction in which a creeper, constructed in the manner shown in the general view, would propel the grain.

The separator, fig. 265, consists of a wire cylinder A B, divided into partitions by a screw C, with long leaves and coarse thread, through which the wheat slowly travels and keeps falling by its own weight, as it moves forward from one extremity of the screen to the other. The space underneath the wire cylinder is divided into chambers by partitions. From *e* to *f*, in the first compartment, the wire of the cylinder is so close as only to admit of dust passing through; from *e* to *g*, in the second, the wire is arranged for only small wheat; and from *g* to *h*, in the third, the large grains of wheat are freely delivered into the spout *k*, from which they are conveyed to the wheat screen. Should there be any stones or large substances in the wheat, they are cast out at *s*. A whalebone brush P B, serves to free the machine from all grains of wheat which fasten themselves between the meshes of the wires. As fast as the brush wears away, its position is regulated and maintained by the adjusting screws *l m*. The cylinder and the screw are connected together, and revolve at a velocity of 20 revolutions per minute.

Fig. 265.

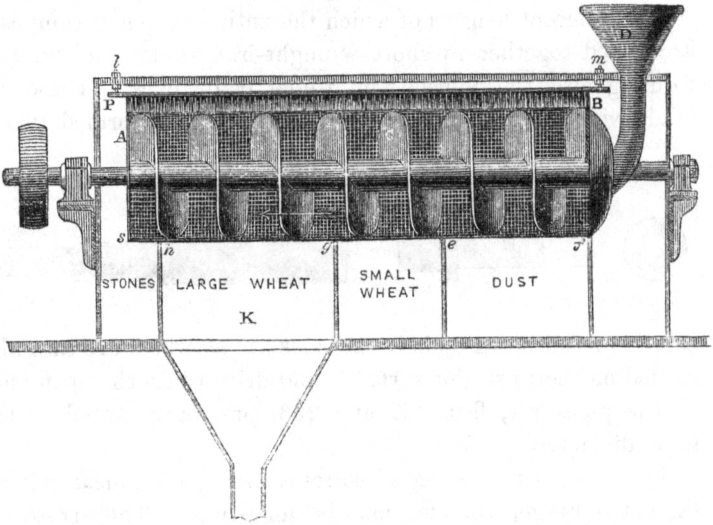

The screening machine is constructed as shown in the drawing, fig. 266. The cylinders *a b* of these machines vary from 4 feet to 8 feet long. The annexed is a small one, 4 feet 2 inches long and 1 foot 4 inches diameter, covered with 6 by 8 wire gauze, and has an inclination of 1 in 2.

Rings of wood *e*, *f*, *g*, $\frac{5}{8}$ thick and $2\frac{1}{4}$ deep, are placed round the cylinder at distances of $4\frac{3}{4}$ inches. To bind the rings together, horizontal planking is used, as shown at *c* and *d*, in the transverse section, by reference to which it will be seen that the cane brushes do not touch the wire gauze, but that $\frac{3}{16}$ of an inch clearance is allowed. The brushes revolve in the cylinder at a velocity of from 350 to 400 revolutions per minute, and by this action the corn, which enters the machine by a wooden spout through the opening *l*, is brushed; the dust passing through the wire gauze falls into the close chamber K, whilst the corn travels to the end of the cylinder *a b*, where a spout *m* is fitted for its reception. On its passage from this spout to the space *o*, it is exposed to the action of a current of air from the fan G, which clears it of dust through the aperture S.

Fig. 266.

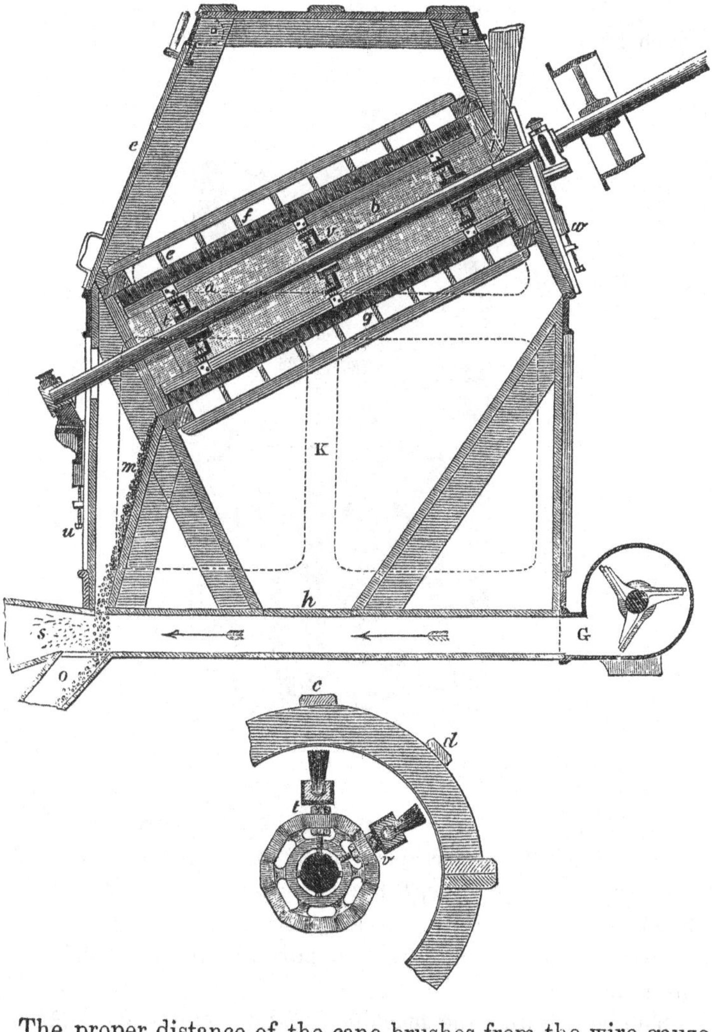

The proper distance of the cane brushes from the wire gauze is maintained, as they wear away, by the screws *t* and *v*, and the shaft on which the brushes are fastened is adjusted by the screws *u w*; *h* is the dust-hole at the bottom of the machine, 12 inches square, through which the dust is taken as often as necessary from the chamber K. By the door *e* convenient access to the machine is afforded, and the dusting and cleaning rendered easy.

Fig. 267 is an outside elevation of the screening machine, which shows the cast-iron frame lined with wood. The screws

Fig. 267.

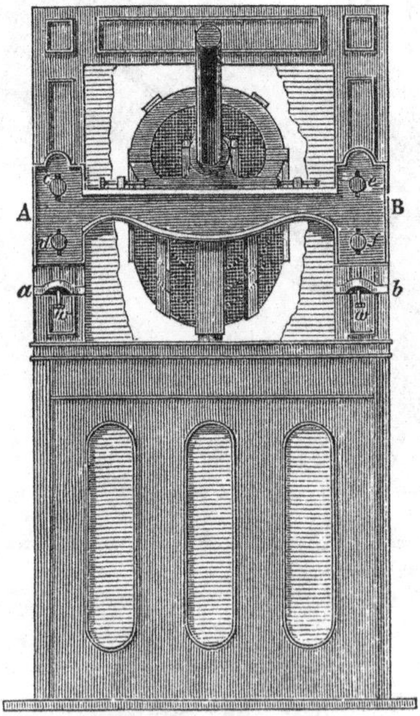

w w, which regulate the shaft on which the brushes are fixed, are screwed into projections *a b* cast on the frame, and by the slots and screws *c*, *d*, *e*, *f*, the whole of the brackets A B can be raised or depressed as necessity requires.

In some mills, besides the separator already described, a smut machine is used.

The corn enters the machine by the spout *a*, fig. 268, and falls upon the cast-iron plate *b c*, which, revolving with the shaft *d*, and the beaters *e*, *f*, *g*, *h*, *k*, *l*, at a velocity of 450 to

500 revolutions per minute, throws the corn to the extremity of the cylinder *m n*, where it comes into contact with the beaters, which, together with the fan *o′*, as the corn passes through, take away the greater portion of the dust and foreign matter. Owing to their not being surrounded by a wooden framing, these machines are placed in a closed room by themselves.

Fig. 268 represents a section of a machine of this kind; the

Fig. 268.

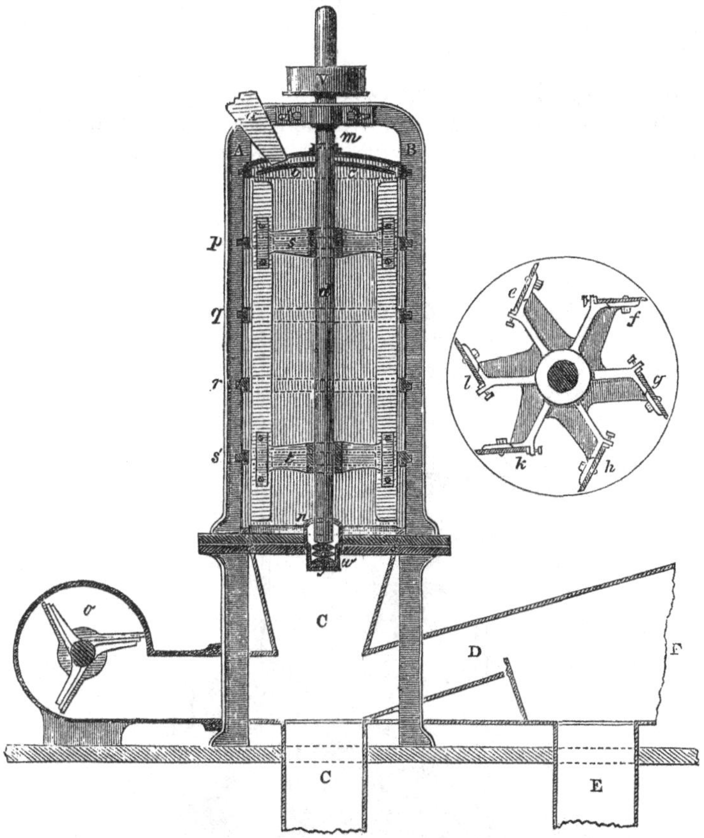

cylinder *m n* is 1′ 7″ diameter, by 4′ 6″ long, covered with 6 by 8 wire gauze, and is surrounded at intervals of 9 inches by rings of iron p, q, r, s'. Inside the cylinder two arms of cast iron s and t are placed, and bolted on these arms are six pieces of iron, as shown in the cross section. The arms are keyed on the shaft d, which is driven by the pulley V, for the bearing of which circular rings α, β, γ, are made and fitted in the box w, and the whole is supported by the framework A, B. By the spout C, C, the corn is conveyed to the screening machine, or the millstones, and by the passage D the hollow grain and heavy dust fall into the spout E, and the dust into a chamber at F.

This description of smut machine is the one most in use at present, having completely superseded the American wheat screen.

Another sort of wheat screen or smut machine has been patented by Mr. James Waltworth.

In consists of a number of cylinders $a\,b$, $c\,d$, $e\,f$, $g\,h$, fig. 269, covered with wire gauze. The outer coating of wire gauze is

Fig. 269.

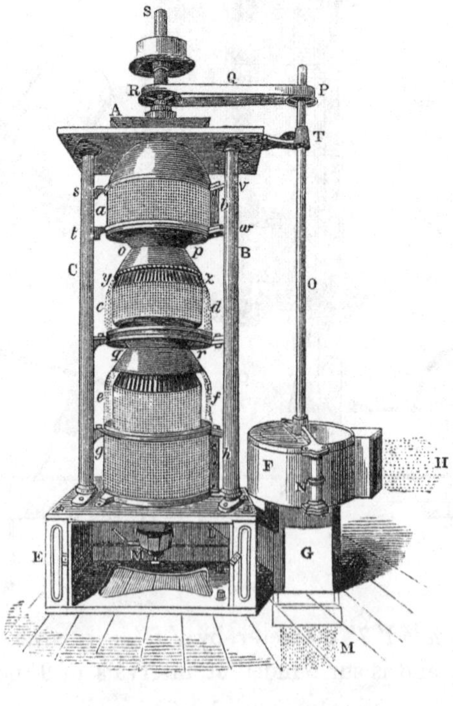

taken away from the cylinders *c d*, and *e f*, in order to show the internal cylinders also covered with wire gauze and grooved plates. These cylinders are supported by rings of cast iron fastened to the projections *s*, *t*, *v*, *w*, &c., on the cast-iron columns C, B, &c.

The wheat enters the machine, which revolves at a velocity of 500 revolutions per minute, by the spout A; it then passes between perforated grooved plates in the first cylinder, like those shown at *o p*, *y z*, in cylinder *c d*. After passing through these perforated grooved plates where it is scrubbed, *r*, the corn descends between the wire gauze of the internal and external cylinder where (the outward cylinder being stationary, and the internal cylinder revolving) it is exposed to a severer brushing, without, however, injuring the grain. Having passed through all the cylinders successively, it falls into the spout L, and on its passage to the part M, it is exposed in the box G to the action of a powerful fan F, which draws up all the light grain dust, &c., and ejects them at H. By the cast-iron box E L, convenient access is afforded to the footstep M. The opening K, admits the air under the perforated plate for the fan. N is a pedestal for the fan shaft O, and T, a bracket which serves at its bearing. The fan is driven by the pulley P, which is connected by the strap Q with the pulley R, keyed on the shaft S.

The value of this machine consists in the large amount of scrubbing surface over which the corn passes. Slight modifications of this machine are made for Egyptian corn, which is first washed, and then run through the cylinders of the machine.

The Framing. — A cast-iron standard or framing A, A, fig. 270, securely bolted to a stone foundation by two holding down bolts, encloses the principal part of the driving and adjusting gearing for each pair of stones. It is made in the form of an oblong box, and is traversed by two horizontal diaphragms or partitions, cast of a piece with it, the upper one for sustaining the footstep of the mill-spindle and its adjusting apparatus, and the lower for carrying the plummer block of the driving shaft. It is surmounted by a large bell-shaped casting B, B, called the cone, firmly bolted, by a flange at its lower end, to the standard, while the upper extremity is expanded, and terminates in a cylinder of a diameter somewhat greater than that of the mill-stones, in which the lower (called the bed-stone) rests, and is secured within it. Two straight and broad flanges are cast at opposite sides of the cylindrical part, for the purpose

Fig. 270.

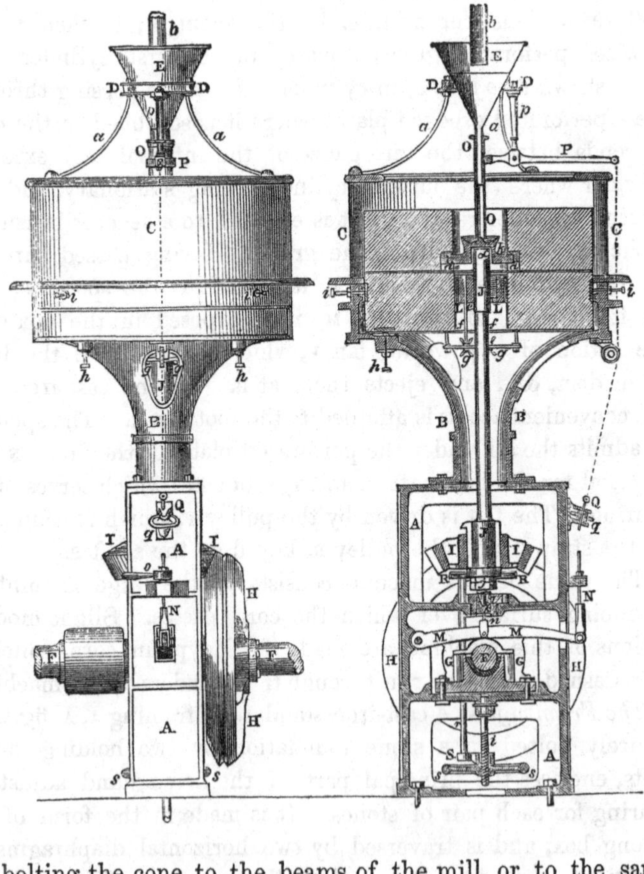

of bolting the cone to the beams of the mill, or to the same part of the framing of the contiguous pairs of stones; while another circular flange passes all round, for sustaining the flooring. Three large openings are left in the upper part of the cone to give access to the interior, and it is provided with suitable arrangements for the reception of the several adjusting screws required for the setting of the lower stone.

Fig. 271 represents a plan of the cone showing the openings and the number and disposition of the adjusting screws above alluded to.

Fig. 271.

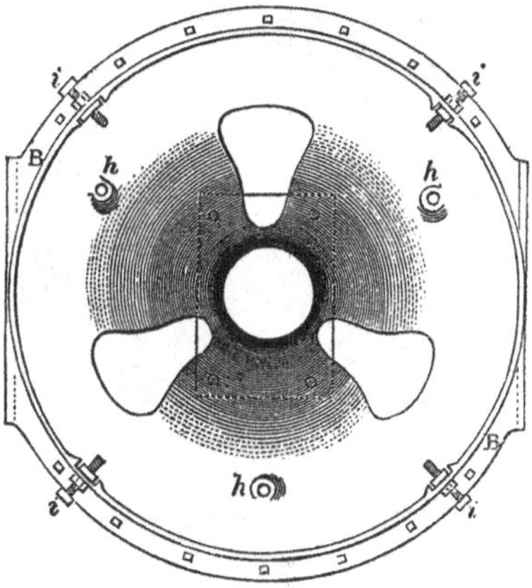

The Stone Case and Feeding-Hopper.—Above the cone, and of the same diameter with the cylindrical part of it, is placed the stone case C, fig. 270, which surrounds the upper stone, and serves to confine the flour which is the result of the grinding. This is simply a cylinder of thin sheet iron, resting upon the stone floor, and having affixed to the top of it a ring of wood, on which the tripod for supporting the feeding apparatus is set. This cover is made open in order to admit the air freely between and around the stones during the process of grinding. A cast-iron ring D, fig. 270, supported by three malleable iron legs *a*, *a*, *a*, forms a sort of tripod in which is placed the hopper E, which receives the grain from the garners above, through the feeding-pipe or spout *b*, and supplies it to the stones by means of the feeding apparatus to be hereafter described. A piece of coarse wire gauze is placed in the hopper, to intercept any foreign body that may descend with the grain.

An enlarged view of the ring D is shown in fig. 272.

The Driving Gear.—The driving shaft F, is part of the line of horizontal shafting which is common to the whole range, and receives its motion from the prime mover, generally through the intervention of a single pair of wheels. The ve-

Fig. 272.

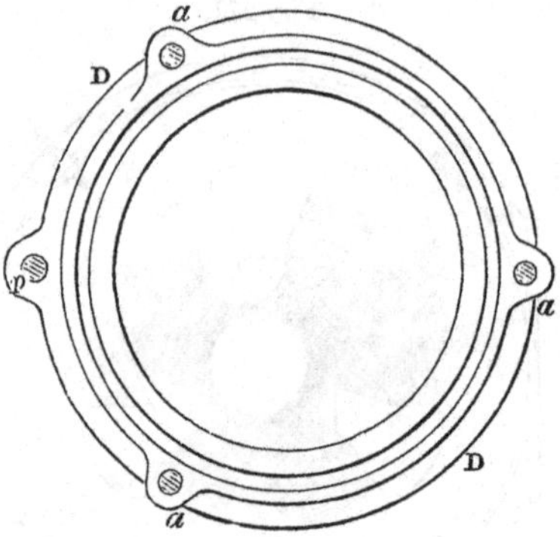

locity of this line of shafting is usually from 70 to 80 revolutions per minute, with stones of the diameter of those in our examples. The different lengths of which it is composed are connected together by couplings of the same description as that described and represented at fig. 213, page 81 of this work. The shaft F revolves in brass bearings, fitted into a plummer block G, fig. 273, bolted to a sole formed, as before noticed, in the standard A, Fig. 274. The strain of the shaft being entirely in a downward direction, this plummer-block requires no cover, the journal being simply protected from injury by a slight brass cap.

Fig. 273.

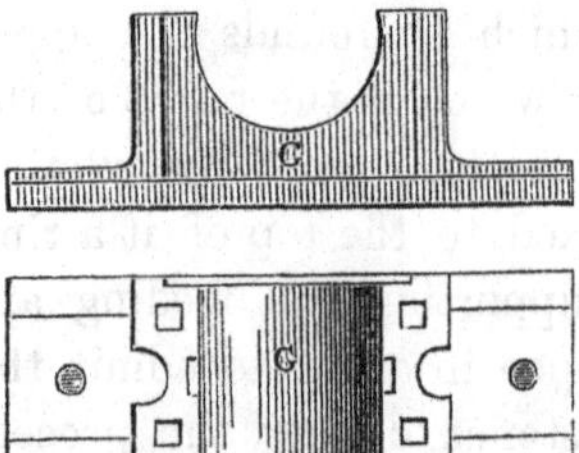

A large bevil mortice wheel H, figs. 270 and 274, working into the pinion I, on the mill-spindle, serves to transmit the motion of the shaft F to the latter. These wheels are made with the greatest possible care and accuracy, so as to work together very smoothly. The pinion is not fixed immovably upon the spindle, but is capable of sliding vertically upon it by means of a sunk feather.

The Mill-spindle and its Appendages.—The mill-spindle

J, J, fig. 270, is made of the best forged iron, accurately turned over its entire length; and rises perpendicularly through the standard A, the cone B, and the lower millstone. It is attached

Fig. 274.

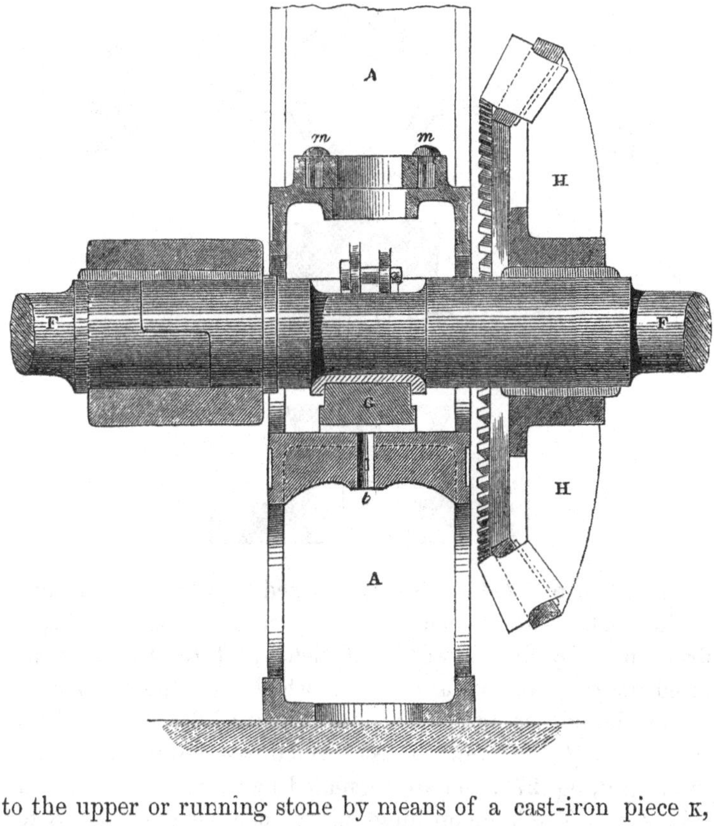

to the upper or running stone by means of a cast-iron piece K, figs. 270 and 275, called the Rhind, which combines this function with that of regulating and delivering the supply of grain to the stones. It will be observed by the drawings, fig. 275, that it forms a species of universal joint, the small steel cross-head *c c*, on the top of the mill-spindle, fitting into corresponding bearings in the rhind, while the projecting tails *d*, *d*, cast upon it at right angles to the former, work in similar bearings formed of small cast-iron pieces sunk into the stone. By this arrangement it will be observed, that the connection between the mill-spindle and the upper stone is complete, while at the same time it admits of the free and unconstrained action of the latter against the grinding surface of the lower stone.

Fig. 275.

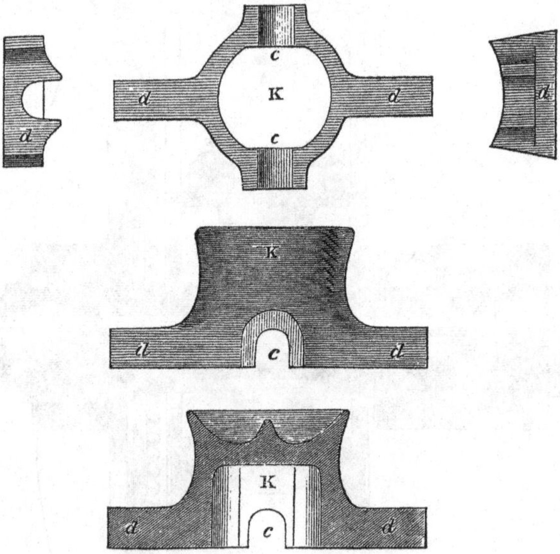

The lower or fixed stone is perforated by a large square hole in its centre, into which the cast-iron block L, fig. 276, is firmly fixed by slips of wood and wedges. Into this block are fitted the three brass bushes *e*, *e*, *e*, which form the upper bearing of the mill-spindle. These are adjusted by means of the wedges *f*, *f*, *f*, the screwed tails of which pass downward through the ring *g*, fig. 277, and are regulated by thumb screws on each side of it. The large openings in the cone, before alluded to, afford access for the working of these screws. Small semi-circular chambers are formed in the socket L, fig. 276, between each bush, and filled with hemp and tallow for the lubrication of the mill-spindle; and the whole is carefully protected from dust by slips of sheet-iron screwed over it.

The Millstones.—The diameter of the millstones most in use at the present day is 4 feet, and their thickness about 12 inches; one half of this thickness is composed of French burr, a very hard, though porous mineral, of a siliceous nature; the other half is made up of plaster of Paris. In consequence of the

Fig. 276.

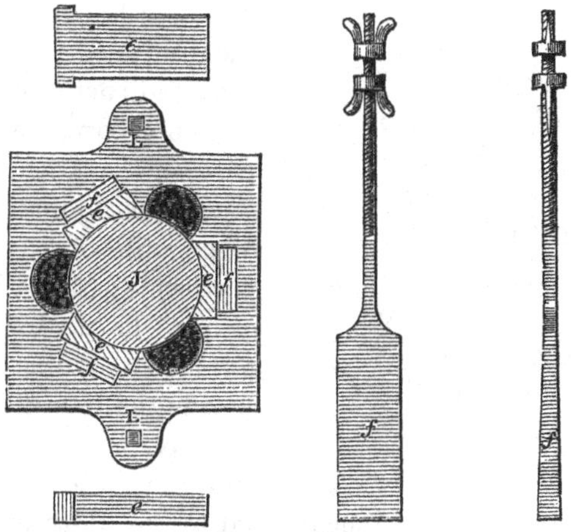

Fig. 277.

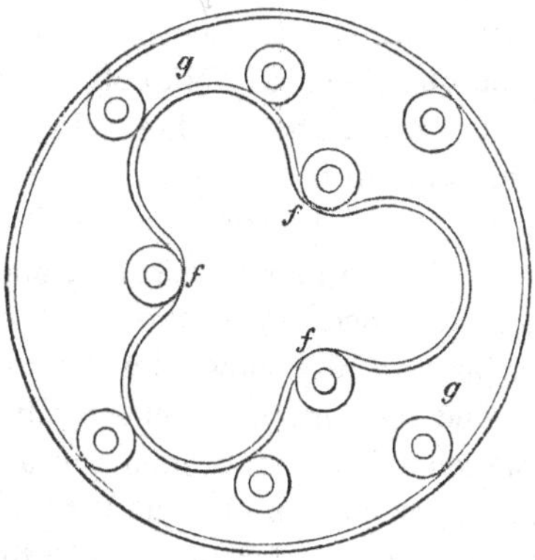

difficulty of obtaining sufficiently large masses of the French stone, it is usual to construct the millstones in segments, which are cemented together, and the whole firmly bound by iron hoops passing round the circumference. The lower stone is, in the first instance, carefully dressed into a perfectly flat, plane surface, but the upper one is made slightly hollow for a small distance from the central aperture, so as to allow the grain to be freely admitted between the stones. Being thus prepared, the circumference of the stone is divided into 11 equal parts, fig. 278; lines are drawn from each division to the centre; these radii determine the limits of the grooves in each compartment. A chord b' c', is then drawn, joining the bounding radii of any two compartments; this chord is, of course, bisected by the intermediate radius J a', in d'. Divide the line d' c', into four equal parts in the points e', f, g', and from these points mark off, on the line d' c', distances equal to the width of the groove to be cut, which is generally from $1\frac{1}{4}$ to $1\frac{3}{4}$ inches wide, then draw through all these points of division lines parallel to the radius J a', terminating them in the radius J c'. These are the outlines of the grooves, which are then to be cut into the stone, perpendicularly on one side, and obliquely on the other, so that each furrow shall have a sharp edge. The direction of the grooves being the same in both upper and lower stones, as they lie on their backs in the position proper for being cut, it is obvious that when the former is reversed and set in motion, their sharp edges will meet each other after the manner of a pair of scissors, (as partially shown by the dotted lines in fig. 278), and thus grind the corn more effectually when it is subjected to the action of the unbroken surfaces between the channels. The land, or portion of stone between each furrow, is cut like the teeth of a file with from 11 to 18 fine grooves to the inch, at an angle of about 45° with the furrows. A brush is placed on the side of the top millstone to clear the cylindrical cover of the flour at each revolution.

Fig. 278.

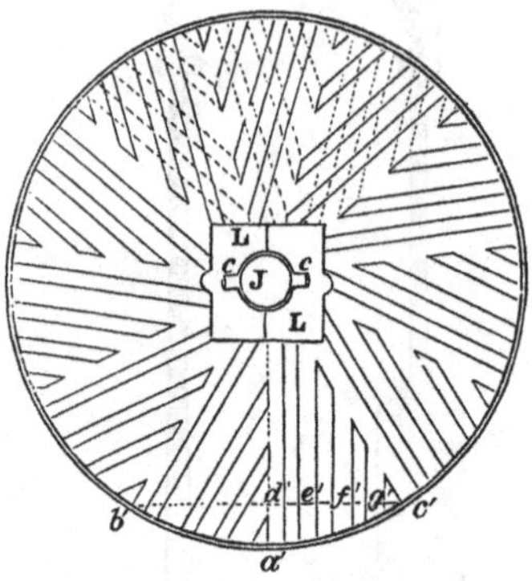

Adjustment of the Lower Stone.—It is essentially important to the proper working of any pair of stones, that the grinding surface of the lower stone should be perfectly level, and that its centre should be exactly perpendicular above that of the lower bearing of the mill-spindle. To secure the former of these conditions, three pinching screws *h, h, h,* figs. 270 and 279, are fitted into the cone, (that number being greatly preferable to four in adjusting the level of any surface), and, bearing against small slips of iron sunk into the stone, it can be raised or depressed by them to any required extent. Fig. 279, *h*, shows the screw for levelling the lower stone on an enlarged scale. The centering of the stone is effected by means of four pinching screws *i, i, i, i,* figs. 270 and 271, acting horizontally upon it. Fig. 279, *i*, shows this screw on an enlarged scale. To secure it against deviating from the truth after having been properly adjusted, all these screws are provided with jam nuts.

Fig. 279.

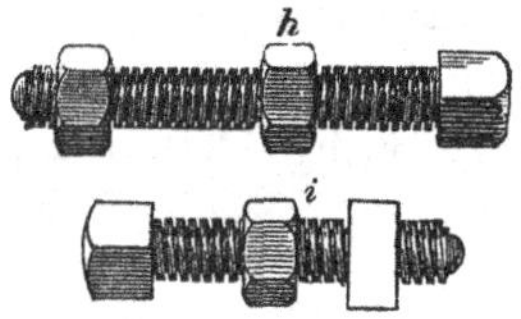

Adjustment of the Mill-Spindle.—The lower bearing or footstep of the spindle J, fig. 270, is also made capable of nice adjustment, both horizontally and vertically. The former is necessary in order to ensure the accurate working of the driving wheel and pinion, and the latter to regulate the pressure of the upper upon the lower stone, and to compensate for the changes produced upon both by the frequent dressings which their grinding surfaces have to undergo.

The footstep *k*, figs. 270 and 280, which is of gun metal, is turned and fitted accurately into a cast-iron socket *l*, resting on the upper diaphragm of the standard A, fig. 270; the hole into which it is inserted, and the annular recess by which it is surrounded, being made of somewhat greater diameter than the corresponding parts of the socket itself. Its exact position is determined and secured by the four radial pinching screws *m, m, m, m,* fig. 281, passing through the ring and working in nuts fitted into recesses cast upon its interior surface, fig. 282.

Fig. 280.

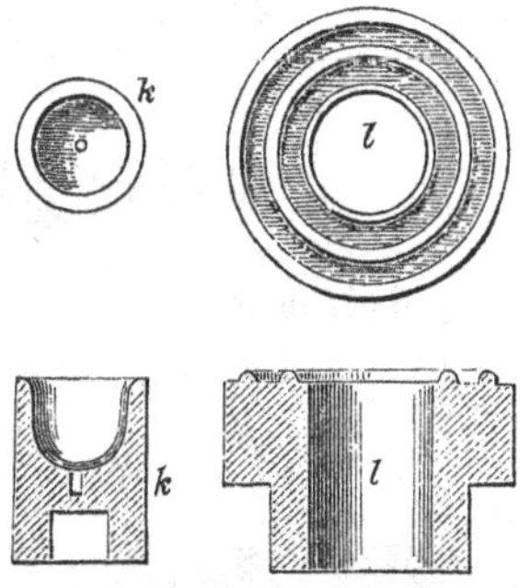

Fig. 281.

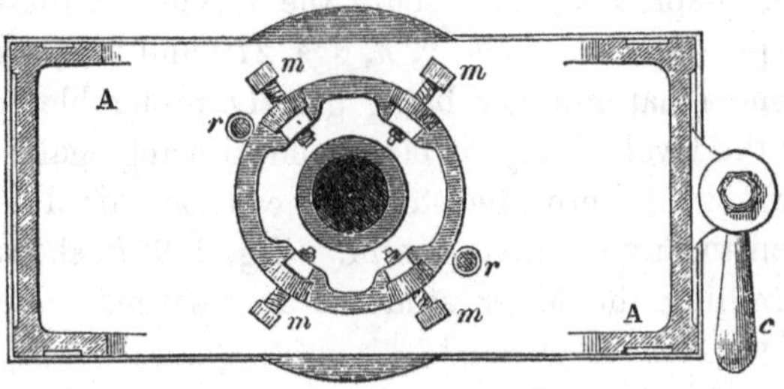

Fig. 282.

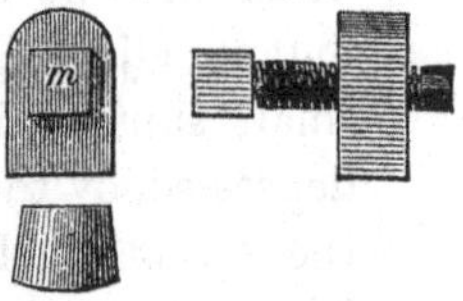

The footstep *k*, is not fixed immovably into the socket *l*, but is capable of sliding vertically into it. Its proper position in this direction is regulated by means of a strong wrought-iron lever M, figs. 270 and 283, having its centre of motion in the

Fig. 283.

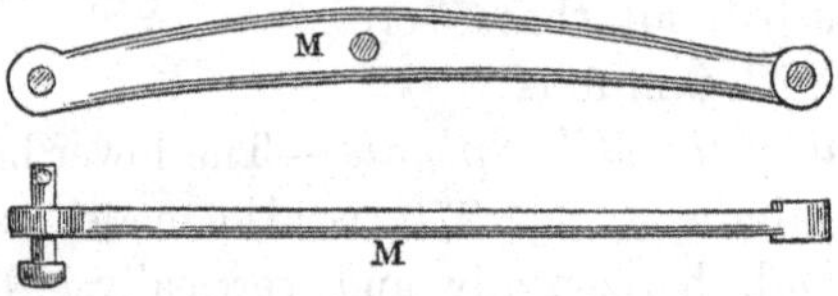

back of the standard A, while its opposite end projects through a slot, and is raised or depressed by means of a screwed rod N, figs. 270 and 284, jointed to it, and passing through a project-

Fig. 284.

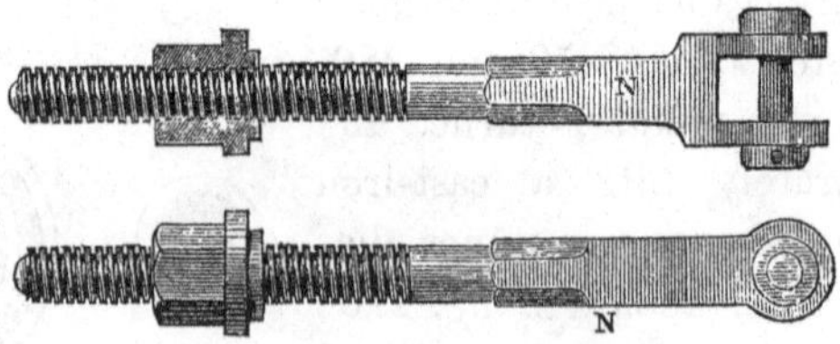

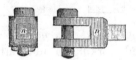

ing shelf cast upon the front of the standard. A small link or saddle *n*, figs. 270 and 285, serves to connect the lever with the footstep *k*; the saddle being provided with a square tail, which is inserted into a similar recess in the under side of the footstep; by which means the latter is prevented from turning in its socket. Thus it will

Fig. 285.

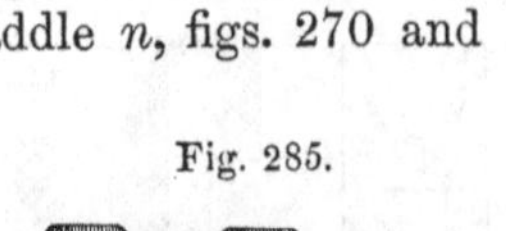

be seen that the entire weight of the upper stone and mill-spindle rests upon the lever M, and that the miller is enabled to vary at pleasure the pressure upon the grain between the stones, and consequently the degree of fineness of the flour produced, by simply turning the nut of the screw N, by means of the key *o*, figs. 270 and 286.

Fig. 286.

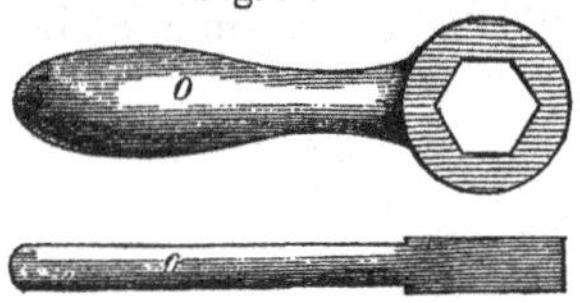

The Feeding Apparatus.—By the old process the stones were fed by a clapper, fixed on the top of the running mill-stone, striking the end of a shoe or trough fixed to the under-side of the hopper, through which the grain travelled, when slightly inclined, into the eye of the stone, every time it received a blow from the clapper. An improvement on this primitive apparatus was effected by the introduction of the damsel, formed of an iron spindle with three blades which acted against the side of the shoe, causing a vibratory motion, and propelling the grain forward in a uniform stream into the eye of the stone, as shown at *a*, fig. 287. The damsel was a

Fig. 287.

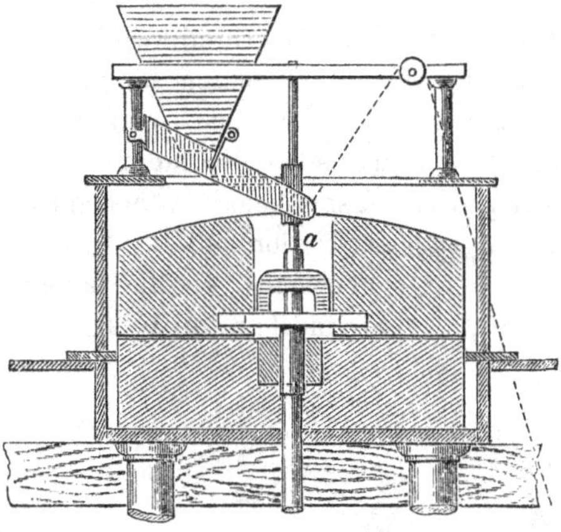

great improvement on the clapper, and continued in general use until the centrifugal or silent feed was introduced.*

* On its first introduction I had to contend against many opponents, who looked upon the new system of feeding as an innovation. Had it been a patent it would probably have come much earlier into general use.

In this arrangement, the supply of grain admitted between the stones is regulated by means of a cast-iron pipe O, figs. 270 and 288, open at both ends, the lower end being brought

Fig. 288.

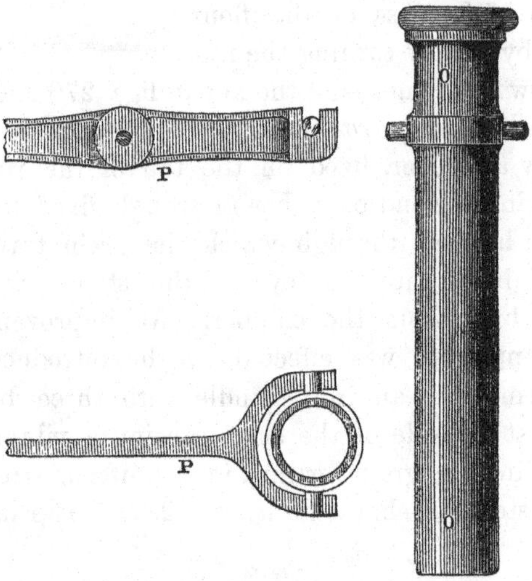

into close proximity with the rhind, while the upper part encloses the pipe in which the feeding hopper E, fig. 270, terminates. It is suspended by means of a cast-iron lever P, which has its fulcrum in the small column *p*, fig. 270, depending from the tripod D. A small chain attached to the end of the lever, and passing over a friction pulley at the bottom of the stone case, serves to connect this feeding apparatus with an ingenious little piece of mechanism Q, figs. 270 and 289,

Fig. 289.

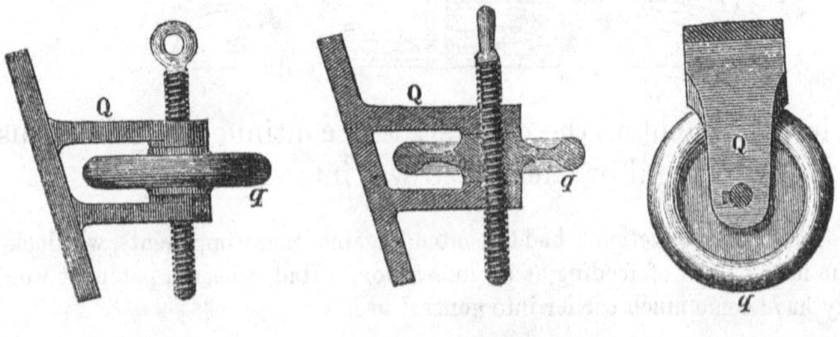

attached to the standard, by which the miller is enabled to regulate the supply with the greatest nicety. This contrivance, as shown in fig. 289, consists of a small hand-wheel *q*, working between the cheeks of a double bracket bolted to the standard A. A small screwed pin forms the axis of this wheel; it passes freely through the cheeks of the bracket, but is screwed into the eye of the hand-wheel, and is prevented from turning with it by means of a feather inserted into the former, and fitting into a groove cut throughout the entire length of the pin, to the upper end of which the chain is attached. By this arrangement it is obvious that by turning the hand-wheel *q*, to the right or left, the small pin will be raised or depressed, and through the intervening mechanism, as shown in fig. 270, the size of the opening between the mouth of the feeding-pipe O, and the rhind cup, will be increased or diminished. On slightly raising the tube O, by the lever P, fig. 270, the grain slides along the surface of the cup or rhind K, which, revolving at a rapid motion, discharges the grain with tangential force over the rim of the cup between the runner and the bed-stone. The grain, as it escapes from under the tube, flies off in tangents from the edge of the cup, and forms a beautiful series of curves, as shown in the drawing, fig. 290. At first this system of feeding, like all other improvements, met with opposition from the millers; but the regularity of the supply, and the absence of noise, soon convinced them of its advantages, and paved the way for its general application.

Fig. 290.

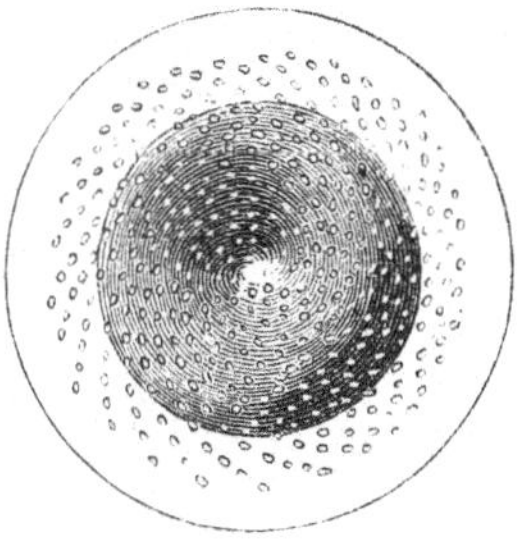

The Disengaging Apparatus.—The driving pinion I, fig. 291, is fitted upon the mill-spindle, so as to be capable of sliding up and down upon a sunk feather. When fully engaged with the wheel H, fig. 270, it rests upon a collar formed on the upper surface of a large brass nut *j*, fig. 292, fitted to a screwed part on the lower end of the spindle, and capable of being fixed by a jam nut, by which the miller is enabled to keep the pinion invariably in its proper position with regard to the wheel, independently of the position of the spindle, which, as we have before had occasion to remark, requires to be slightly lowered

Fig. 291.

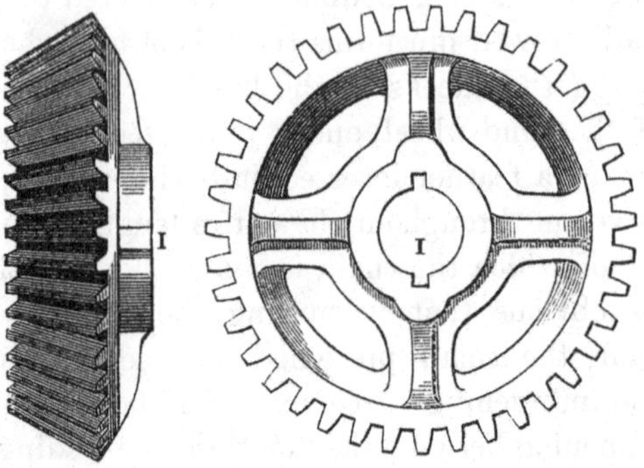

every time the stones are dressed. When properly adjusted, the pinion is secured to the spindle by a taper key.

Fig. 293.

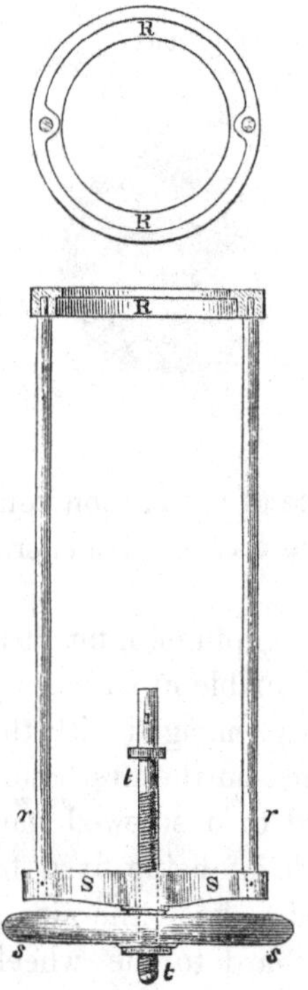

Fig. 292.

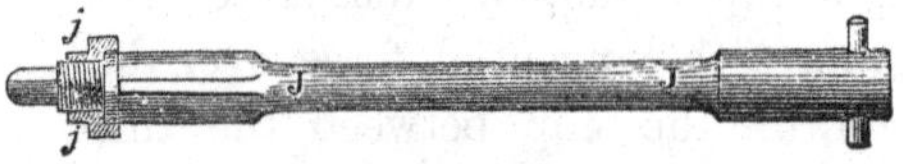

It is, however, necessary to throw each pair of stones, periodically, out of gear with the general range, to admit of their being dressed, &c. For this purpose the tapered key is removed, and the pinion raised out of contact with the teeth of its driving wheel, by means of a species of jack or lifting apparatus attached to the standard, the component parts of which we shall now briefly enumerate.

A cast-iron ring R, figs. 270 and 293, supported upon two upright rods *r*, *r*, is brought in contact with the under surface of the pinion by turning the hand-wheel *s*, which is screwed upon its axis *t*, and carries with it in its ascent the cross-head S, into the ends of which the lower extremities of the rods *r*, *r*, are inserted. The screw *t* is fixed into a socket cast upon the under surface of the lower diaphragm of the standard, and the connecting rods *r*, *r*, which pass through holes formed for their reception in both diaphragms; being set in a diagonal direction in order to clear the lever M, and other important parts of the machinery. Fig. 294 is a plan of the hand-wheel and cross-bar.

Fig. 294.

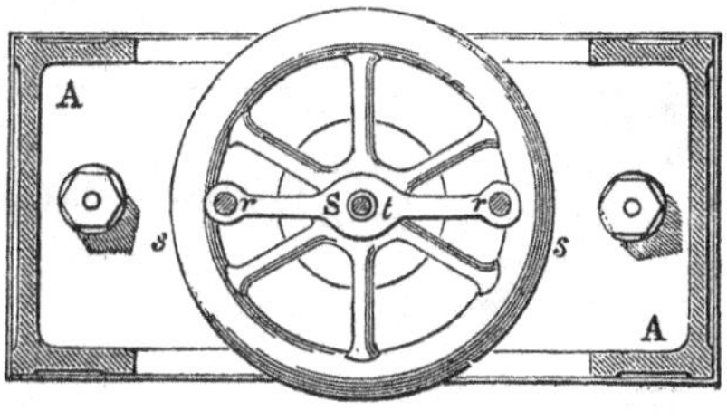

On turning the hand-wheel in a contrary direction, the weight of the pinion again brings it into its working position.

The Stone-lifting Apparatus.

The portable crane or lifting apparatus used for raising the upper stones from their beds, and depositing them on the floor of the mill when they require to undergo the process of dressing, is shown in fig. 295. It consists of a strong malleable iron arm T, T, bent into a form nearly approaching to a quadrant; the lower end works in a cast-iron step *v*, inserted into the stone floor, while its upper extremity is supported by a strong rod U, fitted to rotate upon a stud *u*, fixed into a cast-iron plate, bolted to the beams which support the floor above. The fixed centres, *u* and *v*, are so situated that the machine shall command two contiguous pairs of stones, and the end of the rod U is made of such a form as to admit of its being easily disengaged from the stud; when the entire machine may be removed. A strong screw V, passing through the arm T, and worked by means of a nut formed into a double handle, carries at its lower end the two connecting links W, W, which are attached to the stone by two studs temporarily inserted into it at points diametrically opposite. The links W, W, are bent so as to admit of the stone being inverted while it is suspended in the lifting machine. The running stone is retained in its place in the mill simply by its own weight; it is, therefore, only necessary to raise it out of its bearings when the grinding surfaces require examination or repair.

The sack teagle is very simple, and will be readily understood from the drawing, Plate III. It consists of a barrel R′, provided with a rope of sufficient length to reach to the lower floor of the mill, and fitted to revolve in bearings attached to the roof; it receives motion, when required to be brought into action, from

Fig. 295.

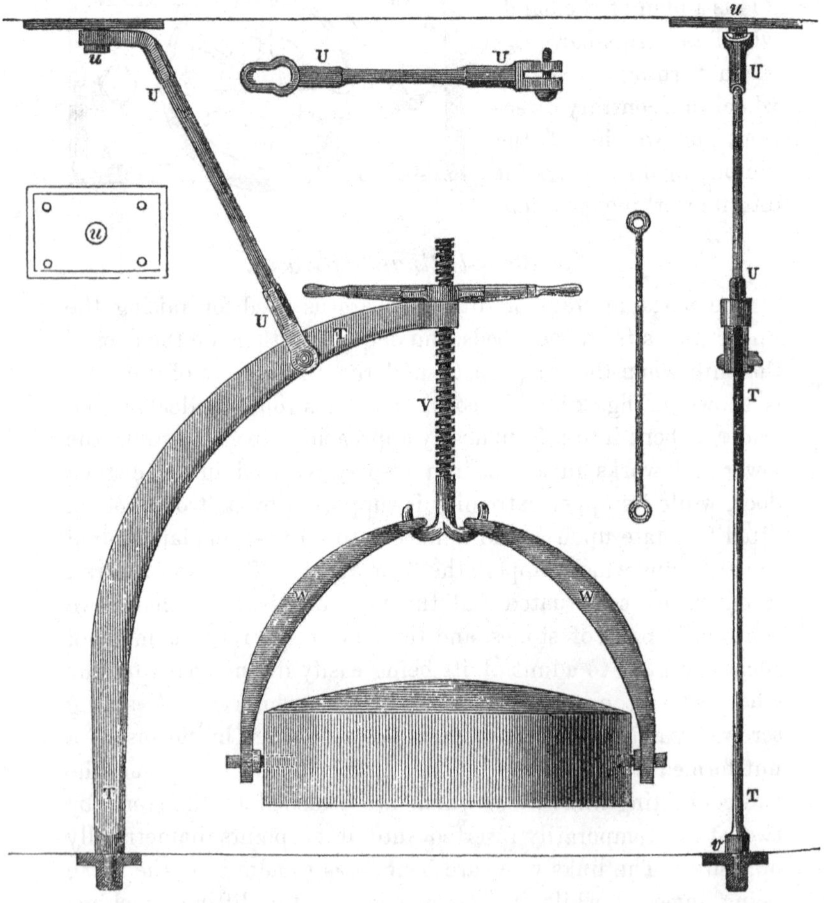

a belt connecting the pulley on its axis with a shaft C′, worked by the engine. The length of this belt is so adjusted that the sack teagle may remain at rest, or be set in motion, according as the long lever (the action of which is to tighten or relax the belt as may be required) is raised or lowered. In the floors of the mill are formed square hatches or openings, through which sacks, &c., may be admitted; and the rope of the sack teagle passing over a guide pulley situated immediately above the centre of these hatches, thus affords a ready means of raising sacks or any heavy articles to the different flats of the mill.

The Dressing Machine.

After passing the millstones, and having been carried by the creeper and elevator to the third storey of the mill, the corn enters the dressing machines; it is here subjected to the last process, and is ready for market.

Fig. 296 represents a dressing machine on a scale $\frac{1}{2}$ inch = the foot, the cylinder $5'.2'' \times 1'.4''$, is of wire cloth, from *a* to *b* is

Fig. 296.

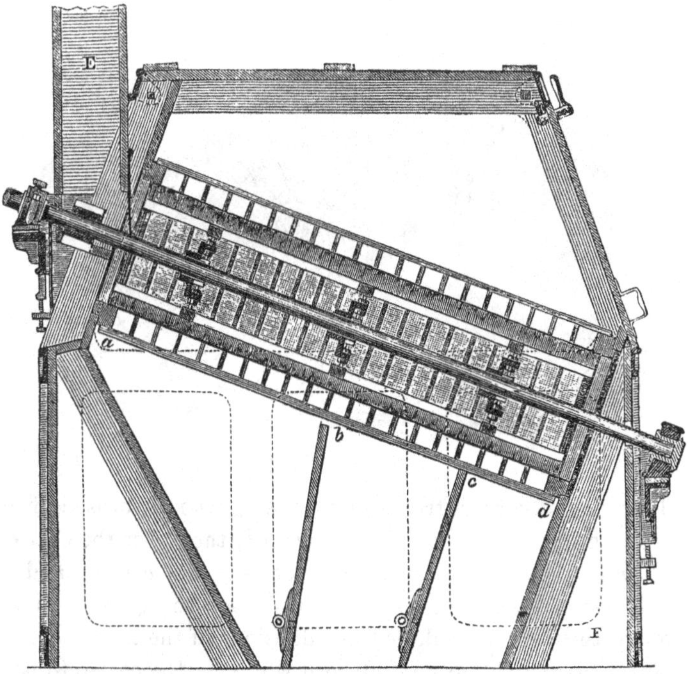

in general up to No. 70 wire gauze, from *b* to *c* is of No. 48 and 50, and *c* to *d* of 30. It seldom happens that wire cloth of

No. 120 wire gauze is used in those machines for the first division, from *a* to *b*, unless it be for the finer description of flour.

An inclination of 3 in 8 is given to the cylinder, and it makes from 560 to 650 revolutions per minute. Rings of wood, $\frac{5}{8}''$ thick, and $2\frac{1}{4}''$ deep, are placed round the cylinder at intervals of $4\frac{3}{4}$ inches, and the whole is bound together by horizontal planking $\frac{7}{8}''$ by $2\frac{1}{2}''$. The same contrivance is used for the adjustment of the brushes and the spindle as in the screening machine; the brushes, however, in this instance touch the wire gauze, and are made of bristle.

Fig. 297.

The ground corn, introduced by the spout E, is brushed first over the finest layers of wire from *a* to *b*, then over the coarser from *b* to *c*, and lastly over the coarsest from *c* to *d*, and is denominated firsts, seconds, and thirds, according as it passes through the first, second, or third divisions of the machine.

That portion of the ground corn which is too coarse to pass through any of the meshes falls out at the end of the cylinder and is called bran.

Some improvements, of late, have been effected in the construction of wire-dressing machines, by substituting an iron instead of a wood framing; by augmenting the speed from 320 to 500, and even 650 revolutions per minute, and by giving an inclination of 45° instead of 20° to the cylinder.

Fig. 297 represents one of these improved iron dressing machines. In its general construction, it somewhat resembles the one already described, fig. 296.

In this machine the ground corn enters the cylinder B B, by the hopper A, containing a patent feed apparatus, which regulates the feed, and thus prepares it for passing through the meshes of the wire cylinder. The internal brushes are secured to the spindle as those in the previous machine, fig. 296, but with a different arrangement for adjustment. The speed of the driving shaft is quickly reduced by the spur wheels, which gear into a wheel on the circumference of the cylinder, thus causing the cylinder B B slowly to revolve. The pulley D is connected with the pulley E by means of the strap F, which causes the external brushes G G to revolve, and thus keep the wire cylinder from being clogged up. By an ingenious arrangement with an eccentric, the brushes are lifted off to allow the cross bars H H, K K, to pass without injuring them, and are reinstated in their position as soon as ever the bars are clear. By twisting the strap, these brushes are made to revolve in a contrary direction, and thus prevent the bristles from taking a permanent set in one direction, and from wearing away unequally. The space under the cylinder is divided into three partitions, or hoppers, by movable boards, and the distance between them is increased or diminished at pleasure by the hand-wheel L. As in fig. 296, that portion of the meal which does not pass through the wire cloth falls out at the end of the cylinder.

The bolting machines are on the Swiss and American principle. Fig. 298 is an end elevation of a bolting machine, with two cylinders, and fig. 299 is a longitudinal section on a smaller scale. The cylinders K K are from 24 to 26 feet long, and 3 feet diameter, and make from 20 to 22 revolutions per minute. At distances of about 4 feet, radiating rods are inserted in the wooden shaft, which forms the reels of the bolting machine, to support the silk covering through which the flour has to pass. This hexagonal wooden shaft terminates in iron pivots which have their bearings in the plummer blocks L, L, bolted to a cross piece of cast-iron M M, which is bolted to the wooden frame-

Fig. 298.

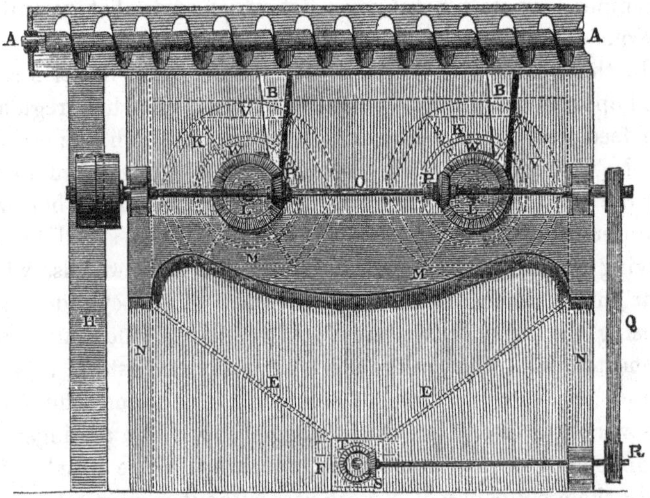

work N N. The machine is driven by a cross shaft O, having two bevil wheels, P P, keyed upon it, which gear into corresponding wheels keyed on the driving shafts of the bolting cylinders. The cross shaft also gives motion through the intervention of the strap Q to the pulley R, which drives a small toothed wheel S, gearing into a wheel T, keyed on the creeper

Fig. 299.

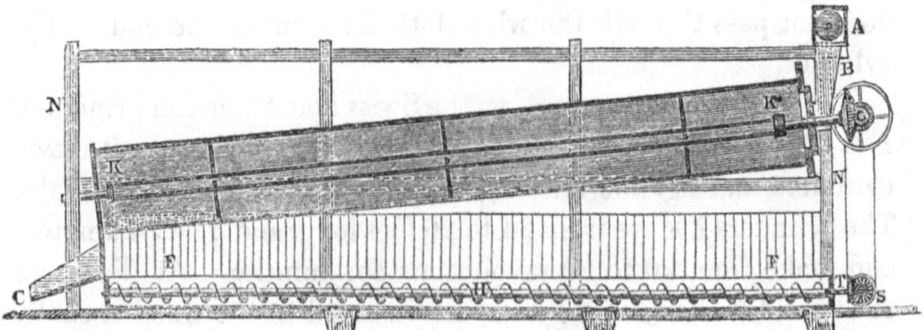

shaft U, which carries the flour along the trough under the cylinders or reels. It will be observed that the cylinder cover, V V is grooved into the end of the cylinder, and moves with the machine, whilst the cover W W is stationary: this enables a constant supply to be given to the machine without waste.

The meal travels from the elevators along the creeper A A, and enters the bolting machine by the hoppers B B; here it makes a progressive onward motion, rising and falling by gravitation from the sides of the reel till the flour has passed through the interstices of the silk and the bran delivered by the spout C at the end of the machine. The flour, after passing through the silk, falls upon the boards E E fig. 298 and F F fig. 299, into the creeper H; and in travelling along this creeper, it is received into bags by the spouts G G G. Should any of the meal have failed to pass into the bolting machine, it is carried on to the spout H′ fig. 298, which conveys it to the elevator from which the creeper A is supplied. The finest flour, chiefly used for confectionery and biscuit-making, is dressed and prepared in this way.

The great advantage of this class of dressing machinery is that the silk-bolting machine dresses the flour direct as it comes from the millstones, as the process is slow, and by the repeated rising and falling of the meal in travelling from one end of the cylinder to the other, it is sufficiently cool for immediate use, or ready to be hoisted above by the sack teagles, to be sacked for the market.

Messrs. John Staniar and Co., Manchester Wire Works, who supply this class of machinery, have lately constructed the dressing and bolting machines, with an improved feed. The meal is conveyed into a hopper with inclined sides, over which a rake slowly revolves, and as the meal travels several times round before it arrives at the feed-spout in the centre, it is exposed to the action of the air, and freed of the greater portion of its moisture. A piece of coarse wire gauze is placed at the top of the spout to intercept any foreign particles which may accidentally have got into the meal.

REFERENCES.

A A, the standard or lower framing of the grinding machinery.

B B, the cone or upper framing.

C, the stone case, of sheet-iron.

D, a cast-iron ring supporting the hopper, carried upon

a, *a*, *a*, three wrought-iron legs, resting upon the top of the stone case.

E, the feeding hopper.

b, a pipe for supplying grain to the hopper from the garners above.

F, the main driving shaft.

G, plummer-block of the shaft F.

H, a bevil mortice wheel, conveying the motion of the shaft F to

I, the pinion of the mill spindle.

J, the mill spindle.

K, the rhind of cast-iron by which the spindle is connected with the upper stone.

c, *d*, the bearings of the universal joint formed by the rhind.

L, the bed-stone box for the upper bearings of the mill-spindle.

e f, bushes and wedges for adjusting the bearings.

g, a thin cast-iron plate, by means of which the wedges *f*, *f*, are adjusted and fixed.

h, *h*, *h*, pinching screws for adjusting the level of the lower millstone.

i, *i*, *i*, pinching screws for centering it.

j, a large brass nut for supporting the pinion I.

k, footstep of the mill spindle.

l, cast-iron socket for the footstep *k*.

m, *m*, pinching screws for adjusting the socket *l*.

n, the saddle or link connecting the footstep *k* with

M, the great lever for supporting, and adjusting the mill-spindle.

N, *o*, screwed rod and key for working the lever M.

O, the movable feeding pipe, of cast iron.

P, lever for adjusting the pipe O.

p, small column forming the centre of motion of the lever P.

Q, *q*, apparatus for regulating the feed.

R, a cast-iron ring for raising the driving pinion out of gear with the wheel.

S, a cast-iron cross-head, being part of the same apparatus.

r, *r*, upright rods connecting the cross-head S with the ring R.

s, *s*, *t*, hand wheel and screw for working the disengaging gear.

T, U, V, W, the several parts of which the stone-lifting machine is composed.

u, *v*, the centres of motion on which it turns.

X, the lower elevator frame.

Y, Y, the creeper of cast-iron.

w, *x*, blocks and studs for connecting the adjacent lengths of the creeper.

y, the brackets in which it revolves.

Z, Z, the creeper box, of wood.

Before closing the treatise on flour mills, it may be interesting for the practical miller to know that an ingenious contrivance for balancing the running millstone has been successfully introduced by Messrs. Clarke & Dunham, of Mark Lane, London. Most persons connected with grinding wheat are aware that millstones are built of blocks of French burr, varying in density. These blocks are cemented and held firmly together by iron hoops, as already described; and the back of the stone is filled in with odd pieces of burrs, and backed up with plaster of Paris. The centre or balance-irons, by which the stone is suspended, are then let into the runner; it is immaterial what sort of iron is used for balancing, as any description will do. The usual custom is to suspend the runner on the stone spindle, and balance it with reference to its gravity alone, thus producing a standing balance by adding the required quantity of lead to produce a stationary equilibrium. This was generally thought enough, and the stone considered fit for work. Had the miller, however, raised the runner to an elevation of half-an-inch from the bed-stone, and rotated it at a speed of from 120 to 130 revolutions per minute, the runner in motion would have dipped or 'wabbled' to the extent, on the average, of a quarter of an inch. The effect of this tilting motion when the stones are at work, is to cause unequal pressure and unequal action upon the face of the stones in contact. The cause of this is the effect of the unequal density of the millstone as a whole, causing in motion an unequal centrifugal action in proportion to the denser parts preponderating above or below the plane of suspension.

In order to obviate this evil, there is introduced into the back of the runner four balance boxes, as shown in fig. 300 at B B. These boxes vary in depth from three to five inches, according to the thickness of the runner; the boxes are fitted with annular weights capable of being adjusted higher or lower, by means of

Fig. 300.

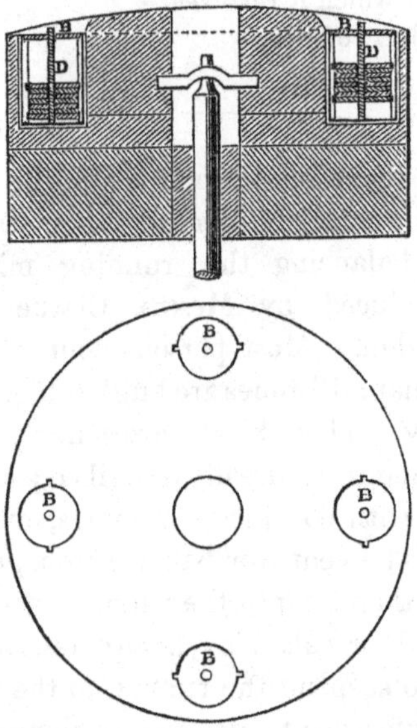

the screws D D, which pass down the centre of the weights. The number of the weights depends on the number required to produce the perfect standing balance. From this it will be seen that the standing balance must be first attained, and then, by raising or lowering the weights in the balance boxes, the running balance is perfectly effected. This is not done by one adjustment only, but by a series of adjustments. Having effected a standing balance, the stone is set in motion, and the dip is found to be on a particular quarter; the stone is then stopped, and the weights in the box opposite to where the dip is are raised higher above the plane of suspension so as to neutralize the variation of the dipping side. This being done, the stones are again set in motion, and the same operation is performed until a perfect balance, at any velocity, is attained. The great advantage of this arrangement is that the weights cannot shift, and the same balance is maintained in good order, and only require altering with the ordinary wear and tear of the stones.

CHAPTER III.

COTTON MILLS.

In our attempts to illustrate the improvements that have taken place in mills and mill-work we have endeavoured to give in detail the present improved state of the machinery for grinding corn, and the next — after what is generally called the staff of life — is the factory system for the manufacture of cotton. Of all the manufacturing interests which the industrial resources of this country present, this is probably the most important, not even excepting the iron and coal trades, and we may readily be excused if we briefly glance at the increase and immense extent to which this manufacture has attained until suddenly arrested by the unnatural war now raging in America. It will be in the recollection of most of our readers, that for the last seventy or eighty years, the mills of Lancashire and those of other parts of Europe, depended almost entirely for their supply of cotton upon the southern states of America, and that the extension of the trade grew up with the facilities of obtaining the means of supply; and although India, Egypt, and other countries, of late years cultivated and exported cotton, yet the chief dependence of Lancashire and other parts was upon the American states. The present miserable war has, however, cut off those supplies, and hence follows the distress and misery which has from this cause overtaken our once comfortable, willing, and industrious population. Our business, however much we may regret this circumstance, is not with the growth or supply of cotton, but its manufacture, and we have now to describe the improved methods and systems adopted for giving motion to the various intricate and ingenious machines now in use.

Until of late years, nearly the whole of our cotton mills were built from five to eight stories in height, with a succession of flats or floors in which the different processes were carried on. Generally speaking, the ground or first floors were appropriated to carding, drawing, and roving, with a separate building for the opening and blowing machines, and these constituted the preparatory process. The rooms above were invariably set apart for spinning, by mules if for fine yarn, but by throstles and mules conjointly if for coarser numbers.

This was the state of the factory system thirty years ago when adapted exclusively for spinning, but the introduction of the power loom and self-acting mule gave a new character to the dimensions and form of factory buildings. In the first instance, it was found that power looms worked better on the ground-floor than those on the upper stories, and that the yarn required a certain degree of moisture to weave freely, which could not be obtained from the heated and dry floors above. These properties peculiar to the ground-floor led to the shed principle, and there is scarcely a cotton mill now in the kingdom where looms are employed that has not a shed attached to the lower story on a level with the ground-floor. Again, it was found after the introduction of the self-acting mule that one man could work, with the assistance of two or three boys, 1,600 spindles with as much ease as he could work 600 spindles by the hand mules. This led to mills of double the width of the old ones, the former reaching from eighty to ninety feet wide. The spinning mills of the present day are therefore more like square towers or large lanterns, with considerable architectural pretensions as compared with the uncouth buildings we have already described. To these square buildings it is usual to attach a weaving shed with all the requisite warehouses and appurtenances for carrying on that additional department of the manufacture.

In order to exemplify our description of a cotton mill, with the steam-engine and transmissive machinery by which it is kept in motion, we might have chosen some of our largest establishments upon the principle referred to above; but having constructed mills for the colonies on a different principle, we have selected for illustration one of those erected in India,

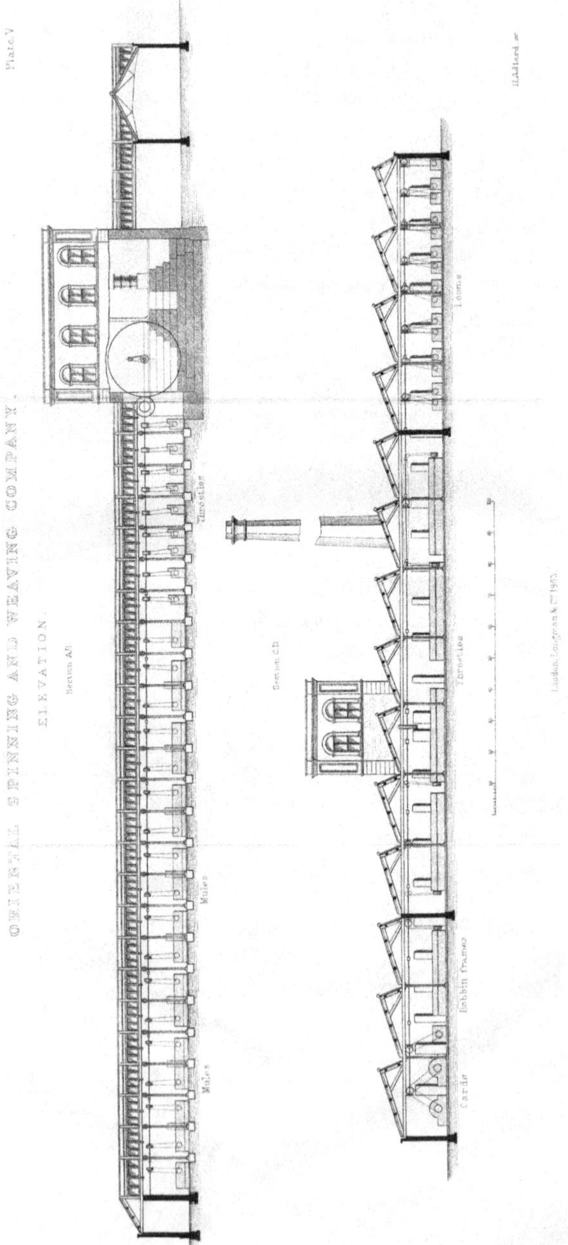

The material originally positioned here is too large for reproduction in this reissue. A PDF can be downloaded from the web address given on page iv of this book, by clicking on 'Resources Available'.

where the whole of the machinery is on the ground floor, and where it is covered by a light roof on the principle of the weaving shed for looms. Messrs. William Fairbairn & Sons have built several of these mills for the Bombay Presidency, and the whole of the machinery being open for inspection can the more easily be traced from the opening of the cotton bales to the finished cloth on the opposite side of the mill.

Description. In the annexed plan and sections it will be observed that the building covers a large space of ground, and is chiefly adapted for the country or small towns where land is cheap. In large cities such as Manchester, buildings of this description are seldom erected, owing to the high price of land and local taxes, from which the country is free; we have therefore most of our cotton mills in the surrounding districts, depending on Manchester as a centre and ready market for the sale and the export of yarn and cloth.

The mill to which we refer, shown in plates V. and VI., was built for India, and is now in successful operation some short distance from Bombay.

Plate VI. is a plan of the buildings showing the position of the machinery and the steam engines at A. The main shafts and gearing are supported on stone or brick pillars through the whole length of the building, receiving in their passage motion from the large pinion at B, which works into the fly-wheel, distributing it to the different lines of shafting on each side.

The steam is supplied to the engines by six boilers, 5 feet 9 inches diameter, and 32 feet long, with internal flues. The engines are each 80 horse-power, collectively 160 horses, 6 feet stroke, 26·8 strokes per minute, and are calculated to work to the full extent of 600 indicated horse-power. The main shafts, which receive motion from the fly-wheel, make 80 revolutions per minute, and are 8 inches diameter for a distance of 70 feet over the throstles and mules, and for a further distance of 35 feet towards the cards, they are $6\frac{1}{2}$ inches diameter, when they gradually taper in both directions to 5 inches at the end over the cards, and to $4\frac{1}{2}$ inches over the looms. The cross shafts over the power-looms are $2\frac{1}{2}$ inches diameter, tapering to 2 inches at the end. The cross shafts over the throstles and mules are $3\frac{1}{2}$ inches diameter, tapering to $2\frac{1}{4}$ inches at the extreme end. Over the slubbing frames and cards they are 3 inches diameter, tapering to 2 inches, the same proportion being observed in the ratio of the power delivered in other parts of the mill.

The cotton is taken from the cotton store, plate VI., and is mixed and sorted for the opening machines at G, where it is thoroughly cleaned, and driven about at great velocity, and freed from husks or seeds where it has not been properly ginned. From this it passes in a fleecy form to the scutchers H, where it is carefully spread upon a travelling cloth in front of the machine, and is carried forward to fluted rollers, by which it is conveyed in uniform thickness to the cylinder containing the beater with three or four arms, which makes about 1,600 revolutions per minute, or travels at the rate of 5,000 feet in the same time. Here it is driven forward into a cylindrical wire case, where it meets with a strong blast of air from a fanner which blows off all the light dust, whilst all the heavier earthy particles fall through the meshes of the wire into a receptacle below prepared for that purpose.*

Most of the improved blowers have two beaters, so that the cotton undergoes a thorough cleansing and opening before it is drawn from the wire cylinders and wound on the large bobbins or beams which form it into a lap ready for the cards. From the blowers the laps are conveyed through the door *a* to the cards, where the object is to clean and straighten the fibres and lay them parallel to each other. This is accomplished by unwinding the laps as they come from the blowers by fluted rollers which bring the cotton in contact with the teeth of the large carding cylinder and the covering flats also lined with teeth. In this way it is carded as the cylinder revolves, but not without being intercepted by the teeth of the flats and rollers, which nearly touching the main cylinder, the cotton is unable to pass without being combed and the filaments straightened in the direction of the two teethed surfaces as they meet each other. After passing through a succession of these flats and rollers, it is taken from the main cylinder by the doffing cylinder, which latter is finally stripped or cleared by a piece

* The scutcher, or blowing machine, is the counterpart of the corn-thrashing machine invented by Andrew Mickle of East Lothian.

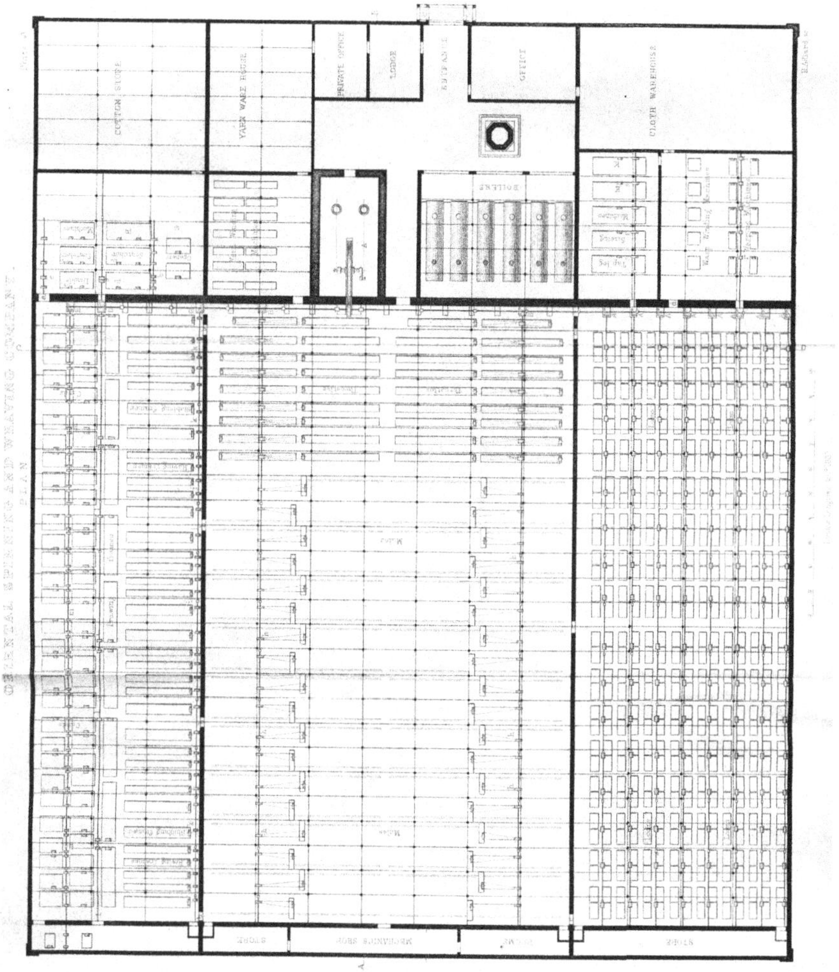

The material originally positioned here is too large for reproduction in this reissue. A PDF can be downloaded from the web address given on page iv of this book, by clicking on 'Resources Available'.

of thin steel, reciprocating by a crank motion on the surface of the cylinder, which forwards the cotton in a fine transparent sheet through a tube to the drawing rollers, where it is wound in circular coils into a can in the shape of a narrow band called a sliver.

From the cards it is removed in the cans, when filled, to the drawing frames, which consist of a series of rollers on cast iron frames, in the line shown at I I in front of the cards, and these rollers are so arranged that the front rollers run about four times the speed of the back ones, and from this increase of speed, it will be seen that the back rollers, whilst they are delivering at a uniform rate, the front ones are rapidly drawing out the fibres and delivering them in a form greatly attenuated to a much smaller sliver than that which first passed between the back rollers. These rollers require to be placed at the proper distance from each other, in order to correspond with the kind of cotton used, and the length of the fibre or staple, as it is frequently called. The object of drawing is to render the whole of the fibres as smooth and parallel as possible, and in the process of drawing as many as from five to six or more slivers in cans are run into one pair of rollers from the cards, and these again are frequently multiplied, drawn and redrawn, according to the quality of the yarn required, before they are fit for the slubbing or roving frames, which is the next process.

The slubbing or roving frame is one of the most ingenious contrivances in the cotton trade, as it not only draws the fibre and elongates the sliver on the same principle as the drawing frame, but it gives it a certain amount of twist, and winds it on a bobbin. For a long series of years this was a difficult process, as the delivering and the twist being the same at all times from the rollers, it requires to be wound on the bobbin one layer upon another, neither too hard nor too soft. If the former, the roving will not wind off the bobbin without breaking; and on the other hand, it must not be too soft, otherwise the fault would be equally objectionable as regards the quantity the bobbin should contain. The great secret therefore is to have it neither too hard nor too soft, but a medium degree of tightness, calculated to hold the exact quantity, and unwind itself freely one layer from off the surface of another, without risk of breakage. This would appear to be a desideratum (in every description of spinning), and to obtain this object with accuracy many ingenious contrivances have been adopted, amongst which we may enumerate sliding straps on conical drums, calculated to the increased circumference of the respective layers as they are wound on, and thus to gain the required degree of tension and compression, as the bobbin continued to increase in diameter, or as each superincumbent layer was wound on. The most important improvement of the roving frame was, however, accomplished by the introduction of the differential motion, which beautiful piece of mechanism effected the object of retarding the motion of the bobbin in the ratio of the increased diameter as it continued to enlarge. This motion gave great exactitude to the process, and enabled the spinner to prepare his rovings with a much greater degree of precision, and in shorter time. After roving, the cotton undergoes a precisely similar process, by passing through what is called a Jack or a finer roving frame. In this it is again drawn with additional twist, and again wound on to similar but smaller bobbins, ready for spinning into yarn, either by the mule or the throstle, as the case may be.

The above description completes, as far as our limits will admit, the preparatory process, until the rovings are in a condition to be handed over to the spinner, to be converted into yarn. Before describing the subsequent processes, we must, however, advert to a most ingenious machine for combing the cotton in place of carding it. This machine was introduced into this country some years since from Alsace, in France, and its operations are performed with such exactitude and precision, as to enable the spinner to produce a superior quality of yarn from an inferior quality of cotton. This machine has received great improvements since its first introduction, from the hands of Messrs. John Hethrington & Co., who have changed several of the motions, introduced others, and rendered it available for the finer descriptions of yarn. It is also extensively used in the preparation of wools of every description, but more particularly those of the alpaca, mohair, &c., as also in the preparatory process for flax, to which we shall subsequently have to refer.

To those familiar with the manufacturing districts the process of spinning is well known, but to the general reader who is not acquainted with those districts, a short account of the two different processes may not be uninteresting. There are two modes of spinning, one by a machine called the throstle, the other by the mule. During the early stages of cotton spinning, the cotton was carded and formed into slivers or rovings by hand cards. They were then placed on a wooden frame behind a row of spindles, fixed in a moveable box, which travelled on wheels, and these again received motion from a wheel and band, and the rovings which passed from the board behind, and delivered by rollers to the spindles, were held fast as the spinner drew the spindles from the rovings to the extent of the stretch; and thus by consecutive movements the rovings were stretched and twisted, and every time the travelling frame was pushed back to the rovings, the yarn previously spun was wound upon the spindles. This was the only method in use before the time of Arkwright, who introduced the cylindrical cards and the water frame, or, as it was subsequently called, the throstle, the noise of the numerous spindles imitating the notes of that bird. It was also designated the water frame, from the circumstance that it could not be worked by hand, but required the power of water to give it motion.

The next process is the throstle, which is on the same principle as the roving frame, but with this difference, that the spindles are smaller and more numerous, and range in rows of 150 or 200 on each side of the frame. It has also this peculiarity, that the bobbin which re-winds the yarn as it is spun is not regulated by the differential motion, but the thread, as it is drawn and twisted from the rollers, is wound on by friction as the frame in which the spindles are fixed rises and falls the length or depth of the bobbin. In this operation it is not necessary to wind the thread on to the bobbin slack, as there is no danger of the layers separating, and the friction given to the bobbin is therefore sufficient to fill it hard and tight.

The rovings are also carried to the mule, which is a totally different machine to the throstle; it has no moveable frame or spindles with bobbins rising and falling; in fact, it is more like the spinning jenny, with a travelling carriage which contains the spindles, and stretches out from the beam or stationary roller frame after the manner of the jenny already described. It, however, combines part of the throstle as well as the jenny, and hence its name of the mule. The mule as left by Crompton possesses many advantages that do not belong to the water frame. It can spin yarn of any degree of softness, and of the finest quality; and since it was made self-acting, it forms its own cop on each spindle, and puts up the carriage, which on former occasions had to be done by hand. These are considerations of vast importance in spinning; and the mule has now attained such perfection that 1,000 spindles can be worked in one carriage with the same certainty and ease as one-third the number could formerly be worked by hand. There is another peculiar property in the mule, and that is the double twist in fine numbers which the yarn receives after the full extent of the stretch is made. When the spindle carriage arrives at this point the rollers become stationary, the motion of the spindles is increased, and the twist required is given according to the quality or purpose for which the yarn is intended. I believe the introduction of the double twist motion is due to the late Mr. John Kennedy, one of our earliest and most successful mule spinners.

Having thus traced the different processes from the bale of cotton to the yarn, our next duty will be to notice the operations of weaving from the yarn into cloth. In our endeavours to accomplish this it will be necessary to glance at the state of the manufacture as it existed previous to the introduction of the power loom, and to show the advantages attained and the enormous increase which these inventions have produced.

From the earliest historical period, the hand-loom has been in use for the purpose of weaving. That of the Hindoos and all other nations have been of the same character, and until the improvement of the flying shuttle, introduced by Kay, we may consider the loom a primitive and unchangeable machine. It is upwards of thirty years since the power-loom was first introduced. After repeated attempts by Major Cartwright, Mr. Shorrocks, and others, to render it available and self-acting, it fell into other hands. These attempts were at first discouraging, but after repeated changes, suggestions, and improvements, it ultimately succeeded in producing a cloth more uniform in character and superior in quality to that of the hand-looms. The result of these improvements was a total change in the cotton manufacture. The hand-looms were thrown out of use, and the hand-loom weavers, who were unable to meet the new state of things, were thrown out of work, and suffered for many years the greatest and most distressing privations. By this transfer from hand to power weaving, the whole system of manufacture was changed, and the manufacture of yarn into cloth was no longer carried on in the domestic cottage, but became a part of the factory system. Large shed buildings were erected for that purpose, and the weavers, chiefly girls, were employed under regulations the same as those in the other parts of the mills.

Before yarn can be woven into cloth, four distinct processes have to be gone through; viz., warping, winding, beaming, and dressing — to prepare the warp for the loom. The first of these, the warping, consists of a large vertical reel, on to which the yarn is wound from the bobbin in measured lengths, several of which, when put together, constitute the warp. It is then, for some qualities of cloth, sized or run through a cistern of liquid flour and water, at nearly the boiling temperature, and from this through rollers which squeeze out the surplus fluid, and leave the yarn saturated with the glutinous substance of flour and water called size. In this state, when partially dried, it is transferred to the loom, where it is woven into cloth. The other process requires more careful manipulation, as the warps have to be formed by winding the yarn from the cop, if it be mule yarn, and from the bobbin if throstle, on to a roller called a beam, and in its passage it is run over a roller about twenty inches diameter, through the divisions of a reed formed of wire, to separate the threads and lay them parallel on to the roller beam.

This done, four or six of the first windings are united on the dressing machine, K K, where they are again passed through reeds at each end, and finally wound upon a large bobbin or beam ready for the loom. In its passage from each end of the machine, it must, however, be observed that it is well brushed or dressed with a pulp of prepared flour and water, which is laid upon the warp, as it passes from the rollers at each end to the beam at the top of the machine, ready for the loom.

The power-loom, although simple in its operations in the first instance, comprises at the present time many important improvements for the manufacture of twills and figure weaving. The revolving shuttle-box, and the changes in colour and form that may be effected, enable it in many cases to compete with the jacquard loom. Many of the beautiful fabrics of mixed goods are woven in this manner, and, judging from what has already been done, we may reasonably look forward to still greater improvements in the quality, as well as the quantity of cloth produced.

It might have been desirable to have noticed the progressive increase of this important branch of industry; but when it is known that a sum exceeding 70,000,000*l.* sterling represents its annual value, we have said sufficient to impress the reader with a desire for its maintenance and cultivation.

We close the chapter on cotton mills with a list of the most approved speeds of the different machines, and a list of wheels, speeds, &c., as now in operation in the mill of the Oriental Cotton Spinning Company:—

Speeds of Machines.

Description of Machine.	Diameter of Pulley.	Number of Revolutions per Minute.
	inches.	
Opener or beater	12	800
Scutcher	$7\frac{1}{2}$	1,600
Rollers	21	270
Cards	18	130
Grinding machines	12	200
Drawing frame	12	226
Slubbing „	13	235
Roving „	11	415
Throstle „	10	562
Winding „	$9\frac{1}{2}$	180
Beam-warping	$14\frac{1}{2}$	50
Tap leg-sizing	13	210
Mules	16	232
Looms, $\frac{9}{8}$ wide	12	140
Looms, $\frac{7}{8}$ wide	11	160

LIST OF WHEELS AND SPEEDS.

Description of Gearing.		Driver. Diameter. ft.	Driver. Diameter. in.	Driver. Revolutions.	Driven. Diameter. ft.	Driven. Diameter. in.	Result. Revolutions per Minute.	
Spur fly-wheel . .	A′	23	4	24·7	7	3¼	80	revolutions of main shaft.
Bevil wheels,	A A A A A A A A	3	0	80	1	11	120	revolutions of cross shaft over looms.
Mortice bevil wheels	B B B	4	7¼	80	2	3¾	160	,, ,, ,, spinning.
,, ,,	G′	4	8½	80	1	7⅜	231·1	,, ,, ,, roving and slubbing.
,, ,,	D	4	0½	80	1	10¾	173·3	,, ,, ,, drawing and blowing.
,, ,,	E	3	6	80	2	3	124·4	,, ,, ,, carding frame.
,, ,,	E	3	8¾	80	2	0½		,, ,, ,, carding frame.
Pulley	*a*	3	0	173	1	8	310	,, counter shaft for opener.
,, ...	*b*	2	7½	310	1	0	800	revolutions of opener.
,, ...	*c*	3	0	173·3	1	4¼	385	,, counter shaft for scutchers.
,, ...	*d d*	2	7½	385		7½	1,600	,, scutchers.
,, ...	*e*	2	10	173·3	1	9	270	,, rollers on double scutcher.
,, ...	*f*	1	3	138·5	1	9	275	,, rollers on single scutcher.
,, ...	*g g g g*	2	10½	160		10	560	,, throstles.
,, ...	*h h h*	2	0	160	1	4	232	,, mules.
,, ...	*i i i*	1	7¾	231		11	415	,, roving frames.
,, ...	*k k k*	1	1¾	231	1	1	235	,, slubbing.
,, ...	*l l l*	1	4	173	1	0	226	,, drawing.
,, ...	*m m m*	1	7	124·4	1	6	130	,, cards.
,, ...	*n*	1	7	124	1	0	124	,, grinding machine.
,, ...	*o*		11	160		9½	160	,, winding machine.
,, ...	*p*	1	0	120	1	0	140	,, looms, $\frac{9}{8}$ wide.
,, ...	*q*	1	0	120		11	160	,, looms, $\frac{7}{8}$ wide.
,, ...	*r r r*		6	120	1	2½	120	,, beam-warping.
,, ...	*s s s*	1	11	120	1	1	120	,, tap leg sizing machine.

CHAPTER IV.

WOOLLEN MILLS.

The difference between cotton and sheep's wool is that the one is a vegetable and the other an animal substance, and the latter being dissimilar in its characteristic properties requires a different treatment in the manufacturing processes. The nature of the fibres of sheep's wool, which curl and hook into each other, is different to most other fibrous substances; some of the early preparatory stages in its manufacture into cloth, however, are the same as in that of cotton wool. The peculiar properties of some of the animal wools is their tendency when worked to entwine the fibres, so as to form a species of cloth called felt, without the aid of spinning and weaving. Hats, horse cloths, and other descriptions of clothing, are made in this way, and that by a process called whipping, which separates the fibres by the vibration of a piece of cord or catgut, drawn tight over the extremities of an elastic bow. With this instrument the fibres are separated by a jerking motion of the hand, and fly off in fine flakes into a receptacle ready for use. It is then worked into a sort of pulp, in a vessel of hot water, to the required thickness. The same principle is observed in fulling blankets, broad cloth, and other fabrics of a similar kind; the only difference being that the former is done by hand in a vessel of hot water, and the latter by the stocks or fulling mill. It is to this latter, and subsequent processes where machinery is used, that our attention is chiefly directed; and for that purpose I have selected a woollen mill erected for the Turkish Government at Izmet, on the Gulf of ancient Nicomedia. I might have taken the large establishment of Messrs. B. Gott & Sons, of Leeds, for illustration; but for the same reason of having all the processes on one floor, as in the previous case of the cotton factory at Bombay, I have deemed it necessary to do the same in that of wool.

There is considerable novelty in the style of the buildings for these works, and the arrangement of the machinery. They were built in 1843, and contained all the improvements of woollen machinery up to that date. They consist of a quadrangular square with a court, A, fig. 301, in the centre; B, the entrance;

Fig. 301.

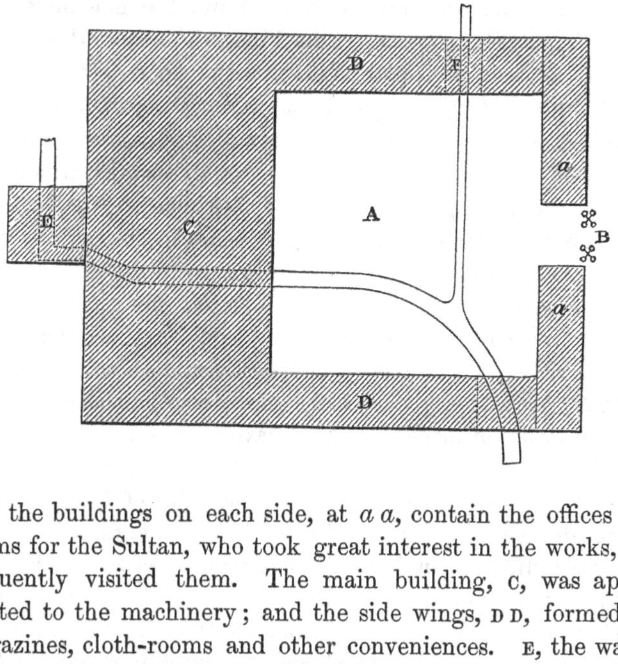

and the buildings on each side, at *a a*, contain the offices and rooms for the Sultan, who took great interest in the works, and frequently visited them. The main building, C, was appropriated to the machinery; and the side wings, D D, formed the magazines, cloth-rooms and other conveniences. E, the water-wheel, in a separate building, gave motion to the machinery in every part of the works. At the end of the wing-buildings on each side were the steam boilers and retorts for gas, and the designs were so arranged as to lock up the whole of the works with one key.

The buildings were erected close to the water-fall on the side of a steep bank, and were designed for the purpose of having the whole of the operations on one floor and within sight. It was arched with iron, and lighted from the top on the bazaar principle, a system of building prevalent for ages in the East. The exterior walls were substantially built of stone, with a portico and ornamental entrance, B, in front. It was originally intended to have constructed another water-wheel, as at F; this was not, however, carried into effect.

Plate VII. is a longitudinal section of the building, showing the piers for supporting the floor, the water-wheel, the machinery of transmission, the dye-house, the position of the machinery, roof, &c.

Plate VIII. is a plan of the woollen mill, showing the arrangement of the different lines of shafting and the machinery, the different passages between the machines, the water-wheel, &c.

It will be seen from plate VII. that the floor of the woollen mill is supported by piers of brick, and resting upon these are the cast-iron columns 20 feet long and 8 in diameter, for the support of the roof. The boiler for the dye-works, which also serves as an heating apparatus, is 7 feet diameter and 24 feet long, and from it cast-iron piping 6 inches in diameter ascends to the woollen mill above.

As the machinery for driving the stocks and gigs is under neath the floor of the woollen mill, it could not well be shown in plate VIII. I have therefore constructed an enlarged view, fig. 302, which clearly shows the position of the shafting, and the description of machinery they have to drive.

Water-wheel and Millwork.—Near the city of Izmet a river of considerable dimensions cascades from a height of 28 feet into the gulf, and on the banks of this stream, in the immediate vicinity of the sea, the mill was erected. The water-wheel is on the suspension principle, and, together with the millwork, was manufactured, and the buildings erected, by Messrs. W. Fairbairn & Sons, Manchester. It is entirely of iron, 30 feet diameter and 13 feet wide, with 72 buckets, 1 foot 6 inches deep, and an opening of 7 inches for the entrance of the water into the bucket. The internal segment, *a*, fig. 302, 28 feet 2½ inches diameter, 324 cogs, 3¼-inch pitch, and 14 inches wide on the cog, drives the pinion, *b*, 5 feet diameter, and gives motion to a strong cross shaft and bevil

Plate VII

WOOLLEN MILL,

FOR

HIS HIGHNESS THE SULTAN

ERECTED AT IZMET

by William Fairbairn & Sons, Manchester

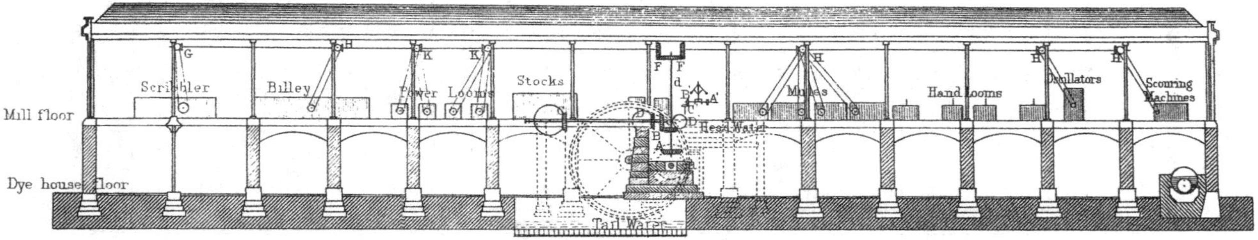

H. Adlard, sc.

London: Longman & Co.

WOOLLEN MILL,

FOR

HIS HIGHNESS THE SULTAN

Erected in 1843.

by William Fairbairn & Sons, Manchester.

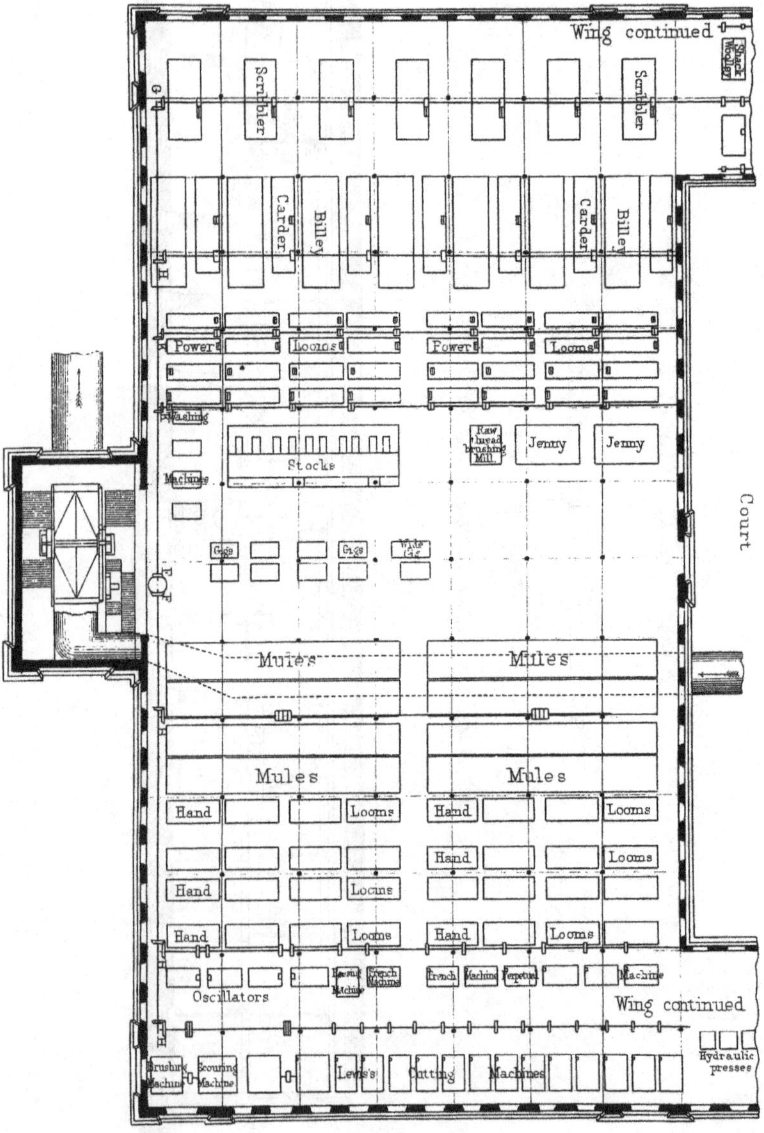

London: Longman & C°.

wheel, *c*, from which the motion is conveyed to the mill by the vertical shaft *d*, and the shaft *f*, driving the gig and stock shafts *h* and *g*.

From this arrangement of the first motion wheels, it will be

Fig. 302.

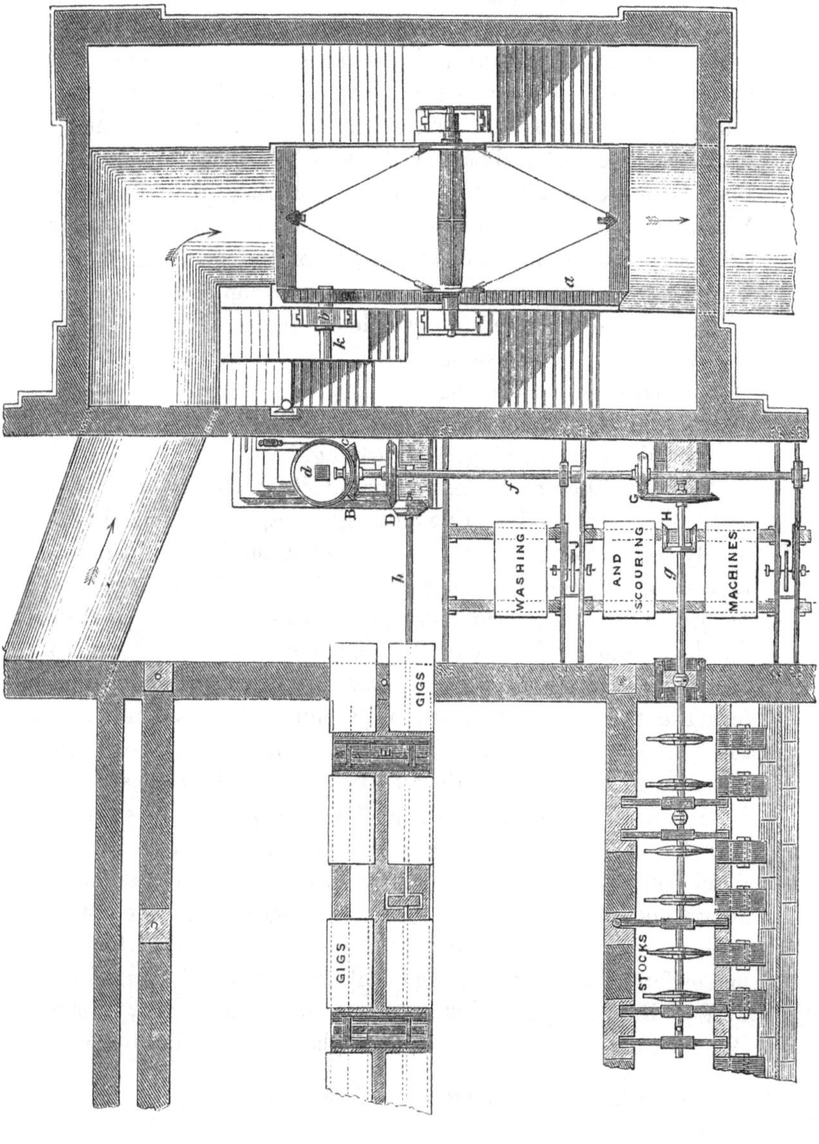

seen (see also plate VII.), that the motive power comes almost direct upon the heaviest portion of the machinery, such as the stocks and gig machines, and by the vertical shaft, *d*, and the horizontal and cross shafts above, it is transmitted to the lighter descriptions of machinery in other parts of the mill. The lower portion of the cast-iron vertical shaft is 11 inches diameter; but after giving off the necessary power by bevil wheels to the horizontal shaft for the stocks, gigs and scouring machines, it is reduced to 10 inches, and tapers to 8 inches at the top. The shafts, *f*, *g*, *h* and *k*, fig. 302, are of cast-iron; *f* is 7½ inches diameter, *g* 8 inches diameter, *h* 5½ inches diameter, and the shaft *k* 12 inches diameter. The main horizontal shafts, plate VIII., are 4½ inches diameter, and taper to 4 inches at their extreme ends. A plan of the position of the stocks, gigs, and washing and scouring machines, is shown in fig. 302.

The processes pursued in a woollen mill are:—

1. Sorting and washing.
2. Teasing and opening.
3. Carding, roving, and spinning.
4. Warping, dressing, and weaving. *

By these different stages of manufacture the wool is converted into cloth. The after-processes may be described very briefly as follows:—After a somewhat similar preparatory process of carding, roving, and spinning, similar to that described under the article cotton, it is taken as it comes from the loom, and submitted to careful washing, by running it over reels through a cistern of water. From this it is transferred to the stocks or fulling mill, and is there submitted to the action of the stock by constant pounding with soap and water in covered boxes, till it attains the required consistency of thickness according to the quality or degree of fineness of the cloth. After this process it is again washed with pure water before it passes to the gigs. Here it is subjected to a severe and almost reverse process; the stocks, by a constant rolling of the cloth in the circular box, give it a thickening or felting character, by the contraction and twisting of the fibres. But the gigs effect a process of separation by teasles (the prickly husk or pod of the plant known by botanists

* For farther information respecting the processes, see Appendix A.

as *Dipsacus fullorum*) fixed in a frame attached to the circumference of a cylinder about 3 feet 9 inches in diameter, run at a velocity sufficient to draw out the fibres, and lay them parallel with the line of the cloth. This it will be observed is the very reverse of the previous process, and by a system of teasle carding, the cloth is now prepared, when dry, ready for the sheering or Lewis frame.

The old process of shearing was effected by stretching the cloth in a frame of convenient length, supported by cushions. On the top of the cloth was fixed two large blades or knives worked by power, and, acting as a pair of scissors, clipped off the projecting fibres, and gave what is called a nap or smooth surface to the body of the cloth.

This operation has, however, been superseded by the Lewis frame, invented by Mr. Lewis about the beginning of the present century. This machine consists of an iron frame with fixed rollers, over which the cloth is drawn, and in its passage a roller with a series of thin steel blades or spiral cutters revolves at great velocity, and cuts off the outstanding fibres, previously drawn into position in line with the cloth by the gig machine. This is an expeditious as well as an accurate process, and the cloth sheared in this manner presents, when finished, a close, a shining, and a smooth texture.

After the shearing the cloth is transferred to the brushing machines, where it undergoes a similar treatment in the dry, as it received from the gig machines in the wet state; it is then well brushed and finished ready for the market.

I have endeavoured in this short description of the different processes to give some idea of the manufacture of woollens, as the whole of the machinery in former times was constructed by the millwright. Now it is in the hands of the machine maker, and, like every other operation of manufacture of modern times, the division of labour, and the organization of separate trades, warrants not only the extended use of machinery for despatch in the manufacture, but greatly increased economy, and much greater perfection in the quality of the cloth produced.

It will be noticed that the manufacture of woollen cloths, such as broad cloths, flannels, blankets, &c., is chiefly derived from the short wools. There is, however, a considerable difference between the woollen and the worsted fabrics, consisting chiefly in the woollen yarn being very slightly twisted, so as to leave the fibres at liberty for the process of felting, whilst the worsted yarn is made from long wool, hard spun, and made into much stronger thread. The worsted manufacture, and all the mixed fabrics of wool, cotton, flax and silk, are made by a different process to that of woollen cloth. In the former, the wool, mohair or alpaca, is cleaned and washed similar to the shorter wool. It is then combed, formerly by hand combs, with rows of long teeth, but now by machines of different constructions, some of them heated by steam of the circular form, and others with the teeth of the combs in line. In combing, a little sprinkling of oil is necessary, in the proportion of a fiftieth or a sixtieth of the weight of the wool, in order to increase the pliancy and ductility of the filaments, and to straighten them in parallel layers as they are drawn from the teeth of the comb, and formed into a roving or sliver. From the combing machine the slivers go to the drawing and roving; and from thence to the spinning machines, in every respect similar to the throstles or water-frames used for cotton. In some cases, and for some description of wools, carding is substituted for combing, and the usual subsequent process of drawing, roving, &c., are gone through, as in many other operations where the wool travels from its raw state to that of yarn.

In both the woollen and the worsted manufactures the processes are probably more complex than those of most other textile fabrics, and in that of power weaving the difficulties to be encountered have been considerable, owing to the softness of the material and want of twist, or hardness in the yarn forming both weft and warp. From this cause, the power-loom has made slower progress in the woollen manufacture than in that of worsted and all kinds of mixed stuffs, where a stronger yarn and more rigid material has to be dealt with.

The woollen and worsted trades, like most other manufactures, have increased in an accelerated ratio from the commencement of the present century up to the present time. We have no reliable returns of the state of the manufacture at the beginning of the present century; but in 1857, according to the factory returns, there were 806 woollen and 445 worsted mills at work, and the total value of the exports of woollen and worsted goods and yarn was 13,645,175*l.*, or about one-fifth of that of cotton.

List of Wheels and Speeds.

Description of Gearing.		Driver. Diameter.	Driver. Revolutions.	Driven. Diameter.	Result. Revolutions per Minute.	
		ft. in.				
Segments	a	28 2½	2·2	5 0	12·8	revolutions of main cross shaft.
Wheels bevil	A	9 8	12·8	4 10	25·6	„ upright shaft.
Wheels mitre	B	5 3	25·6	5 3	25·6	„ main cross shaft for driving gigs and stock shaft.
Wheels bevil	C	3 9¼	25·6	7 0	14	„ stocks shaft.
Wheels „	D	6 0	25·6	3 0	51·2	„ gigs shaft.
Wheels spur	E	4 0½	51·2	1 10½	108	„ gigs.
Wheels bevil	F	6 0	25·6	3 0	51·2	„ cross shaft overhead for driving longitudinal shafts.
Wheels „	G	3 8¾	51·2	2 0½	92	„ longitudinal shafts over scribblers.
Wheels „	H, H, H	3 6	51·2	1 10½	92	„ longitudinal shafts over mules, billeys, and carders, &c.
Wheels „	I	3 0	14	1 6	28	„ cross shaft to washing machines.
Wheels spur	J	4 8½	28	2 4	56	„ washing and scouring machines.
Wheels mitre	K K	2 4	51·2	2 4	51·2	„ longitudinal shafts over power-looms.
Pulley	A′	0 9¾	92	2 0¾	36	revolutions of governors.
Worm	B′	0 4.	36	1 1	1·2	„ short shaft through the mill.
Wheel	C′	1 6	1·2	1 6	1·2	„ upright shaft.
Worm	D′	0 6½	1·2	2 0¾	·031	„ rack shaft.

List of Pulleys and Speeds of Machines.

	Driver. Diameter.	Driver. Revolutions.	Driven. Diameter.	Result.	
14 Pulleys	1 10¼ × 7	92	2 0	85	revolutions of scribblers, two pulleys for each scribbler.
14 „	1 10¼ × 7	92	2 0	85	„ carders „ „ carder.
1 „	1 11½ × 16	92	1 9½	100	„ teaser hook woolley.
2 „	2 6 × 9	92	1 6½	153	„ counter shaft over the shack woolleys.
2 „	2 4 × 14	153	1 2	306	„ shack woolleys.
8 „	1 7 × 13	92	0 11½	152	„ mules.
4 „	1 11½ × 12	92			oscillators.
1 „	2 6 × 12	92			cross rassing machine.
4 „	2 2½ × 9	92	2 6	80	„ brushing mills.
7 „	2 5½ × 4¼	92	0 7	400	„ French machines and perpetual machines.
15 „	2 5½ × 4	92	0 5	540	„ Lewis cutting machines.
32 „	1 3¼ × 10	51·2	1 5	46	„ looms or number of picks per minute.

CHAPTER V.

FLAX MILLS.

FLAX mills are contemporary with those for the manufacture of cotton and wool, and, in fact, it may be said, that this and almost every other branch of industry received its impetus from the introduction of the steam engine and the improved machinery of Arkwright, which speedily found its way with certain modifications to the manufacture of other textile fabrics besides cotton. Messrs. Marshall and the late Matthew Murray, machine makers of Leeds, are entitled to the merit of having been the first to introduce machinery for the spinning of flax; and the mills of Messrs. Marshall have been in operation since the close of the last century to the present time. Belfast and the linen districts of the north of Ireland were for many years supplied with linen yarn almost exclusively from Leeds, and nothing was done by the Irish manufacturers in the shape of spinning their own flax until 1824, when the first mill was built by Messrs. Mulholland of Belfast. Since that time flax mills have increased to an extent which rivals if it does not exceed the manufacture of Leeds.

The eastern districts of Scotland, Fifeshire and Dundee, were early in the field, and even as far north as Aberdeen the manufacture of flax has been in a flourishing state. The Scotch flax specimens are, however, chiefly employed in the sail cloth and shirting manufacture, and the facilities afforded for the import of flax from Russia and the Baltic are powerful inducements for the extension of the manufacture in that part of the united kingdom. They also spin, as well as the Irish, considerable quantities of home-grown flax, and the mills in both countries give employment to an active and industrious population.

Mr. I. G. Marshall states in a paper read before the British Association for the Advancement of Science, 'That the first essay in flax spinning in Leeds was made at a small mill driven by water, called Scotland mill, about four miles from Leeds, by my late father, John Marshall, in partnership with Samuel Fenton of Leeds, and Ralph Dearlove of Knaresborough. This was in 1788 and 1789.

'The wonderful success and large profits attending the introduction of Arkwright's invention into cotton spinning had about this time attracted general attention to mechanical improvements applied to manufacturing purposes. The spinning of flax by machinery was a thing much wished for by linen manufacturers. It attracted the attention, amongst others, of Mr. Marshall, who was so strongly impressed with the advantageous field for invention and enterprise offered by flax spinning, that he devoted himself entirely to the new enterprise.

'It appears that some attempts at flax spinning had already been made on a small scale at Darlington, and some other places, as the first spinning machines used at Scotland mill were on a patent plan of Kendrew & Co. of Darlington. This did not answer; experiments were made, and a patent taken out for a plan of Matthew Murray's, then foreman of mechanics with Mr. Marshall.

'In 1791, a mill was built in Holbeck, Leeds, and at first driven by one of Savery's steam engines, in combination with a water-wheel; but in 1792, one of Boulton and Watt's steam engines of twenty-eight horse power was put down. In 1793 there were 900 spinning spindles at work. We may take this small item as our first statical datum of flax spinning in Leeds.'

Dr. Ure, in his Philosophy of Manufactures, speaking of flax, states that 'flax is the bark or fibrous covering of the stem of the well-known plant called by botanists *linum*, because it constitutes the material of linen cloth. The spinning filaments are separated from the parenchymatous matter either by steeping the plant in water, or by exposing it for some time to the action of the air and weather. The former, which is the commonest or safest method, is called water retting; the latter is called dew retting.* Both act by a slight degree of fermentation in the

* Since Dr. Ure's work was written new chemical processes have been adopted for separating the bark or husk from the flaxy fibre, by which a great saving of time is effected, the new process occupying only a few days, whereas six weeks were required by the old steeping process.

substance which attaches the flaxy filaments to the vegetable vessels and membranes. The crude flax is dried by being spread on the grass, and is then subjected to an instrument called the brake, which breaks and separates the boon or core from the true textile flax.'

As considerable interest is attached to the retting process, so as to cause the bark or straw to separate freely from the filaments, it is necessary to ascertain that the retting is perfect and well dried, after which they are exposed to the action of the double breaking machine, and to this and subsequent processes where machinery comes into requisition I shall more exclusively confine my attention.

To illustrate this subject, I might have selected the large mills of Messrs. Marshall, with whose works I have been professionally connected for a number of years, but as these extensive works do not weave the yarn into cloth, and as some of the machinery is in mills five stories high, and the new mills are under brick arches lighted from the top, it was found more convenient to select a smaller establishment, where the whole of the processes are carried on, from the flax as it comes from the grower till it issues from the mill in the shape of cloth. To attain these objects I have chosen a mill erected some years since at Narva, in Russia, for the Baron Stieglitz, the description of which is as follows:

Fig. 303 is a longitudinal section, showing the water wheel, the machinery of transmission, the brick arching for rendering the building fireproof, the pillars for supporting the floors, roof, &c.

Fig. 304 is a plan showing the form of the building, the entrance, arrangements of the power looms, heating apparatus, &c.

The mill was built on the side of a steep bank, and a system of arching, not shown in the drawing, was found necessary for the purpose of levelling the surface. On the piers of these arches, the cast-iron pillars for the support of the iron beams and arches were placed, and over the water wheel is a strong wrought-iron girder, which supports two of the columns and the arched floors above, rendering the whole structure perfectly fireproof. On reference to the plan, fig. 304, it will be seen that the building is rectangular, with a wing A′ at one end. A bleach house is attached to the end of the mill next the water-wheel, and a shaft running parallel to the wall BB′ Fig. 304, connected by a cross shaft extending the whole length of the bleach house, gives motion to the machinery of that part by a pair of bevil wheels through the opening B′. These wheels are marked K and L in the list of wheels and speeds. The lower, or first floor, contains the looms for weaving. On the second floor, Fig. 303, the preparatory process is carried on, and the top or third floor is exclusively employed for spinning. The heating apparatus consists of a boiler, 5 feet diameter and 12 feet long, from which cast-iron piping 6 inches diameter ascends to the various rooms. The mill is lighted with gas by piping from another mill near at hand, belonging to the same company. The roof consists of wrought-iron principals, to which red deal planking $1\frac{1}{2}$ inches thick, stretching from one principal to the other, is securely bolted. This planking was covered with wrought-iron sheet-plating over the entire roof, as in general use in St. Petersburg and Moscow.

During the building of the mill, it was, however, determined to carry the walls 6 feet higher than they are shown in Fig. 303, and by this elevation a large room, 223 feet long and 71 feet wide, was attained, without pillars, for reeling through its entire length. At C′ Fig. 304, is the entrance; and over the staircase a water cistern, 40 feet long by 12 feet wide and 6 feet deep, was placed in the attic story for supplying the frames with water and for various other conveniences of the mill.

The motive power consists of a water wheel with ventilated buckets, 24 feet diameter and 20 feet wide. It contains 56 buckets, with an opening of $7\frac{3}{4}$ inches between them for the entrance of the water. The internal segments are 20 feet $4\frac{3}{8}$ inches in diameter, with 192 cogs, 4 inch pitch, and 14 inches wide on the cog. Into this wheel the pinion A, 4 feet $6\frac{3}{4}$ inches diameter, with 43 cogs, gears. On the pinion shaft a wheel, 12 feet 1 inch diameter, is keyed, to give motion to a second pinion 4 feet $6\frac{3}{4}$ inches diameter, which is keyed on the shaft for driving the upright at an accelerated speed; and on this shaft the bevil wheel A, which drives the upright,

FLAX MILL FOR BARON STIEGLITZ,

Erected at Narva, Russia.

Fig. 303.

Longitudinal Section.

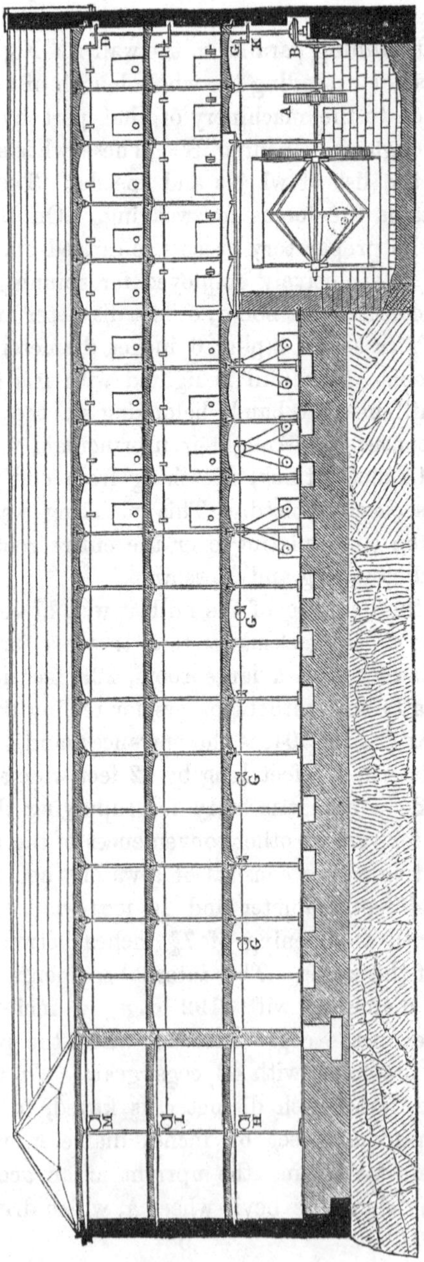

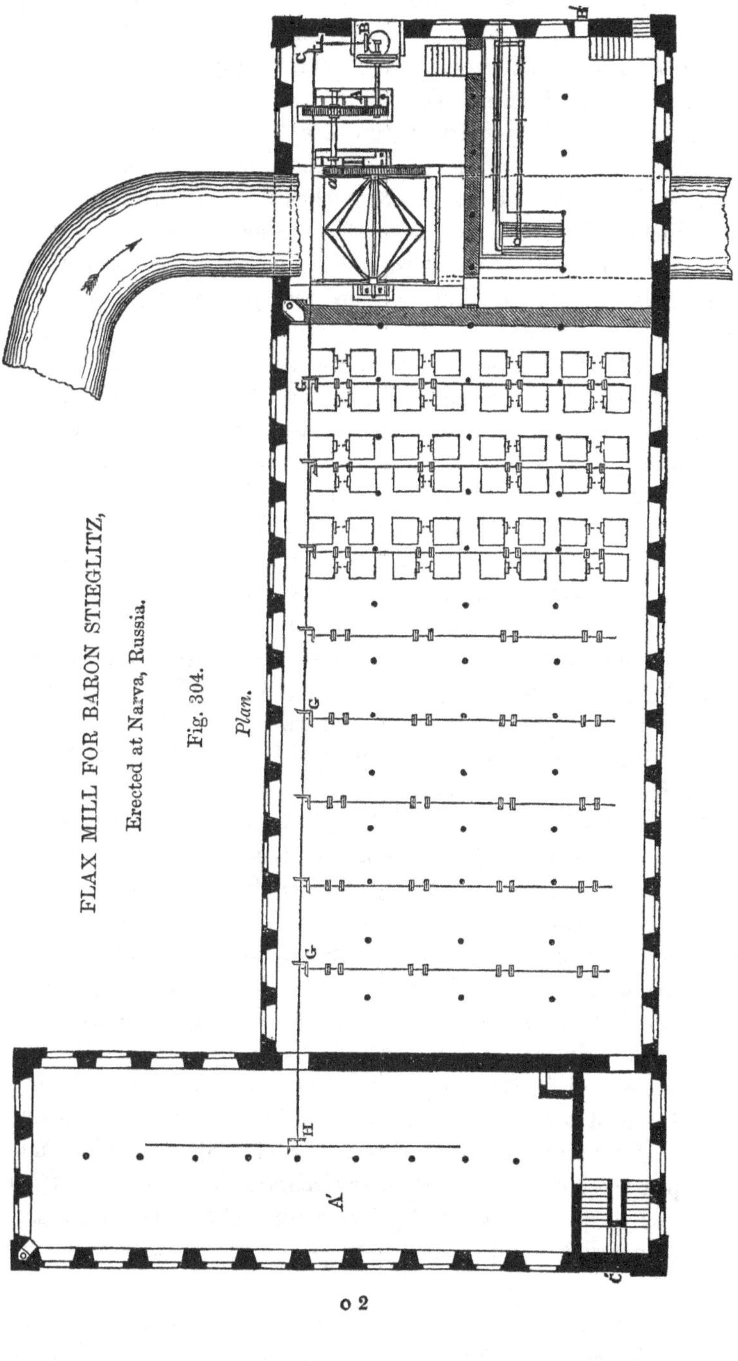

FLAX MILL FOR BARON STIEGLITZ,

Erected at Narva, Russia.

Fig. 304.

Plan.

is also keyed. By reference to the list of wheels and speeds, page 203, the form and dimensions of the other wheels will be found. The vertical shaft B is 8¼ inches diameter, and tapers to 6½ inches. The main horizontal shaft over the looms and in the preparing room are 3½ inches diameter, and taper to 2¾ inches at the end. In the third, or spinning-room, the main horizontal shaft is 3¼ inches diameter, and tapers to 2¼ inches at the extreme end. The minor cross shafts are of the same description as those in cotton mills, already noticed at pages 173 and 174, and therefore need no further description.

The Process.—Preparatory to the flax being scutched it is passed over a machine called a breaker. This machine consists of several coarsely-fluted rollers of cast iron, which are heavily weighted, and as the flax is passed through between them, the straw or boon, which has become quite brittle by the process of retting, is broken and partly separated from the fibres of flax. The stricks, or handfuls of flax, are now passed on to the scutching machines, to have the broken straw separated from the fibre. This is effected by causing a series of rapidly-revolving blades or beaters to come in contact with the pendent stricks of flax, and by repeated blows to clear off the straw and give a softness to the flax fibre. There is necessarily a portion of fibre carried off by these beaters. This is afterwards taken out from amongst the shives or broken straws, and is partly cleaned. It is called the scutching tow, and is spun into yarns of the coarsest quality. The scutched flax is now made up into bundles, or, as they are called, heads of flax, and in this state sold to the manufacturer.

As the roots and top parts of the flax are of a quality inferior to the middle portion, it is sometimes found advantageous to cut the flax into three lengths: this is generally done with the finest description of flax from Courtrai and Flanders. The machine used for this purpose is a breaking machine, consisting of 4 pair of grooved rollers and a rapidly-revolving plate about 20 inches diameter, with a number of projecting diamond-shaped steel cutters on its outer circumference. These cutters occupy a space of about 1½ inches in width. The workman takes a strick of flax, and places it so that it will meet the cutter at the exact place where he wishes the strick to be broken, and allows the ends to pass on each side between pairs of grooved rollers, which slowly revolve. These carry forward the flax and hold it firmly, whilst the revolving plate with the cutters gradually breaks through the whole of the fibres, and the strick is severed in two. The same process breaks off the outer end. These machines are generally made double with four pairs of grooved rollers, i. e. two pairs on each side of the revolving cutters, so that two men can be employed on the same machine at the same time.

The flax, whether divided or in its full length, must now pass in succession through the various processes of heckling, spreading, drawing, roving, and spinning.

I may here describe an important difference between the state in which the raw material flax is presented to the spinner and that in which cotton, wool, or silk is found previous to being manufactured. The fibres of cotton, wool, and silk are supplied by nature already in their finest state of subdivision; they require merely to be straightened and formed into a continuous thread. In raw flax, on the other hand, the ultimate fibres, which are very fine, are united by a gummy matter into broad strips or ribands, and a very operose process called hackling is required to subdivide the material into finer fibres before the spinning process can begin.

Heckling, the first of these processes, consists in effectually completing the process commenced in scutching, and in splitting the fibres so as to make them equal in size, and also capable of being spun to as fine numbers as possible. Heckling machines are various, according to the quality of the flax to be operated upon; a description of one machine, however, will suffice to show the manner in which the process is carried out: take for example Baxter's Street Heckling Machine for Long Line, and it will be found that this machine consists generally of six gradations of heckles (although that number may be increased if required), fastened upon a strong leather sheet 8 feet wide, each heckle being 16 inches long. This sheet runs at a quick speed over two rollers, and inclines downwards from the part where the heckles first strike the flax. The flax is divided into stricks or handfuls, and being spread into holders is there firmly compressed by bolts, a little more than one half of the whole length of fibre being allowed to hang from the bottom bite of the holder. The holder with the pendent strick of flax is now introduced into the head of the machine, which is moveable up and down, and by a self-acting motion the holder is pushed forward so as to arrive exactly over the first tool or set of heckles when the head is at the highest; consequently, the heckles commence operating upon the ends of the flax at first, and gradually enter deeper into the strick as the head descends. When the head has attained its lowest point, the bite of the holder is quite close upon the points of the heckle-pins. The head dwells in this position for a short time, to give the flax a proper amount of heckling, and then rises again to the top, pushing the holders forward to the next tool, then turning them round so as to present the other side of the flax to the pins. The same thing is repeated to each holder, until it has passed over the whole of the six tools. It will be understood that each tool contains heckles finer and closer set than its predecessors, so that the action upon the flax is very gradual, and that there is at all times a strick of flax upon each tool, so that there is one delivered at each rise of the head. When the holder and flax is delivered from the machine, the heckled portion of the flax, which, it will be observed, is little more than half, is spread into a holder and fastened as before; the other end is liberated from the holder, and is in its turn subjected to the process of heckling. The short, loose, and weakest fibres, which are taken out by the heckle teeth, are called *tow*, which is also spun into yarn, but of a quality inferior to that produced from line or heckled flax.

The first machine over which the heckled flax or line is passed, is the spreader or first drawing frame, the object of which is to transform the stricks into an endless ribbon or sliver.

The stricks of 'line' are subdivided by the attendant into still smaller portions, of an equal weight, and laid upon a travelling sheet, each small portion being slightly elongated by the hands, and laid so as to overlap about three-fourths of the length of that which was spread before it, it being desirable in spreading to have as equal a thickness over the whole sheet as possible. This travelling sheet carries forward the flax to a pair of rollers called retaining rollers, which revolve at the same surface speed as the feed sheet. At a distance (regulated by the length of the fibre) from the retaining rollers are another pair of rollers called drawing rollers, and the intervening space is filled with iron bars or fallers, on which are fastened a series of gills or heckles. These fallers rise up close in front of the retaining rollers, and the pins of the heckles enter into the flax as it is passed through by these rollers, and carry it forward to the drawing rollers, the speed of the fallers being calculated to take the flax exactly as it is delivered by the retaining rollers. The drawing rollers revolve at a surface speed, varying according to the material, from 20 to 60 times that of the retaining rollers and gill bars. The flax is consequently drawn out by these rollers to a length from 20 to 60 times what it was when originally laid upon the sheet.

A spreader is generally composed of two feed sheets, and to each sheet there are two gills; there are therefore four slivers formed on the machine at one time, all of which, by means of a condensing plate with four diagonal holes placed in front of the drawing roller, are doubled into one, and passed through a delivering roller into a can. Each can is calculated to hold a certain length of sliver, and the ringing of a bell, connected with the delivery roller, informs the attendant when that length has been delivered.

Next comes the drawing frame. A certain number of cans as they come from the spreader are taken, weighed, and formed into sets of 12, 16, or 24 per set, as the case may be, the heavy and the light cans being arranged so that the total weight of each set is the same. The slivers from these cans are then conducted, 4 or 6 together, between a pair of retaining rollers as in the spreader, drawn by drawing rollers through gills, in the same manner, and all the slivers of the set are again doubled and delivered into one can. The draught between the retaining and drawing rollers is not, however, so great in the drawing frames as in the spreader, varying only (according to the material) from 8 to 24. The process of drawing and doubling is again repeated until the sliver is of an equal thickness in all its parts, care having been taken to regulate the draft of the machines, so that the number of yards of sliver in one pound is suitable for the size of the yarn to be spun.

The roving frame is the next machine over which the sliver is passed. The process of elongating the sliver in this machine is the same as in the drawing frames, but instead of its being doubled after passing the drawing rollers, each sliver is slightly twisted by a flyer, and laid upon a wooden bobbin. Most roving frames are now made self-regulating, in the same manner as those used in the cotton manufacture, the only difference being that for most kinds of flax the bobbins are larger than for cotton. The 'line' is now prepared ready for the spinning frame.

The 'tow' or short fibres thrown out by the heckling machines has also to undergo a process of preparation but slightly different from that to which the line is subjected.

The first operation consists of cleaning and straightening the tangled fibres by means of a carding engine. This machine is composed of a large cylinder 4 or 5 feet in diameter, and 6 or 8 feet in width, the whole of its circumference being covered with teeth set closely together, projecting about $\frac{3}{8}$ths of an inch, and slightly inclined in the direction in which the cylinder turns. The cylinder revolves at about 150 or 200 revolutions per minute, and is surrounded by several rollers, also covered with pins, which assist in straightening and equalising the tow. The first of these rollers are the feeders, a pair of rollers about $2\frac{1}{2}$ inches diameter, which, being fed from a creeping sheet (similar to that employed in the spreader), pass the tow slowly to the cylinder, but, having their teeth set at an angle so as to retain the tow, the cylinder can only take a small portion at a time. After the feeders come several pairs of rollers, 6 or 7 inches diameter, called workers and strippers. The workers revolve slowly with the cylinder, but with their teeth set at a keen angle, and pointing so as to take the material that does not lie straight upon the cylinder. The worker is then cleared by the stripper, which revolves much more rapidly, but not so fast as the cylinder, which in its turn clears the stripper. The tow is thus passed on through several pairs of workers and strippers, each succeeding pair being set closer to the cylinder and to each other until it arrives at the doffers, which are the last of the rollers upon the card. A card has generally 2 or 3 doffers, the last being, of course, set closer to the cylinder than the first; these cylinders are about 14 inches diameter, and have the teeth bent in the same manner, and revolve in the same direction as the workers. The tow is stripped from the doffers by the rapid strokes of a knife, which rises and falls close to the pins on the face of the roller, and the fibres are conducted and delivered by rollers in the form of a continuous sliver. As it is generally necessary to pass the tow over two cards, the slivers from the first card are formed by a lapping machine into a ball about 20 inches diameter, and from this fed to a second card, which is called the finisher. The construction of the finisher card is the same as the first or breaker card, but the pins are finer and set still more closely. There is also attached to the front of the most improved finisher cards a drawing head, in which all the slivers are drawn over a rotary or porcupine gill, doubled over a plate, and delivered into one can.

The cans from the finisher card are now arranged into sets, and the slivers are doubled, drawn, and made into rove in the same manner as 'line.'

The most important machine that has been for some years introduced into the flax trade, is that known as Flishmann's Tow-combing machine, similar to that used in the manufacture of cotton. By passing tows over this machine, they are cleared from all impurities, as well as from the little buttons and knots formed both in heckling and carding, and thus nothing is left but clear fibre. Yarns can be spun from the tow to the same fineness, and having as good an appearance as the yarns produced from the heckled line. As yet, tow-combing is only carried on by a few of the most advanced houses in the trade, but there is every probability that in a few years it will become much more general. Each of the various kinds of combing machines now in use for wool, and several designed expressly for the purpose, have been tried on tows.

The great desiderata in each case are:—

1. The thorough cleaning of the tows from all buttons, shives, &c.

2. The attainment of the above object with the smallest possible amount of loss in waste.

3. The passing of as much weight of material as possible per day.

The machine which has been most successful in obtaining the first-mentioned results, is that known as Flishmann's.

Spinning.—There are three modes of spinning flax; viz. dry, with cold water, and with hot water. Dry spun yarns are chiefly of the coarsest quality, and are used in the manufacture of sail cloth, canvas, sacking, &c. Cold water spun yarns are used principally for shoe threads, and for making twines. The processes of dry and cold water spinning are identical, except that in the latter the fibres as they pass from the drawing roller are damped, which gives the yarn a smoother and more regular appearance. The finest qualities of yarn are the hot-water spun. In this process the rove, before it reaches the retaining rollers to be drawn out, is completely saturated by passing through a trough containing water heated by steam. The hot water macerates the fibres, and dissolves a portion of the gummy matter contained in the flax, and so renders it capable of being spun to a much greater degree of fineness. The length of the fibre is, by this process, very materially shortened, the distance between the retaining and drawing rollers being only from 3 to 4 inches, whilst for the dry spun yarns, the distance is about 18 inches for line, and 8 or 9 inches for tow. The strength of the yarn is not, however, at all impaired by this, and it is much smoother and more even than that spun in the full length.

Reeling, the next operation, is the winding of the yarn from the bobbins round a barrel 2½ yards in circumference. 120 revolutions of this barrel make a lea, or 300 yards, which the attendant ties up separately,—10 leas constitute a hank, and 20 hanks a bundle. It is in bundles that the yarn is generally made up, and a small machine called a 'Bundling Press' is used to compress the yarn into as small a space as possible, previous to its being sent to market.

The above description of the flax process has been kindly furnished by Messrs. Fairbairn & Co. of Leeds, to whom and the late Sir Peter Fairbairn the flax trade is greatly indebted for the introduction and working out of the screw gill and other preparatory machinery.

It will not be necessary to enter into the process of weaving, as the power-loom has made slow progress in the manufacture of linens, which are chiefly woven on the hand-loom. For the coarser descriptions of cloths, the power-loom has been adopted, but not successfully, in fine linen, as there appears to be great difficulty in dressing and preparing the warps, and I believe at the present time most of the Irish linens are manufactured by the hand-loom weavers of Ulster.

In Manchester and other parts of the manufacturing districts, a description of mixed goods called domestics are manufactured for shirtings, and are wove on the power-loom the same as cotton. The late firm of Messrs. Leys, Mason & Co. of Aberdeen were at one time large manufacturers of the coarser linens by power, and several attempts at linen power-loom weaving have been made in other establishments, but not successfully, either as regards fine linens or cambrics. Most of the flax mills in Leeds confine their operations to spinning, with the exception of Messrs. Marshall, who of late years have paid considerable attention to weaving by power, I believe, successfully, but the difficulties have not been altogether surmounted, and several years may yet elapse before this important desideratum in the manufacture of flax into cloth is attained. Much has already been done, but the results so far have not been attended with complete success, and can therefore only be looked upon as experimental.

The bleaching, beetling, polishing, and finishing of linen is carried on extensively in the north of Ireland and at Barnsley, in Yorkshire. It is an important branch of industry in both countries, and to effect these objects, machinery for boiling in lea, washing, drying, calendering, winding, packing, pressing, &c., are requisite. In some descriptions of goods the beetling process is adopted, and this consists in winding the cloth on to an iron cylinder of about 20 inches diameter, placed under a row of stampers 4 inches square, made of beech, which by a revolving tappet-shaft raises the stampers and allows them to fall in succession upon the cloth, until it attains a beautiful polish and wave-like appearance.

Annexed is a list of wheels and speeds, and the number and size of the pulleys, with the velocity of the different machines.

List of Wheels and Speeds.

Water-wheel, 24 feet diameter. Velocity on periphery = 4·4 feet per second = 3·5 revolutions per minute.

Description of Gearing.		Driver. Diameter.	Driver. Revolutions.	Driven. Diameter.	Result. Revolutions per Minute.	
		ft. in.		ft. in.		
Segments . . .	a	20 $4\frac{3}{8}$	3·5	4 $6\frac{3}{4}$	15·6	revolutions of cross shaft.
Wheels	A	12 1	15·6	4 $6\frac{3}{4}$	41·1	„ shaft driving upright.
Wheels	B	7 $11\frac{3}{4}$	41	3 $11\frac{3}{4}$	82·3	„ upright shaft.
Mitre wheels . .	C C	3 2	82	3 2	82	„ cross shaft No. 1.
Bevil „ . .	D D	4 0	82	2 $2\frac{3}{8}$	149	„ cross shaft in Nos. 2 and 3 room, preparing and spinning-rooms.
Mitre „ .	E E	2 5	149	2 5	149	„ longitudinal shaft in No. 2 preparing-room.
Mortice bevil wheels	F F	3 $5\frac{5}{8}$	149	2 1	248	„ „ shafts in No. 3 spinning-room.
Mitre wheel . .	G	3 2	82	3 2	82	„ „ shaft in No. 1 weaving-room.
Bevil wheels .	G G G G	2 2	82	2 0	88	„ cross shafts in No. 1 weaving-room.
Mitre „ . .	H	2 4	82	2 4	82	„ cross shaft in wing building, No. 1 room.
Mitre „ . .	I	2 4	149	2 4	149	„ shaft in wing No. 2 preparing-room.
Mitre „ . .	J	1 8	88	1 8	88	„ cross shaft, No. 1 winding-room.
Bevil „ . .	K	2 3	82	2 $0\frac{1}{8}$	90	„ cross shaft to bleach-house.
Mitre „ . .	L	2 0	90	2 0	90	„ shafts in bleach-house.
Bevil „ . .	M	1 $1\frac{1}{2}$	248	2 3	124	„ cross shaft, No. 3 wing, mechanic's shop.
Governor Speeds.						
Pulley	A′	0 $9\frac{1}{4}$	82	1 9	36	revolutions of governors.
Spur wheel . . .	B′	0 8	36	2 0	12	„ cross shafts.
Spur „	C′	0 5	12	1 9	2·8	„ shaft across cistern.
Bevil „	D′	1 $1\frac{1}{2}$	2·8	2 3	1·4	„ worm shaft.
Worm „ . . .	E′	0 $6\frac{1}{2}$	1·4	2 $0\frac{3}{4}$	0·38	„ rack shaft.

LIST OF PULLEYS AND SPEED OF MACHINES.

Number of Pulleys.		Driver. Diameter. ft.	in.	Driver. Width. ft.	in.	Driver. Revolutions.	Driven. Diameter. ft.	in.	Result. Revolutions per Minute.		
32	Pulleys	1	5	1	2	88	1	2	106	revolutions of	heavy looms.
8	,,	1	6	1	2	88	1	2	113	,,	light looms.
8	,,	1	5	1	2	88	1	4	93	,,	light looms.
2	,,	2	1	0	8½	88	1	10	100	,,	pumps.
2	,,	1	10	0	7	88	1	11	84	,,	calender.
1	,,	1	0½	0	10	88	1	6	60	,,	cauderoy.
2	,,	1	0½	0	10	149	0	10	200	,,	shaft-minding.
4	,,	1	5¼	0	9	149	1	4	160	,,	warping-machine.
4	,,	1	8	0	8	149	1	4	180	,,	dressing-machine.
2	,,	1	10½	0	8½	149	1	6	180	,,	spreader.
5	,,	2	2½	0	9	149	1	8	200	,,	finisher-card.
1	,,	1	10¼	0	7	149	1	6	180	,,	line 1 drawing.
2	,,	2	0	0	9	149	1	6	200	,,	line 2 drawing.
1	,,	2	3	0	7	149	1	6	225	,,	line 3 drawing.
6	,,	1	5¼	0	9	149	1	2	180	,,	tow-drawing.
2	,,	1	6½	0	8	149	1	6	150	,,	roving.
1	,,	2	0	0	9	149	1	6	200	,,	roving.
2	,,	1	10¼	0	7	149	1	6	180	,,	tow-roving.
3	,,	2	2½	0	9	149	1	8	200	,,	break card.
1	,,	1	8	0	8	149	1	4	180	,,	break card.
1	,,	2	0	0	9	149	1	6	200	,,	teaser.
1	,,	0	10	0	10	149	1	8	120	,,	lap.
2	,,	1	3¼	0	6	149	1	2	160	,,	heckling.

CHAPTER VI.

SILK MILLS.

Dr. Ure states, in his 'Philosophy of Manufactures,' that 'the silk-worm is a precious insect, which was first rendered serviceable to man in China, about 2,700 years before the Christian era. From that country, the art of rearing it passed into India and Persia. It was only at the beginning of the sixteenth century that two monks brought some eggs of the silk-worm to Constantinople, and promulgated some information on the growth of the caterpillars. This knowledge became, under the Emperor Justinian, productive of a new source of wealth to the European nations. From Greece, it spread into Sicily and Italy, but did not reach France till after the reign of Charles VIII., when the white mulberry-tree and a few silk-worms were introduced into Dauphiny by some noblemen on their return from the conquest of Naples. No considerable result took place till, in 1564, Traucat, a common gardener of Nismes, laid the first foundation of a nursery of white mulberry-trees, with such success as to enable them to be propagated within a few years over all the southern provinces of France.' The cultivation of the mulberry, according to this statement, dates from an early period, but the silk manufacture, like most other branches of industry, was first introduced into this country by emigrants from France and Italy, during the persecution of the Protestants, and shortly before the Edict of Nantes. At that time, everything was done by hand, from the cocoon to the web, as it left the hands of the weaver. The winding, throwing, spinning and weaving, were all effected by manual labour, until the first silk mill, driven by power, was introduced by Mr. John Lombe into Derby, about the year 1716. For a series of years this mill and others, constructed from the same model, did nearly the whole of the spinning, and the throwsters, as well as the hand spinners, were, in consequence, superseded by the greater economy, accuracy, and despatch of the new manufacture. The neighbourhoods of Spitalfields, Coventry, and Macclesfield, became the chief seats of the silk manufacture, and for a considerable time they continued to enjoy the monopoly without little, if any, change or improvement in the machinery. The trade seemed to languish rather than improve until after the peace of 1815, when it was introduced on a greatly increased scale into Manchester, where it underwent the same changes and improvements as most others of the textile fabrics. The machines for spinning, doubling and throwing silk continued for nearly a century the same as they were when first introduced into Derby. They consisted of a wooden frame, about 3 feet 6 inches wide and 7 feet high, with two or three tiers of spindles, each tier being driven by an upright shaft A and large drums, with tightening pulleys, as shown at *a*, *a*, &c., Fig. 305.

Fig. 305.

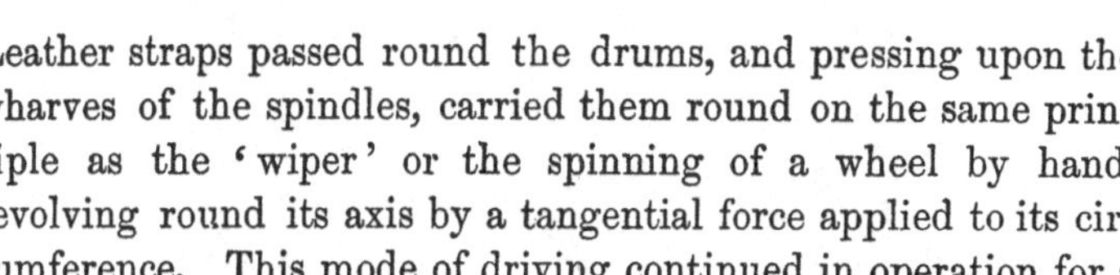

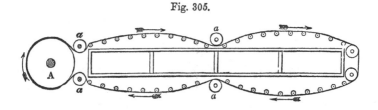

Leather straps passed round the drums, and pressing upon the wharves of the spindles, carried them round on the same principle as the 'wiper' or the spinning of a wheel by hand, revolving round its axis by a tangential force applied to its circumference. This mode of driving continued in operation for a great number of years without variation, until the late Mr. Vernon Royle built a large silk mill in Manchester, when the whole of the machinery underwent a total change, wooden frames and friction traps were removed, light cast-iron frames and cotton bands were substituted for driving the spindles, and the whole machine was remodelled on the same principle as the throstle for spinning cotton. These improvements were introduced by the late firm of Fairbairn & Lillie, and the result was a great improvement in the motion of the spindles and a great increase of speed, by which one half more yarn per spindle was produced than what could be obtained from the old frames. The alterations were further improved by the late Mr. Ritson and

Messrs. Wren & Hopkinson, to whom I am indebted for the description of the present improved process in the manufacture, and to whom silk manufacturers may be referred as the best makers of this description of machinery.

The following section of the improved machine is taken from Dr. Ure's work on the 'Philosophy of Manufactures:'—

'The machine for twisting the single threads of silk, either before the doubling or after doubling, is called the spinning-mill, sometimes also the throwing-mill, though the latter term

Fig. 306.

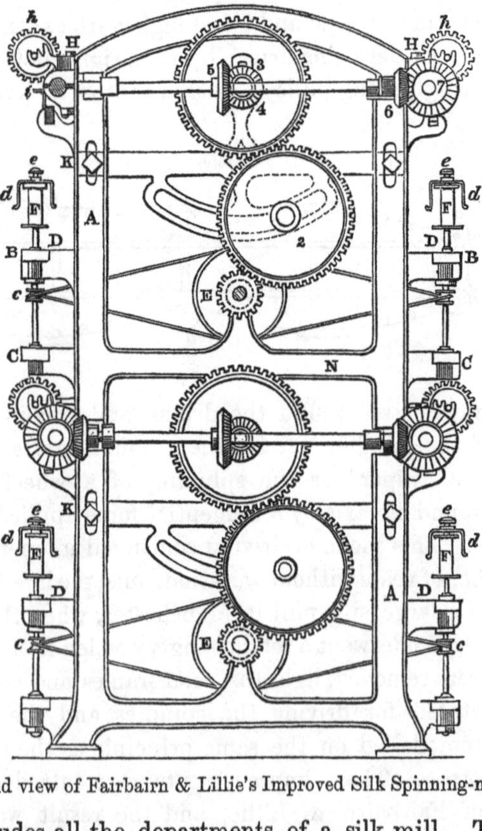

End view of Fairbairn & Lillie's Improved Silk Spinning-mill.

often includes all the departments of a silk mill. The section of this apparatus in Fig. 306, shows four equal working lines, namely, two on each side of the frame, one tier being over the other. In some spinning-mills there are three tiers, but the uppermost is a little troublesome to manage, as it requires the attendant to mount a stool or steps.

'A A are the end frames or uprights, bound with cross-bars N N; and two or more similar uprights are placed immediately between the ends. They are all connected at their sides by beams B and C, which extend through the whole length of the machine. D D are the spindles, having their top bearings fixed in the bar B, and the bottom or step bearings in the bar C. These two bars together are called by the workmen the spindle-box: *c c* are the wharves, turned by cords passing from the horizontal tin cylinders E, which lie along the middle of the mill, midway between the ranges of spindles. F F are the bobbins with the double silk, which are fixed on the tapering spindles by pressing them down; *d d* are little flyers, or forked arms of wire attached to a disc of wood or washer, which revolves loosely upon the top of the said bobbins F F and round the spindles, one of their arms being sometimes bent upwards to serve as a guide to the thread; *e e* are pieces of wood pressed on the top of the spindles, to prevent the flyers from being thrown off; *h h* are the ends of the winding bobbin-shaft, laid in slots near H, as in the former machines. The winding bobbins are driven by toothed wheels cast on one end of their square iron axes, in the line of *h*, which wheels are turned by toothed wheels on a bar in the line of the bevel wheel 7. On these bobbins, fig. 307, which are of considerable diameter, the silk is wound, and distributed

Fig. 307.

Bobbins of spirally wound silk.

diagonally by a peculiar differential mechanism. K K are the guide bars, with the guides *i*, through which the silk passes, being pulled by the winding bobbins on their horizontal in the line of *h*, and delivered by the flyers *d d*, from their vertical twisting bobbins and spindles F. By the revolution of the tin cylinder E, driven by a steam-pulley fixed on its end, motion is communicated immediately through the cords to the wharves *c*, and their spindles; and mediately through the plate-wheels 2 and 3, and the bevel-wheels 4, 5, 6, 7, to the rest of the machine. The toothed wheel at E is called the change-pinion, because, by changing it for another of a smaller or a larger size, the speed of the plate-wheel 2 and 3 may be changed. The axis of the plate-wheel 2 lies in a curvilinear slot, in which it can be shifted to suit the size of the change-wheel put in at E, and to keep it in proper gearing, after which it is fixed by a screw-nut.'

We have selected for illustration a silk mill erected some years since in the south of England, driven by a water-wheel 22 feet diameter and 10 feet wide inside the bucket. The internal segments are 20 feet 6 inches diameter, 2½ inches pitch and 8 inches wide on the cog. The segments communicate motion to a spur wheel 4 feet diameter, and by the shaft A, fig. 308, on which a second spur wheel is keyed, the motion is conveyed to the line of horizontal shafting on the ground floor, and also to a line of vertical shafts which communicates with the rooms above. The bottom room F, figs. 308 and 309, is filled with spinning frames, and driven by the small cross shafts *b*, *b*, *b*, &c.; the second floor, C, fig. 309, is occupied by doubling machines and throwsters, and the upper floor H contains the lighter description of machinery, and is chiefly employed in preparing the hanks for the throwing and spinning and winding the yarn as it comes from the machines below.

The building consists of three stories and an attic, at one end of which is the water-wheel and staircase, and the adjoining buildings contain the boiler for heating, gas works, &c., &c. For some years this mill was worked exclusively by water, but subsequently it was found necessary to have steam as an auxiliary, and hence followed a considerable extension of the mills, and a corresponding increase of machines.

Fig. 308.

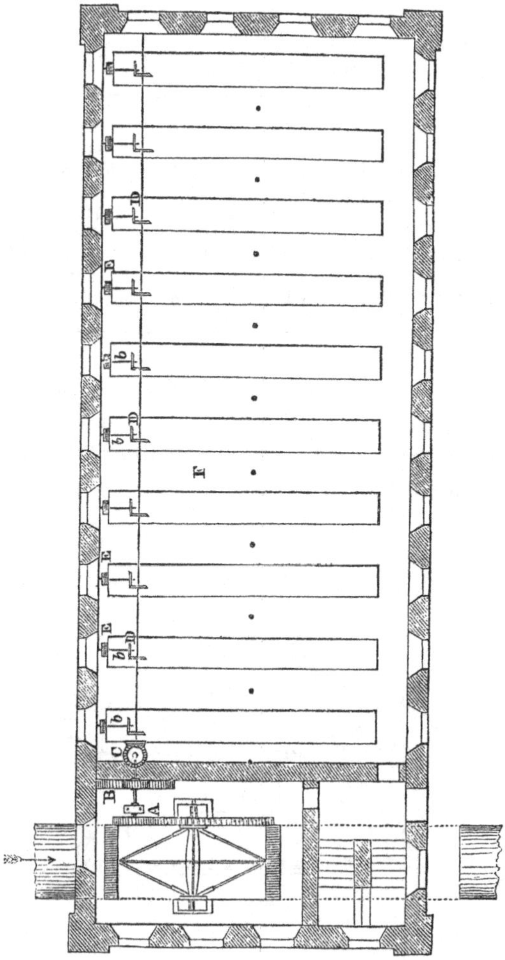

The following references and calculations exhibit the speed of the water-wheel and the shafting, spinning frames, &c., in the mill:—

SPEEDS OF SHAFTS, WHEELS, &c.

Water-wheel 22 feet diameter; velocity of periphery 4 feet 2½ inches per second, = 3·7 revolutions per minute.

Description of Gearing.	Driver.		Driven.	Result.
	Diameter.	Revolutions.	Diameter.	Revolutions per Minute.
Spur segments A	20 6	3·7	4 0	18·96 revolutions of cross shaft.
Spur wheels B	9 0	18·96	3 1	55·17 revolutions of horizontal shaft.
Bevel wheels C	3 9	55·17	2 6	62·75 revolutions of upright shaft.
Bevel wheels D D D, &c.	1 10½	55·80	1 1	96·53 revolutions of cross shafts.
Pulleys on shaft do.	2 6	96·53	0 8	361·98 revolutions of frames.

The speeds of the throwsters, &c., in the second room are the same as the spinning on the ground floor, and the winding machines are driven slow to suit the quality of the silk as it is drawn from the hanks.

Fig. 309.

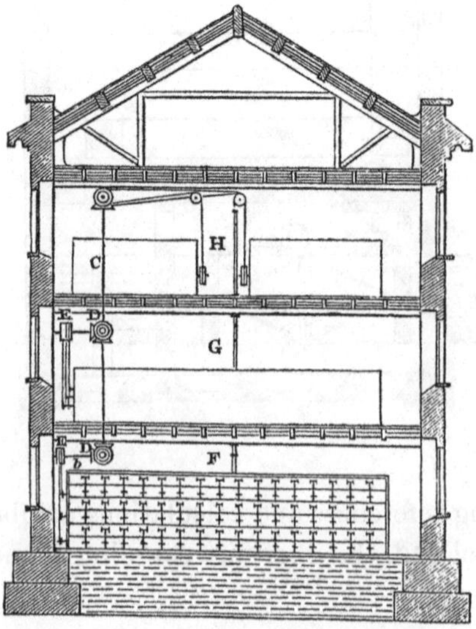

The *Raw Silk Spinning Machinery* is used for the winding and twisting of silk as imported into this country in hanks: the thread being already formed in the cocoon, no drawing process, as in cotton, is needed, and the skill of the manufacturer is exercised in freeing the imported hanks from knots, lumps, and entanglement, 'sizing' or matching the strands of silk and spinning them together, so that the twist shall be regular and perfect, and for this purpose the machines in general use are usually named winding, cleaning, spinning, throwing, dyed-silk, pirn-winding, &c.

The hanks being of different lengths, two sizes of winding machines are used—one suitable for hanks imported from China and Persia, and about 120 inches long; and the other for Indian and Italian, about 72 inches long; but the principle of action being alike in both, one description will suffice. The hank is extended on a swift, which is constructed of wood, being a small centre with metal pivots, with slender arms of lancewood radiating in pairs from this centre; each pair of arms has a string tied round them, so that the hank is distended into a hoop, or rather hexagon, which readily revolves and unwinds; the bobbin which takes up the silk from the 'swift' lays horizontally, and is rotated by friction-rollers, so that when through entanglement or otherwise the swift ceases to revolve and give off the thread of silk, these rollers slip, and the bobbin stops without breaking the thread, when the attendant can adjust the work and remove the impediment to motion. Sometimes this winding is done upon a machine which also 'cleans' the silk, but usually there is a separate machine for this purpose, where the bobbin from the winding frame is placed on a shelf near the floor, and from it the thread is un-wound and re-wound upon a bobbin on a spindle on the top of the frame; this spindle is in the same position and rotated by friction rollers precisely as in the winding frame, but the silk in its way from one bobbin to the other is passed through a 'cleaner,' which is two knives of steel, of which the edges are set parallel to each other, and a minute distance apart; according to the fineness of the silk so is this distance made more or less by means of adjusting screws, and the object of this is to remove from the surface of the silk thread all knots or other excrescences which may interfere with the regularity of twisting, and of the woven goods. Sometimes this cleaning is produced by passing the silk thread between two steel rollers about $\frac{5}{8}$ of an inch diameter, also adjustable by screws, with the same object of arresting knots and lumps, until the attendant removes them from the thread. After the silk has undergone the process of cleaning it is ready for the spinning machine, which twists the single thread so as to give it strength and increased elasticity for the manufacture of sewing silks and the warps of woven fabrics, and this is a machine similar to that described at page 207, consisting of two or three tiers of upright spindles rotated by a tin roller turning them by cotton banding, or in some machines by an endless leather belt revolving horizontally round the machine, and rubbing against each spindle in its circuit; on these upright spindles the bobbins of silk from the cleaning machine are placed, so as, whilst each spindle is rapidly rotated, the thread is being drawn off, and again wound upon another spindle placed horizontally, and turned by small toothed gear, or by friction rollers. These machines have the usual arrangements of change pinions to suit the required twist per inch in the thread, and it is of the highest importance that the delivery and taking up of the thread shall be uniform, otherwise the defective and irregular twist will seriously deteriorate the beauty and value of the manufactured fabrics. This machine fulfills in the economy of the silk manufacture the function of the throstle or the mule in the cotton trade, and sometimes the spinning, together with the processes of 'doubling' and 'throwing,' is done at one operation, as in 'Shute's patent,' where the larger spindle carries round with it two or more smaller bobbins upon spindles, which being revolved in a contrary direction, by rubbing against a stationary band, spin, double, and throw upon one machine. Other varieties of machines have been adopted for the same end, but on account of the diminished speed of the spindles, increased loss in waste, and greater cost in wages, the plan most commonly adopted is to double and 'to throw' on distinct and separate machines. The doubling machine is similar in form to the cleaning machine. The bobbins from the spinning machine are placed upon a shelf near the floor, and the ends of silk from two or more bobbins, according to the sort of work done, are wound together in one cord or strand on a bobbin rotating horizontally, and moved by friction rollers, as in the winding and cleaning frames; but it is essential that this cord or strand shall in every part of its length be composed of the same numbers of the ends of silk laid evenly, and with the same amount of tension together. Before the silk arrives at the bobbin on which it is to be wound each fibre or end passes through the eye at the end of a light wire lever, which, whilst all is going on properly, is upheld by it, but should one of the fibres of silk break, then its wire lever drops upon a second lever, and overbalancing it causes its further end to rise up, and arrest, by means of a ratchet-wheel, the motion of the winding-on bobbin: thus, without the stoppage of the machine in general, that particular bobbin waits motionless for the attention of the operative, when the broken end of the fibre is re-pieced, and, the levers being restored to position, the silk proceeds as before. After the silk has thus been doubled, or several threads laid evenly together, it is taken to the throwing machine to be again twisted, for the doubling machine does not, as in the cotton manufacture, double and twist at one time; and this twisting, as in cotton, is effected by the spindle revolving in a contrary direction to that of the spinning machine. This process is almost a repetition of the spinning, except that, in place of winding on to bobbins, this is done upon reels of 43 to 44 inches circumference; indeed, many silk throwsters 'throw' their silk upon the spinning machine by placing the spindle bands so as to rotate the spindles in the contrary direction, and then reel off the silk into hanks ready for the dyer. The silk when dyed is re-wound from the hank upon bobbins of tin or wood, by machines named soft or dyed-silk winding frames, similar in principle and action to the winding machines for raw silk, and the bobbins are now delivered to the weaver for warping and winding upon pirns for weft. The warping machines are of the usual form, with a large wooden fly as in cotton warping, and the weft is wound on pirns by girls, with the simple hand wheel (forming one at a time), or in the most modern mills by the pirn winding machine, which contains 40 to 100 spindles, under the care of one attendant: the pirn is formed upon bobbins specially shaped for the purpose, sometimes by running in a metal internal cone, which as the pirn fills gradually forces it upwards, until its spindle is out of gear from the driving power, and then it remains motionless until the attendant re-adjusts the position of the spindle and places on it an empty bobbin. The same effect is produced by three small conical formed rollers pressing on the outside of the bobbin, but both these varieties of machine have been found to be injurious to the delicate shades of colour in the dyed silk, as by the compression and friction the thread is flattened and glazed, and thus rendered unequal in appearance when in the piece goods; and to obviate this serious defect the plan used by the best manufacturers is to wind the pirn without external pressure upon the bobbin, which is placed on a spindle, which by toothed gear gradually sinks down in the machine, until its driving-band arrives at a loose pulley, which then allows the spindle to rest until the full pirn is removed; or by another mode, the traverse or winding-on rail rises gradually, and effects the same object. In this description of the silk manufacture, I have not gone minutely into a description of the mechanical arrangements necessary to produce these beautiful and costly fabrics; in point of fact, the machines are not intricate in construction, but they require careful workmanship. The traverse rods for forming the bobbins are moved by the well-known appliances called the sun and planet crank motion, the simple crank, the heart, the oval wheels, the mangle wheel motion, &c.; and whilst one manufacturer for the peculiarities of his business may adopt one or more of these, another may prefer the application of others to his machines. Many manufacturers, in this advanced age of sewing silks, in Leek and elsewhere, spin and throw with the simple hand-wheel, where a boy (as in twine or rope making) carries the ends of silk and makes them fast at the other end of a room, and the man called the twister rapidly whirls the hooks on which the silk is tied, gradually moving up the wheel as the twisting proceeds to accommodate the shortening of the thread; then, after his judgement tells him that the thread is sufficiently twisted, he fastens two or more ends of the twisted silk upon one hook, reverses the direction of revolution in his hand-wheel, and 'throws' all into one strand.

Lately, there have been introduced to the silk trade several novel machines which, though they have not hitherto obtained extensive use in this country, may prove useful aids to silk industry. One is a mode of winding the silk from the cocoon, and spinning it on the same machine. This has been done by Mr. Chadwick and others of Manchester, and beautiful work produced, but the difficulties in a new machine of turning off a paying quantity of work, joined to the want of commercial facilities for obtaining from abroad an adequate supply of cocoons, has hitherto impeded the success of the experiment. Another is a mode of 'sizing' or measuring the thickness of the silk thread, by passing it between two or more rollers nicely adjusted, and so arranged that when a part of different thickness occurs, the rollers move a system of levers which either stop the winding-on bobbin, or else transfer the thread to another bobbin. This operation is also accomplished by taking paper spools exactly alike in weight, and winding upon each of them a definite number of yards of silk, then with a delicate balance assorting them, placing those of like weight in distinct lots, and thus obtaining a number of spools with equal lengths and weights of silk to be put together on the doubling machine; for this matching is essential to the regularity of the twist in the silk spinning, as when threads of unequal diameters are 'thrown' together it is very difficult to prevent its being unevenly done, and harder twisted in one place than another, or 'corkscrewed,' as it is technically called. At present, it is the office of a manager or operative of approved skill to 'size' or match by the eye or touch the various bobbins of raw silk, before placing them on the spinning or doubling machines. To meet this purpose, there has also been invented in France a doubling system which, in place of taking several distinct strands or threads of silk, and winding them together as in the doubling machine first described, only deals with one thread of silk, which in an ingenious manner is doubled or rather tripled upon itself into three strands, by means of a traversing carriage like that of a cotton mule, putting at the end of each traverse a loop in the silk, doubling, and thus, so to speak, matching the silk with itself, with the same view of attaining an improvement in the manufacture of the thread when twisted together. In the silk dye-house, a very useful machine from America has recently been introduced. To the present time, the silk, after being dyed, has been 'stringed' or glossed by means of the severe hand labour of men twisting it in the hank with sticks to and fro, so as to rub the strands together, and produce the beautiful lustre so characteristic of the material; this required strong and skilful men, and with every exertion a workman could not finish much per day; but the machine in question entirely dispenses with the great physical exertions of the workman, and enables him to produce a larger amount of polished silk. This operation is performed in a box of cast iron with a steam-tight door, in which the silk can be placed on two rollers, the upper one adjustable to suit the varying lengths of the hanks, and the lower roller is fixed upon the head of a piston rod attached to a piston moving downwards in a steam cylinder; when the silk is placed upon the rollers, the door is shut, and the high pressure steam introduced, thoroughly saturating the silk; then by another valve the steam enters the steam cylinder, and the silk, whilst immersed in steam, is strongly stretched by the pressure applied to the piston; during the process, the silk hank is slowly revolved upon the rollers by gearing communicating with the outside of the box, so that the shades of colour shall not be varied at those parts of the hank which bend round the rollers, and a few seconds suffice for the completion of several hanks.

This class of machinery is, with slight modifications, adapted for the manufacture of sewing silks, warp and weft for piece goods, and for crape; but for inferior fabrics in silk the waste or spun silk is used, and this manufacture is very similar to that of flax or fine cotton. The waste silk from the cocoon winding factories or the raw silk manufactories is combed or heckled, then cut into lengths of from 1½ to 5 or 6 inches, then placed in the opening machines, afterwards sewn up in small bags and boiled, to cleanse it from gum and other impurities, and after drying, is batted or opened with sticks like fine cotton; from this it is passed through breaker and finisher carding engines, drawing and roving frames, and mules constructed with rollers of spaces and adjustments, adequate for fibres of this length, when it is reeled and dyed as the raw silk before described.

The introduction and description of the different preparatory processes in the manufacture of the textile fabrics—although not bearing directly on the subject of millwork, for which these volumes were originally written—are nevertheless analogous, and work so closely into each other that a treatise on mills and millwork would not have been considered complete unless the processes of manufacture for which the mills were designed were introduced. This must, therefore, be my apology for the introduction of matter which appears to belong more to the machinery of manufactures than to the wheels, shafting, and pulleys by which they are driven.

On the question of what is the province of the millwright, and what exclusively belongs to the machinist, we are yet, notwithstanding the great advances made in the division of labour, unable to state where the millwright ends and the engineer and mechanist begins. There is no correct nor definite line of demarcation between the one or the other; and it is a curious fact that the industrial mechanical progress of the last half century has not from that period marked any reliable principle of organisation by which one mechanical operation is distinguished from another. They seem to run into each other without any definite outline of distinction, and the millwright of the present day appears to maintain as in past times his original character of a 'Jack of all trades,' and there are none of the varied forms of mechanical manipulation pursued in this country in which the millwright is not employed. This is the state in which he appears to germinate, alternately changing from the mill to the machine, and from the machine to the equally important duties of the civil engineer. We have many instances of these transmigrations in the history of Bundley, Smeaton, and Rennie; and I believe there is no lack of them at the present day, as many examples may be adduced of men who have risen to distinction from the humble origin of a working millwright.

To show the intimacy which exists, and the claims which the millwright has upon almost every mechanical profession, I may instance that of corn mills, in which not only the moving power, whether wind or water, but all the machinery is exclusively devoted to the skill of the millwright; and it is for this reason that I have endeavoured to describe more minutely the various machines and movements in that department of manufacture than in those of the textile fabrics. In the latter, the trades are divided and subdivided into machine makers, smiths, fitters, turners, &c., but the millwright stands alone as an operator in all these trades; and although the name may be lost in the changes which have taken place in the organisation of the different callings, there is, nevertheless, a sprinkling of the old trade requisite to give character and consistency to the industrial progress of the country. Having thus described the mills for corn, cotton, and other textile manufactures, I have now to revert to paper mills, oil mills, powder mills, &c., in which the millwright undertakes the construction of the whole or the more prominent parts of the machines of which these establishments consist.

CHAPTER VII.

OIL MILLS.

THE means adopted for extracting oils from seeds and nuts in the early stages of civilisation were of a very primitive kind, consisting simply of a few poles driven into the ground supporting two horizontal cross-bars, between which a bag containing the seed was placed. A lever was then brought to bear against one or both of the horizontal levers, causing severe pressure upon the seed from which the oil was expressed. This rude apparatus, long in use in India and Ceylon, was necessarily slow and inefficient, but an improvement was introduced by Mr. Herbert, whose object was to construct what he considered a powerful and effective machine. It consisted of an upright

Fig. 310.

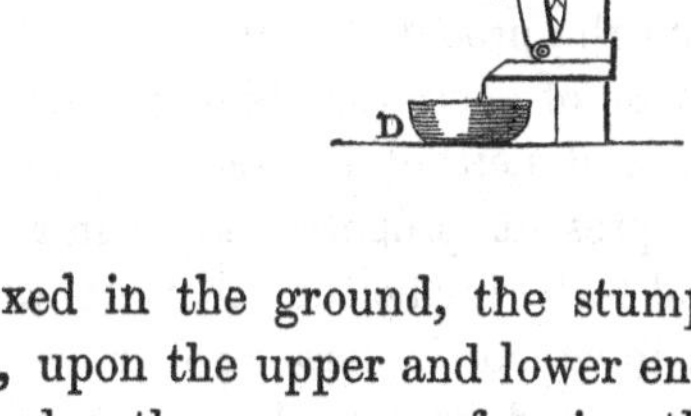

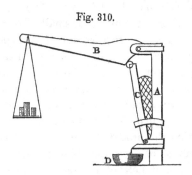

post A, firmly fixed in the ground, the stump of a tree being frequently used, upon the upper and lower ends of which were fixed levers B and C, the upper one forming the fulcrum of the horizontal lever B, and the lower one the joint of the vertical lever C. At the top of this was fixed a roller to diminish the friction of the lever B. The pressure was obtained by the weight of a man applied to the end of the lever B, and that, acting on the vertical lever C, brought it in contact with the bag containing the seed which expressed the oil — on the principle of a pair of nut-crackers — into the vessel D below. Machines of this description with double levers came into use, but all of them were very imperfect, and until the Dutch stamper-press was introduced from Holland, there was no machine in this country entitled to consideration as an oil mill. The screw and the hydraulic press are recent inventions of this country; but, before considering their comparative merits, it will be necessary to refer, generally, to the course of operations to be performed previous to the compression of the seed, which is the last of five operations to which it is subjected.

The *first* operation consists in passing the seed through a flat screen or shaker, which is kept in a constant state of agitation to clear it of all foreign matter and to prepare it for the *second*, which is to pass the seed through a pair of crushing rollers. In this operation the two rollers are of unequal diameters, the larger one being 4 feet diameter, and the smaller 1 foot diameter, the breadth of both being 16 inches, or 14½ inches on the face. The larger roller makes 56 revolutions per minute, driving the smaller one by friction. The seed is supplied through a hopper by means of a small roller very slightly grooved, which is made to revolve for the purpose of feeding the main rollers, being driven by a strap from the larger roller passing over a pulley outside the hopper. The amount of feed is regulated by a moveable plate adjusted by a screw. Underneath the rollers are placed scrapers kept in contact with them by weights, for the purpose of removing any seed adhering to the surfaces during the process of crushing. These rollers for a long time were made of equal diameters; but it was found that they crushed the seed neither so well nor so expeditiously as they do in their present proportions. After the equal sized rollers were found to be inefficient, that known as the Ipswich mill was adopted, in which the larger roller was 6 feet diameter, and the smaller 1 foot diameter; but experience proved that, when any hard substance got between the rollers, the leverage over the journals was so great that it caused much wear and tear upon those parts. Seed crushers have, therefore, by degrees adopted the medium sized rollers, which are found to be exceedingly effective and not liable to derangement. A pair of rollers, such as described, will crush upon an average about 4½ tons of seed in 11 hours, which is sufficient for two sets of hydraulic presses.

The *third* operation consists in grinding the seed under a pair of edge stones, weighing together about 7 tons, and making about 17 revolutions per minute. The stones, if of good quality, and the seed pure, require to be faced every three years, and will last for a great length of time. One pair of edge stones will grind sufficient seed for two double hydraulic presses; the time of grinding being about twenty-five minutes, when it is ready for transfer to the *fourth* operation, which is to heat the seed in a double steam-kettle of the annexed form, fig. 311,

Fig. 311.

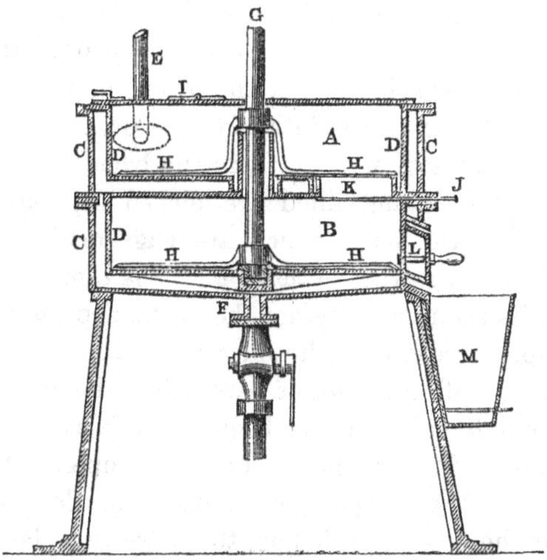

which represents a vertical section. The kettle consists of two cylindrical chambers A and B, one above the other, each of which is composed of an external casing C, which surrounds the internal or inside kettle D, with a sufficient space left between the two round the sides and bottom to allow a free circulation of the steam. The steam is admitted by the pipe E, and the condensed water passes off at F from the bottom of the kettle. The shaft G gives motion to two arms or stirrers H H, in each chamber, revolving at the rate of 36 revolutions per minute, which keep the seed constantly agitated, so that every particle of it may come in contact with the heated sides and bottom of the kettle. The upper chamber A is covered with a sheet-iron lid I, through which the kettle is charged. In heating the seed, the upper chamber A is filled first, and the seed is allowed to remain in it from 10 to 15 minutes; the slide J is then withdrawn, and the seed falls through the opening K into the lower chamber B, where it remains until it is required to be taken to the press; the door L is then opened, and the whole of the seed is discharged from the chamber B by the action of the revolving stirrers H. The seed falls through a funnel M, under which is placed a bag of suitable dimensions to contain a sufficient quantity of seed to make a cake weighing 8 lbs. after the oil is expressed from it. Each of the chambers in the heating kettle will contain sufficient seed for charging one single press; the heating of the seed is therefore a continuous operation of first charging the upper chamber A, and then allowing the seed to pass into the lower one B, in which it is heated to 170° Fahr., and is then withdrawn and placed in the bags.

The bags after being filled are placed separately between what are called the hairs, which are bags made of horsehair with an external covering of leather. The same description of bags and hairs are used, whether the oil be expressed by means of the stamper, screw, or hydraulic press.

The last operation is that of expressing the oil by pressure through the interstices of the bags, and this is done either by a square-threaded screw press, or by stampers which are of Dutch origin. The stamper press consists of a long rectangular box open at the top; at each end there are two plates, between which one bag of seed is placed, yielding a cake of 9 lbs.; next to one of the inner plates is a filling-up piece, then an inverted wedge, then another filling-up piece, after which is introduced a vertical driving-wedge, and, lastly, another filling-up piece is inserted between the driving-wedge and the other inner plate. As soon as the bags have been placed vertically in the press-box, a stamper made of hard wood, about 16 feet long and

8 inches square, with a descent of about 22 inches in the final stroke, is allowed to fall at the rate of 15 strokes per minute for a period of about 6 minutes upon the head of the driving-wedge, which is sufficient to drive it down level with the top of the press-box, the stamper being worked by two cams, or wypers, on a revolving shaft. Side by side with the first stamper is a second one, immediately above the inverted wedge, which is held suspended at a fixed point by means of a lever, while the first stamper is in action; but, as soon as it is time to remove the bags, the first stamper is raised by means of a lever above the point at which the cams come in contact with it, and by the same means the other stamper, which was previously suspended, is allowed to fall upon the inverted wedge, driving it downwards and thereby releasing the working wedge, so that the attendant may remove the bags and repeat the operation. A press like this will not do more than 12 cwt. of cake per day.

The last mode of expressing the oil is by means of the hydraulic press, which may fairly be said to be the most approved system that has yet been adopted. This press is simply Bramah's press arranged specially for the purpose of expressing oil, and appears to have been in use for this work more or less for thirty years, although the earlier presses were very defective as compared with those in use at the present time.

One of the first hydraulic presses applied to oil mills was constructed by Messrs. Martin Samuelson & Co. of Hull, to whom I am indebted for the description of the machinery.

In this arrangement, only one press and one set of small pumps was introduced. The box A, which receives the seed, is in one piece, and runs upon a small tramway for the purpose of withdrawing it from the press to remove the cake and replenish the bags; each time, therefore, that the press is put into operation, the entire box has to be withdrawn, in order to empty and replenish it, and it has then to be replaced upon the ram B, after which it is lifted bodily upwards so as to bring it into contact with the press-head C, which fits accurately in the press-box A, and acts as the point of resistance when the pressure is upon the ram. The constant withdrawal and lifting of this heavy box must evidently be a great loss of power and time. Presses of this description have been at work at Deptford until within the last few weeks, but they have now been removed and replaced by those known as Blundell's presses, which are now universally admitted to be the most efficient appliance for the purpose.

Blundell's double hydraulic press is shown in the annexed cut, fig. 312, which shows a vertical section of the press with an elevation of the pumps. The double hydraulic press consists of two distinct presses A and B, supplied by two pumps C and D, one of which, C, is $2\frac{1}{2}$ inches diameter, and the other, D, 1 inch diameter, both connected to each distinct press cylinder by means of hydraulic tubing E. The stroke of each pump is 5 inches, and they make 36 strokes per minute; the larger

Fig. 312.

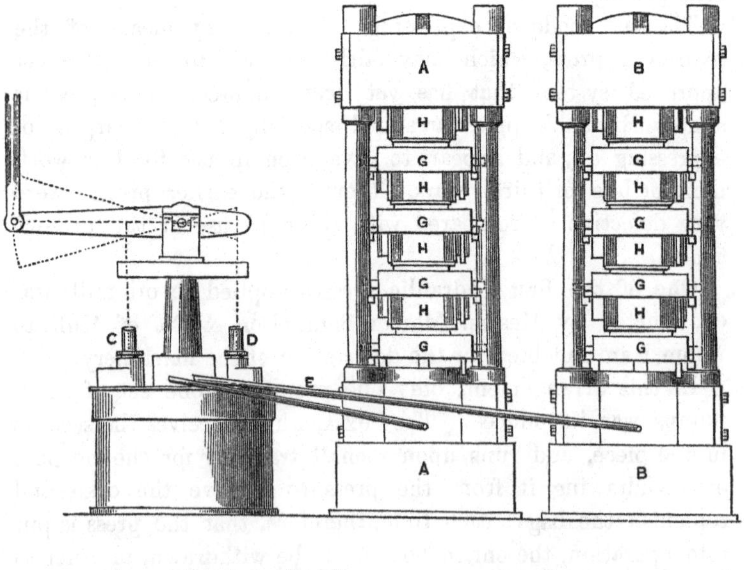

pump C is weighted to 740 lbs. per square inch pressure, and the smaller, D, to 5,540 lbs. per square inch. The diameter of the press rams is 12 inches, and the stroke 10 inches. Each press is fitted with four boxes G, G, and receives four bags of seed in the spaces H, H, producing in all a weight of 64 lbs. of cake at each operation. After the heated seed has been removed from the heating kettle, and placed in the canvas and hair bags, which is done as speedily as possible, so that it may retain its heat, the attendant first fills one press A, and opens the communication between the large pump C and the charged press A, by means of the valves I, which causes the ram to rise until there is a total pressure of about 40 tons exerted on the press; the safety-valve connected with the large pump C then rises and is kept open by means of a small spring catch. Whilst this operation is going on in the first press A, the second press B is being filled in the same manner; the communication is then opened between the large pump C and the press B by means of the valves I, the safety-valve of the pump C having been replaced in its original position; the ram of the second press B is then raised to a corresponding position with that of the first press A, when the safety-valve of the pump C rises a second time. The communication between the large pump C and the press B is then closed, and at the same time a communication is opened by the valves between the small pump D and the presses; and the extreme pressure exerted by the small pump D, amounting to about 300 tons, is allowed to remain upon the rams for about 7 minutes from the time they were first brought into action; this, together with 3 minutes allowed for emptying and charging the press, is the full time required for expressing the oil in the most effectual manner. The oil in leaving the seed passes through the canvas bag, and then through the hair bag, where it finds a free exit at the edges; thence it runs into a channel or groove, which passes round the upper portion of each press-box G; a communication is made from one box to another by means of piping, so that the oil passes from the upper boxes through the lower ones, and thence into the cistern, which is called the spell-tank, being just large enough to hold the produce of one day's work. These presses are not worked with water; it has been found that oil which is not of a glutinous nature works much better, and keeps both the pumps and presses in a better condition. It is scarcely possible, if the presses are properly constructed, that they should meet with any accident: this can only occur through carelessness, when excessive weight is placed upon the safety-valve levers, and the valves themselves are allowed to stick through want of cleanliness, from the attendant not taking care to remove the oil, which sometimes becomes clotted round the valves. Each of these presses is capable of producing 36 cwt. of cake per day of 11 hours, and the yield of oil may be taken at about 14 cwt. in the same time; this, of course, depends much upon the quality of the seed. The cake is trimmed, or pared at the edges, by means of a small paring knife, after which it is put into a kind of rack to allow it to cool and dry, so that it will not become mouldy when stacked. The oil is pumped from the spell-tanks into larger tanks, capable of holding from 25 to 100 tons, where it is allowed to remain for some time for the purpose of settling, previous to being brought to the market in that condition, or to undergoing various other processes, such as refining, &c.

For all practical purposes, the screw press is quite unfit as compared with either the stamper or the hydraulic press, from the objection that it is constantly liable to break-downs when driven by steam-power, there being no portion of the machinery that will yield, if the pressure is not relieved in time, either by the attendant or by some self-acting contrivance, the best of which are very uncertain in their action; whereas in the case of the stamper press, the stamper being loose and independent of the press-box, any risk of breakage by an overstrain or excess of pressure is in a great measure avoided by the stamper recoiling and leaving the wedge at a fixed point after it is tightly driven home.

The following comparative results of the stampers and the hydraulic presses, as exhibited in Messrs. Earles & Carter's oil mills at Liverpool, may be interesting. On the same area of floor, 121 feet by 30 feet, where 12 stampers were previously employed, the largest quantity of seed crushed in one year, working during the day, was 13,000 quarters; but with 8 hydraulic presses, working day and night, they have produced 52,000 quarters; this would be equivalent to about 30,000 quarters for day-work only; and with 10 hydraulic presses, which now form the complement, they will crush 65,000 quarters of seed, working day and night. With the stampers they were compelled to work the seed twice over, whereas with the hydraulic presses it is only necessary to work it once, in both instances yielding the same quantity of oil, and the consequent saving of labour of nearly 25 per cent. The difference in point of wear and tear between the two modes of crushing is also found to be considerable in favour of the hydraulic presses, while the cost of wedges with the stampers is very considerable. Altogether there is an important saving in the new method, either as regards the amount of labour or the power required to work the hydraulic presses.

The practical conclusions to be drawn from the results appear to be that, in the same sized mill, the hydraulic presses produce about three times as much oil as the stampers. They do this with less wear and tear, at a considerable reduction of labour (since the seed has only to be handled once), and the general expenses of working the hydraulic mill as compared with the stampers, owing to the increased production, is much less per quarter of seed crushed in the former than in the latter process. The consumption of coal, general charges, and interest on capital, plant, &c., is the same whether 13,000 quarters per annum are crushed by the stampers, or three times that quantity by the hydraulic presses.

In conclusion, we may observe that it appears, from official returns in 1841, that the quantity of seed imported into this country for the oil manufacture was 364,000 quarters; in ten years it increased to 630,000 quarters, and in 1856 it was 1,100,000 quarters, producing about 144,000 tons of cake and 56,000 tons of oil.

CHAPTER VIII.

PAPER MILLS.

Dr. Ure, in his 'Dictionary of Arts, Manufactures, and Mines,' states, in his article Paper, that 'it is much to be regretted that in tracing the origin of so curious an art as that of the manufacture of modern paper, any definite conclusion as to the precise time or period of its adoption should hitherto have proved altogether unattainable. The Royal Society of Sciences at Göttingen, in 1755 and 1763, offered considerable premiums for that especial object, but, unfortunately, all researches, however directed, were utterly fruitless. The most ancient manuscript on cotton paper appears to have been written in 1050, while Eustathius, who wrote towards the end of the twelfth century, states that the Egyptian papyrus had gone into disuse but a little before his time.'

Speaking of the origin of the manufacture, it may be stated that the Chinese were early in the field, and probably gave birth to the art of making paper from vegetable matter reduced to pulp long before its introduction into Europe. Dr. Ure observes that, 'Several kinds of their paper evince the greatest art and ingenuity, and are applied with much advantage to many purposes. One especially, manufactured from the inner bark of the bamboo, is particularly celebrated for affording the clearest and most delicate impressions from copper-plates, which are ordinarily termed India proofs. The Chinese, however, make paper of various kinds, some of the bark of trees, especially the mulberry tree and the elm, but chiefly of the bamboo and cotton tree, and occasionally from other substances, such as hemp, wheat, or rice straw. To give an idea of the manner of fabricating paper from these different substances, it will suffice (the process being nearly the same in each) to confine our observations to the method adopted in the manufacture of paper from the bamboo — a kind of cane or hollow reed, divided by knots, but larger, more elastic, and more durable than any other reed. The whole substance of the bamboo is at times employed by the Chinese in this operation, but the younger stalks are preferred. The canes being first cut into pieces of four or five feet in length, are made into parcels, and thrown into a reservoir of mud and water for about a fortnight, to soften them; they are then taken out and carefully washed, every one of the pieces being again cut into filaments, which are exposed to the rays of the sun to dry and to bleach. After this they are boiled in large kettles, and then reduced to pulp in mortars, by means of a hammer with a long handle; or, as is commonly the case, by submitting the mass to the action of stampers, raised in the usual way by cogs on a revolving axis. The pulp being thus far prepared, a glutinous substance extracted from the shoots of a certain plant is next mixed with it in stated quantities, and upon this mixture chiefly depends the quality of the paper. As soon as this has taken place the whole is again beaten together until it becomes a thick viscous liquor, which, after being reduced to an essential state of consistency, by a farther admixture of water, is then transferred to a large reservoir or vat, having on each side of it a drying stove, in the form of the ridge of a house, that is, consisting of two sloping sides touching at top. These sides are covered externally with an exceedingly smooth coating of stucco, and a flue passes through the brickwork, so as to keep the whole of each side equally and moderately warm. A vat and a stove are placed alternately in the manufactory, so that there are two sides of two different stoves adjacent to each vat. The workman dips his mould, which is sometimes formed merely of bulrushes, cut in narrow strips and mounted in a frame, into the vat, and then raises it out again, the water passing off through the perforations in the bottom, and the pulpy paper-stuff remaining on its surface. The frame of the mould is then removed, and the bottom is pressed against the sides of one of the stoves, so as to make the sheet of paper adhere to its surface, and allow the sieve (as it were) to be withdrawn. The moisture, of course, speedily evaporates by the warmth of the stove, but before the paper is quite dry it is brushed over on its outer surface with a size made of rice, which also soon dries, and the paper is then stripped off in a finished state, having one surface exquisitely smooth, it being seldom the practice of the Chinese to write or print on both sides of the paper. While all this is taking place the moulder has made a second sheet, and pressed it against the side of the other stove, where it undergoes the operation of sizing and drying precisely as in the former case.'

With respect to the time when the manufacture of paper was first introduced into England, we have no reliable data. The earliest trace of a paper mill is supposed to have been erected at Stevenage, in Hertfordshire, about the year 1498 or 1499; also in Scotland, about the middle of the seventeenth century, when a company was formed for the manufacture of white writing and printing paper. But, in fact, little was done in the way of perfecting the manufacture till the middle of the last and the beginning of the present century. Up to the latter period the only machinery then in use was the rag engine, and the moulds and felts as practised by hand in single sheets from the liquid pulp. It is a curious fact that, notwithstanding that paper has been made in this country and other parts of Europe from two to three centuries, few if any improvements, till of late, have been effected in the shape of machinery for the purpose of increasing the quantity and reducing the cost of the manufacture. Such was the imperfect state of the manufacture when a working model of a continuous machine was introduced into this country from France, from the paper manufactory of Monsieur L. Didot, at Essonne. This occurred in 1801, when the invention was purchased by Messrs. Fourdrinier, whose experiments and labour in perfecting the machine were never rewarded. Subsequently it passed into the hands of the late Mr. Donkin, of Bermondsey, of whose improved machine and the drying and cutting machine attached we give a sketch.

During a recent alteration of the paper duties, a long and vigorous opposition was maintained on the part of the manufacturers against the Legislature. It was maintained that the price of rags from the continent would be greatly increased, and that serious injury would be the result. These fears have not, however, been realised, as a number of resources were at hand, and the variety of substances which enters into the formation of pulp and the manufacture of paper is so great, that I cannot refrain from again quoting Dr. Ure.

The Doctor says that, 'Silks, woollens, flax, hemp, and cotton, in all their varied forms, whether as cambric, lace, linen, holland, fustian, corduroy, bagging, canvas, or even as cables, are or can be used in the manufacture of paper of one kind or another. Still, rags, as of necessity they accumulate and are gathered up by those who make it their business to collect them, are very far from answering the purposes of paper making. Rags to the papermaker are almost as various in point of quality or distinction as the materials which are sought after through the influence of fashion. Thus the papermaker, in buying rags, requires to know exactly of what the bulk is composed. If he is a manufacturer of white papers, no matter whether intended for writing or printing, silk or woollen rags would be found altogether useless, inasmuch, as is well known, the bleach will fail to act upon any animal substance whatever. And although he may purchase even a mixture in proper proportions, adapted for the quality he is in the habit of supplying, it is essential, in the processes of preparation, that they shall previously be separated. Cotton in its raw state, as may be readily conceived, requires far less preparation than a strong hempen fabric; and thus, to meet the requirements of the papermaker, rags are classed under different denominations, as, for instance, besides fines and seconds, there are thirds, which are composed of fustians, corduroy, and familiar fabrics; stamps or prints (as they are termed by the papermaker), which are coloured rags, and also innumerable foreign rags, distinguished by certain well-known marks, indicating their various peculiarities. It might be mentioned, however, that although by far the greater portion of the materials employed are such as have already been alluded to, it is not from their possessing any exclusive suitableness — since various fibrous vegetable substances have frequently been used, and are, indeed, still successfully employed — but rather on account of their comparatively trifling value, arising from the limited use to which they are otherwise applicable.'

The same authority goes on to state, that 'almost every species of tough fibrous vegetable, and even animal substance, has at one time or another been employed; even the roots of trees, their bark, the bine of hops, the tendrils of the vine, the stalks of the nettle, the common thistle, the stem of the hollyhock, the sugar-cane, cabbage stalks, beet-root, wood shavings, sawdust, hay, straw, willow, and the like. Straw is occasionally used, in connection with other materials, such as linen or cotton rags, and even with considerable advantage, providing the processes of preparation are thoroughly understood. Where such is not the case, and the silica contained in the straw has not been destroyed (by means of a strong alkali) the paper will be invariably found more or less brittle; in some cases so much so as to be hardly applicable to any purpose whatever of practical utility. The waste, however, which the straw undergoes, in addition to a most expensive process of preparation, necessarily precludes its adoption to any great extent. Two inventions have been patented for manufacturing paper entirely from wood. One process consists in first boiling the wood in caustic soda lye, in order to remove the resinous matter, and then washing to remove the alkali; the wood is next treated with chlorine gas or an oxygenous compound of chlorine in a suitable apparatus, and washed to free it from the hydrochloric acid formed; it is now treated with a small quantity of caustic soda, which converts it instantly into pulp, which has only to be washed and bleached, when it will merely require to be beaten for an hour or an hour and a half in the ordinary beating-engine, and made into paper. The other invention is very simple, consisting merely of a wooden box enclosing a grindstone, which has a roughened surface, and against which the blocks of wood are kept in close contact by a lever, a small stream of water being allowed to flow upon the stone as it turns, in order to free it of the pulp, and to assist in carrying it off through an outlet at the bottom. Of course, the pulp thus produced cannot be employed for any but the coarser kinds of paper. For all writing and printing purposes, which, manifestly, are the most important, nothing has yet been discovered to lessen the value of rags, neither is it at all probable that there will, inasmuch as rags, of necessity, must continue accumulating, and before it will answer the purpose of the papermaker to employ new material, which is not so well adapted for his purpose as the old, he must be enabled to purchase it for considerably less than it would be worth in the manufacture of textile fabrics; and, besides all this, rags possess in themselves the very great advantage of having been repeatedly prepared for paper-making by the numerous alkaline washings which they necessarily receive during their period of use.'

In considering the various processes or stages of the manufacture of paper, we have first to notice that of carefully sorting and cutting the rags into small pieces, which is done by women, each woman standing at a table frame, the upper surface of which consists of very coarse wire cloth, a large knife being fixed in the centre of the table nearly in a vertical position. The woman stands so as to have the back of the blade opposite to her, while at her right hand on the floor is a large wooden box, with several divisions. Her business consists in examining the rags, opening the seams, removing the dirt, pins, needles, and buttons of endless variety, which would be liable to injure the machinery or damage the quality of the paper. She then cuts the rags into small pieces, not exceeding 4 inches square, by drawing them sharply across the edge of the knife, at the same time keeping each quality distinct in the several divisions of the box placed on her right hand. During this process much of the dirt, sand, and so forth, passes through the wire cloth into a drawer underneath, which is occasionally cleaned out. After this the rags are removed to what is called the dusting machine, which is a large cylindrical frame covered with similar coarse iron wire-cloth, and having a powerful revolving shaft extending through the interior, with a number of spokes fixed transversely, nearly long enough to touch the cage. By means of this contrivance, the machine being fixed upon an incline of some inches to the foot, the rags which are put in at the top have any remaining particles of dust that may still adhere to them effectually beaten out by the time they reach the bottom. The rags, being thus far cleansed, have next to be boiled in an alkaline lye or solution, made more or less strong as the rags are more or less coloured, the object being to get rid of the remaining dirt and some of the colouring matter. The proportion is from four to ten pounds of carbonate of soda with one-third of quicklime to the hundredweight of material. In this the rags are boiled for several hours, according to their quality. The method generally adopted is that of placing the rags in large cylinders, which are constantly, though slowly, revolving, thus causing the rags to be as frequently turned over, and into which a jet of steam is cast with a pressure of something near 30 lbs. to the square inch.

Before considering the subsequent process the material has to go through before it issues from the mill in the finished state of paper, it may be interesting first to describe the position of the machinery in order to show how the work is accomplished.

The following sketches, figs. 313, 314, and 315, represent plans and sections of the machinery part of a paper mill, but none of the outbuildings, such as dry houses, bleach works, and finishing rooms, which are generally attached to the buildings containing the moving power.

Fig. 313 is an elevation of the mill, showing the water-wheel A, and the principal shafts F′, B, C; the vat F, containing the pulp; the rag-engines R; the machine L, with endless wire; the drying machine K; the cutting machine H; the rag machines D in the upper story (which are used for cutting and preparing the rags for the boiling and bleaching process); the pillars for supporting the floor and roof; the roof, &c.

Fig. 314 is a plan showing the position and gearing of the longitudinal shafts; *a*, *a*, the pillars for supporting the floor; W, W, the hydraulic presses; T, the bench for sorting the paper; V, the staircase; *b*, *c*, *c*, the method generally adopted for gearing the shafting to the paper machine L; the drying machine K; and the cutting machine H, fig. 314, &c.

Fig. 315 is a section showing the position of the rag-engines R, R, the upright B, &c.

The building is rectangular, as shown at fig. 314, with a powerful water-wheel A at one end. The mill is 125 feet long, two stories high, and 50 feet wide. It contains dust and rag machines in the sorting-room above; eight engines for converting the rags into pulp; a continuous paper, drying, and cutting machine; hydraulic presses; glazing machines, &c. It will be seen that the water-wheel gives motion to a line of strong cast-iron shafts, varying from 10 to 8 inches diameter, a little above the ground floor. On these shafts, next to the wall, is a bevel-

Fig. 313.

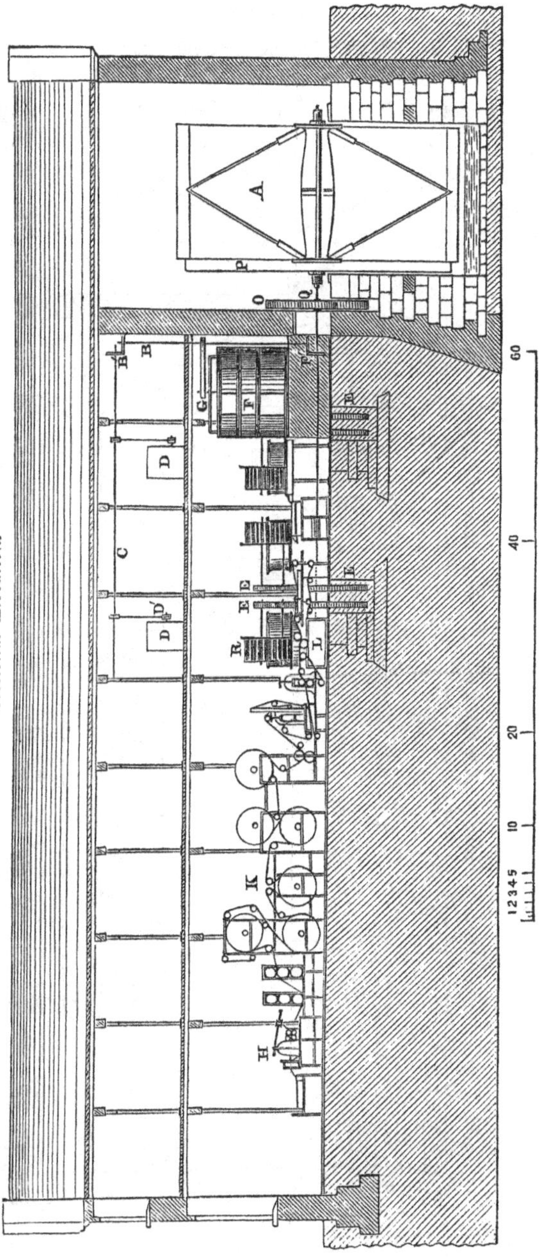

PAPER MILL.

Sectional Elevation.

Fig. 314.

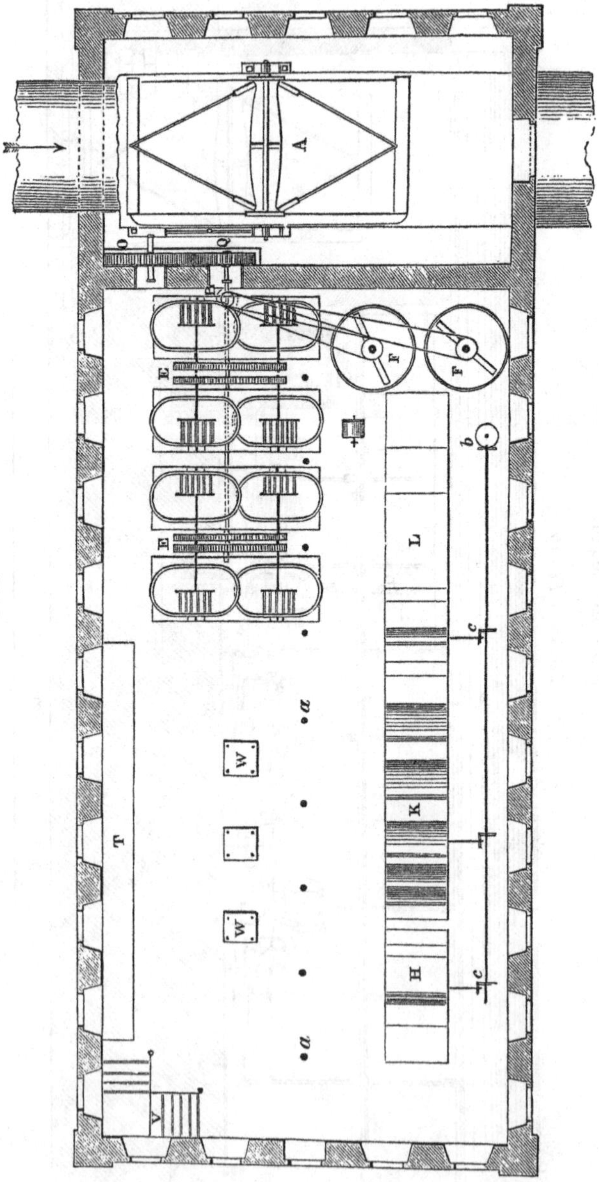

PAPER MILL.

Plan.

wheel for driving the upright B, of cast iron, 5 inches in diameter, and wrought-iron shaft C′, 3½ inches diameter, which drive the rag machines D, D in the room above. On the line of shafts which passes under the rag or pulp engines are four large spur-wheels E, E, E, E, which give motion to eight pinions on the axes of the spindles, 4½ inches diameter, and cylinders of the rag engines which contain steel cutters fixed all round their periphery. These cylinders, which are about 3 feet diameter, revolve at the rate of 170 revolutions per minute, and, coming into close contact with similar knives placed at a slight angle in a fixed iron block below, the rags are torn or macerated until they are reduced to the state of pulp. After this process the liquid is run from the higher to the lower engines, where it is comminuted into an exceedingly fine pulp, which in proper time is discharged into a vat or cistern below. From this it is

Fig. 315.

Cross Section.

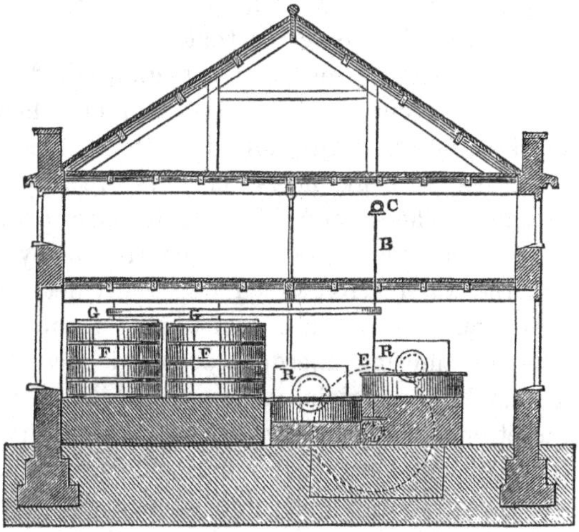

raised by a pump to the vats F over the paper machine, where it is kept in motion by a revolving agitator G, and from thence is run in measured quantities on to the frame and wire-cloth shaker of the continuous paper machine, where it is finished, dried and cut into sheets.

Our space does not admit of a full description of the paper-making machine, with all its motions, strainers, rollers, blanket-carriers, &c.; we may, however, direct attention to an elaborate description of these ingenious contrivances in Dr. Ure's 'Dictionary,' pages 391, 392, to which we have already referred.

The quantity of machinery in a paper mill is not considerable when compared with mills for the manufacture of the textile fabrics, and, with the exception of the paper machine, the drying, cutting, and sizing machines, a mill for the manufacture of paper, so far as its mechanism is concerned, is almost exclusively a piece of well-constructed millwork; the other parts of the manufacture belong to the chemist, both before and after the mechanical operations have been effected.

Connected with the manufacture of paper, there is one point of considerable interest and importance, and that is, what is commonly, but erroneously, termed the *water mark*, which may be noticed in the Bank of England notes, cheques, and bills, as also in every postage and receipt label of the present day.

On this point Dr. Ure observes that, 'Water marks have at various periods been the means of detecting frauds, forgeries, and impositions, in our courts of law and elsewhere, to say nothing of the protection they afford in the instances already referred to, such as bank notes, cheques, receipt, bill, and postage stamps. The celebrated Curran once distinguished himself in a case which he had undertaken by shrewdly referring to the water mark, which effectually determined the verdict. And another instance, which may be introduced in the form of an amusing anecdote, occurred once at Messina, where the monks of a certain monastery exhibited, with great triumph, a letter as being written by the Virgin Mary with her own hand. Unluckily for them, however, this was not, as it easily might have been, written upon the ancient papyrus, but on paper made of rags. On one occasion a visitor, to whom this was shown, observed, with affected solemnity, that the letter involved also a miracle, for the paper on which it was written was not in existence until several centuries after the death of the Virgin.'

There is, perhaps, no description of manufacture, where machinery is not extensively employed, that embraces a greater variety of forms and quality in its products than paper. Chemical combinations have probably as much, if not more, to do with the manufacture than applied mechanics; and, assuming that we are able to enumerate the different kinds of manufacture, such as writing, printing, and warping papers, which come from the mills, we should find in the first five different sorts, known as cream-wove, yellow-wove, blue-wove, &c.; in printing, two sorts, laid and wove; and in warping four, blue, purple, brown, &c. Exclusive of these there are other varieties, such as blotting and filtering papers, which are rendered absorbent by an admixture of woollen rags introduced into the process of preparing the pulp, and to these again may be added the capabilities to which paper may be moulded under the well-known name of *papier-maché*, which constitutes such a variety of articles of utility and ornament in both the fine and useful arts. This material, when moulded and pressed, can be formed into models, busts, tables, trays, antique candelabra, and a vast variety of useful and classical forms, which, forcibly compressed, is capable of retaining the impressions as originally produced.

The paper manufacture of this country has undergone, like most other manufactures, many changes and improvements. Little more than a century ago the whole manufacture was only equal to two-thirds the consumption, now there is a considerable surplus exported free of duty. In 1835 the weight of manufactured paper that paid duty was . . . . 70,000,000 lbs.
In 1845 it had risen to . . . . 124,000,000 ,,
In 1855 it stood at nearly . . . . 167,000,000 ,,
In 1859 it was about . . . . 218,000,000 ,,
and in six more years it may reach . . . . 300,000,000 ,,

Such is the importance of this valuable branch of industry, that to every appearance its increase is only circumscribed by the supply of rags and the material employed in the manufacture.

The following Table exhibits the speeds of the different shafts and machines shown in figs. 313, 314, and 315 :—

List of Wheels and Speeds.

Water wheel 30ft. diameter. Velocity on periphery 4ft. per second, equal to 2·5 revolutions per minute.

Description of Gearing.	Driver. Diameter.	Driver. Revolutions.	Driven. Diameter.	Result. Revolutions per minute..
	ft. in.		ft. in.	
Segments P .	29 0½	2·5	4 6½	16 revolutions of cross shaft.
Spur wheels O	10 6½	16·0	5 6½	30·6 " " longitudinal shaft.
Spur wheels E	11 8	30·6	2 0½	174 " " rag engines.
Bevil wheels F′	3 8¾	30·6	2 0½	56·3 " " upright shaft.
Bevil wheels B′	3 6	56·3	1 10½	103·7 " " longitudinal shaft C.
Pulley D′ . . .	2 0	103·7	1 3	165·9 " " rag-cleaning machine.
Pulle y . . .	1 0	56·3	3 0	18·7 " " agitater.

It will be observed, as noticed in the first part of this work, that the speeds at which water wheels are driven vary according to the heights of the falls; in this wheel, as in most others adapted to high falls, the speed seldom exceeds 4 feet circumferential velocity, unless there is an unlimited supply, when the speed may be slightly increased.

CHAPTER IX.

POWDER MILLS.

THE manufacture of gunpowder is always a precarious operation, and this is not surprising when we consider the properties and chemical combinations of the ingredients of which it is composed. An admixture of sulphur, nitre, and charcoal in due proportion when fired by a spark, whether accidentally or otherwise, will produce the phenomenon of explosion, and cause serious injury to surrounding objects within range of its destructive effects.

This well-known composition is used for every description of fire-arms, and its safe application depends upon a knowledge of the fact that, at the moment of ignition, violent deflagration takes place, accompanied by the evolution of a large quantity of gas.

The quantity produced by explosion is about 900 times the volume of the powder; but, owing to the high temperature it attains as it passes from the solid to the gaseous state, and the space it occupies at the moment of formation, it is probably three times that amount, or 2,700 times the actual volume of the powder from which it is evolved. This immense increase of volume, and corresponding amount of expansive force, will account for the effects of explosive substances, and the velocity with which a projectile is discharged from a gun. There are many fulminating substances, chiefly compounds of nitrogen and chlorine, besides gunpowder, which explode with great rapidity, the whole mass to every appearance being instantaneously converted into gas. Now this is not the case with gunpowder, as time is an element in its combustion, and it frequently happens that, in a gun loaded with an extra charge of powder, the combustion is incomplete; and hence follows the necessity of limiting the quantity to the power of the gun, in order that the projectile may have the full force of the exploded powder. This is well known to artillerists, and, also, that no composition fulfills the requisites required in fire-arms so effectually as a well-proportioned mixture of nitre, sulphur, and charcoal. It is this composition which constitutes the propulsive force of projectiles, and the more pure these substances are when properly mixed, the more perfect is the powder.

In powder mills, as in others where chemical operations are carried on, it would be foreign to the objects of this treatise to enter upon the various processes by which a certain article is manufactured from the crude materials of which it is composed; but simply to describe the mechanical operations by which the objects are attained, and that more particularly when it comes within what has been considered the province of the millwright. In treating of this particular manufacture, and the machinery by which its ingredients are mixed, ground, and sifted, which from the time of its invention has been entrusted to the hands of the millwright, it may be interesting to trace, as simply as possible, the preparatory process by which the ingredients already described are manipulated in combination before they arrive at what is called the finished state of gunpowder.

We have noticed that nitre forms one of its principal ingredients, and this is prepared from its crude state, as nitre of commerce, by solution in hot water and crystallization. By this process, after being carefully washed and every impurity removed, it is in a pulverulent state ready for combination.

Sulphur is prepared by fusion in an iron pot, at a temperature of about 230°; the impurities are then removed by skimming, and the denser parts allowed to sink to the bottom, from whence they are discharged into a receptacle or vessel to cool.

Charcoal, of the three ingredients of gunpowder, is the most important, and much depends on the quality of the wood used. I give the article as it appears in the last edition of Dr. Ure's 'Dictionary of Arts and Manufactures.'

He states that, 'Woods which are best adapted for the production of pyroligneous acid are not fitted for the manufacture of gunpowder; the charcoal must therefore be prepared specially. The following are the essential properties of good charcoal for powder:—1. It should be light and porous. 2. It should yield little ashes. 3. It should contain little moisture. The woods yielding good powder charcoals are black alder, poplar, spindle tree, black dogwood, and chesnut. Hemp stalks are said to yield good charcoal for gunpowder.

'The operation of preparing the charcoal naturally divides itself into three processes. 1. The selection of the wood. 2. Preparation of the wood previous to carbonisation. 3. The carbonisation.

'In selecting the wood care is to be taken to avoid the old branches, as the charcoal made from them would yield too much ashes. The bark is to be rejected for the same reason. The wood is to be cut into pieces from 4½ feet to 6 feet long. If the branches used are more than ¾ of an inch in diameter they are to be split. If the wood be too large, great difficulty will be found in uniformly charring it.

'There are two methods employed in the charring of wood for gunpowder. In one, the operation is conducted in pits; but the process more commonly resorted to is distillation in cylindrical iron retorts. There are certain advantages in the pit process, but they are more than counterbalanced by the convenience and economy of distillation. The stills used are about 6 feet long, and 2 feet 9 inches in diameter. The ends of the cylinders are closed by iron plates, pierced to admit tubes of the same metal. Some of the latter are for the introduction, during the carbonisation, of sticks of wood, which are capable of being removed to indicate the stage of the decomposition, while another communicates with the condenser. The more freely the volatile matters are allowed to escape, the better the quality of the resulting charcoal. If care be not taken in this respect, especially as the distillation reaches its close, the tarry matters become decomposed, and a hard coating of carbon is deposited on the charcoal, which greatly lowers its quality. The process of burning in pits is considered to yield a superior coal, owing to the facility with which the gases and vapours fly off.

'The degree to which the burning or distillation is carried materially influences the nature of the resulting powder. If the operation be arrested before the charcoal becomes quite black, so that it may retain a dark-brownish hue, the powder will be more explosive than it would be if it were pushed until the charcoal had attained a deep black colour. When it has been found that no more volatile products are being given off, the fire is damped, and in a few hours the contents of the cylinders are transferred to well closed iron boxes to cool.'

The three ingredients having been carefully prepared are now mixed so as to effect a thorough incorporation necessary for the production of good powder. The original method was by stampers shod with brass beaters working in wooden mortars, the stampers in this case being raised by a revolving shaft and tappets worked by a water-wheel, or horses, as most convenient.*

The government powder mills at Waltham Abbey, and other private establishments, have not, until of late years, undergone any material improvement since their erection. They could not, however, escape the changes which for the last twenty years have been in operation in almost every other kind of manufacture. New contrivances and improvements upon old ones have been applied to powder mills as well as to those of a different character, and although the system of grinding by edge stones has not been superseded, great improvements have nevertheless been introduced, by substituting turned cast-iron rollers or runners in place of edge stones, and these revolving on turned cast-iron beds give greatly increased accuracy to the movements, and less danger from sparks or small crystals from the runners than when composed of stone. As this kind of machinery has been renewed at Waltham Abbey, a descrip-

* In the year 1839 the author visited Constantinople, at the request of Sultan Mah'moud, for the purpose of inspecting and reporting upon the then existing state of the founderies, firearms' manufactory, powder mills, &c.; and during this investigation he found the powder mills driven by relays of horses, working in the interior of large wooden wheels, the same as a dog-spit. About the same time steam engines were introduced for working the machinery at a distance by compressed air. The engine, boilers, air cylinder, &c. were in this case at a distance of 200 yards from the machinery, in order to prevent accident. This arrangement was subsequently changed, as it was found that more than one half the power of the engine was lost, by friction, in forcing the air through the pipes, and working it over again into cylinders giving motion to the machinery at that distance from the motive power.

tion of the arrangements and improvements introduced by Mr. Anderson, of Woolwich, and others, may not be without interest.

Two entirely new establishments were constructed under the direction of these gentlemen by Messrs. W. Fairbairn & Sons, of Manchester, and Messrs. B. Hick, of Bolton, the former executing the water wheels, edge runners, hydraulic presses, and granulating machines, the latter a 40 horse-power steam-engine and six pairs of edge runners.

In the construction of powder mills, it is a question of much importance to have the grinding and other processes separate from each other, as in case of explosion in any one department it should not communicate to the others. This precaution has been considered essential in every well-regulated powder mill, and to attain that object, most establishments have their mills at 100 to 150 yards distant from each other. Where this arrangement is found inconvenient, and the mills have to be nearer together, they are then separated by butts or mounds of earth, at a considerable height, and tapering at the top like the roof of a house. The more recent erections are however

Fig. 316.

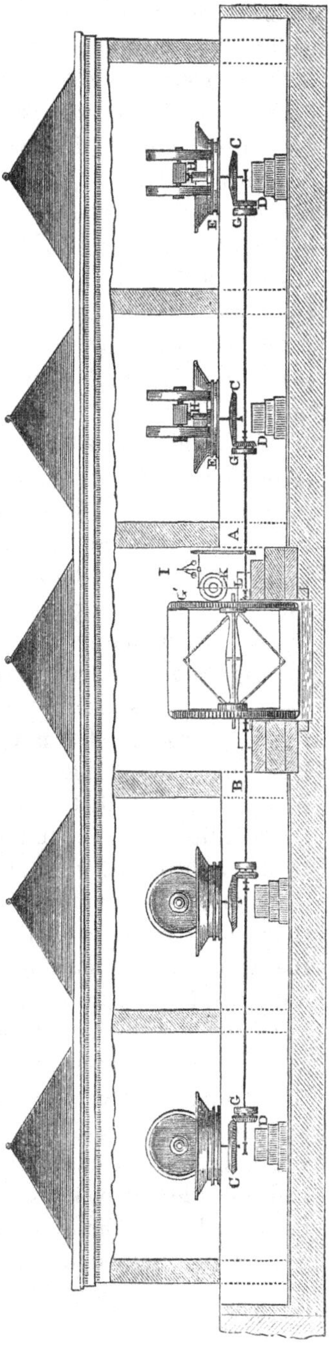

different since the introduction of the cast-iron runners, as may be seen by the preceding arrangement, fig. 316, where the mill is driven by a water wheel.

On this plan it will be observed, that the water wheel is 14 feet diameter, 12 feet wide inside the buckets, with segments on the shrouds, giving motion to two pairs of runners on each side of the water wheel. Each pair of runners is in a separate house, covered with a light iron roof, and partitioned from each other by a thick wall, intended in case of explosion to save the adjoining mill. How far this arrangement will answer the purpose has not been determined, as no explosion has taken place since the mills were finished, and there is therefore no proof of the amount of security afforded, nor of the direction of the explosion, which, it is expected, would be vertical through the roof. It will be observed that the water wheel gives motion to a line of shafts A and B, fig. 316, one on each side of the water wheel placed in an underground tunnel, and by the bevil wheels C, D, gives motion to the vertical spindle which passes through a tight brass stuffing box (see also fig. 317) in the large bed plates E and F, and gives motion to the edge runners above; the spindle in this case being steadied by the conical standard H, into which is inserted a brass bush for that purpose. The pinion on the tunnel shaft is bored and keyed upon the sliding clutch G, which works by friction on the principle described at page 92 of this work. All the wheels in these mills have wood and iron teeth, as iron working into iron is always attended with danger, where they are liable to come in contact with particles of powder. The bed plates on which the runners revolve rest upon the side walls of the tunnel, and the apparatus for starting and stopping the runners (which consists of a worm wheel and crank) is worked by a wheel and handle from the outside.

In order to illustrate more in detail the method of constructing the mixing mill, and the connection between the vertical shaft and the runners, fig. 317 has been introduced. E, is the bed plate which is grooved out to admit the conical standard H. The inclined side R is bolted to the bed plate E, which contains the brass bush *a*. The standard is bolted to the bed plate with countersunk headed bolts, and a brass bush *e* is firmly inserted in its upper extremity to steady the vertical spindle. A box *b*, planed in the inside and keyed on to the vertical spindle in which another square brass box is allowed freely to slide, forms the connection between the spindle through the runner and the vertical shaft. A brass bush *g* is driven into the runner, which is prevented from sliding off the spindle by a collar through which a split pin is driven. A brass cap *f*, bolted to the collar of the brass bush, completes the mill. The object of this cap is to prevent any grease from falling into the powder.

Fig. 317.

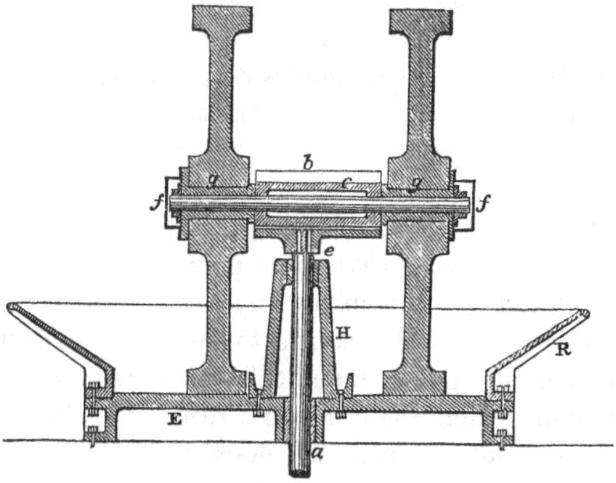

Scale $\frac{1}{4}''=1$ foot.

The water wheel is on the ventilated principle, applicable to a variable fall, which does not exceed 3 feet in low ebbs; at other times, when the river is flooded, there is a large supply of water and a considerable increase of head. The runners are regulated by the governor I, which acting upon the worm wheels K, L, communicates motion to the shuttle, which moves upon the back surface of the cast-iron breast, and increases or diminishes the supply to the wheel, as may be necessary to maintain a steady uniform motion in the machines. It will be observed that the iron runners are not equidistant from the vertical spindle which carries them round on the surface of the bed plate. This is for the purpose of grinding the ridges which are thrown up by the outer edge of the central runner, and with the aid of scrapers the powder is brought immediately under the runners, the more effectually to incorporate the mixture of the ingredients. During the grinding, the powder is kept moist by a little water at proper intervals, to make the particles adhere together, and form the mass into a sort of paste.

Another water wheel is employed to give motion to the hydraulic presses, crushing, sifting, and granulating machines, and these operations are carried on by a series of machines, adapted to the different kinds of powder which the government require for small arms and cannon. Other methods, besides those of edge stones or cast-iron runners, have been called into use for effecting the thorough incorporation of the ingredients for the production of good gunpowder, but none of them have been found to answer so well as the cast-iron runners or edge stones; the only objection to the edge stones being their liability to strike fire.

Previous to the grinding and mixing process under the runners, the ingredients have to be pulverised, which is accomplished by means of a number of revolving drums and balls. The pulverised materials require to be sifted in cylinders in order to be properly mixed, and this completes the preparatory processes, and renders the compound ready for grinding.

The subsequent processes as described by Dr. Ure are as follow: —

'The cake produced by the action of the stones is ready for graining or corning. For this purpose the cake is subjected to powerful pressure, by means of an hydraulic press. The mass is then broken up and transferred to a species of sieve of skin or metal pierced with holes. A wooden flail is placed on the fragments, and the sieves are violently agitated by machinery. By this means the grains and dust produced by the operation fall through the holes in the skin or metal discs, and are afterwards separated by sifting. Sometimes the machinery is so arranged that the graining and separation of the meal powder is effected at one operation. The meal powder is reworked, so as to convert it into grains. The next operation to which the powder is subjected is glazing. Its object is to render it less liable to injury, by absorption of moisture or disintegration during its carriage from place to place. The glazing is effected by causing the grained powder to rotate for some time in a wooden drum or cylinder, containing rods of wood running from end to end. The grains, as they rub against each other and against the wooden ribs, have their angles and asperities rubbed off, and at the same time the surface becomes harder and polished. It is finally dried by exposure to a stream of air, heated by means of steam.'

The following Table shows the composition of the various gunpowders in use amongst the different nations of Europe and America.

TABLE OF THE COMPOSITION OF VARIOUS GUNPOWDERS.

	Nitre.	Sulphur.	Charcoal.
English war powder	75·0	10·0	15·0
" sporting powder	77·0	9·0	14·0
French war powder	75·0	12·5	12·5
" sporting powder	76·9	9·6	13·5
" blasting "	62·0	20·0	18·0
" " " (another kind)	65·0	20·0	15·0
United States war powder	75·0	12·5	12·5
Prussian war powder	75·0	11·5	13·5
Russian " "	73·8	12·6	13·6
Austrian " "	75·0	10·0	15·0
Spanish " "	76·5	12·7	10·8
Swedish " "	75·0	16·0	9·0
Chinese " "	75·7	14·4	9·9

It will be seen from the above that the different nations do not materially differ in the proportions at which they have arrived by experience. They approximate nearly to each other, and the differences which exist may be accounted for by the superior or diminished purity of the substances of which they are composed. We regret to remark that the limits of our treatise preclude us from taking advantage of the drawings of the granulating machines, presses, &c. in our possession.

For the speed of the runners, and of the shafting and governor gear, we beg to refer the reader to the annexed list of wheels and speeds, which enter into this description of Powder Mills.

List of Wheels and Speeds.

Description of Gearing.	Driver.		Driven.	Result. Revolutions per minute.
	Diameter.	Revolutions.	Diameter.	
	ft. in.		ft. in.	
Water wheel .	14 0	—	—	6·68 revolutions per minute, equal to 4·89 feet per second.
Segments G′ .	12 9	6·68	3 6½	24·2 revolutions of main horizontal shaft.
Bevil wheel C .	2 3	24·20	5 4⅝	10 revolutions of runner.
Governor Gear.				
Pulley H . .	1 9	24·20	1 2	36 revolutions of governors.
Worm wheel K	0 4	36·00	0 8¼	1·79 ,, vertical shaft.
Worm wheel L	0 5	1·79	2 0	0·031 ,, rack shaft.

It might have been interesting to have noticed in this place a new process of manufacture which to some extent is superseding the use of gunpowder, and that is guncotton, first invented by Professor Schonbein, and now greatly improved by General von Linz and the Austrian Government; but as that manufacture is now under the consideration of the authorities at Woolwich, and as experiments are instituted for the purpose of analysing its constituents, and testing its powers as compared with our best qualities of gunpowder, it would be premature in this stage of investigation to venture an opinion upon its comparative merits as a substitute for the present compounds.

CHAPTER X.

IRON MILLS.

THE mechanical operations connected with the manufacture of wrought-iron consist of shingling, hammering, rolling, &c., to which we may add the forging of 'uses;' that is, the forging of those peculiar forms which constitute the frames and parts of steam engines, iron ships, railway carriages, locomotives, &c., required in those important constructions.

In tracing the whole of the processes in the manufacture of wrought-iron bars and plates, it will not be necessary to dwell on those practices which have been superseded by more modern and improved machinery. Suffice it to observe, that the puddled balls have to be fashioned into oblong slabs or blooms by the blows of a heavy forge hammer. During this operation, the scoria and impurities contained in the balls are separated from them in the shape of scoria by the force of impact, and by a continued series of blows the iron is rendered malleable, dense, and compact. The slabs are then passed through a series of grooved iron rollers, which reduce them to the form called puddle bars. These are again cut up and piled regularly together or faggotted, and brought to a welding heat in the reverberatory furnace, when they are a second time passed several times through grooved rollers, and by this latter process are made into bars or rails ready for the shears. In the manufacture of plates the same process of heating and piling is observed, with this difference, that the pile at a welding heat is passed through plain surface rolls until the required thickness and dimensions are attained.

In order to arrive at a clear conception of the mechanical operations employed in the manufacture of iron, it will be necessary to describe more at length the processes as at present practised, with the improved and powerful machinery now employed; and as much depends upon the application of the motive power the steam-engine claims the first notice. Until of late years the vertical steam-engine was invariably used for giving motion to the forge hammer and rolling mill, which were placed on one side of the fly-wheel and the hammer on

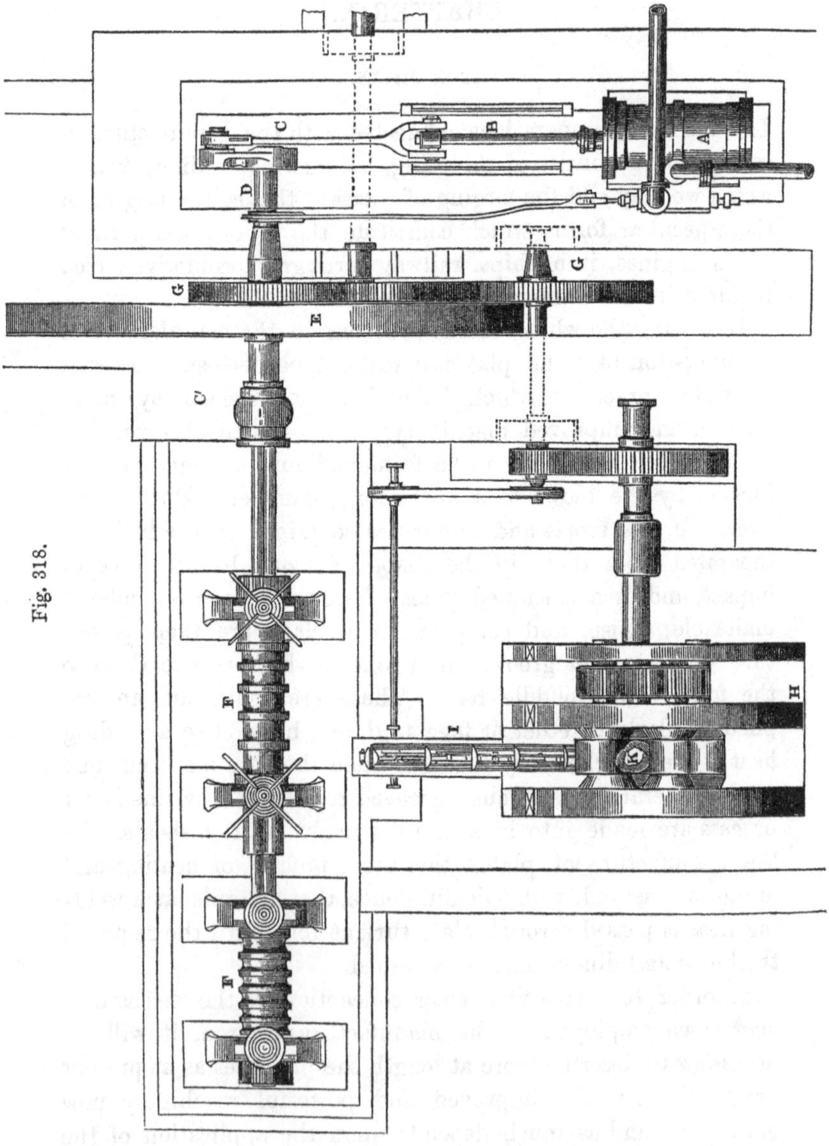

Fig. 318.

the other; but the high pressure non-condensing engine is found to be decidedly preferable, as the waste heat passing from the puddling and heating furnaces is quite sufficient to raise the steam for working the rolls and one of Brown's bloom squeezers, as shown in the following drawing:—

In this arrangement the cylinder A (fig. 318) is placed horizontally, and is supplied with steam from boilers near the puddling furnaces. The piston rod and slides B, and connecting rod C, give motion to the crank shaft D, on which is fixed a heavy fly-wheel E. The puddling rollers F F are driven direct from the end of the fly-wheel shaft, being attached to it by a disengaging coupling C′; the bloom squeezer H is driven by a train of spur wheels G G. Under the lower rolls of the squeezers a Jacob's ladder or elevator I is fixed, for raising the block, which, deprived of its impurities, is reduced to an oblong shape by passing between the rollers of the squeezers. The block on leaving the rollers is carried in front of one of the projecting divisions of the ladder I, and thrown on to the platform in front of the rollers F F; the workman then seizes it with a pair of tongs, and forces it into the largest groove in the rolls; it is then passed in succession through the other grooves till it attains the required form of the bar. This squeezer process is substituted for the old method of preparing and shingling the puddle balls, still much practised from its simplicity in reducing them to shape by a heavy hammer called the forge hammer or helve, shown in fig. 319. It consists of a

Fig. 319.

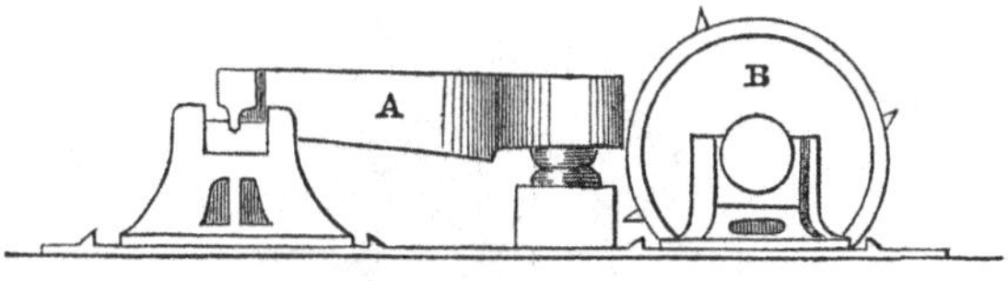

heavy mass of iron A, resting on a pivot at one end, and lifted by projecting cams on a revolving wheel B, at the other; between these points, and nearer the front, is the anvil on which the puddler's ball is thrown to receive a rapid succession of strokes, which force out the impurities, and reduce it to a form suitable for insertion between the rolls.

Other squeezers have also been used for the same purpose, one of them consisting of two massive jaws worked by a lever and crank, between which the ball is moulded by severe pressure to the necessary form. The squeezer is however alleged to have the effect of lapping up cinder in the iron to a greater extent than in the case of the forge hammer. This instrument is sometimes called the alligator, from its resemblance to the mouth of that animal, and it will be observed that the puddle ball is reduced in size by being rolled by the puddler to the back part of the jaws, where the leverage is more powerful as its diameter decreases.

One of the most perfect machines of this class is Brown's bloom squeezer already referred to. The heated ball of puddled iron K, thrown on the top of the machine, is gradually compressed between the revolving rollers as it descends, and at last emerges at the bottom, where it is thrown on to the moveable 'Jacob's ladder' I, fig. 318, by which it is elevated and discharged in front of the rolls, as already described. This machine effects a considerable saving of time; it will do the work of twelve or fourteen furnaces, and may be kept constantly going as a feeder to one or two pair of rollers.

There are two distinct forms of this machine, one in which the bloom receives only two compressions; and the other, which is much more effective, where it is squeezed four times before it leaves the rolls and falls upon the Jacob's ladder.

There is another machine for preparing the blooms by com-

Fig. 320.

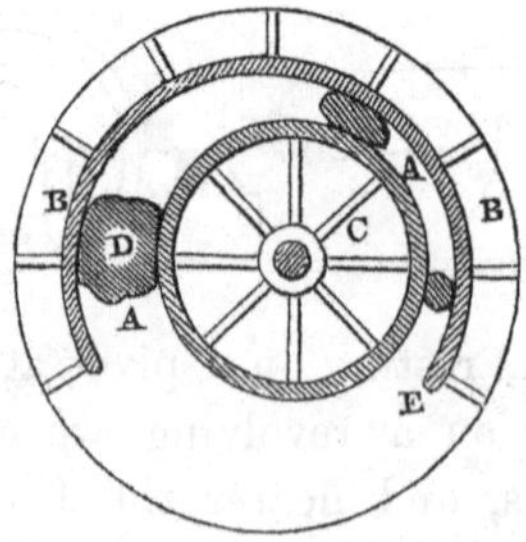

pression, namely, a table firmly embedded in masonry, as shown at A A, fig. 320, with a ledge rising up from it to a height of about 2 feet, so as to form an open box. Within this is a revolving box C, of a similar character, much smaller than the last, and placed eccentrically in regard to it. The ball or bloom D is placed between the innermost revolving box C and the outer case A A, where the space between them is greatest, and is carried round till it emerges at E, compressed and fit for the rolls.

The bloom, after leaving the hammer or squeezer, is at once placed in the rolling mill, fig. 321. This consists of massive

Fig. 321.

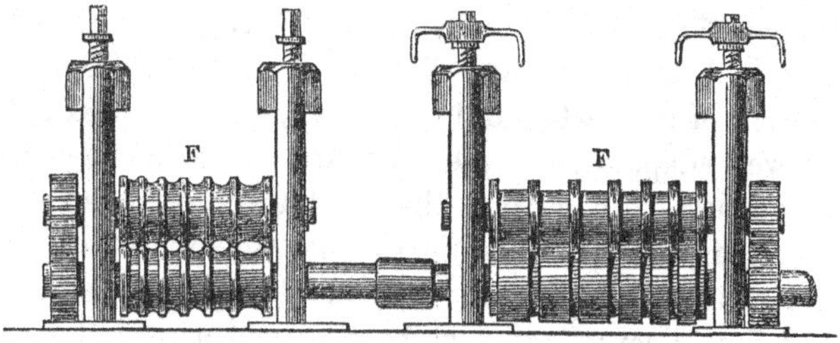

grooved rollers connected by toothed pinions, and put in motion by the steam-engine. The rollers are fixed in massive framing, which has to support a prodigious strain, as the bloom is drawn in, and compressed and elongated as it passes through. The bar so formed is passed through a succession of similar grooves, decreasing in size till it is reduced to about 4 inches wide, three quarters to an inch thick, and 10 or 12 feet in length. In this state it is cut to pieces, piled, heated, &c., as already described, and a second time rolled. The bars produced by this second process are called merchant bars; or the bloom may be rolled into plates. Again, instead of being rolled, it may be brought under the steam hammer and forged into *uses*, or those variously shaped masses of wrought-iron which are employed by the engineer, millwright, or iron ship builder, where these *uses* are extensively employed. We have stated that the horizontal, non-condensing steam-engine, from its compact form and convenience of handling, is admirably adapted for giving motion to the machinery of iron works. For this object it is superior to the beam engine, as its speed can be regulated with the same facility as the condensing engine by simply moving the lever of the steam valve, so as to suit all the requirements of the manufacturer, under the varied conditions of pressure, and the power required for rolling heavy plates or bars. It is also much cheaper in its original cost, and all its parts being fixed upon a large bed plate, requires a comparatively small amount of masonry to render solid and secure.

In regard to the manufacture of the rollers for the boiler plate, and merchant train, the greatest care must be observed in the selection of the iron and the mode of casting. In Staffordshire there are roller makers, but in general the manufacturer casts his own; and as much depends upon the metal, the strongest qualities are carefully selected and mixed with Welsh No. 1 or No. 2, and Staffordshire No. 2. This latter description of iron, when duly prepared, exhibits great tenacity, and is well adapted, in either the first or second melting, for such a purpose. In casting, the moulds are prepared in loam, and when dry are sunk vertically into the pit to a depth of about 5 feet below the floor. The moulding box is surrounded by sand firmly consolidated by beaters, and a second mould or head is placed above it, which receives an additional quantity of iron to supply the space left by shrinking, and retain the roll under pressure until it solidifies, and thus secures a great uniformity and density in the roller. The metal is run into the mould direct from the air furnace by channels cut in the sand; and immediately the mould is filled, the workman agitates the metal with a rod, in order to consolidate the mass and get rid of any air or gas which may be confined in the metal. This stirring with iron rods is continued till the metal cools to a semifluid state, when it is covered up and allowed slowly to cool and crystallize. This slow rate of cooling is necessary to favour a uniform degree of contraction, as the exterior closes up like a series of hoops round the core of the casting, which is always the most porous and the last to cool. In every casting of this kind, it is essential to avoid unequal contraction; and this cannot be accomplished unless time is given for the arrangement of the particles by a slow process of crystallization. Rollers for boiler plates and thin sheet iron are difficult to cast sound, on account of their large size. They are subjected to very great strain, and require to be cast from the most tenacious metals. The bearings or neck should be enlarged, or turned to the shape shown at A A, and the cylindrical part B for plate rolls should be slightly concave, because, when the slab is first passed through the rollers, it comes in contact with a small portion only of the revolving surface. The central parts of the roller thus become highly heated, whilst their extremities are perfectly cool. The consequence is, that the expansion of the roller is greatest in the middle, and unless this be provided for by a concavity in the barrel, the plates become buckled, that is, both warped and uneven in thickness, and, consequently, imperfect and unfit for the purposes of boiler making. Bar rolls are generally cast in chill, and great care is required to prevent the chill penetrating too deep, so as to injure the tenacity of the metal and render it brittle.

Fig. 322.

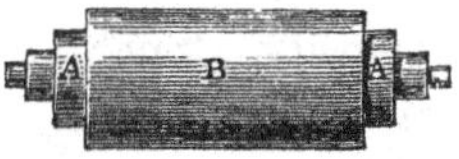

There are different kinds of rolling mills used in the iron manufacture, and they vary considerably in their dimensions, according to the work they have to perform. The first, through which the puddled iron is passed, we have already described as puddling rolls. There are others for roughing down, which vary from 4 to 5 feet long, and are about 18 inches diameter; those for merchant bars, about 2 feet 6 inches to 3 feet long, and 18 inches in diameter, are in constant use. The boiler plate and black sheet iron rolls are generally of large dimensions; some of them for large plates are upwards of 6 feet long, and 18 to 21 inches in diameter; these require a powerful engine and the momentum of a large fly-wheel to carry the plates through the rollers; and not unfrequently, when thin wide plates have to be rolled, the two combined prove unequal to the task—and the result is, the plates cool and stick fast in the middle. The greatest care is necessary in rolling plates of this kind, as any neglect of the speed of the engine or the setting of the rolls results in the breakage of the latter, or bringing the former to a complete stand.

The speed of the different kinds of rolling mills varies according to the work they have to perform. Those for merchant bars make from 60 or 70 revolutions a minute, whilst those of large size, for boiler plates, are reduced to 28 or 30.

Others, such as the finishing and guide rollers, run at from 120 to 400 revolutions a minute. In Staffordshire, where some of the finer kinds of iron are prepared for the manufacture of wire, the rollers are generally made of cast-steel, and run at a high velocity. Such is the ductility of this description of iron, that in passing through a succession of rollers, it will have elongated to ten or fifteen times its original length, and, when completely finished, will have assumed the form of a strong wire of $\frac{3}{8}$ to $\frac{1}{4}$ of an inch in diameter, and 40 to 50 feet in length.

A high temperature is an indispensable condition of success in rolling. The experience of the workman enables him to judge, from the appearance of the furnace, when the pile is at a welding heat, so that, when compressed in the rolls, the particles will unite. Sometimes it is necessary to give a fine polish or skin to the iron as it leaves the rolls; but this can only be done when the iron cools down to a dark red colour, and by the practised eye of an intelligent workman.

Another description of rolling for giving a fine skin or polish to the iron is to roll it cold, and this process not only adds to its appearance, but, it increases its tenacity and renders it suitable for many purposes, such as slender shafts where stiffness is required.

Before closing this description of the manufacture of iron as relates to the rolling mill, the whole process would be incomplete if the new appliance for rolling of large masses into enormous plates, were omitted. The casing of ships of war with enormous large and thick plates is a new discovery in the art of war, and its appliance not only to ships, but forts and citadels, presents an entirely new era in the history of naval and military tactics. To what extent its future developement may attain time alone can determine; but of this it is quite evident, that a change of construction, and a new system of operations in the laws of defence and attack, is imminent. We have, therefore, to prepare new mills and new manufactories upon a scale commensurate with the wants and requirements of the changes now in progress. To what extent the present transition may be carried is unknown, as everything depends on future developements in the manufacture of guns and the antagonistic force of iron in its powers of resistance to the increased power of projectiles at high velocities. But however this may be, the enterprise, power, and energy of the British manufacturers, as instanced in the works of Beale and Brown, is fully able to meet the emergency, and to roll plates of any dimensions required for the purposes of defence.

As an example of this determined spirit we may notice a series of experiments which took place in presence of the Lords of the Admiralty at Messrs. John Brown and Co.'s works, Sheffield, in April last. For some time previous both Messrs. Beale and Brown had been making preparations for the rolling of these plates, and in order to put the reader in possession of the experiments at Messrs. Brown's works we have given in Appendix II. a graphic description of the whole process as taken from the correspondent of the 'Times,' who was present on the occasion.

APPENDIX.

I.

THE following extracts, taken from a paper by Mr. EDWARD BAINES, M.P., on the Woollen Manufacture of Leeds, read at the Meeting of the British Association for the Advancement of Science in 1858, will be found interesting :—

In giving a description of the different processes, Mr. Baines states—'That the processes of the Woollen Manufacture are more numerous and complex than those of any other of our textile manufactures. In one of those complete and beautiful establishments where fine cloth is both manufactured and finished—as that of Messrs. Benjamin Gott and Sons of this town, which has long ranked with the first woollen factories of any country—the spectator who may be admitted to it will see all the following processes, namely :—

1. Sorting the wool; no less than ten different qualities being found in a single fleece.

2. Scouring it with a ley and hot water, to remove the grease and dirt.

3. Washing it with clean cold water.

4. Drying it; first in an extractor—a rapidly revolving machine full of holes—and next, by spreading it and exposing it to the heat of steam.

5. Dyeing, when the cloth is to be wool-dyed.

6. Willying, by revolving cylinders armed with teeth, to open the matted locks and free them from dust.

7. Teasing, with a teaser or devil, still farther to open and clean.

8. Sprinkling plentifully with olive oil, to facilitate the working of the wool.

9. Moating, with a moating-machine, to take off the moats or burs, i.e. seeds of plants or grasses which adhere to the fleece.

10. Scribbling, in a scribbling-machine, consisting of a series of cylinders clothed with cards or wire-brushes working upon each other, the effect of which is still further to disentangle the wool and draw out the fibres.

11. Plucking, in a plucking-machine, more effectually to mix up the different qualities which may remain in the wool.

12. Carding, in a carding-machine, resembling the scribbler, but more perfectly opening the wool, spreading it of a regular thickness and weight, reducing it to a light filmy substance, and then bringing it out in cardings or slivers about three feet in length.

13. Slubbing, at a frame called the billy, generally containing sixty spindles, where the cardings are joined to make a continuous yarn, drawn out, slightly twisted, and wound on bobbins.'

[By a new machine, called the Condenser, attached to the carding-machine, the wool is brought off in a continuous sliver, wound on cylinders, and ready to be conveyed to the mule, so as to dispense with the billy.]

' 14. Spinning on the mule, which contains from 300 to 1,000 spindles per pair.

15. Reeling the yarn intended for the warp.

16. Warping it, and putting it on the beam for the loom.

17. Sizing the warp with animal gelatine, to facilitate the weaving.

18. Weaving at the power-loom or hand-loom.

19. Scouring the cloth with fuller's earth, to remove the oil and size.

20. Dyeing, when piece-dyed.

21. Burling, to pick out irregular threads, hairs, or dirt.

22. Milling or fulling, with soap and warm water, either in the fulling-stocks or in the improved milling-machine, where it is squeezed between rollers.

23. Scouring, to remove the soap.

24. Drying and stretching on tenters.

25. Raising the nap of the cloth, by brushing it strongly on the gig, with teazles fixed on cylinders.

26. Cutting or shearing off the nap in two cutting-machines, one cutting lengthwise of the piece and the other across.

27. Boiling the cloth, to give it a permanent face.

28. Brushing, in a brushing-machine.

29. Pressing in hydraulic presses, sometimes with heat.

30. Cutting the nap a second time.

31. Burling and drawing, to remove defects, and marking with the manufacturer's name.

32. Pressing a second time.

33. Steaming, to take away the liability to spot.

34. Folding or cutling for the warehouse.

These processes, as has been said, are greatly more numerous than those required by any other textile manufacture, and they are performed by a much greater variety of machines and of work-people.

It is pretty obvious that there must be proportionate difficulty in effecting improvements which will tell materially on the quantity or the price of the goods produced.'

In another part of his paper, Mr. Baines gives an interesting account of the Shoddy Trade, which is chiefly carried on at Batley, near Dewsbury. He says that 'He must now explain a new branch of the trade, which has risen up with great rapidity and attained extraordinary dimensions, to which, indeed, we are compelled to ascribe much of the present prosperity and extension of the Yorkshire trade. Its origin dates as far back as 1813, but it was long regarded with disapprobation as a dishonest adulteration. It consists in mixing with wool, in the course of manufacture, a very inferior species of wool, made from the tearing up of old woollen and worsted rags, and to which the names have been given of *shoddy* and *mungo*. Shoddy is the produce of soft materials, such as stockings, flannels, &c., and mungo of shreds or rags of woollen cloth. The latter is of very superior quality to the former, being generally fine wool, which, after being once manufactured and worn, is torn up into its original fibres (by cylindrical machines armed with teeth), only shorter and feebler, and not susceptible of being dyed a bright colour. Both shoddy and mungo give substance and warmth, and the latter will receive a fine finish; but, from the extreme shortness of their fibre, the cloth made from them is weak and tender. If cloth made of these kinds of rag-wool is expected to have the tenacity of goods made from new wool, it will utterly disappoint; but there are immense quantities of goods where substance and warmth are the chief requisites, and where strength is of no importance. Among them are paddings, linings, the cloth used for loose and rough great coats, office coats, and even ladies' capes and mantles. Broad cloth may be made with a large admixture of these cheap and inferior materials to look almost as well as that made of pure wool; but the goods for which they are more properly adapted are what are called pilots, witneys, flushings, friezes, petershams, duffels, honleys, druggets, as well as blankets and carpets.

The price of shoddy varies from $\frac{3}{4}d.$ per pound to $5d.$, and the white shoddy from $2d.$ to $10d.$ per pound. The proportions of these materials used in this district are about one-third mungo and two-thirds shoddy. Some goods, such as low-coloured blankets and pea-jackets, are made with only one part of pure wool to six parts of shoddy; but in the whole district perhaps one-third of wool may be used with two-thirds of shoddy or mungo.

It is one of the objects of improvements in the useful arts to give value to that which possessed no value, to utilise refuse, to economise materials, and, as it were, to prolong their existence under different forms to the latest date. The waste swept up from the floor of the cotton mill is made into beautiful paper. The oil washed out of woollen cloth is now extracted from the muddy liquid which formerly ran to waste, and is saved for fresh oleaginous uses. Scraps, shavings, dust, the contents of sewers, are all made valuable. Why, then, should not the wool of the sheep undergo a second manufacture? If the cloth made of shoddy and mungo is sold for what it really is, no one is deceived. It may, indeed, be fraudulently sold for what it is not, and the man who does so ought to be branded as a cheat. But if the use of shoddy and mungo will answer nearly as well as wool for a vast variety of purposes, and will enable the consumer to obtain two or three yards of cloth where he formerly obtained only one, it should be received as a lawful and valuable improvement in manufacture.

The place where shoddy was first used in this manner was Batley, by Mr. Benjamin Law, and the first machines for tearing up the rags were set up by Messrs. Joseph Jubb and J. and P. Fox. The manufacture has forced its way, and made Batley, Dewsbury, and the neighbourhood the most prosperous parts of the woollen district. There are now in Batley alone fifty rag-engines in thirty-five mills, producing no less than 12,000,000 lbs. of rag-wool per annum (after deducting for loss of weight in the manufacture); and I am assured, on good authority, that three times this quantity is made in the district. The rags are gathered from all parts of the kingdom, as well as imported regularly from the continent, America, and Australia. There is now a considerable manufacture of the shoddy, or rag-wool, in Germany, and it is believed that no less than 9,000,000 or 10,000,000 lbs. weight was imported last year.

How profitable this trade is to the workmen may be inferred from evidence which has been obtained, to the effect that 5,408 operatives in Batley received, 812*l.* of weekly wages, or an average of 14*s.* 1*d.* each.

Another method of cheapening cloth has also been extensively introduced in the woollen manufacture, though by no means to the same extent or with the same success as in the worsted—namely, the use of cotton warps. This also was regarded as a great deterioration of the fabric, and to some extent it is so. The cloth is not so warm as when made all of wool, and it has a certain harshness of feel; but it is not, like shoddy cloth, tender: on the contrary, it is stronger than if made entirely of woollen yarn. Many kinds of goods of great beauty are thus made, among which may be mentioned the tweeds used for trousering, and grey cloths used for ladies' mantles and other purposes. Cloths with cotton warps are generally called union cloths.

With respect to the value of the woollen manufacture of the United Kingdom, Mr. Baines's estimates are as follows:—

Estimated Annual Value of the Woollen Manufacture of the United Kingdom, 1858.

(1) Raw material—		
lbs.		£
75,903,666	foreign and colonial wool . . .	4,717,492
80,000,000	British wool at 1*s*. 3*d*. per lb. . . .	5,000,000
	Shoddy and mungo—	
45,000,000	30,000,000 lbs. shoddy at 2½*d*. per lb. 15,000,000 ,, mungo at 4¾*d*. ,,	609,370
	Cotton and cotton warps, $\frac{1}{50}$ of the wool	206,537
200,903,666		10,533,399
(2) Dye wares, and soap		1,500,000
(3) Wages—150,000 work people at 12*s*. 6*d*. per week		4,875,000
(4) Rent, wear and tear of machinery, repairs, coal, interest on capital, and profit—20 per cent. on the above		3,381,680
	Total . .	£20,290,079

II.

The visitors were then conducted through the extensive new and old mills and workshops, where some 3,000 hands were busily engaged in melting, bending, hammering, and twisting great masses of seething iron into every conceivable form its stubborn nature could be made to take. It was really a wonderful sight. On every side, amid thick smoke and deafening clamour, the blazing rites of Moloch—the furnace god of old—were being celebrated. Great furnaces blaring in the fierce white glare which shone from their crevices were stuffed to the mouth with monstrous cranks and shafts and uncouth bosses of red-hot metal. Every now and then some one of them was opened, with a flash that filled the smoky atmosphere with a glare as from snow, and a mass of metal, seething and spluttering in a blaze of sparks, was dragged off and moulded, like so much wax, under the blows of steam hammers that made the earth tremble and the whole building to jump and chatter under the stroke, as if from the shock of a little earthquake. It was wonderful to see the skill with which the groups of workmen, uniting all their individual exertions in a series of violent efforts like a weird species of dance, contrived to hedge and move about the great masses on the anvils, so that the hammer struck only where and how they chose. While the heat lasted in the mass, and that was for a long time, they never paused or slackened in their work, and though literally almost scorched by their proximity to red heaps, they kept on toiling till the work was done, and the lump that a quarter of an hour before was almost melted iron was picked up by some huge crane that came travelling along the smoky walls, and carried off, glowing through the gloom, a finished piece of work. At other places there were tilt and lever hammers, wearying the very air with the clattering din of their tremendous strokes. At others great ingots of steel were cast by the Bessemer process—small plates were rolled and roughly cast aside in great red slabs to cool, or hurried backwards and forwards in iron trucks, scorching even the hardened workmen out of their tracks as they came burning past. On every side there were furnaces and smoke and red-hot metals, while in out-of-the-way nooks men in steel caps and wire vizors, and cased below in rough steel leggings, like jack-boots of iron, fought in a crowd like so many salamanders round some rough mass that was dangerous in its fierce heat, and which sent back aggressive spurts of red-hot metal in return for every blow. Such fiery combats as these were going on in all directions; the 'Sheffield carpet' of the factory—iron plates—was hot and painful to the feet; the air was arid with a sulphury warmth that was like the glow of an over-heated stove. When we have said thus much, and added that there were roaring pipes of steam mounting into the air, side by side with great iron trumpet-shaped piles of chimneys, out of which jets of red flame roared and flapped into the smoke above like gigantic flambeaux—that lower down long lines of lathe bands flew noiselessly in all directions, and that the background was filled in with glimpses of ponderous fly-wheels whirling their arms through the smoke and turning rolling-mills or lapping-hammers, or shearing down with noiseless might the great lumps of iron that were brought in to be cut up,—we have said enough to indicate the view which met visitors on their first introduction to this glowing scene of industry. Though not the first, yet by far the most important process which their Lordships were shown, was the operation of rolling the great plate—by far the largest single plate that has ever yet been rolled in the world. This took place in what are called the New Mills of the Atlas Works, which were used on Thursday for the first time, and where great ranges of furnaces have been erected, with their mouths opening on the iron tramway which leads direct to the double rollers through which the plate passes. One may guess at the solidity required for mills of this kind when it is stated that some of the rolls used at this mill on Thursday have a first foundation of no less than 60 tons of solid iron, resting on masonry carried far below the earth. The rolls themselves are 32 in. in diameter and 8 ft. wide, and are turned by an engine of 400 horse-power, putting in motion a fly-wheel large enough, apparently, to make a world rotate if only well balanced on its axis. A powerful screw, applying its force through compound levers, allows the distance between the rollers to be adjusted to the fraction of an inch, so that the plate which on its first rolling is forced through an interval of—for instance, 12 inches apart—is on its next wound through one of 10, next through one of 8, and so on till the required thickness has been carefully and equally attained by tremendous compression through every part of the metal. There were a great many visitors to see the rolling of this formidable mass, which was fortunate, as one would certainly be frightened to witness the terrible process alone. After some delay and quick glimpses made by the most hardened workmen, who, rushing up to the door of the furnace, got a half-blinded glance into its white interior, it was decided that the mass was ready, for, strange as it may seem, an armour-plate requires more than mere heating, and has to be cooked and watched in its cooking with as much care as if it was an omelette, and the plate that is drawn before it is 'done to a turn,' generally remains a permanent ornament of the unlucky manufacturer's workshop, which no one will have at any price. When at last this eventful moment had arrived, on Thursday the door of the furnace was slowly raised, and a colossal pair of pincers with very long handles, fastened to a chain drawn by machinery, was swung in. For an instant some men rushed forward, and, shielding their faces from the deadly heat that shot from the furnace, adjusted the bite of these forceps on the plate, and then ran back as the chain began to tauten, and the great inmate of the blazing den was slowly dragged forth on to the long iron trucks in front of the door, and there lay in its huge length and thickness a mass of living fire, which none could approach, or scarcely even look at, so fierce was its glow and terrific heat. The chains that should have pulled it forthwith to the rollers were too slack, and then arose shouts and cries and commands, as the men did battle with this mass of fire, coming so near it, in their attempts to gather up the slackened chains, that one literally almost expected to see them fall, scorched and shrivelled, on the ground. In its great glare they fought and struggled with the chains till at last all was adjusted, and the great pile of angry fire began to move slowly downwards towards the mills, the men following it with hoarse shouts and directions, now hid in steam, as buckets of water were dashed over the mass, and the next moment standing in an atmosphere of white light, to which the light of the day around was mere dusk. The rollers did not bite directly the mass came to them, and when they did the engine was almost brought to a standstill by the tremendous strain upon it; but at last the soft plate yielded, and the rollers seemed to swallow it as they wound it slowly in, squeezing out jets of melted iron like squirts of fire, that shot about dangerously as the pile was compressed from 19 inches to 17 inches thick by the irresistible force of the rollers. *Ce n'est que le premier pas qui coûte*, and the victory was certain when the mass had once passed through the mill, and both visitors and workmen gave a tremendous cheer at the success. From this time it was kept rolling backwards and forwards, the workmen sweeping from its face the scales of oxide that gathered fast upon it with long-handled besoms that, though soaked in water, caught fire and blazed up as fast as they were used. With every time it was passed through the rollers were screwed closer and closer together, as we have already mentioned, till at the end of about a quarter of an hour, after leaving the furnace an almost melted mass, it was passed through for the last time, and came out opposite the furnace door it had so lately left, no longer shooting forth spiteful sparks, but shorn of half its heat, subdued and moulded to its proper form—a finished armour-plate, weighing 20 tons, 19 ft. long, nearly 4 ft. wide, and exactly 12 in. thick throughout from end to end. This is the most signal triumph that any rolling mills have yet achieved.

Other smaller plates were then rolled with a quickness and certainty that proved the skill already gained in this new and most important branch of manufacture. One plate was 17 feet long by 4 feet broad and 5½ inches thick; one 19 feet long by 4½ feet wide and 4½ inches thick; one we have already alluded to 41 feet long by 3 feet 10 inches broad and 4½ inches thick. A lesser plate was also rolled 18 feet long, 5 feet wide, with a thickness of 6 inches on one edge and 3 inches on the other. The method of converting cast iron by the Bessemer process into the tough soft Bessemer metal, a combination of the qualities between soft steel and tough wrought iron, was next shown. It is needless now to enter on a description of the very beautiful and very terrible process, to witness, which the metal goes through in the converter as it is stimulated to a white heat by the passage of the air blown by force-pumps upwards through the mass. No fireworks can surpass the brilliancy of the display this process affords as it approaches its completion, and the stream of violet flame and clouds of burning sparks pour from the mouth of the converter as from a gigantic squib. Nor is it necessary here to enter into a detail of the now well-known process, which was a subject of such controversy a few years since, but which is now being so generally and advantageously adopted throughout England and the continent. Suffice it to say, that in twenty minutes from the time of putting in the charge of cast iron, it was, without any expenditure of labour, poured out into the mould, an ingot of soft tough steel weighing three tons. This metal, after undergoing hammering, is now most extensively used for steel rails at stations, points, and junctions, where the wear is great, and in these trying situations it seems almost indestructible. A great deal has also been used in making Blakely rifled guns in this country for both Federals and Confederates. These are the ordnance which the Americans always speak of as Parrott guns, and by them they are more highly prized than those of either Armstrong or Whitworth. Yet it is stated that the Ordnance Select Committee have refused even to try these guns at Shoeburyness. After these processes were over, and the various planing and filing shops had been duly examined, the visitors were entertained by Mr. Brown at a most sumptuous *déjeûner*.

weighing 20 tons 13 [illegible], nearly 4 ft. wide, and [illegible] 12 inches thick [illegible] out from end to end. This is the [illegible] of any rolling mills [illegible] yet achieved.

Other smaller plates [illegible] been rolled [illegible] thickness and certainty [illegible] proved the skill [illegible] and [illegible] of [illegible]. One [illegible] feet [illegible] by 4 feet broad and 5½ inches thick [illegible] 13 feet long [illegible] feet wide [illegible] inches thick [illegible] feet long by 3 [illegible] inches [illegible] 4½ inches thick. A [illegible] 18 feet long [illegible] thickness [illegible] edge and 3 inches [illegible]. The method [illegible] by the [illegible] process into [illegible] metal, a [illegible] of the qualities between soft steel and [illegible] wrought iron, [illegible]. It is needless [illegible] a description of the [illegible] process, to [illegible] which the metal goes through in the [illegible] white heat by the passage of the air [illegible] fireworks can [illegible] the [illegible] of the display this process affords as it [illegible] its completion, and [illegible] of [illegible] and clouds of burning [illegible] of the [illegible]. [illegible] here [illegible] into detail of the [illegible] a process, which [illegible] of such [illegible] years since, but which is now being so generally and [illegible] adopted throughout England and the continent. [illegible] to say, it is [illegible] putting in the charge of [illegible] labour, poured [illegible] weighing three tons [illegible] underground [illegible] is now most extensively used [illegible] and [illegible], where the wear is great, [illegible] these trying situations [illegible]. A great [illegible] has also [illegible] in making [illegible] in this country [illegible]. These are the [illegible] which the [illegible] always speak of as [illegible], and by them they are [illegible] of [illegible] Armstrong [illegible] Whitworth. [illegible] that the [illegible] have [illegible] to try these guns at Shoeburyness. After [illegible] processes were seen, and the [illegible] plates and [illegible] shops had been duly examined, the visitors were entertained by Mr. Brown at a most sumptuous déjeûner,

INDEX.

PRINTED BY SPOTTISWOODE AND CO., NEW-STREET SQUARE, LONDON